T3-BOB-112
3 0301 00075699 5

Archaeological Chemistry IV

ADVANCES IN CHEMISTRY SERIES **220**

Archaeological Chemistry IV

Ralph O. Allen, EDITOR
University of Virginia

Developed from a symposium sponsored
by the Division of the History of Chemistry
at the 193rd Meeting
of the American Chemical Society,
Denver, Colorado,
April 5-10, 1987

American Chemical Society, Washington, DC 1989

Library of Congress Cataloging-in-Publication Data

Archaeological chemistry IV.
(Advances in chemistry series; 220)

Bibliography: p.
Includes index.

1. Archaeological chemistry—Congresses.

I. Allen, Ralph O. II. American Chemical Society. Division of the History of Chemistry. III. Series.

QD1.A355 no. 220 [CC79.C5] 540 s [930.1'028] 88-7953
ISBN 0-8412-1449-2

PRINTED IN THE UNITED STATES OF AMERICA

Advances in Chemistry Series

M. Joan Comstock, ***Series Editor***

1988 ACS Books Advisory Board

FOREWORD

The ADVANCES IN CHEMISTRY SERIES was founded in 1949 by the American Chemical Society as an outlet for symposia and collections of data in special areas of topical interest that could not be accommodated in the Society's journals. It provides a medium for symposia that would otherwise be fragmented because their papers would be distributed among several journals or not published at all. Papers are reviewed critically according to ACS editorial standards and receive the careful attention and processing characteristic of ACS publications. Volumes in the ADVANCES IN CHEMISTRY SERIES maintain the integrity of the symposia on which they are based; however, verbatim reproductions of previously published papers are not accepted. Papers may include reports of research as well as reviews, because symposia may embrace both types of presentation.

ABOUT THE EDITOR

RALPH O. ALLEN is Professor of Chemistry and the Director of the Office of Environmental Health and Safety at the University of Virginia in Charlottesville, Virginia. He received a liberal arts education (B.A., 1965) at Cornell College in Mt. Vernon, Iowa. After beginning his research on trace elements in meteorites at Argonne National Laboratory, he completed his graduate studies at the University of Wisconsin in Madison (Ph.D. 1970) where he helped analyze the first lunar samples. In 1970 he began his academic career at the University of Virginia but remained involved in the lunar analysis program through his association with Argonne National Laboratory.

His research in trace element geochemistry lead to the application of nuclear techniques to lithic archaeological materials and to forensic evidence. In addition to being widely published in these fields, he has conducted research and has published in the fields of hazardous waste management and safety. He has been a Fellow of the Royal Norwegian Council for Scientific and Industrial Research (1977–1978) and the Norwegian Marshall Foundation (1983). He has been the faculty liaison for forensic science and training at the Federal Bureau of Investigation's (FBI) National Academy since 1981, a member of the Operations and Planning Committee for the FBI Forensic Science Research and Training Center, and was presented the Erikson Award by the FBI. He is the chairman of the ACS Subdivision of Archaeological Chemistry, a part of the Divsion of the History of Chemistry.

CONTENTS

ART OBJECTS

ORGANIC MATERIALS

FIBERS

INDEXES

PREFACE

MOST CHEMISTS DO NOT HAVE THE OPPORTUNITY to experience quite the same sense of awe and excitement as was experienced by archaeologist Howard Carter when he first opened the tomb of the boy king Tutankamun. However, a growing number of chemists have shared in the excitement of discovering the past in a more subtle manner. The application of chemical techniques to the study of archaeological materials has brought chemists, archaeologists, anthropologists, and historians together to enhance our knowledge of the past. The same techniques that help chemists characterize materials in their efforts to improve the future have also helped us understand the earlier uses of materials found in nature.

Archaeological chemistry has not only allowed chemists to share in the study of the past, but has strengthened archaeologists' understanding of procedures that can be used to characterize archaeological materials. The chapters included in this volume show the interdisciplinary nature of archaeological chemistry. The symposium upon which this book is based included 1 day devoted to the growing contributions that chemists and biochemists have made to the study of proteinaceous materials in archaeological samples. Another day was devoted to the important contributions that have been made by nuclear techniques.

RALPH O. ALLEN
Department of Chemistry
University of Virginia
Charlottesville, VA 22901

1

The Role of Chemists in Archaeological Studies

Ralph O. Allen

Department of Chemistry, University of Virginia, Charlottesville, VA 22901

This chapter is an overview of the wide variety of archaeological studies conducted by chemists. From the earliest stone artifacts to the artistic manuscripts and textiles of the more recent past, the studies presented in this volume show the wide range of materials that have been studied by chemical techniques. The field keeps expanding as chemists help provide information valuable in the interpretation of archaeological sites and artifacts. Besides helping to detect fraudulent artifacts and artistic objects in museum collections, chemists have studied the physicochemical deterioration processes that destroy the monuments and objects of the past. Thus, the role of chemists is more than just discovery of the past; it includes investigation that may help preserve the artifacts for future generations to enjoy and study.

HUMAN CULTURE CAN BE OBSERVED in the physical artifacts of the past. Even without recorded histories, the partial record of materials that have survived the ravages of time provides us with insight into ancient times. The oldest surviving materials are human bones and the simple stone artifacts that are evidence of ancient workmanship. The stone (or lithic) tools can be used to describe early culture. Whether we study the rough chipped-stone implements of the Paleolithic era or the finer microblade tools of the Neolithic era, it is clear that early humans knew the mechanical properties of many natural materials. It was not until much later that the methods of chemical transformation were learned.

Evolution of culture can be traced in the gradual refinement of the stone tools produced and the materials used by early humans. Eventually, objects

0065–2393/89/0220–0001$06.00/0

were made of native metals like copper and gold. With time, and presumably through experimentation, fire was used to alter the physical characteristics and even the composition of certain materials. Thus, artifacts like pottery, mortar, glass, and metal alloys have chemical compositions that reflect the chemistry of the raw natural materials used, as well as the changes introduced by new and evolving technologies.

Differences in chemical compositions usually provide new information about artifacts. By differentiating between the sources of the raw materials used to produce objects, it is possible to infer cultural contacts. For some artifacts, detailed studies of compositional differences can also help us understand production methods. The remains of the humans themselves may also be analyzed to provide useful information. The contributions that chemists have made in the study of archaeological materials have gone far beyond the simple chemical analysis of the materials. This volume gives but a small part of the great contributions that chemists have made toward the understanding of ancient materials and technologies.

The use of careful chemical analysis to enhance the understanding of prehistoric technologies is not new. Archaeological chemistry was pioneered by many of the earliest chemists, including the seventh president (1882) of the American Chemical Society, John W. Mallet. Mallet's doctoral dissertation at the University of Gottingen was the first account of a chemical investigation of prehistoric Celtic objects, including precious stones, glass beads, pigments, bronzes, and gold ornaments. Although analysis of the objects was difficult by today's standards, the results were accurate enough to provide new insights into these Celtic artifacts. For example, Mallet used the chemical compositions to define the sources of the raw materials. On the basis of his geological and chemical knowledge, Mallet concluded that during the early Christian times, the Celts used only native gold (which they graded by color) and had not yet discovered the means of extracting silver from ores.

Since these early endeavors, the analytical techniques have improved, and the potential role of the chemist in archaeological studies has increased. Greater sensitivity in the analysis of atomic and molecular species has helped scientists differentiate between artifacts. Improvements in selectivity for the measurement of molecular species have generated more information about decorative objects and those organic materials that have survived. The ability to measure isotopic compositions accurately has provided some of the most valuable information for archaeological studies. For example, the analysis of carbon-14 has revolutionized the dating of archaeological sites. With the improvement of methods and the resulting increases in detailed chemical studies of archaeological materials, the problems of data management and interpretation have increased. Fortunately, the increased availability of computers to manage data bases has enhanced the ability of the chemist and the archaeologist to deal with all of the data that is produced.

One of the goals of this book is to illustrate how chemists use their knowledge and new techniques to provide further information on the artifacts of past cultures. Another goal is to acquaint archaeologists with chemical techniques that may be useful to them in their studies, and to help them appreciate some of the problems in the interpretation of chemical data. Archaeologists and chemists must, for example, understand the role of proper sampling in applying chemical methods to archaeological problems. Archaeologists must appreciate that, even though the chemist can measure very low concentrations with great precision, the differences between samples may not be significant. The chemist's knowledge of the materials themselves must be used to help the archaeologist know which variations are significant.

Archaeological chemistry is a marriage between two disciplines and requires ongoing cooperation and interaction. As a result of this interaction, our knowledge of the past and of materials (both natural and synthetic) increases. Many of the chapters in *Archaeological Chemistry IV* show how chemists and archaeologists can work together and grow in their understanding of each other's disciplines.

Trace Elements in Lithic Artifacts

Glimpses of the past are revealed by the few artifacts that are found. Care is required to obtain as much information as possible from these artifacts. In most cases, the context and even orientation in which the object is found will help in the eventual interpretation of an archaeological site. However, chemical analysis of all artifacts is both impractical and unwarranted. The decision on whether to study an artifact and what techniques to use should be based upon an understanding of the material from which the object is made. This decision can be illustrated by considering stone artifacts. Although early people must have used many natural materials, those like the reeds and wood have decayed, whereas the lithic or stone objects have survived. Many of the extant objects were made from fine-grained hard stone like flint. Physical examination of these stone tools suggests that the styles and types of tools did not vary much over long periods of time and over large geographical areas. The geological sources for high-quality raw materials were widely distributed and probably did not require a sophisticated barter or exchange system.

Chemical methods can sometimes be used to distinguish between stone artifacts that are made with very similar materials. Lithic raw materials were chosen because of properties, such as hardness, that depend upon crystallinity and chemical composition. Geochemical processes determine both the chemical composition and physical properties of the rock. In many cases, lithic artifacts do not require chemical analysis to show that they are made from different materials. A simple visual examination is frequently sufficient.

In Chapter 2, Hancock, Pavlish, and Sheppard give an example of a case in which visual examination of stone tools was not adequate to differentiate between lithic artifacts that were produced from rocks that were very different in their origins. During the Mesolithic and early Neolithic times, the inhabitants of what is now Portugal used a variety of materials. Although most of the stone tools were classified by the archaeologists as sedimentary cherts, Hancock concluded that many tools were made of volcanic rhyolite.

The basis for Hancock's differentiation between rocks of different geological origin was the analysis of the impurities (trace elements) in the silica-rich rocks. The trace element concentrations in a particular rock depend primarily upon the minerals that are present in the rock. Although there are many factors to consider, generally trace elements in minerals are ionic species that substitute in a silicate crystal lattice. The degree of substitution depends on how much of the impurity is present and on how well it is matched in ionic size and charge to the major ions in the mineral. The silica lattice (quartz) itself does not accommodate impurities, as there are no major cation sites at which substitution can occur. Thus, trace element concentrations tend to be very low in silica-rich rocks like quartz and chert. Chapter 2 describes a number of silica (quartz)-rich rocks, but suggests that some that look like chert were of volcanic origin, and thus contained larger amounts of feldspar. Because feldspars are aluminosilicate, the presence of feldspar in the rhyolite is apparent from the higher concentration of Al (along with associated Na, K, and Ca) in those artifacts made of rhyolite.

Some of the most successful provenance studies have been for obsidian, the glass formed from the rapid solidification of volcanic lava. Obsidian was available only in limited areas of volcanic activity, but because it was a natural glass, it proved to be an excellent material for making the sharp tools (e.g., knives) that are often found hundreds of miles away from known sources of obsidian. Matching an artifact to a distant obsidian deposit was possible because the chemical compositions of these natural glasses are often quite different. The composition of volcanic lavas depends on how much rock has already crystallized from the lava. Because the trace elements are removed from the lava by substituting for major elements in a crystal lattice, the composition of the lava and the resulting glass also depends upon the degree of crystallization (or likewise partial melting of rocks). The glass from a single lava flow is fairly homogeneous; therefore, sampling is not as great a problem as it is in the case of a rock that is a mixture of several different minerals.

Of the many studies of obsidian, Blackman's study of obsidian artifacts in Iran from the period of 3500 to 1800 B.C. is typical (*1*). The results showed that obsidian from a single region varied in composition because volcanic glass was produced at several different times throughout geological history. Thus, the grouping of artifacts by using a hierarchical aggregative clustering methodology did not necessarily identify distinct geographical sources.

As useful as the chemical analysis of certain lithic artifacts has been in elucidating patterns of resource procurement, care must always be taken in interpreting the results. For example, in the early work on soapstone (steatite) artifacts, it appeared that the rare earth elements (REE) were particularly useful as "fingerprints" for the geological sources (2). Soapstone was a widely used soft rock, but it is a relatively rare geological material formed by local or regional metamorphism. Analysis of material from prehistoric quarries has shown the variability in the composition of the soapstone from a particular outcrop. In many cases, the differences between source outcrops is greater than the variability within a geological body. In cases of regional metamorphism where the geological formation is very large, only small and gradual compositional differences were found in outcrops that were hundreds of miles apart (3). The use of trace elements to determine patterns of resource use is not as helpful as when distinct outcrops have been identified.

Analysis of Ceramic Pottery

Pottery is produced by the conversion of sedimentary clay (produced by the weathering of rocks) into hard rocklike objects. The clay minerals, which were formed by the chemical decomposition of certain rock-forming minerals, contain trace elements. The sediments in which these clays are found, however, also contain fragments of the primary minerals from the parent rock (including grains of silica sand). These detrital components, which result from the physical and chemical breakdown of minerals, are often accompanied by authigenic minerals that are chemically precipitated from aqueous solutions. In some ceramics, additional components were added as temper during production.

If a clay sediment containing a high concentration of authigenic calcite ($CaCO_3$) was used, the pottery produced was generally more vitreous (glassy). Where a sediment contained very low amounts of Ca, there were technological advantages to adding Ca as shell or calcite temper. In Chapter 3 by Allen, Hamroush, and Hoffman, the importance of Ca in the production of Predynastic Egyptian pottery is discussed. Chapter 3 also describes some of the difficulties in characterizing the river sediments that were used for pottery production. Sediments tend to be variable in composition and clay content because during deposition from the river water there is sorting of grains according to size and density.

Early potters selected materials that were rich in clay, but in some cases, they separated coarser components from the sediments. In other cases, they added temper. In Chapter 4, Bishop and Neff discuss the effect of temper on the chemical analysis of pottery. They point out that the concentration of an element measured in a ceramic artifact can be represented mathematically. Bishop and Neff show that the analysis of pottery sherds alone can give rise to groups with different compositions, because of the temper used rather than the source of the clay. Precise instrumental neutron activation analysis (INAA) data was used to show how variability in the chemical composition of ceramics can be accounted for by the addition of volcanic ash. In this case, the clustering routines of the statistical analysis did not show differences in the sources of the clays, but differences in the production methods. Unless there are some geological and compositional differences between the clay sediments used, the analysis of pottery will not show different origins and could give misleading results when differences due to such things as temper occur. Chapter 4 shows clearly how the speed and versatility of various statistical or pattern-recognition approaches to data analysis can outstrip the logic of the analysis.

In Chapter 5, Olin and Blackman explain that differences in the chemical compositions of pottery are caused by both the use of temper and by chemical and mineralogical differences in the source of the clay. Olin and Blackman report on the continuation of their studies of majolica (a common earthenware pottery) from the Spanish Colonial period in Mexico. They used INAA as well as microscopic examination of the minerals to show that majolica produced in Spain could be distinguished from that produced in Mexico. Volcanic temper was present in the ceramics produced in Mexico, and the chemical analysis of these local ceramics suggested different production centers in Mexico. The discovery of a chemically distinct group of sherds added to the typological classifications of this pottery.

Many times it is impossible to identify and sample the sources of clay used by ancient artisans. In these cases, ceramic studies must include a large number of samples. In Chapter 6, Hancock and Fleming give an example of this kind of study. A large number of samples of Neolithic Iranian ceramic sherds (207 from three time periods) were analyzed as the first step in determining whether there were any real compositional differences in the pottery. The normalization approach described in this chapter has a geological basis because minerals like quartz (sand) can act as dilutants. Normalization helps avoid misinterpretation of the analytical results for ceramics. These authors also address some of the analytical problems, including contamination and leaching of the sherd when it is buried.

Many trace element studies of archaeological samples have used neutron activation analysis (NAA). Although this technique is not useful for all elements, it is very sensitive for many of those that have proved to be valuable indicators of geochemical processes (e.g., the rare earth elements). The precision of the actual measurements is usually high and easy to determine. Samples can be irradiated with little or no sample preparation, so there are few chances of contamination during the analysis. However, the limited number of nuclear reactors severely limits access to this type of analysis. When samples are sent to a distant laboratory for analysis, the critical interaction between archaeologist and analyst can be lost.

In Chapter 7, Foust, Ambler, and Turner report on studies of from the southwestern United States using atomic absorption (AA) and inductively coupled plasma (ICP) emission spectroscopy as alternatives to the much more expensive NAA. Dissolving lithic samples to prepare the solutions necessary for this type of analysis is a problem, but now the methodology is much more readily available to archaeologists. In this study, three clays could be distinguished on the basis of the 8 elements they measured. Although many of the samples could be assigned to a particular clay source, the analysis of the Anasazi pottery sherds gave results that were somewhat ambiguous. The mixing of clays or the addition of temper is described as the reason that high levels of certain trace elements have been found in some of the sherds. For ceramic studies where large numbers of samples must be analyzed, the less expensive AA procedure may make chemical analysis a more routine part of archaeological studies.

One aspect of ancient ceramic technology was the addition of various tempers to the clays. In some cases, the addition of salts was used to affect the colors and texture of the pottery (*see* Chapter 3). In Chapter 8, Mitchell and Hart discuss a somewhat different approach to the study of ceramic technology. They looked at the mineralogical changes that clays undergo when they are heated in the presence of other minerals (especially those containing Ca). The mineralogy, as determined by using X-ray diffraction, varied as a function of temperature. Although such changes have the potential to be used to determine firing temperatures, it is often necessary to confirm these temperatures by other methods. Nevertheless, this research is an example of the basic studies in ceramics and mineralogy that have helped elucidate the earlier technologies.

Composition of Metal Artifacts and Art Objects

The history of the use of various metals is preserved as a part of the archaeological record. The age of a metal artifact can be determined either by carbon-14 analysis of organic remains associated with the artifact or by its association with ceramics of known age. In many cases, the technology used for working the metal or extracting it from ores can be obtained from careful chemical and microscopic analysis of the metal artifact itself. Most of the metals that were used by early people are not very abundant in the earth's crust. For example, the average abundance of Cu in the crust is only about 45 ppm. Fortunately, the processes of geochemical differentiation and mineralization tended to concentrate specific metals in localized deposits. These deposits were often readily recognizable by their color and texture. Whereas color may have been important to the discovery of metals and ores, the thermodynamic stability of the metal compounds (the ease with which the ores could be reduced to metals) determined the order in which they were used. It is interesting that only 8 of the 70 metallic elements (Fe, Cu,

As, Sn, Ag, Au, Pb, and Hg) were recognized and used before the 18th century.

The analysis of metal artifacts has been used extensively to differentiate materials by sources. X-ray fluorescence and neutron activation analysis have both proved valuable in determining elemental concentrations. Native metals, such as gold, contained impurities that could, in some cases, be used to characterize their sources. However, the smelting of ores to recover the metals often changed the concentrations of impurities. Later, as alloys (e.g., bronze and brass) were produced, the compositions were intentionally altered and controlled. In some cases, the re-use of materials or the lack of quality control made the alloy composition quite variable (especially in terms of the trace components).

A promising approach to this problem has been the use of lead isotope ratios to characterize sources. Chapter 9 by Gale and Stos-Gale is an example of this type of study. The isotopic ratios of lead are variable because some of the isotopes are the daughters from the radioactive decay of uranium and thorium (*4*). Even though the amount of lead in bronze artifacts is small, Gale has been able to distinguish between sources of the ore on the basis of the ratios of the various lead isotopes. The sources of silver, lead, and copper in the Bronze Age Mediterranean are discussed.

Chemical analysis, especially detailed examination of inclusions with scanning electron microscopy (SEM), has provided considerable insight into metal-working technology. Chapter 10, by Manea-Krichten, Heidebrecht, and Miller is one example of a technological study. The interpretation of early smelting practices is complicated by the contamination from the ceramic crucibles. This contamination is indeed a problem as some of the fragments (from the archaeological site at Tel Dan, Israel) that this group investigated were metal-fused to ceramic crucibles. Although these results are preliminary, the study does indicate some of the problems in interpretation that can result when the samples are not chosen to answer a specific question. The chapter includes a discussion of the sources for the tin used at this site.

Although Chapter 10 describes a study aimed at elucidating early technological methods, the research presented in Chapter 11 by Carter and Razi shows how the analysis of coins can provide historical political information. This chapter is the latest in a series of chemical studies of Roman coins that have been included in the earlier volumes of the *Archaeological Chemistry* series. These studies have shown how coins were produced and how early Roman mints functioned. Chapter 11 shows how the political fortunes of the Roman Empire affected the composition of the coins. Debasement of the coins reflected periods of political turmoil.

Coins were valuable as money, but they were also artistic endeavors. Metals have been used in numerous ways in art objects. The great value of certain art objects has increased the need for authentication, which is a different aspect of archaeological chemistry. In some cases, the process of authenticating art objects has provided chemists with knowledge about the artists' materials and techniques. Chapter 12 by Williams, Hopke, and Maguire describes the X-ray fluorescence analysis of medieval Limoges enamels. Although this is a preliminary report on these gilded copper ecclesiastical objects, the goal was to determine the metals used to create the different colored enamels. X-ray fluorescence was used to analyze the different colored enamel regions. Although one goal of this research was to understand how the colors were produced, clearly another aspect of this work was to provide a basis to discriminate between different materials of similar appearance.

Another type of study that looks at the decorative, artistic objects of the past is found in Chapter 13 by Demortier. Early gold jewelry was produced from physically working native metals. Intricate objects were fashioned by brazing or soldering individual pieces together. Demortier describes attempts to determine how the brazing alloy used in ancient jewelry was produced. Demortier used X-rays excited with a beam of high-energy protons (PIXIE) to analyze microscopic portions of jewelry thought to be over 2000 years old. This detailed analysis revealed four different kinds of brazing or soldering. Although several sources for this solder have been suggested by later descriptions in the historical literature, Demortier's analysis suggests that Cd was used. The possibility of having a Cd-based solder in antiquities from Iran and Syria is discussed because it has been argued that Cd solders indicate modern forgeries.

Another example of detailed microchemical analysis is given in Chapter 14. Orna, Katon, Lang, Mathews, and Nelson examined the colored portions of several medieval manuscripts by using infrared microspectroscopy. In one example, a forgery was detected on the basis of a blue ink containing ferric ferrocyanide that was not used until 500 years after the manuscript was supposed to have been produced. In many cases, this type of examination can help authenticate a particular object and can aid in the development of greater understanding of the techniques used by the artists. In some instances, inorganic salts were used to produce colored inks. These inks were analyzed by using X-ray diffraction. For tiny samples of organic inks, a microscopic attachment to a Fourier transform infrared spectrometer (FTIR) was valuable. Besides showing which materials were used for sizing and tanning the parchments, FTIR helped Orna et al. identify the source of the red ink as some species of insect.

The importance of preserving archaeological monuments is discussed by Burns and Matsui in Chapter 15. Chemical information not only helps our understanding of early technologies, but can help answer questions about how best to preserve an object. Burns and Matsui describe their studies of very large art objects: the richly decorated tombs of the Egyptian pharaohs. Deterioration of these monuments has resulted from physicochemical processes that are site-specific. Deterioration often depends on such things as the composition of the plaster or the rock from which statues are carved. Chapter 15 describes experiments using X-ray photoelectron spectroscopy and electron microprobe analysis to determine why the painted murals on the walls of tombs have deteriorated so rapidly. The goal of understanding the complex mechanism of deterioration (which involves the plaster, the pigment, and the atmosphere) is to devise conservation measures to protect and restoration practices to revitalize these ancient works of art.

Proteinaceous and Organic Artifacts

Besides the stone or lithic tools, bones represent another large class of artifacts that have survived because of their mineralogical nature. The physical characteristics of these bones have provided much information about human evolution. There have been extensive studies using light and electron microscopes to elucidate the structures of fossil shells, bones, and teeth. In some of these studies, various fossil components seen with the scanning electron microscope were analyzed chemically to ascertain whether the structures were compositionally related to their modern counterparts. These experiments revealed detailed microscopic and compositional similarities. With the advent of improved methodologies, it became clear that bones and similar mineralized portions of living organisms could preserve some of the proteinaceous material of the organism.

Although proteinaceous material did survive, it was altered over time. Some studies have concentrated on the assignment of time scales to the archaeological record on the basis of the changes in the protein composition of bone and teeth over time. Other studies have attempted to discern paleopathological data and dietary information from the analysis of the calcified tissue. However, in all cases, relatively little is known concerning the basic biological processes of postburial change in these proteinaceous materials. Thus, the study of proteins in archaeological materials continues to focus on understanding how the postburial environment affects chemical and physical changes in the proteins themselves, as well as attempting to answer specific archaeological questions relating to diet, age, etc. At the 8th Symposium on Archaeological Chemistry in Denver, a number of papers dealt with the important research on proteinaceous matter in archaeological materials. These papers are being published elsewhere, but some of the research in the field must be summarized for the sake of completeness.

In the pioneering work on the occurrence and stability of proteins and amino acids in fossils, Abelson determined that thermally unstable amino acids such as threonine and serine were either much reduced or absent in fossils, whereas more stable amino acids, such as glycine and alanine, were still present (*see* ref. 5). The total concentration of amino acids decreases dramatically with time. von Endt and Erhardt reported their results concerning the differential chemical disintegration of amino acids in compact faunal bones from three archaeological sites (*6*). They also reported on the composition and state of preservation of proteinaceous material in the calculus from prehistoric human teeth. The results of amino acid analysis of similar bones from each site indicated that there was significant within-site variability, and that the most chemically reactive amino acids varied the most. The stability of bone decreases in proportion to depth of burial and in relation to the severity of the soil environment. The amino acid content of dental calculus differs greatly from that of bone. Amino acid analysis of calculus from three distinct American Indian populations differed by more than 10% between groups.

Isomeric forms of amino acids result from different arrangements of the different atoms or groups of atoms around an asymmetric carbon atom or chiral center. Although laboratory-produced amino acids are usually a mixture of the L and the enantiomeric D isomers, those produced in biological systems are virtually all present in the L configuration. Isoleucine has two chiral centers and can thus exist as the D,L-isoleucine pair, or as the closely related D,L-alloisoleucine pair. Rearrangement of atoms on the chiral carbons of the naturally occurring protein amino acid L-isoleucine results in the formation of the nonprotein amino acid D-alloisoleucine. This rearrangement process, called epimerization, must occur in the proteinaceous material of fossils because alloisoleucine is observed (*see* ref. 7). Indeed, in older fossils the amount of isoleucine decreases and the amount of alloisoleucine increases.

Amino acids such as alanine, aspartic acid, and isoleucine may change from L to D configurations over time by the process of racemization. These changes found in fossils led Hare and Abelson to suggest that epimerization and racemization might be used in techniques for dating shell and other proteinaceous material (*8*). However promising these results appeared, it soon became apparent that there were complications in applying this theory of amino acid dating. For example, the half-life for the epimerization of isoleucine in cow bone was calculated to be 110,000 years in bone at 20 °C and 290,000 years at 15 °C. This finding shows how sensitive the reaction is to temperature, a factor that is often not known for archaeological specimens (*9*). Some problems can be overcome if radiocarbon dating of bone from the same site can be used to "calibrate" the racemization rates of amino acids.

Problems remain, and differing lines of evidence suggest that racemization dating of bone is, at best, tentative, and that many of the "early" dates for human skeletons are wrong. For instance, the Sunnyvale skeleton (racemization dated at 70,000 years B.P.) is morphologically indistinguishable for skeletons dated between 400 and 1600 years B.P. (*10*). Another important and controversial skeleton was described in Denver. Bada and Masters presented the results of a detailed study of the general composition and extent of racemization of amino acids isolated from the Del Mar Man skeleton (*11*).

These remains were found eroding from the face of a coastal bluff in Del Mar, CA, in 1929. A variety of dates have been obtained for these remains, depending upon which materials were analyzed and which dating method was used. Bada and Masters reported on new dates based upon the new and more sensitive C-14 dating technique. The results, obtained on amino acids isolated from the bone, indicated that the skeleton was 5400 years old. These results were made possible by the more recent important advances in carbon dating that are described in Chapter 16.

In Chapter 16, Harbottle and Heino address the particular application of carbon isotopic dating to textile fibers. The discussion of the accelerator mass spectrometric (AMS) technique, which allows the accurate measurements of isotopic ratios in very small amounts of carbon, shows why this technique has had such an important impact on the field. Chapter 16 summarizes the progress in this very important methodology that remains the most accurate absolute measure of the ages of artifacts.

Other isotopic measurements may also be useful in providing information on proteinaceous material. Because amino acids contain nitrogen, an analysis of the nitrogen remaining in prehistoric bones may provide an approximate indication of the quantity of protein remaining in the bone. Perhaps of greater importance, the nitrogen that remains may also provide insight into diet. In the papers presented in Denver by Hare, Fogel, Stafford, and Hoering (*12*), as well as in the paper by Bada and Masters (*11*), the isotopic analysis of carbon and nitrogen were described. Hare noted that most of the amino acids in collagen from bones of animals that had been reared on controlled diets were enriched in C-13 and N-15 as compared to the corresponding amino acids in their food. Threonine was an exception and was depleted in N-15 relative to that in the food source. Unlike most amino acids, threonine does not derive its nitrogen via transamination from the cellular nitrogen pool, but rather inherits it from the threonine that is consumed. The isotopic fractionation of N-15 in threonine has been suggested as a means to distinguish between herbivores and carnivores at different levels in food chains. The patterns of stable isotope content in amino acids from bones of fossil herbivores and carnivores were reported to be well-preserved.

There can be little question about the the value of absolute techniques for determining the ages of artifacts. As the ability to measure isotopic ratios has improved, not only has the impact of carbon dating improved, but a new technique has been proposed. In Chapter 17, Taylor, Slota, Henning, Kutshiera, and Paul discuss the feasibility of using a long-lived isotope of calcium (Ca-41) as a means of determining the absolute age of a bone. The isotopic abundances are low so the accelerator mass spectrometric approach (AMS) has been used to make the measurements of Ca-41. Where the application of carbon dating to the determination of the age of bone requires the survival of some of the proteinaceous matter (e.g. collagen), the inorganic portion of the bone, which contains the calcium, survives more readily. Contamination of the amino acid fraction in bone samples is another problem with carbon dating which, it was reasoned, should not be as great a problem when the inorganic mineral phase is used to determine the age. Finally, one of the reasons that radiocalcium dating can complement carbon dating is that the half life of Ca-41 is longer ($t_{1/2}$ of about 10^5 years compared to C-14 $t_{1/2}$ of 5730 years). Taylor et al. discuss the current status of the radiocalcium dating method in terms of the potential problems including environmental and diagenetic effects.

Some of the potential problems with using the inorganic portion of bones for age determination and for the discovery of the constituents of ancient diets are discussed in Chapter 18. This chapter by El-Kammar, Allen, and Hancock shows some of the changes that occur when bones are buried. The scanning electron microscope (SEM) and X-ray diffractometer were used to show how new mineral phases, including some that contain calcium, can fill the voids left in bone as the organic fraction decomposes. Contact between the bone and soil not only affects the major elements like calcium, but the trace elements as well. Comparisons between bones buried in the Nile delta and bones of mummies that were protected from direct contact with the soil show very clearly that diagenesis and contamination are problems when the inorganic portion of the bone is analyzed. The photomicrographs show clearly that even the interior regions of a bone can be affected if ground water is present.

Part of the problem with using the more developed radiocarbon dating techniques with bone samples is that, in many cases, the bones being studied are too old. The longer half-life for the Ca-41 will make it more valuable for determining the age of the paleolithic bones that are of great interest to anthropologists. In Chapter 19, Robins, Sales, and Oduwole describe another technique that appears to be valuable for extending the dates of bone samples to materials older than can be obtained with the radiocarbon method. This is not, however, an absolute technique based upon changes in the isotopic ratios, but a technique based upon changes in the electron spin resonance (ESR) signal. The chapter describes some of the problems with using this technique, including the effects of temperatures on the change in the ESR spectrum.

Organic Residues and Fibers

In a number of instances, organic materials other than those in bones, shells, and teeth survive the ravages of time. One of the more interesting examples is the nearly rocklike organic resin called amber. Long noted for its attractive color, this material had great value. Because amber is a resin from certain plants, differences in its composition reflect the original plants. In Chapter 20, Lambert, Lee, Welch, and Frye report on their continuing investigation that is aimed at "fingerprinting" amber from different sources. This chapter describes the use of C-13 nuclear magnetic resonance (NMR) spectroscopy for the characterization of amber. Chapter 20 reports that amber from two Mexican sites are similar to each other. These results differ from results obtained for Baltic and Dominican amber.

The focus of most archaeological ceramic studies has been on provenance or technology. There is also a growing body of specific evidence on how the pottery was used. Chapter 21 by Beck, Smart, and Ossenkop describe the organic tars used to line ancient Mediterranean amphoras. Chapter 21 includes a description of how the residues from amphora contents can be analyzed. As in most cases where unknown organic materials are encountered, the most powerful analytical technique is gas chromatography with mass spectrometric detection (GC–MS). This technique is expensive for the analysis of large numbers of samples.

Beck et al. show the advantages of using simpler, less expensive methods like FTIR and thin layer chromatography (TLC). These techniques provide speed and ease of analysis. When used to investigate complex mixtures like the natural products examined by Beck et al., more extensive methods may be required to validate the findings.

One of the most exciting new techniques in analytical chemistry has been the development of sensitive methods based on immunological responses. One of the main advantages of immunological techniques is their high selectivity, even in complex matrices. This approach has made a dramatic impact on clinical chemistry. As new probes are developed, they will probably be applied to ancient proteinaceous materials.

Radioimmunoassay techniques are capable of measuring as little as 10^{-12} g of protein in fossils. Because bone collagen is somewhat species-specific by virtue of its nonhelical ends, the system has potential for clarifying the phylogeny of humans. Lowenstein has developed antisera in rabbits capable of establishing the immunological distance (I.D.) between humans, pygmy and common chimp, rhesus monkey, guinea pig, dog, cat, calf, and mouse (*13*). By using the collagen I.D. Lowenstein suggested that humans and chimps are close to each other, as are rats, mice and guinea pigs. These techniques were then used to test antihuman and antichimp sera against a series of six hominid fossils ranging in age from 1000 (Hungary) to 1,900,000 (Omo) years old. It was discovered that immunological activity decreased with time and changed from being humanlike to being chimplike in the time period between Cro-Magnon and Neanderthal. These experiments demonstrated that immunologically reactive collagen fragments persist in 2-million-year-old fossils and exhibit species-specific antibody binding.

Radioimmunoassay techniques may aid in clarifying genetic relationships if sufficient immunological activity is maintained in archaeological bones. In Chapter 22, Herr, Benjamin, and Woodward discuss some new immunological tests that can distinguish between blood and tissue of human and animal origin. Although the probe described by Herr has not been used to examine any archaeological samples, it is included in this volume because it provides a valuable overview of the field. The review of this rapidly developing field will be valuable to archaeologists as new probes are developed that could provide new insight into archaeological problems. This technique may soon provide more specific genetic information, even with small amounts of preserved DNA in hair and in other proteinaceous material that survives as part of the archaeological record.

In Chapter 22, Herr et al. describe a monoclonal antibody used to test for human albumin in body fluids. The development of this monoclonal antibody was aimed at forensic evidence, but if (and it must be tested) degradation processes do not destroy the antigenic site that this antibody recognizes, it could be used for archaeological samples. Certainly as these techniques are developed they will prove valuable in the studies of objects that are suspected of containing blood. For example, these newer methods could help further define an artifact like the Shroud of Turin, which is described in Chapter 23.

In Chapter 23, Dinegar, Adler, and Jumper present a summary of the known history and the earlier scientific examination of the famous religious artifact known as the Shroud of Turin. On the basis of the previous interpretations of the changes in the cellulose fibers and the blood stains, a plan is presented for more specific testing using improved techniques. The authors propose that other tests be made at the same time that samples of the Shroud are being analyzed. Although these other tests may help clarify the changes in the fibers of this historical textile, the real test of its antiquity will be the radiocarbon analysis. Nevertheless, the continued progress on the analysis of natural fibers will provide more detail on the changes that account for the body images and stains on the Shroud. (The image is interpreted as having been the result of an oxidation–dehydration reaction.) In addition, study of the Shroud may aid conservationists as they attempt to better understand natural aging processes and find ways to preserve these more fragile artifacts.

Ballard, Koestler, Blair, and Indictor discuss some of the problems with the preservation of silk in Chapter 24. Their work was aimed at understanding how various components added to the silk during the manufacturing of a number of historical flags affected the embrittlement and decomposition of the material. They found that colorants and chemicals added to weight the natural silk fibers could be detected by using X-ray fluorescence. They correlated the inorganic additives to the deterioration of the flags.

On the other hand, organic silk itself is the subject of Chapter 25 by Hersh, Tucker, and Becker. A number of historical silk fibers were examined in order to understand the chemistry of the degradation process. The roles of photochemical processes as well as heat are discussed. The chemical processes of aging are complex and slow. Either historical samples must be studied, or methods must be found to reliably accelerate the aging process in the laboratory. The eventual object of the studies presented in Chapters 24 and 25 is to provide better ways to preserve the silk artifacts.

In Chapter 26, Jakes and Angel show that the analysis of elemental concentrations can also aid in the study of ancient textiles. In this chapter, the inorganic components and their distributions in the fibers were the basis of a method to identify the fibers. Jakes and Angel used a scanning electron microscope to image the fiber's morphology, and energy dispersive X-ray spectrometry to determine elemental distributions. This approach is suitable for the very small samples often encountered as archaeological fibers.

In earlier work, Jakes and coworkers have shown that it is not always necessary for the organic fibers to survive in order to obtain information on the textile. Chapter 27 by Jakes, Sibley, Kuttruff, Wimberley, Malec, and Bajamonde discusses the imprints or pseudomorphs of fabrics that are made when metal salts replace the decomposing organic matrix. In this research the metal pseudomorphs were from bronze weapons that are over 3000 years old. Jakes et al. used photomicrographs to observe the manner in which the yarn was woven to produce the original fabric. The statistical methods that were used to identify and correlate the patterns in these textile pseudomorphs are very similar to those originally used in provenance studies of the less fragile stone and pottery artifacts. This study shows the importance that computers and statistical methods have attained in aiding in the organization of the extensive data sets generated as a part of archaeological studies.

Conclusions

These examples of how chemists contribute to archaeology are not intended to provide a complete review of the field. The purpose of this chapter was to indicate how the field has matured. In some of the earlier work, the analysis of inorganic artifacts took more of a "shotgun" approach. It was thought that the probability of differentiating between materials was enhanced by analyzing for a large number of elements. For this reason, multielement techniques with proven accuracy like neutron activation analysis were favored. Grouping artifacts on the basis of chemical similarities led to increased use of various statistical protocols (computer programs). However, as different types of artifacts were studied (by chemical means), it became increasingly clear that a more detailed understanding of the materials themselves made it more likely that useful archaeological information could be obtained.

More detailed studies, using an ever wider array of analytical techniques, have increased our understanding of many natural materials and of the early technologies. This greater understanding of the materials provides a better rationale for the choice of samples and techniques to be used. The chapters in this volume reflect the diversity of materials being studied. In each chapter, there has been an attempt to understand the material itself as well as to increase our knowledge of the past.

Literature Cited

1. Blackman, M. J. In *Archeological Chemistry III*; Lambert, J., Ed.; Advances in Chemistry Series No. 205; American Chemical Society: Washington, DC, 1983; pp 19–41.
2. Allen, R. O.; Pennel, S. E. In *Archeological Chemistry II*; Carter, G., Ed; Advances in Chemistry Series No. 171; American Chemical Society: Washington, DC, 1978; pp 230–257.
3. Allen, R. O.; Hamroush, H.; Nagle, C.; Fitzhugh, W. In *Archeological Chemistry III*; Lambert, J., Ed.; Advances in Chemistry Series No. 205; American Chemical Society: Washington, DC, 1983; pp 3–18.
4. Gale, N. H.; Stos-Gale, Z. *Sci. Am.* **1981,** *244,* 176–192.
5. Abelson, P. H. In *Organic Geochemistry;* Berger, I. A., Ed.; McMillan: New York, 1963; pp 431–455.
6. von Endt, D. W.; Erhardt, W. D. Chemical Studies on the Proteins of Archaeological Bone and Teeth, Abstracts *of Papers*, 193rd Meeting of the American Chemical Society, American Chemical Society: Washington, DC, 1987.
7. Hare, P. E.; Mitterer, R. M. *Year Book, Carnegie Institution of Washington* **1967,** *65*, 362–364.
8. Hare, P. E.; Abelson, P. H. *Year Book, Carnegie Institution of Washington* **1968,** *66*, 526–528.
9. Bada, J. L. *Earth and Planet Sci. Let.* **1972,** *15*, 223–231.
10. Gerow, B. A. *Soc. Calif. Arch., Occasional Papers* **1981**, *3*, 1–12.
11. Bada, J. L.; Masters, P. M. Amino Acids in the Del Mar Man Skeleton, Abstracts *of Papers*, 193rd Meeting of the American Chemical Society, American Chemical Society: Washington, DC, 1987.
12. Hare, P. E.; Fogel, M. L.; Stafford, T. W.; Hoering, T. C.; Mitchell, A. D. Stable Isotopes in Amino Acids from Fossil Bones and Their Relationship to Ancient Diets, Abstracts *of Papers*, 193rd Meeting of the American Chemical Society, American Chemical Society: Washington, DC, 1987.
13. Lowenstein, J. M. In *Biogeochemistry of Amino Acids*; Hare, P. E.; Hoering, T. C.; King, K. Jr., Eds; John Wiley and Sons: New York, 1980; pp 41–51.

RECEIVED for review February 17, 1988. ACCEPTED revised manuscript August 5, 1988.

LITHIC AND CERAMIC MATERIALS

2

Lithic Material from the Mesolithic and Early Neolithic Periods of Portugal

Instrumental Neutron Activation Analysis

R. G. V. Hancock[1], L. A. Pavlish[2], and P. J. Sheppard[3]

[1]SLOWPOKE Reactor Facility and the Department of Chemical Engineering and Applied Chemistry, University of Toronto, Toronto, Ontario, Canada M5S 1A4
[2]Archaeometry Laboratory, Department of Physics, and Department of Anthropology, McLennan Physical Laboratories, 60 St. George St., Toronto, Ontario, Canada M5S 1A7
[3]Department of Anthropology, University of Auckland, Auckland, New Zealand

Lithic artifacts from Portugal were analyzed by rapid instrumental neutron activation analysis (INAA). Concentrations of 15 short-lived isotope-producing elements were determined in both archaeological and geological samples. The archaeochemical objectives were the identification and source determination of materials used in prehistoric stone tool production. Elements measured in the materials were present in low concentrations, and many were attributable to contamination and diagenesis. INAA results demonstrate the heretofore unknown use of rhyolitelike rock in the Portuguese coastal Mesolithic period and suggest as its source a volcanic zone 20 km inland.

Members of the Canadian–Portuguese archaeological project are engaged in a program of archaeological research on the Holocene prehistory of southern Portugal. The ultimate goal of this project is the identification of the causes and effects of the transition from a hunting-and-gathering economy to a more settled Neolithic way of life (i.e., hunting–gathering, pastoralism, and agriculture). The project focuses on the effects of this change on diet, health, and demography. These factors may

0065–2393/89/0220–0021$06.00/0

be particularly useful in evaluating theories concerned with the relationship between developing territorialism and the advent of megalithic structures along the Atlantic coast (*1*).

One aspect of this project is the analysis of the stone tools and manufacturing debris recovered from excavated sites. In addition to standard typological and attribute analysis, a program of identification and source determination for lithic materials using instrumental neutron activation analysis (INAA) and petrography was carried out. The purpose of this research program was to provide insights into questions about ancient human behavior: Where, when, why, and how did prehistoric peoples use the natural resources in their environs (*2*)?

Archaeological Background

Lithic artifacts were collected from three sites south of the industrial town of Sines along the western coast of the southern Portuguese Alentejo province (*see* Map I). Two of these sites, Samouqueira and Palheiroes do Alegra, are 30 km apart on cliffs above the modern sea shore; the third site, Fiais, is 12

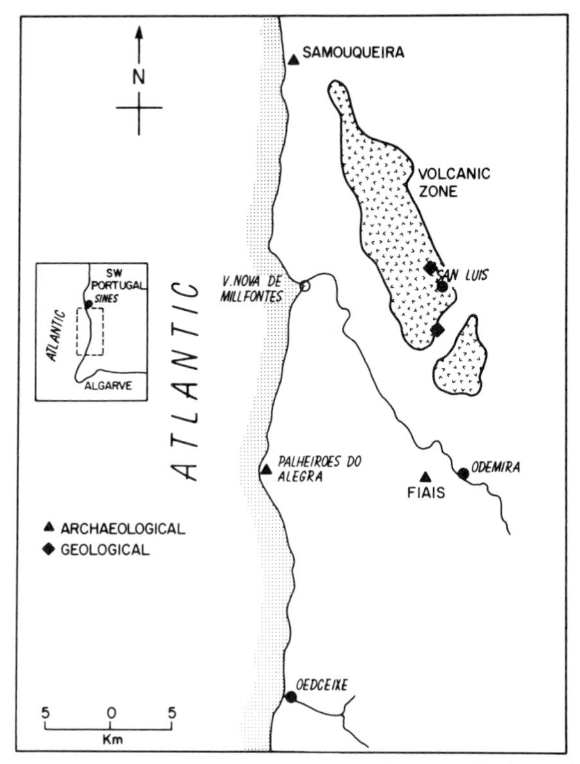

Map I. Map of the region of southern Portugal from which the archaeological and geological samples were taken.

km inland from Palheiroes do Alegra near the town of Oedmira. The site of Samouqueira, which is dated at 6370 ± 70 B.P. (TO-130)*, yielded the remains of people who used both local marine (limpets, mussels, and fish) and terrestrial (red deer, wild boar, and rabbit) resources. A comparison of stable carbon isotope data on human bone from Samouqueira with similar data from older Mesolithic and younger Neolithic sites indicated that the individuals from Samouqueira had a varied "Mesolithic" diet with food obtained from hunting and gathering (*1*).

Fiais and Palheiroes do Alegra have been tentatively dated on typological grounds. Fiais is assigned to the Early Aceramic Neolithic period (Stage IB, ca. 6000–5000 B.P.), and Palheiroes do Alegra is assigned to the Early Mesolithic period. Data on fauna (predominantly red deer) and the lack of any permanent structures at Fiais suggest that it represents a large hunting camp. Palheiroes do Alegra is a disturbed site in a wind-blown erosional depression with minimal preservation of faunal material. However, its location and tool assemblage suggest the inhabitants may have had an economy and way of life similar to that indicated for Samouqueira. No evidence was found at these sites of a sedentary life based on domesticated plants and animals (*3*, *4*).

Raw Material Collection and Classification

Over 3000 artifacts from the three sites have been visually sorted and analyzed into categories according to color, texture, and inferred general rock type. Table I shows the rock types that were used to make stone tools. These types include a wide variety of very fine-grained rocks with conchoidal fracture, which were classified as chert, as well as quartzite, graywacke, massive milky quartz, rock crystal quartz, mudstone, and miscellaneous coarse-grained igneous rock. These rock types are mostly SiO_2 and have low concentrations of elemental impurities that permit differentiation.

At the coastal sites, graywacke and quartzite cobbles were used extensively in the production of large crude chopping tools, whereas the fine-grained materials, including rock crystal quartz, were used to produce microlithic assemblages. These small tools are generally assumed to have been hafted as barbs or projectile points.

The graywacke and quartzite cobbles used for tool making were probably obtained locally from the storm beaches immediately below the coastal sites. Conversely, chert or quartz crystals have not been found on coastal beaches or adjacent regions, despite intensive survey. Although potential chert-bearing calcareous sediments are located inland, no source is known to exist in this region of southern Portugal (*5*).

* TO-130 is a radiocarbon data log number that permits any interested party to contact the Toronto ISOTRACE Lab and request data on date 130.

Table I. Visual Field Classification of Lithic Materials by Rock Type and Site

	Samouqueira		*Pal. do Alegra*		*Fiais*	
Rock Type	%	x^a	%	x	%	x
Fine-grained chert	44.8%	43	58.1%	22	78.1%	31
Quartzite	19.2	9	5.0	1	3.6	1
Graywacke	34.0	6	26.4	8	1.7	—
Milky quart	—	—	7.4	1	6.5	1
Rock crystal quartz	2.0	1	0.3	—	9.3	1
Misc.	—	—	2.8	—	0.8	—
Number of samples	1340	59	994	32	1025	34

[a]Number of INAA samples taken.

At the inland site of Fiais, the absence of a local source of river or beach cobbles probably accounts for the limited importance of quartzite and graywacke at the site. Large cobble chopping tools made of these materials are absent from the site. Flaked rock crystal quartz is common at Fiais, a fact suggesting that these crystals may occur naturally in the area (*6*).

Procedure

Samples (156) were taken from 54 reference lithic pieces that represented five rock types. These samples were analyzed at the SLOWPOKE Reactor Facility of the University of Toronto. They were irradiated for 1 min at 2 kW, or for 1 or 2 min at 5 kW (depending on their radioactivity level in preliminary tests). Upon removal from the reactor, the samples, which weighed between 0.1 and 0.3 g, were left to decay for 18 min and were counted for 5 min with a Ge(Li) γ-ray detector coupled to a multichannel analyzer. Trace element concentrations were calculated with the comparator method (*7*). The 15 elements examined were barium, titanium, sodium, aluminum, potassium, manganese, calcium, uranium, dysprosium, strontium, bromine, vanadium, chlorine, magnesium, and silicon. The first seven of these elements were the most useful in the differentiation of major rock types.

Analytical Results

Rhyolites. Figure 1 is a ternary diagram showing standardized, scaled elemental abundances from INAA for potassium, sodium, aluminum, and titanium obtained from the samples. The diagram shows the separations of the various rock types. Figure 2 indicates feldspar-rich compositions by showing the relationship between the aluminum concentration and the potassium and sodium concentrations. These materials, therefore, were most probably associated with the formation of either small, igneous, quick-cooling extrusive lava flows or near-surface, intrusive, dikes or sills. Figures 1 and

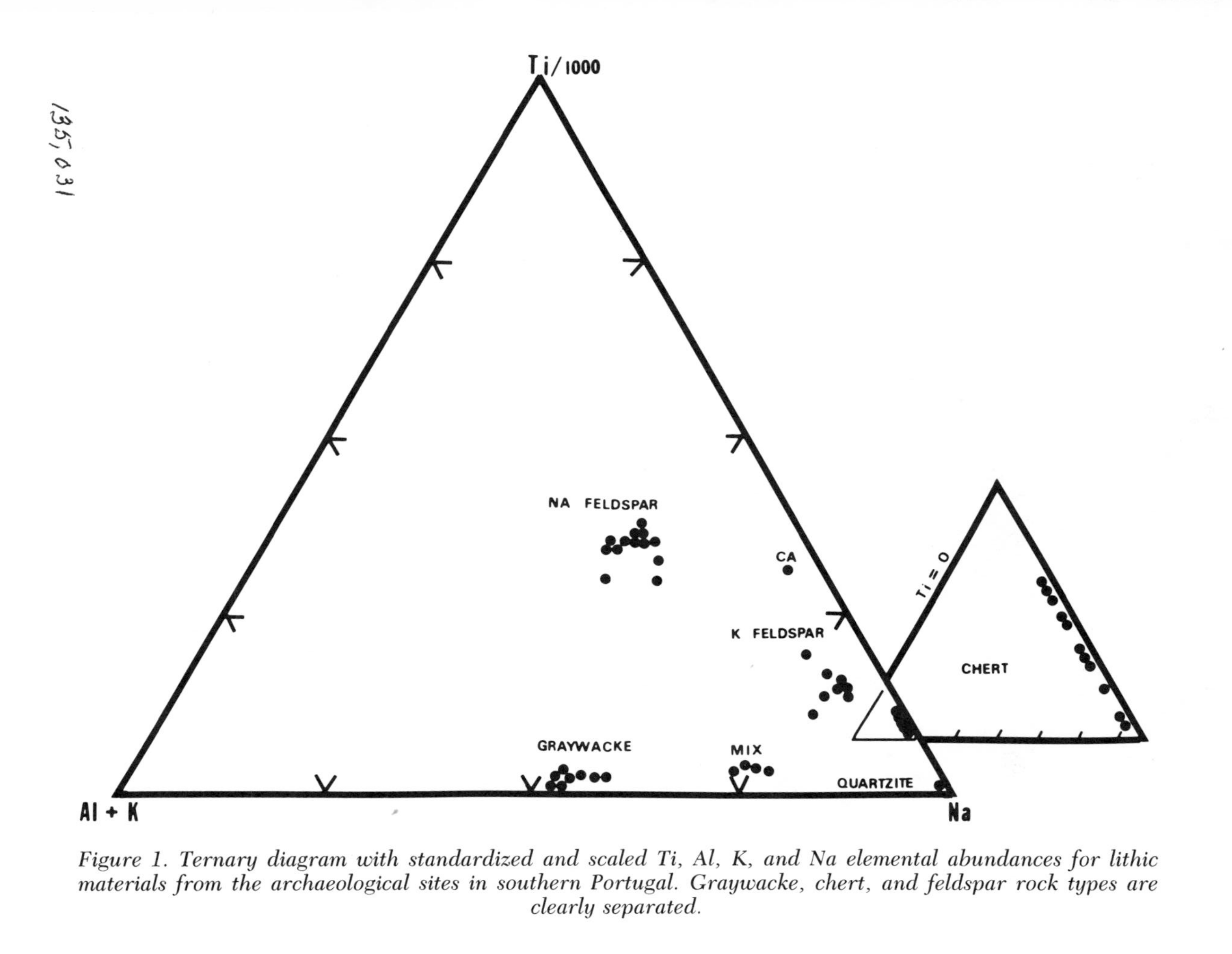

Figure 1. Ternary diagram with standardized and scaled Ti, Al, K, and Na elemental abundances for lithic materials from the archaeological sites in southern Portugal. Graywacke, chert, and feldspar rock types are clearly separated.

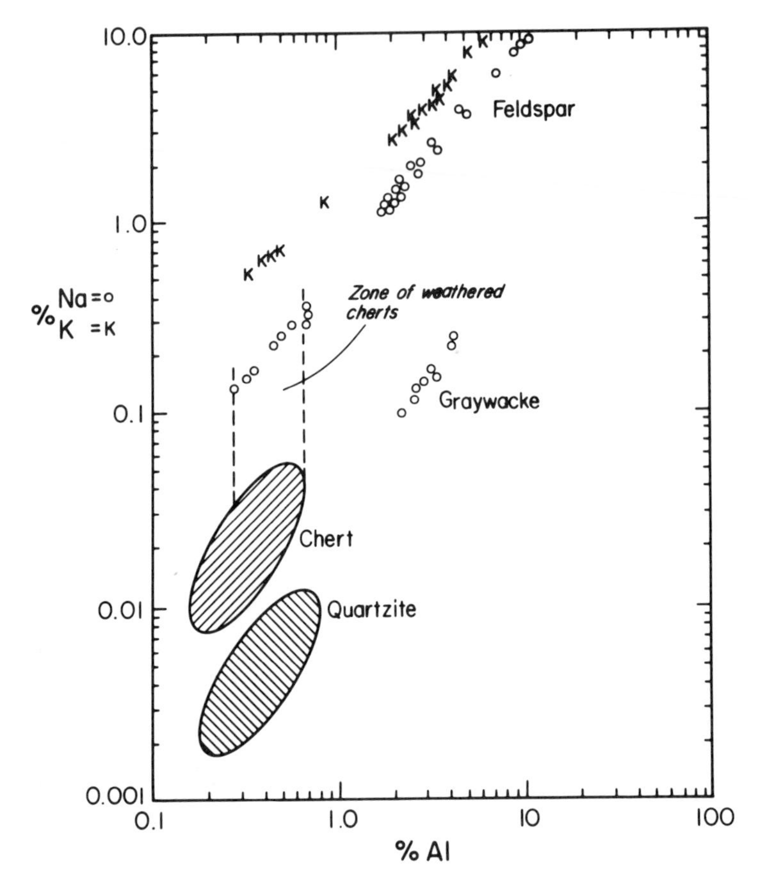

Figure 2. Plot of the relationship between Al and K or Na in the feldspar samples. The graywacke, chert, and quartzite also have unique Al/Na values.

2 show that the relative abundance of various elements is inversely related to the degree of SiO_2 purity in the rock specimens.

Subsequent examination of petrographic thin sections confirmed that the rock was igneous with a rhyolitelike composition made of a fine-grained quartz crystal matrix with feldspar inclusions. This material formed a large percentage of the fine-grained material from both coastal sites. This material constituted, however, only a very small percentage of the material from the inland site of Fiais (Table II).

The INAA data showed that 42.4% of the samples visually identified as chert in the field were rich in feldspar minerals (*see* Table II), and confirmed that the reference collection was made of quartzite, (K–feldspar-rich rock and Na–feldspar-rich rock), graywacke, and chert (8).

A series of Portuguese geological rhyolite samples were collected from the area around San Luis (*see* Map I) for comparison to the archaeological samples. These geological samples are visually similar to some rhyolitelike rocks from archaeological sites. The INAA data indicate that although there are general similarities between the geological and archaeological materials, the two materials are not identical (Table III).

Cherts. Our analyses of siliceous materials from the three sites clearly separate cherts and quartzite from the rhyolitelike rocks (Figure 2). Unweathered and relatively pure cherts and quartzites have Al concentrations in the range of 0.15% to 0.30%. [Although aluminum concentrations may increase because of weathering, the aluminum "base value" is generated by the neutron–proton reaction on ^{28}Si and provides a standard check for SiO_2 purity (cf. Figure 2).] Other elements (Ti, Ca, Na, K) were also present in small concentrations. Table IV shows analytical results for Al, Na, K, Ca, Ti, and Mn from cherts. On the basis of their impurity levels, the cherts were categorized as very clean, clean, or dirty. The impurities were, for the most part, caused by the presence of mineral phases other than quartz (*9*, *10*).

Table II. The Reclassification of Fine-Grained Chert into Chert and Rhyolitelike Rock on the Basis of INAA Results

Rock Type	*Samouqueira*	*Pal. do Alegra*	*Fiais*
Chert	42.4	69.0	99.7
Rhyolitelike	57.6	31.0	0.3
Number of samples	600	578	801

NOTE: Results for chert and rhyolitelike rock are given in percents.

Table III. Representative INAA Data from Samples of Geological and Archaeological Rhyolitelike Rock

Rocks	*Al* (%)	*Na* (%)	*K* (%)	*Ba* (*ppm*)	*Ti* (*ppm*)	*Mn* (*ppm*)	*Number of Samples*
			Geological				
N. San Luis	6.0	0.10	8.0	5000	100	0–7	16
S. San Luis	4.0	0.10	6.0	400	400	3–8	15
			Archaeological				
Samouqueira	5.0	3.0	1.5	500	500	0–25	25
P.d. Alegra	9.0	7.5	0.0	0.0	4000	0–100	9

Table IV. Representative INAA Data for Different Chert Qualities

Element	*Very Clean*	*Clean*	*Dirty*
Al (%)	0.25 ± 0.05	0.30 ± 0.05	0.45 ± 0.05
Na (%)	0.01–0.02	0.02–0.03	>0.03
K (%)	<0.005	<0.02	>0.02
Ca (%)	0.0–0.15	0.0–0.50	0.0–1.0
Ti (ppm)	<20.0	20.0–65.0	65.0–250
Mn (ppm)	0.0–20.0	0.0–150	0.0–200

Before provenance work can be carried out, it is necessary to establish the actual degree of weathering and potential diagenetic alteration that the chert materials have undergone. The majority of chert types from Palheiroes do Alegra and Fiais may be susceptible to weathering because they contain Ca. Samouqueira's major chert types appear to be genetically related to volcanic deposits (meta-volcanics) with low impurity levels and less susceptibility to weathering.

Figure 3 shows that, on the basis of Na and Cl concentrations alone, a clear distinction can be made between the coastal site of Palheiroes do Alegra and the inland site of Fiais. If one or the other of these elements had not been analyzed, the data may have been improperly interpreted. The NaCl appears to be a diagenetic factor in the coastal materials. The cherts from coastal Palheiroes do Alegra were deposited in sand dunes and seem to have been weathered by Ca leaching. The presence of Ca indicates that the chert is probably of geologically calcareous origin. The salt from the ocean subsequently percolated through the relatively clean sand and some was deposited in the chert.

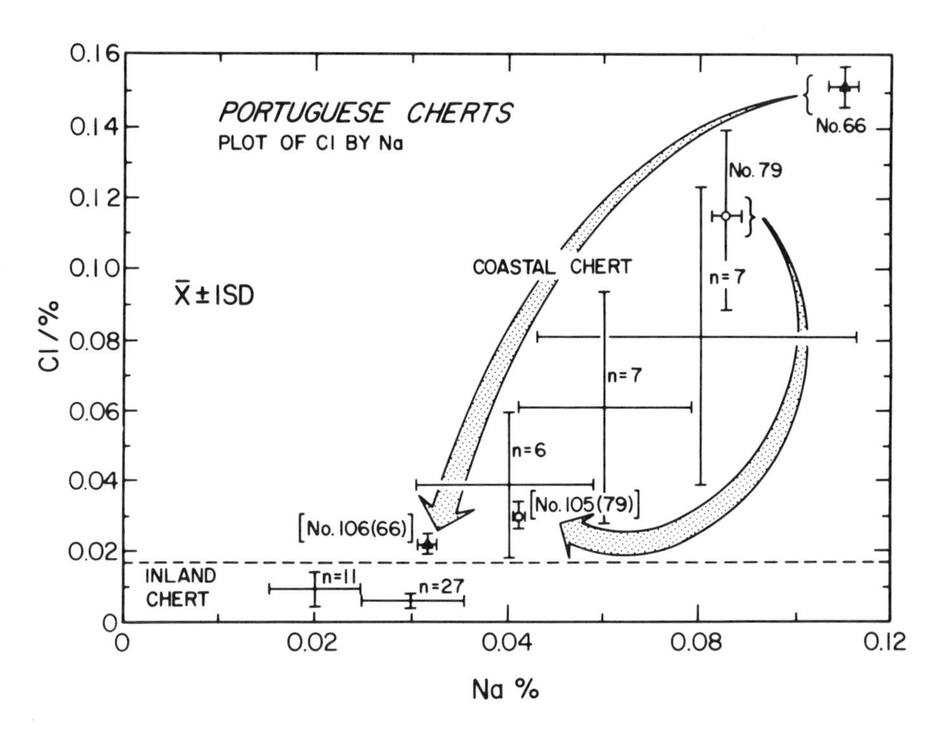

Figure 3. The Na and Cl results obtained from samples at both coastal and interior sites. The average values and standard deviations are plotted for the samples. Salt removal is demonstrated for two samples.

To check this hypothesis, two samples were soaked in distilled water for 24 h. Sample No. 106 (resampled 66) was crushed, and sample No. 105 (resampled 79) was not. After reanalysis, the only significant changes were in the Na and Cl concentrations (*see* Figure 3). These two elements will be of little use in provenance work but can provide an indication of weathering.

The majority of cherts from the inland site of Fiais were also from calcareous deposits. Figure 4 shows that Al and Ti increase at the interior site, which is buried in a clay-rich soil. It is admittedly difficult to determine whether the increases in such elemental concentrations are caused by diagenetic factors or by inhomogeneity of the materials. These changes stand out in sharp contrast to the majority of cherts from Samouqueira, which on the basis of thin-section analyses, are of a meta-volcanic origin. These cherts are less susceptible to weathering and appear to have experienced little

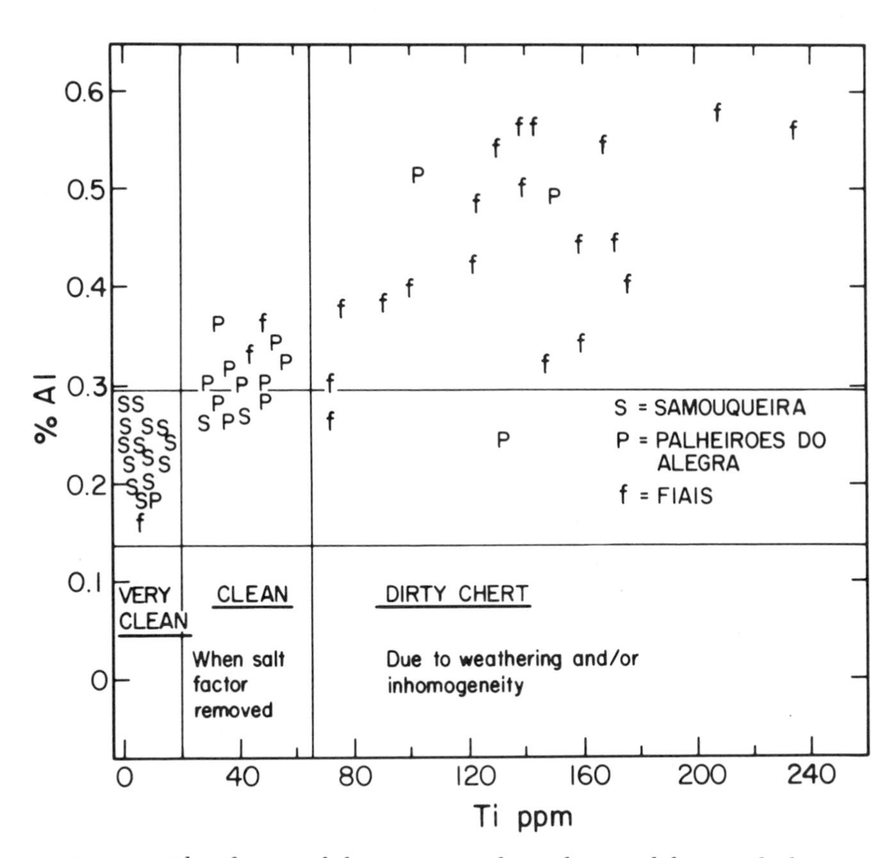

Figure 4. Classification of chert into very clean, clean, and dirty on the basis of relative amounts of Al and Ti.

weathering or diagenetic interference. Cherts from a volcanic setting may provide a greater opportunity for source determination than those from a sedimentary environment.

Conclusions

INAA results indicate that rock types can be efficiently separated when macro–visual characteristics are ambiguous. The provenance of both volcanic and chert materials is made difficult when weathering, diagenesis, and the quality of the material are considered. We have demonstrated that similar materials weather differently in varying archaeological contexts. It is likely that parent materials weathered in situ also will have unique diagenetic constraints. Therefore, neither geologically weathered nor unweathered materials can be assumed to be comparable to "similar" materials recovered from archaeological sites. As a consequence, clean cherts from both a geological and archaeological context will probably provide the most reliable source material for provenance-oriented analytical programs.

Acknowledgments

The authors thank Cathy D'Andrea and Debbie Ross for their undaunting assistance in this work. The study was partially funded by a grant from the SSHRC (Social Sciences and Humanities Research Council of Canada) and the analysis was made possible through an infrastructure grant to the SLOWPOKE Reactor Facility at the University of Toronto, Canada.

References

1. Lubell, D. Progress report on Social Sciences and Humanities Research Council of Canada (SSHRC) Operating Grant 410-84-0030: Archaeological Research in Portugal, Report on file with SSHRC, Ottowa, 1985.
2. Earle, T. K.; Erickson, J. E. In *Exchange Systems in Prehistory;* Earle, T. K.; Erikson, J. E., Eds.; Academic: New York, 1977.
3. Sheppard, P. J. "Studies of Lithic Material: Technological and Stratigraphic Observations at Samouqueira." Report on file with the Portuguese Geological Survey, Lisbon, 1985.
4. Lubell, D. personal communication.
5. Agar, D. V. *The Geology of Europe;* McGraw-Hill: England, 1980.
6. Sheppard, P. J. "Studies of Lithic Materials from Portugal." Report on file with the Portuguese Geological Survey, Lisbon, 1988.
7. Hancock, R. G. V. *J. International Institute of Conservation, Canadian Group* **1978,** *3*(2), 21-27.
8. Pavlish, L. A.; Sheppard, P. J.; Hancock, R. G. V. "NAA: A Method of Determining Feldspar Types and Their Mixtures in Fine-Grained Rhyolites." Paper presented at the workshop Analytical Tools in Archaeometry 1; Oct. 7–9, 1985, Newark, Delaware.

9. Luedtke, B. E. *American Antiquity* **1978,** *43*, 413-423.
10. Luedtke, B. E. *American Antiquity* **1979,** *44*, 744-757.

RECEIVED for review May 11, 1987. ACCEPTED revised manuscript August 30, 1988.

3

Archaeological Implications of Differences in the Composition of Nile Sediments

Ralph O. Allen[1], Hany Hamroush[1,3], and Michael A. Hoffman[2]

[1]Department of Chemistry, University of Virginia, Charlottesville, VA 22901
[2]Earth Sciences and Resource Institute, University of South Carolina, Columbia, SC 29208

The ability to distinguish Nile sediments from different time periods and different sources allows identification of sediments at different archaeological sites. Nile river sediments representing a range of geological ages from Predynastic sites at Hierakonpolis were analyzed by instrumental neutron activation analysis. Some Nile sediments and the Egyptian pottery produced from them can be distinguished on the basis of the relative distributions of the lanthanides. Other groups of sediments are chemically similar but have differences due to "dilution" by sand (mostly quartz). Fractions of Nile sediments separated on the basis of size showed that most of the trace elements were in the fine-grained fraction, but the trace elements in the sand-size fraction seemed to reflect the relative contributions of the different Nile tributaries and the geology of the drainage basins.

THE USE OF FIRE TO TRANSFORM CLAY PASTE into solid ceramic vessels may have been one of the earliest efforts at chemistry. Since the earliest times, pottery has been made by using clays formed by the weathering of rocks. Sedimentary deposits containing clay minerals also contain fragments of other minerals that are broken from the source rocks as they weather. The chemical composition of the sediments used as a clay source determined some of the characteristics of the pottery that was produced.

[3]Present address: Geology Department, Faculty of Science, Cairo University, Giza, Egypt

0065-2393/89/0220-0033$06.00/0

With high-calcium (calcareous or marl) sediments found north of the Dead Sea, potters produced highly vitrified (glassy) wares (*1*). At the same time (ca. 3500 B.C.), potters along the Nile River used firing conditions similar to those used for high-calcium sediments, but because the sediments generally contained less than 3% CaO, the pottery was far less vitrified than that produced from marl sediments. Pottery from these two locations differed in color as well as in the degree of vitrification. The relatively iron-rich Nile sediments were used to produce a range of red, buff, brown, orange, and even black pottery, in contrast to the white, gray, and black wares produced from low-iron calcareous sediments.

Mineralogical composition was also important. Sediments containing too much coarse-grained material (e.g., quartz or other minerals) made poor pottery unless the coarser grains were removed to enrich the clay content of the sediment. Fine-grained clay sediments needed coarser grained sand or other temper added to aid in working the paste and firing the pottery. In some cases, the mineral fragments that occurred naturally with the clay minerals served as a natural temper, and as such, may provide a basis for distinguishing the source of the sediment. The identification of clay sources and the observation that additional components were added to produce a particular type of pottery helps enhance our understanding of the technological practices of the ancient potters.

In a geochemical sense, sediments tend to represent an average of the geological terrain from which they are derived. The clay minerals are formed by the chemical weathering of rock-forming minerals like feldspars and ferromagnesian silicates. Other minerals, like quartz, remain unaltered but are gradually broken into small fragments that mix with the clays. Thus, a river sediment is a mixture of weathered minerals (clays) and relic minerals derived from the surface rocks in the drainage basin.

The chemical composition of a river sediment is a reflection of the average chemical composition of rocks in the drainage basin. Sediments from different rivers might differ in chemical and mineralogical composition, whereas sediments from a single river should be fairly uniform in chemical composition if similar grain-sized fractions are compared. Sediments deposited at different locations along the same river can differ in mineralogical and chemical composition. These differences result from the differential sedimentation or deposition of individual mineral grains from the running water that transports the grains downstream. The patterns of deposition depend upon the grain size (which is related to the mineral's hardness), the density of the particles, and the water's flow rate.

The Nile River provides a good example of the similarities and differences between sediments from the same river. The common origin of these Nile sediments is confirmed by the high degree of correlation in the elemental compositions between different Egyptian Nile sediment samples (*2*).

These samples are similar in composition, but they have subtle chemical differences.

To understand the sediments deposited along the Nile, it is helpful to realize that as the river flows through Egypt, it is flowing through a very large delta. Sediments from two distinct drainage basins are carried by the Blue Nile and the White Nile until they merge and form the Nile. Before the construction of the Aswan High Dam, millions of tons of suspended matter was carried by the Nile as it emptied out onto this delta and deposited as sediments along the banks of the river. In the geological past, these sediments were deposited on top of the earlier sedimentary deposits that reflected earlier environments such as smaller deltas, alluvial fans, flood plains, and even ocean floors. The resulting complex sequence of sedimentary deposits is often difficult to interpret.

Part of the work described in this chapter was aimed at studying the Nile sediments in one locality to determine whether the complex deposits could be differentiated by chemical means.

Identifying the sources of sediments chemically would be valuable in answering a number of archaeological questions. For example, the similarity in the compositions of the modern Nile sediments suggested that the compositions of early Egyptian pottery made from Nile clays would be similar and of little use for provenance studies (*see* ref. 3); therefore, it was somewhat unexpected when earlier (*4–5*) studies of Predynastic Egyptian pottery suggested that there were some geochemically significant differences between ancient Nile sediments, deposited some 40,000 years ago, and those deposited more recently.

As a part of ongoing studies at the important Predynastic Egyptian sites at Hierakonpolis, we have attempted to understand how the sediments deposited in this area have varied spatially and over time. The results of these geochemical studies, summarized in this chapter, indicate how a detailed understanding of the sediments can enhance our knowledge of the environmental evolution of the local landscape, help to relate separate archaeological sites in an area, and provide a basis for comparisons of the pottery produced in the area.

Archaeological Significance of Hierakonpolis

Since the time of the Greek historian Herodotus (484–425 B.C.), scholars have speculated on the role of the Nile in the origin and development of Egyptian civilization. One of the most important sites at which this role can be investigated is Hierakonpolis, where both ancient legends and archaeological evidence suggest the first leaders of unified Egyptian state emerged in the fourth millennium B.C. (*6*).

The name Hierakonpolis (Greek for City of the Hawk) was originally used in a more restricted sense to refer to the Archaic–Old Kingdom (ca. 3100–2230 B.C.) walled town of "Nekhen," which is located in the modern Nile flood plain along the west bank of the Nile river about 650 km south of Cairo. In this chapter, the name Hierakonpolis will be used to embrace a larger archaeological and geographical region surrounding the town of Nekhen, including archaeological sites ranging in age from Lower Paleolithic to Greco–Roman (ca. 250,000–31 B.C.).

The best-known and most important aspects of this area are the extensive Predynastic (ca. 4000–3100 B.C.) settlement and cemetery complexes. Long before and during the unification of the Egyptian state (ca. 3100 B.C.), Hierakonpolis was a town of religious and political distinction and may have served as the Predynastic capital of Upper Egypt. Whereas little was known about the nature and distribution of sites in the Nile flood plain, the large areas of Predynastic occupation (stretching for about 1.5 km along the low desert on the edge of the modern cultivation zone) have been the subject of considerable investigation (e.g., refs. 6–8). The Predynastic settlements and cemetery complexes also extend about 3.5 km west of Nekhen into the western desert along an ancient drainage course known as the Great Wadi or Wadi Abul Suffian.

The "royal" cemetery complex (Tombs 1 and 2), where a considerable amount of fine Plum Red Ware (PRW) was found, was located in the Great Wadi. The Plum Red Ware pottery appeared to have been fired at sites along the northern side of the Great Wadi (localities 39 and 59) on the upper beds of ancient (Cretaceous) sediments (variegated shales and sandstone) that are a part of the Nubian formation.

The Nubian sedimentary formation has a different origin than the Nile sediments, and the two types of sediments can be easily distinguished on the basis of trace element contents (*4*). As Figure 1 shows, these ancient variegated shales and ferrugineous sandstone beds are exposed in parts of the low desert surface west of the cultivation zone, as well as in the high desert areas that border the Great Wadi.

In most areas of the wadi floor and the low desert area, the Nubian formation is covered by an average of 5–7 m of Pleistocene Nile silts (*9*). These older Nile sediments represent different episodes in the evolution of the Nile River (*10–13*). The oldest Nile sediments (called Protonile) in the area, exposed at high (about 125 m above sea level) Pleistocene terraces, were deposited during the Lower to Middle Paleolithic period (*10*).

The more recent (Neonile) sediments in the area include the Masmas formation (clay and sandy silts deposited some 40,000 years ago). Neonile sediments called the Sahaba formation are younger (ca. 20,000 B.C.), and, as seen in Figure 1, these sediments cover much of the area bordering the modern flood plain on which most of the Predynastic sites are located. Some chemical differences are seen between these two Neonile sedimentary units

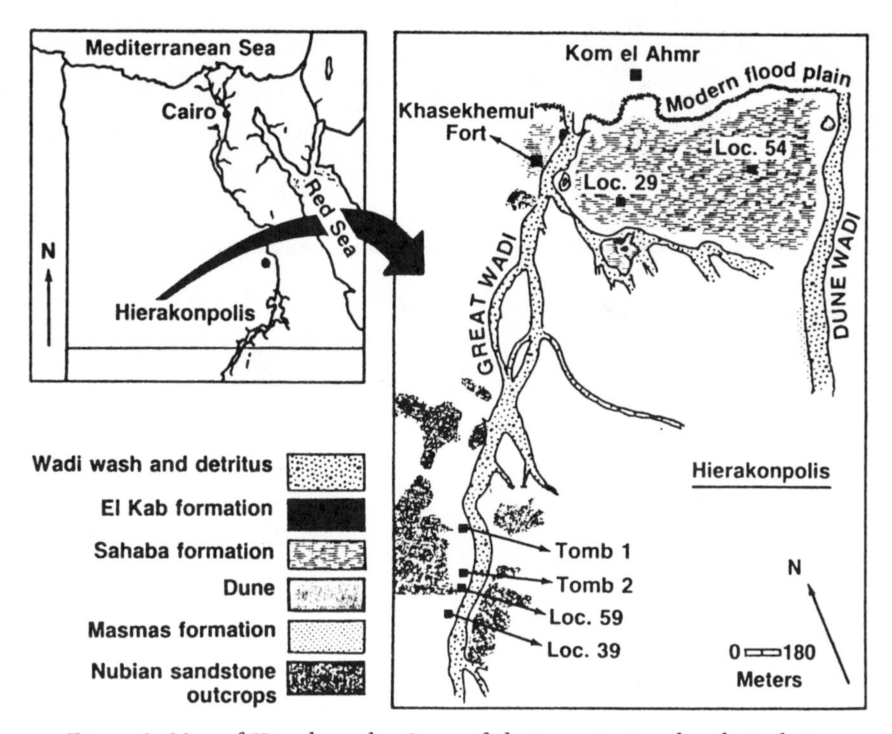

Figure 1. Map of Hierakonpolis. Some of the important archaeological sites are shown along with the sedimentary formations exposed in the area. Tombs 1 and 2 mark the Predynastic "royal" cemetery.

(*4–5*). A final Neonile sediment is the younger (ca. 9000 B.C.) El-Kab lithozone that is exposed in only one area near locality 24 (*14, 15*). It was anticipated that most of the El-Kab formation sediments were covered by the modern Nile sediments on the flood plain and would be found directly under these cultivated soils.

Experimental Details

Samples. The Archaic–Old Kingdom walled town of Nekhen was in the modern flood plain. This region (known locally as Kom el Ahmr), first dug by the Englishmen J. E. Quibell and F. W. Green in the late 1890s, was investigated by an interdisciplinary team in 1984. Several trenches were excavated along a small canal between the site and the edge of the nearby desert. A manual auger was used to sample the sediments in the 16 locations shown in Figure 2. Visual examination of the sediments and the artifacts (pottery) found at each level showed that the area around Nekhen was more stratigraphically complex than anticipated.

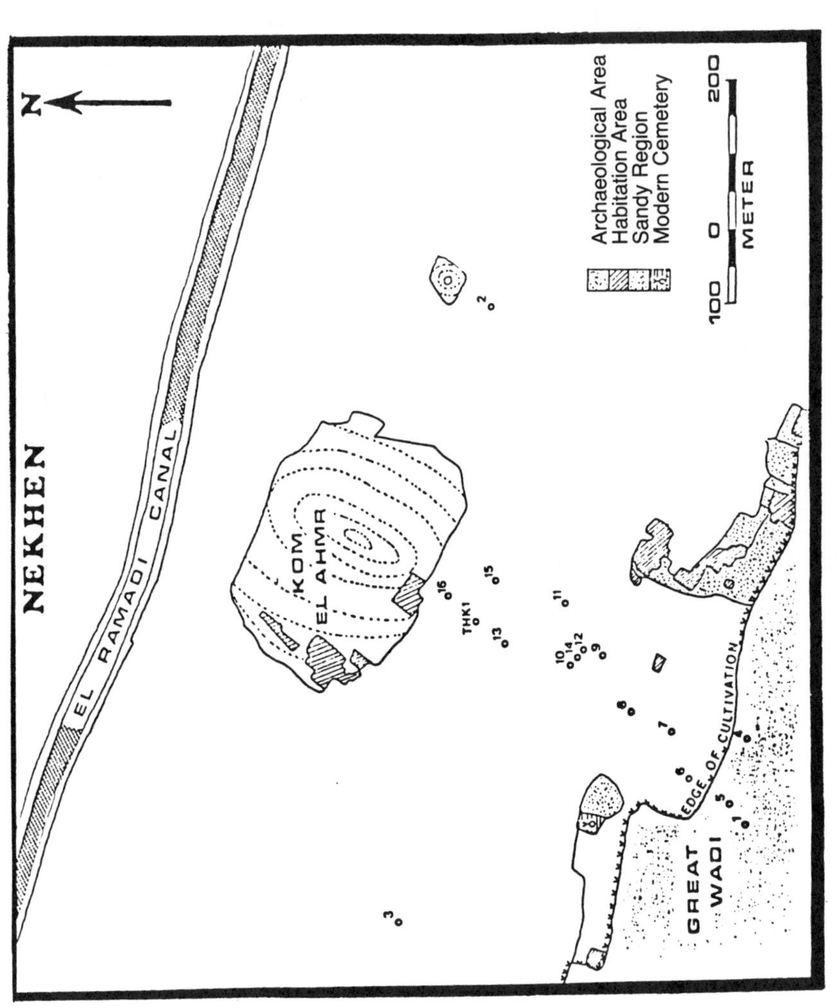

Figure 2. Map of Kom el Ahmr area at Hierakonpolis showing the partially excavated archaeological areas, including the old temple and the numbered locations from which test cores were taken.

Below the sediments disrupted by cultivation, a unit (Unit A) was found that was about 120 cm thick and that contained a disorderly mixture (almost an inverse time sequence) of Roman, Pharaonic, and Predynastic sherds. The next 15 cm or so was a clay-rich Nile silt (Unit B) in which there was a normal ceramic sequence with sherds dating from about 300 B.C. to about 2500 B.C. Table I describes the typical core sample. The layers of sediments were compacted and, in some cases, partially hardened by a calcareous cement, but in all cases, the samples could be easily crumbled.

Below Unit B a silty sand layer (Unit C) was encountered in the cores taken between the mouth of the Great Wadi and the center of the Kom el Ahmr. Unit C contained a sequence of Old Kingdom to late Predynastic artifacts dating from about 2500 B.C. to 3200 B.C. Below Unit C, a very compact, well-sorted thick layer of Nile clay and sand was encountered (Unit N). No cores or trenches reached below this sedimentary deposit. Unit N contained occasional Predynastic ceramic and flint artifacts. Samples were collected from each 10–15-cm auger cut within each sedimentary unit in each core or trench. In addition, numerous samples of the Neonile deposits and other sediments from the nearby low desert and Great Wadi were taken for analysis.

Nilotic sediments were sampled across the river from Hierakonpolis from the Nekheb formation at El-Kab (78.5–80 m above sea level), which was thought to have been deposited beginning about 11,000 years ago (*15*). In addition, a series of 27 Nilotic sediments were collected from along a 350-km stretch of the Nile Valley in Upper Egypt. Because the Nile sedi-

Table I. Definition of Sedimentary Units from a Typical Core Sample from Vicinity Between Kom el Ahmr and the Edge of the Cultivation Zone

Unit	*Appropriate Thickness (cm)*	*Approximate Dates*	*Approximate Period*	*Cultural Sediment*
Overburden	~100	1968–present	Modern	Cultivation Zone
A	120	Ca 220–320 B.C.	Ptolmaic	Anthropic Nile[a]
B	15	to 2500 B.C.	New Kingdom	Nile
Sharp contact				
C	~100	2500–3100 B.C.	Archaic (Dynasty 1–2)	Wadi Deposits
	~50	3100–3200 B.C.	Protodynastic	
Gradual contact				
N	~75	3200–3400 B.C.	Gerzean	Nile
	~75	3400–3700 B.C.	Amratian	Nile
	at least 150	3700 – ?[b]	Badarian	Nile

NOTE: Units are in order of depth from surface (at 82.23 m above sea level) to the bottom of the core (about 5 m below the surface).

[a]This layer appears to be the result of leveling the site for agricultural purposes (ca. 320–220 B.C.) as described in ancient writings.

[b]On the basis of earlier conclusions (*14*) drawn from site at El-Kab across the Nile from Nekhen, the deposition of this series of Nile sediments, which is called the Nekheb lithozone, probably began about 9000 B.C.

ments originate from different geological terrains in Africa, samples of fine silts were obtained from the White Nile, the Blue Nile, and the region of Khartoum where these major rivers join to form the Nile before it flows north through Egypt (*16*).

Analytical Methods. Bulk samples of the Nile sediments (200–500 mg) were analyzed with instrumental neutron activation analysis (INAA). The basic procedure has been described elsewhere (*4, 17*). Samples and appropriate geological reference standards were irradiated for 1 h in the University of Virginia research reactor at a flux of 1.2×10^{13} neutrons cm^{-2} s^{-1}. The samples were counted with a relatively thin Ge(Li) detector that provided high efficiency and resolution for low energies (0.86 keV full width at half maximum (FWHM) for a 122-keV Co-57 peak), but was sufficiently efficient for counting high-energy γ-rays (1.1% efficiency for 1.332-MeV Co-60 γ-ray relative to 3- $\times$ 3-in. NaI crystal). All samples were counted 4–7 days after irradiation and again 30–40 days after irradiation.

As will be discussed later, the initial analysis suggested that the bulk samples were very similar in composition. Thus, to understand the differences that were observed in the Neonile sediments, a more detailed investigation of the different units from the region around Kom el Ahmr was begun. Dried bulk samples were examined under a binocular microscope to detect organic matter and to determine the grain size distribution. This examination was supplemented with scanning electron microscopy (SEM) to determine the lithology.

Depending upon the relative proportions of sand and clay, representative samples of between 10 and 50 g were stirred with deionized distilled water for 10 min to remove any water-soluble salts (e.g., NaCl). After allowing the sediment to settle for 5–6 h, the water was decanted and the water wash cycle was repeated five more times. Following the final wash with water, the carbonate salts, which cemented some of the grains together, were removed by slowly adding a 5% HCl solution until there was no effervescence. It was anticipated that this acid would also remove iron oxide coating from the mineral grains. After rinsing six times with distilled water, a sample (about 1 g) was taken for analysis.

The remaining material was warmed in a 10% H_2O_2 solution and allowed to stand overnight to remove all organic matter. After washing with distilled water, the sediment was suspended in 425 mL of a solution that contained 2.35 g of sodium hexametaphosphate (Calgon) and stirred overnight. Within 30 s after stirring was stopped, the sample was filtered through a 63-μm (4-phi mesh) screen to separate the sand-sized grains (0.063–2 mm) from the finer particles. Although this fraction is described in this chapter as mud, it is a mixture of what sedimentologists would call the silt and clay fractions. After washing with distilled water five or six times, separate grain size fractions were dried (at 50 °C for 6 h) and sampled for analysis.

A portion of the sand-sized fraction was mixed with 1,2-tetrabromoethane (specific gravity of 2.9). The less dense grains of quartz and feldspar floated on the surface and were separated from the more dense (heavy) minerals (mostly pyroxenes, amphiboles, opaque minerals, and epidotes). Both fractions were washed with acetone 20–25 times, washed with distilled water five or six times, and finally dried overnight at 80 °C. For each sediment, both fractions of the sand-sized material were then examined with binocular and polarizing light microscopes to determine the efficiency of the separations. The less dense fraction was white or clear and composed of fragments of quartz and feldspar. This fraction showed very little contamination of the grains by the darker mineral fragments that were found in the heavier fraction.

The separation of specific minerals was not the goal of this study, so there was no effort to remove the small number of white quartz fragments from the dark heavy minerals. By making these separations, it was possible to analyze the relic minerals free from the clay minerals that carry most of the trace elements in these river sediments. During the analysis of these fractions, the concentration of Br was used to detect contamination of the grains. The results indicated that the extensive washing was sufficient to remove the tetrabromomethane.

Results and Discussion

Nile Sediments. Throughout the millions of years of its history, the Nile has carried sediments to and through Egypt. On its nearly 7000-km course from central and eastern Africa to the Egyptian Delta, suspended matter is carried by the Nile River. Table II contains a summary of data that suggests that the compositions of the sediments that are deposited by the Nile along its banks are very similar. When the 27 samples of modern sediment from along a 350-km stretch of the Nile in Upper Egypt were analyzed by INAA, there were differences of as much as 30% in the concentrations of the 17 elements measured (nine of the rare earth elements, Th, U, and the elements in Table II). The ratios of the various trace elements, however, were very similar for each sample, a fact suggesting that the major differences were due to variable amounts of quartz sand in the sediments.

Quartz typically yields low concentrations of the elements when measured by this procedure, and therefore does not contribute significantly to the overall trace element composition of the bulk sediment samples. Thus, the presence of quartz in sediments acts as a dilutant to their overall trace element content. Unfortunately, Si cannot be measured by this INAA pro-

Table II. Average Concentrations (by Weight) of Some Elements in Bulk Sediment Samples from Hierakonpolis

Sediment Formation[a]	Fe_2O_3 (%)	Na_2O (%)	*Co* (*ppm*)	*Sc* (*ppm*)	*Cr* (*ppm*)	*Hf* (*ppm*)	*La* (*ppm*)
Nile River Sediment							
Composite (27)[b]	8.5	1.5	28	19	132	6	36
S.D.	±1.6	±0.2	±6	±2	±17	±1	±4
El Kab (3)	10.8	2.0	33	21	122	8	43
Sahaba (5)	9.3	2.8	29	21	244	6	33
Masmas (5)	9.3	1.6	32	18	148	6	33
Protonile (5)	5.0	1.0	13	7	62	6	23
Wadi (4)[c]	5.0	1.0	15	10	78	5.5	23
Nubian							
Sandstone (4)	0.6	0.05	6	1	n.a.	3.2	6

NOTE: All samples were measured by INAA. The elemental concentrations of Fe and Na were measured but have been calculated as the weight percent oxide, although this calculation probably does not represent the exact chemical nature of these species.

[a]The number of separate samples of each type shown in parentheses.

[b]The average for 27 modern sediment samples from along a 350-km length of the Nile in Upper Egypt. The standard deviation for the NRSC samples is typical of that observed for all the other groups.

[c]Mixture of materials (Old Nile and Nubian Formation) washed from the Great Wadi.

cedure. An inverse correlation between Si and the trace elements would prove that dilution by sand was the source of the observed variations in these Nile sediments.

Averages for the 27 bulk samples are given in Table II for some of the elements measured. These averages, called Nile River Sediment Composite (NRSC), are very similar to the averages for modern Nile sediments compiled by Tobia and Sayer (2). This similarity clearly confirms that, other than the variations due to grain size (sand content), the modern Nile silts are quite uniform in composition.

Recognition of the mechanisms by which trace elements are partitioned into minerals suggests the importance of looking at the relative distributions of groups of elements that have similar chemical behavior. The rare earth elements (REE), or lanthanides, have been particularly useful because they usually occur as trivalent cations that differ from each other only in ionic size. Each mineral, as it is formed, partitions the REE and other trace elements into its crystal lattice on the basis of ionic size and charge. The REE are distributed in minerals on the basis of size, and the total concentration in a rock depends upon the minerals that are present. In some cases, there is an "anomaly" in the behavior of Eu, which can be separated from the others when it is reduced partially to the 2+ oxidation state.

The relative distributions of REE in geological materials are often represented by plotting the normalized* REE concentration (concentration of element in the rock divided by the average concentration of that element in chondritic meteorites) as a function of atomic number (which is inversely proportional to the radius of the 3+ ion) as shown in Figure 3.

In Figure 3 the REE concentrations in the Nile sediments are all similar, so the REE patterns A, B, and C are offset from each other. These are semilog plots. Pattern D is the NRSC, an average of 27 modern Nile sediments from Upper Egypt. Pattern A is the average of five samples from the Masmas formation (the oldest of the Neonile sediments). Pattern C is the average of five samples from the Sahaba formation, and Pattern B is the average of three samples from the El Kab formation all from Hierakonpolis. Pattern E is the average of five samples of the much older Protonile sediments, and Pattern F is the average of four samples of the even older (non-Nile) Nubian formation sandstone. The typical modern Nile sediment (shown as the NRSC) shows an enrichment of La relative to Lu and the negative Eu anomaly (concentration relative to Sm and Gd) that is typical of the earth's crust.

The results for the average Nile sediment of Upper Egypt (NRSC) compare quite well with the averages for the older Nile sediments (El-Kab and Sahaba) found at Hierakonpolis (Table II), a result suggesting that there

* This normalization process helps highlight the effects of geochemical processes on the separation of this group of very similar elements (*12*, *19*).

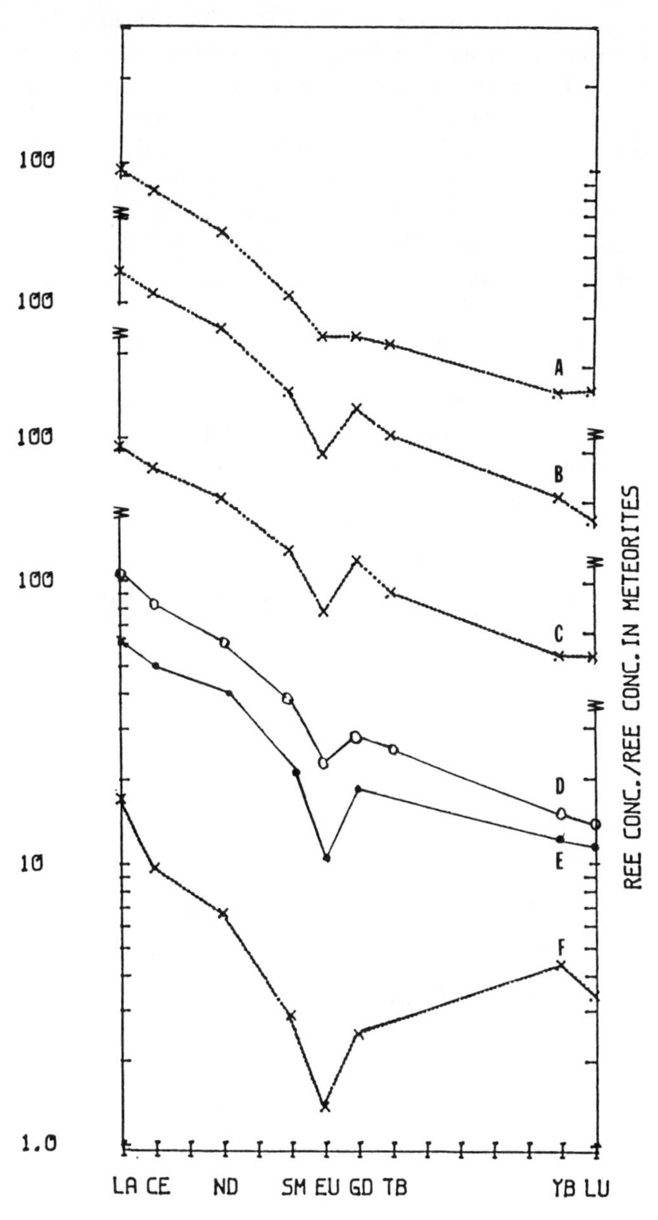

Figure 3. Average rare earth element (REE) distributions in sediments from Hierakonpolis.

has been little change over the last 20,000 years. However, if the relative concentrations are examined, there appear to be some differences in the REE distribution patterns (Figure 3) for the older Masmas formation. Although the concentrations of La, Lu, and Eu are very similar in the Masmas sediment, the Eu is not anomalous but has a concentration that would be expected in minerals where it was mostly present in the +3 oxidation state. In the older Protonile sediments at Hierakonpolis, the REE patterns are also different and have a pronounced negative Eu anomaly and a lower La/Lu ratio than the Masmas formation (Figure 3). The differences between Protonile and the younger Nile sediments are also clear from the other elements in Table II, and correlation diagrams like that in Figure 4 for Co and Sc in the bulk samples.

The results from the Nubian sandstone are also shown in Table II and Figure 3 for comparison to the various Nile sediments at Hierakonpolis. These ancient Nubian formation sediments, which contain large amounts of quartz, are clearly distinguishable. Material washed from the Great Wadi is a mixture containing Nubian formation sediments (mainly gravel-sized) and the older Nile sediments (mainly mud and sand-sized) that had been deposited on the ancient wadi floor. The similarity between the average wadi sediment and the Protonile sediments (Table II) suggests that the contributions to the finer sand and mud-sized material (older Nile sediments) by

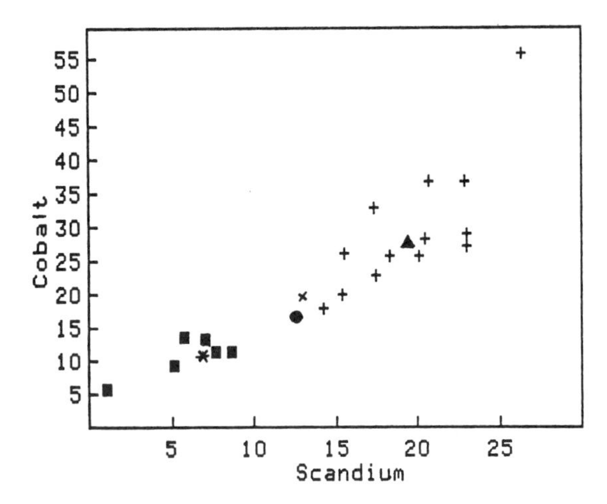

Figure 4. Concentrations of Co and Sc in bulk samples of sediments found at Hierakonpolis. The Nubian sandstone (■) and the Protonile sediments contain far less Co and Sc than the Neonile (Masmas, Sahaba, and El Kab) sediments (+). For comparison the average concentrations for Units B (●), N (×), C (*), and the NRSC (▲) are shown. Although sediments from Units B and N are Neonile, the low concentrations suggest higher proportions of sand (dilutant) than in the older Neonile sediments from the low desert area.

the Nubian sandstone is small. The REE distribution patterns suggest that the wadi deposits contain some Masmas formation material.

Considering the great similarities in recent Nile sediments, it was not surprising that the average concentrations of trace elements from the various levels in the Kom el Ahmr (Nekhen) area were very similar. Although Units B, C, and N appear to be similar to each other (Table III), especially in the relative distributions of the REE (Figure 5), there are some differences between these and the typical Nile silt (NRSC) for some elements. If Unit B is compared with NRSC and the El-Kab average, most of the differences can be accounted for on the basis of the dilution by sand.

The dilution effect of the coarser grained material (sand-sized fraction) can best be seen in Table III, as the mud (made up mostly of clay minerals) fraction (<63 μm) contains much higher trace element concentrations. The mud-sized fractions are very similar for Units B and N, but the concentrations in the mud from Unit C are significantly lower. This distinction of Unit C is even more pronounced when the REE distributions are compared (Figure 6). Although the bulk samples of all three units show characteristic negative Eu anomalies (Figure 5) in the mud fraction (Figure 6), only Unit C shows this Eu anomaly. The negative Eu anomalies in the bulk samples of Units

Table III. Average Concentrations of Some Elements in Sediments from the Kom el Ahmr (Nekhen) Area

Fraction[a]	Fe_2O_3 (%)	Na_2O (%)	*Co* (*ppm*)	*Sc* (*ppm*)	*Cr* (*ppm*)	*Hf* (*ppm*)	*La* (*ppm*)
Sediment Level B							
Bulk (3)	5.6	1.3	18	13.5	137	8.4	32
Acid Washed (2)	4.0	1.0	13	10.0	81	7.7	8
Mud (3)[b]	13	—	32	29.5	255	9.6	32
Sand (7)[b]	1.5	—	5	2.8	11	3.3	7
Heavy (3)[b]	29	—	67	72.6	674	72.6	56
Light (4)[b]	1.1	—	4	1.8	9.2	1.3	6
Sediment Level N							
Bulk (3)	5.0	1.0	17	11.0	106	6.0	29
Acid Washed (3)	5.2	1.0	13	10.0	64	6.6	13
Mud (3)	11.1	—	26	24.5	197	8.1	26
Sand (9)	2.1	—	11	3.3	32	3.3	8
Heavy (3)	36	—	75	59	802	56	91
Light (3)	0.9	—	7.6	1.4	5.9	1.1	7
Sediment Level C							
Bulk (3)	4.0	0.7	12	7.5	72	8.1	36
Acid Washed (4)	3.0	0.6	11	6.4	45	7.6	21
Mud (3)	5.8	—	15	13	100	7.0	31
Sand (3)	0.3	—	3.7	0.5	6	0.6	8
Heavy (3)	36	—	69	81	811	115	121
Light (2)	0.5	—	1.3	0.8	3.3	1.0	5

[a]The number of separate samples analyzed is indicated in parentheses.
[b]Heavy- and light-density fractions of sand (2 mm–63 μm) separated with 1,2-tetrabromoethane. Mud-sized fraction is <63 μm.

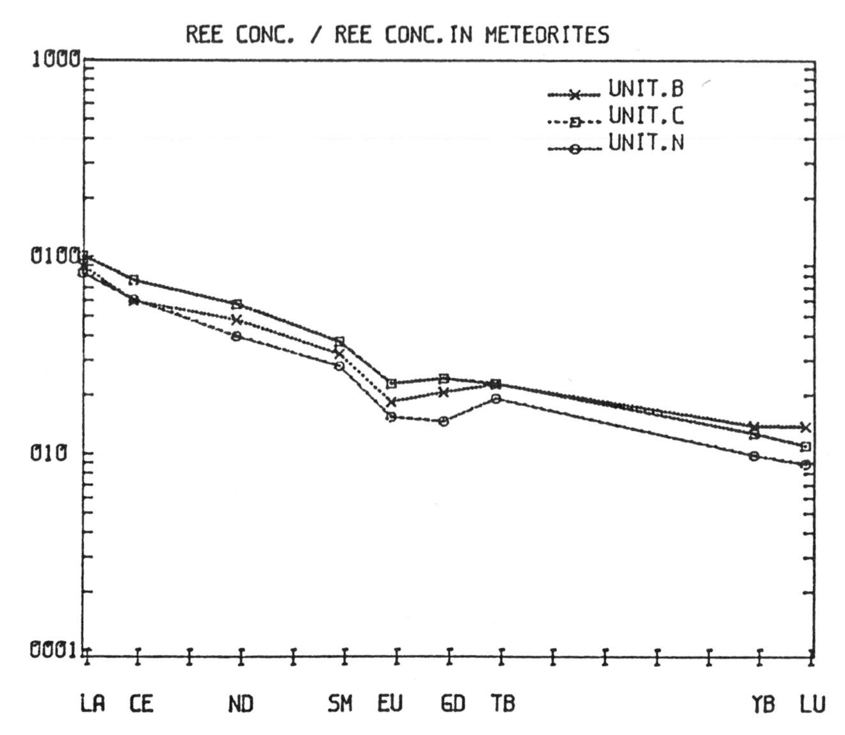

Figure 5. Average REE distributions in sedimentary layers below the cultivation zone in the Kom el Ahmr region. Unit B (×) and Unit N (○) were deposited directly by the Nile flooding. Unit C (□) sediments were washed from the Great Wadi.

B and N are the result of the large amounts of the coarser minerals that have a negative Eu anomaly.

The bulk samples from Nekhen (Units B, C, and N) contain less Fe than the other Nile sediments in the area. Many of these sediments were examined by using X-ray powder diffraction in an attempt to determine which phases contained iron and how the oxidation state and mineralogy changed during the production of pottery. There was no clearly defined crystalline iron oxide phase present. Although Fe may be present as an impurity in many of the minerals, it appears that, in these sediments, the Fe occurs as amorphous or fine-grained mineral phases (e.g., geothite, coating the surface of other grains). As shown in Table III, even though the Fe is dispersed, it is not readily soluble in dilute acid.

With the possible exception of Cr, the acid wash removes only relatively small amounts of these trace elements. Although it was not measured in all samples, the Ca content was low (1.5–3% CaO) and much was easily removed in the acid wash. There are, however, layers of crystalline white salt in the older Nubian sediments (such as the Esna or variegated shales) along the

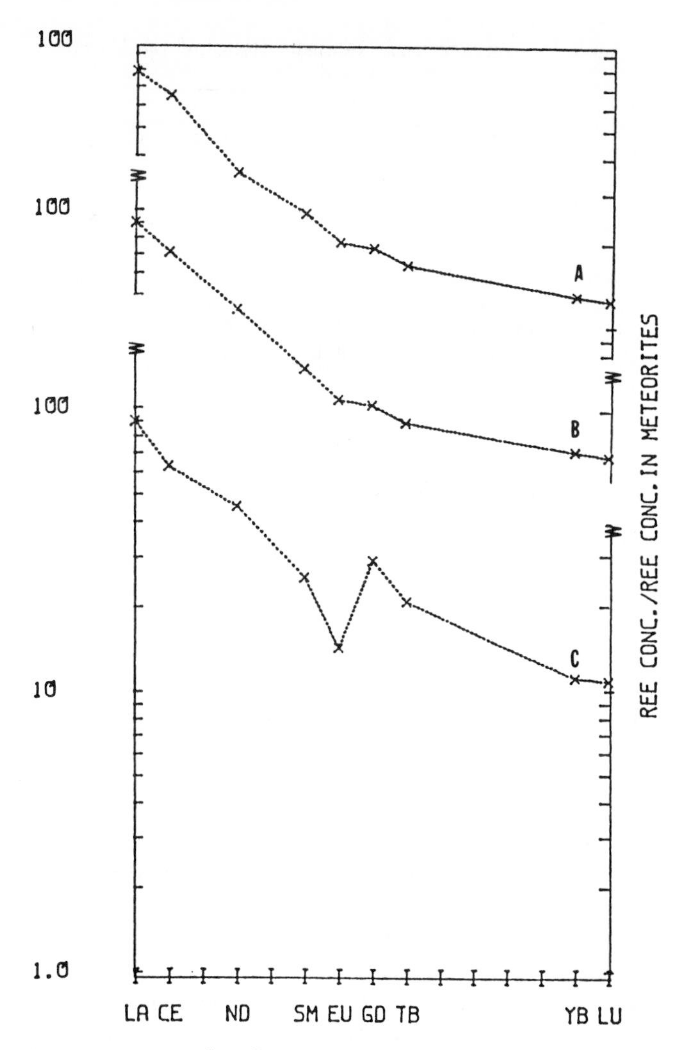

Figure 6. Average REE distributions in mud sized (<63 μm) fractions from Units N (designated as A), B, and C in the region of Kom el Ahmr. Number of samples analyzed is given in Table III. Note that patterns for Unit N(A) and B are offset for clarity.

sides of the Great Wadi. X-ray diffraction showed these salts to be $CaCO_3$ and $CaSO_4$, which could be mobilized by water.

The presence of soluble Ca salts in the sediments of Unit C was particularly obvious in our observations using an energy-dispersive X-ray detector with the SEM to compare the intensities of characteristic X-rays. The results in Table IV were obtained for the bulk samples. The loss of Ca from Unit C sediments (probably as $CaCO_3$) upon washing with dilute HCl is

Table IV. Average X-ray Intensity Ratios of Ca/Al and Ca/Si

Fraction	*Ca/Al*	*Ca/Si*	*% CaO*[a]
Bulk of Unit C	0.43	0.18	3.9
Unit C sample after washing with acid[b]	0.37	0.04	n.a.[c]
Bulk of Unit N	0.40	0.25	2.0
Unit N sample after washing with acid[b]	0.56	0.23	n.a.
Sahaba Formation	0.36	0.13	3.9
Straw Tempered Ware	0.36	0.13	2.3
Plum Red Ware	0.48	0.17	3.4
Hard Orange Ware	1.40	0.53	11.7

NOTE: These values are not elemental concentrations.
[a]Measured with INAA with a 10-min irradiation procedure.
[b]Washed with 5% HCl.
[c]n.a. means not available.

shown by the substantial decrease in the Ca/Si ratio. The same pattern was found for wadi sediments, which, although they are old Nile sediments, contain contributions from the Nubian formation materials, and probably include redeposited $CaCO_3$.

Unit C at Nekhen is clearly different from the sediments above and below it. Comparing all features (including the location of its occurrences), it is clear that Unit C is material washed from the wadi and deposited in a small "alluvial fan" during the period from 3200 to 2500 B.C. At this time, Nile flooding must not have reached the site, but there were substantial amounts of water to wash sediments from the wadi. This evidence supports other paleoclimatic studies that suggest periods of moist conditions on the Egyptian desert (*18*).

The shape and characteristics of this alluvial deposit provide a possible explanation for the location of the Nekhen Temple at the mouth of the Great Wadi. During earlier periods (7000 B.C. or earlier), material washed from the wadi may have created a "wadi fan" that was the high ground in or near the Nile flood plain that existed at that time. This high point during Nile floods had religious significance to the early Egyptians who believed that life first arose from such an area in the midst of the water (*19*). This area also had a more practical attraction for Predynastic settlers wishing to remain near the Nile but above the average flood level (*20*).

Units B and N are clearly Nile sediments of similar composition, despite being deposited at different times. Besides the varying relative content of clay (mud) and the coarser grained minerals, the only reason one would expect to see variations with time would be if the drainage patterns had changed. From a geological standpoint, the drainage basins of the Blue Nile (and Atbara) and White Nile basins are significantly different. The Precambrian undifferentiated metamorphic rocks of the Blue Nile basin are covered by volcanic rocks. The older (Late Cretaceous–Tertiary) occurrences are mainly as flood basalts, with interbedded trachytes and rhyolites near the top. The younger (Pliocene–Quaternary) volcanic rocks, exposed in the Lake Tana Basin and in several tributary gorges of the Blue Nile, occur as basaltic cones and craters with some tuffs and alkali rhyolites at the top. The White Nile basin is underlain by Precambrian igneous and high-grade metamorphic rocks such as gneisses, mica schists, granulites, granodiorites, serpentinites, and quartzites. The contribution of the sedimentary cover to both basins is very limited (*16*).

During the modern times before the dam, the Blue Nile and Atbara were thought to contribute about three-fourths of the sediments that traveled to Egypt. If this ratio differed in the past, then any differences would best be observed in coarse-grained mineral fractions. Although this fraction was typically 90% quartz and feldspar, it does contain the heavy minerals characteristic of the geological terrains. Shukri (*21*) first observed that sediments from the Blue Nile contained a higher frequency of pyroxene minerals, whereas the White Nile contained a higher frequency of epidotes (*21*). Such differences in the frequency distributions of heavy minerals could result in distinctive differences in the trace element distributions. For example, elements like Cr are generally more concentrated in the volcanic rocks drained by the Blue Nile, whereas elements like Th and Hf might be expected in the minerals from the metamorphic rocks drained by the White Nile. In the sand-sized fraction of the samples from Units B and N (Table III), the concentrations vary considerably from sample to sample, especially within Unit N, and there is considerable overlap in the concentration ranges for the two units.

In a number of cases, geochemically similar elements are well correlated to each other in the sand fractions from each unit (e.g., Sm and Eu in Figure 7). This correlation is not surprising if the trace elements are associated with the heavy minerals. Small variations in the amount of a particular heavy mineral will cause large differences in the trace element content measured, if the sand fraction is mostly quartz and feldspar, which contain so few trace impurities. Figure 7 shows the relationship between Sm and Eu. There is a correlation between these elements in the samples from each of the two levels. Figure 7 suggests that there is a different Sm/Eu ratio for the sands from these two Nile sediment deposits.

Similar relationships are found for other elements like Sc and Cr. In the sand-sized fractions, the REE distribution patterns show a more pronounced negative Eu anomaly and higher La/Lu ratio for the material from Unit N as compared to samples from Unit B. Interpretation of these observations is difficult because the heavy fractions contain as many as 10 different minerals. Pyroxene (mainly augite), amphiboles (mainly hornblende), iron oxides, chromite, ilmenite, rutile, and zircon were identified by using a polarizing microscope and energy-dispersive X-ray analysis. However, the

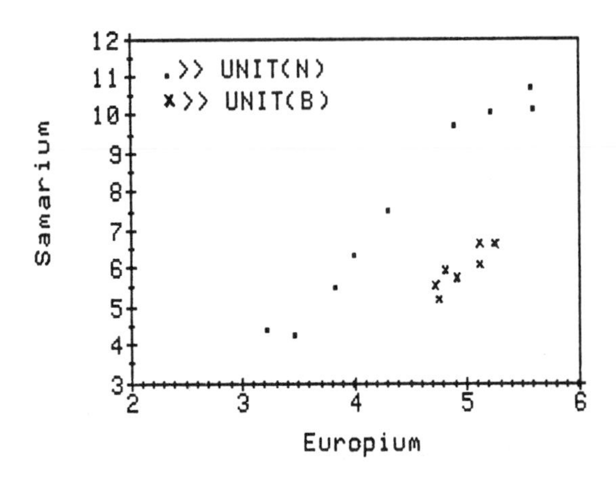

Figure 7. Relationships between the normalized Sm and Eu concentrations (ratio of concentrations of each element to concentrations in chondritic meteorites) for the sand-sized fractions (2 mm to 63 μm) of the Nile sediments of Units B (×) and N (*) in the Kom el Ahmr region.

trace element data suggests that there must be differences in the relative proportions of the various heavy minerals that are present.

Shukri (*21*) indicated that the frequency distributions of minerals from the Blue and White Niles differ, so differences in the mineralogical composition of the heavy sand fraction could result from varying the contributions from the two sources. The geological differences in the two drainage basins had suggested that Cr might be an element that was characteristic of the volcanic rocks of the Blue Nile source. To help confirm this suggestion, Ahmed El-Kammar of Cairo University analyzed silt (very fine-grained sediments) from the Blue and White Niles (collected early this century) by using instrumental neutron activation analysis. The Cr/Th ratio in silt from the Nile north of Khartoum was 20, which compared well with the NRSC, which was 17, and the ratios for the mud fractions of Units B and N which were 23 and 15, respectively. For the silt from the Blue Nile (above Khartoum), the Cr/Th ratio was about 43 with a Cr concentration that was only 20% higher than that in the Main Nile silt. For the White Nile silt, the Cr was substantially lower and the Th only slightly higher than for the Blue Nile. The Cr/Th ratio in the White Nile silt was about 10 (*16*). Although this result does indicate the geochemical differences we expected, this size fraction (primarily clay minerals) has become homogenized and does not preserve the characteristics of the geological source as well as the unaltered mineral fragments in the sand-sized fraction (*21*).

Figure 8 shows the results for the individual analysis of the sand-sized fractions from Units B, N, and C as a triangular correlation diagram for Cr, Hf, and Th. Ignoring the samples from Unit C, which is a wadi sediment that may be representative of very old Nile deposits, the material from Unit

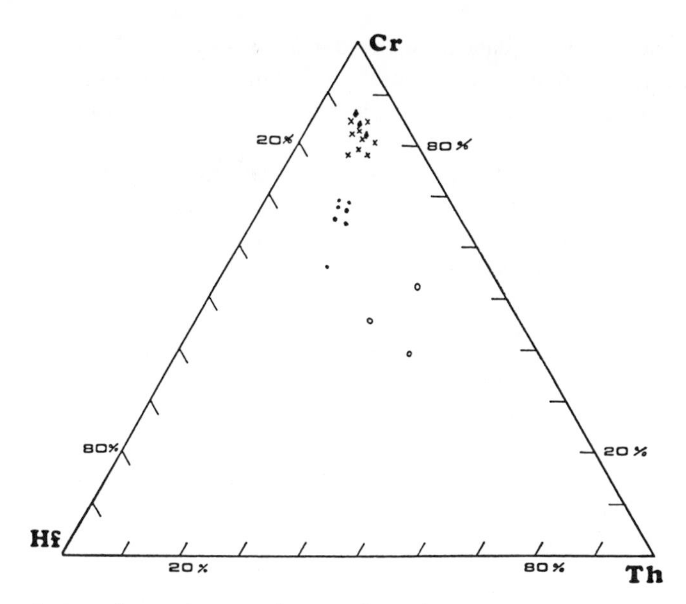

Figure 8. Correlation diagram for Cr, Th, and Hf in the sand-sized (2 mm to 63 μm) fractions from Units N (×), B (*), and C (○) at Hierakonpolis and the Nekheb lithozone across the Nile at El-Kab (♦). This diagram is made by summing the concentrations of Cr, Th, and Hf in a sample and calculating the fraction which each element contributes to this summation. The % Cr in the sum is on the right side (100% at the top) and % Hf is on the left.

N clearly contains a higher proportion of Cr than Unit B. This information suggests that the contributions from the Blue Nile (East African source) were higher during the deposition of Unit N than during the deposition of Unit B. This difference shows up only in the sand-sized fraction where the unaltered fragments of minerals from the different geological terrains are found.

Although not as many samples analyzed, the same trend holds when data for the heavy mineral portion of the sand fraction are similarly plotted. The fine-grained fraction, which contains primarily the clay minerals (products of the chemical alteration of the rock-forming minerals), generally contains much higher concentrations of these trace elements. The compositions of the "mud" fractions are very similar for all of the samples from Units B and N. In fact, the concentrations in the bulk samples are essentially determined by the fraction of "mud" in the sample. The overwhelming domination of the fairly homogeneous mud fraction shows up when the data from this fraction and from the bulk samples are plotted in the same manner as in Figure 8. All of the samples (mud fraction and bulk samples) from Units B, C, and N clump together on such a plot.

The Cr/Th ratio is somewhat higher in the heavy minerals from Unit N than from Units B and C, but it appears that an important factor in Figure 8 is the relative proportion of light and heavy minerals in the coarse fraction.

A mass balance using data from the density fractions and the entire sand fraction suggests that heavy minerals containing Cr, Th, and Hf represent about 1–2% of the sand-sized material from Unit B and 4–5% from Unit N.

Archaeological Interpretation of Geochemical Findings. In addition to suggesting changes over time in the relative contributions of the Nile's two drainage basins to its sediment load, our geochemical research clarifies a number of related sedimentological, climatic, and archaeological issues.

SEDIMENTOLOGICAL ISSUE. First, the analysis of the sand-sized fractions gave a reliable method to correlate the various Nile sediments in the vicinity of Hierakonpolis. Because the sediments of Unit N, which were deposited during the Predynastic period up to about 3200 B.C., could be distinguished, we compared them to sediments from the archaeological sites on the east bank of the Nile at El-Kab.

Vermeesch (*15*) recorded a series of sediments under the cultivation zone at El-Kab that included the Nekheb lithozone (fine sand and clay), that was assigned an age earlier than the Predynastic period. When sand-sized fractions of Nekheb material from El-Kab were analyzed, it became clear that the Nekheb formation and Unit N at Hierakonpolis were the same Nile sediments. In Figure 8, the correlation of Cr, Th, and Hf in the Nekheb formation is the same as for Unit N. The REE distribution patterns for these two groups of sediments also match (Table V). This approach not only allowed similar Nile sediments to be distinguished (Units B and N) but also allowed us to match deposits from different locations in the area (El-Kab and Nekhen).

Because one portion of the sediment formation (Unit N) can be dated by the archaeological context (pottery dating to ca. 3200 B.C. on the top of Unit N), this data can be applied to date the geological horizons at other locations where the sediments are found. By comparing both sides of the Nile, it appears that the episodes of flooding represented by the El-Kab formation predated the Predynasitc habitation of Hierakonpolis, and the Nekheb lithozone represents material that was deposited by the periodic flooding that occurred during much of the Predynastic period.

CLIMATIC ISSUE. Second, analysis of sediments around the Kom el Ahmr throws new light on local climatic conditions that affected the human patterns of settlement observed by archaeologists. Long before the large-scale Predynastic occupation of the region (ca. 4000–3100 B.C.), rich Nile sediments (Masmas, Sahaba) were deposited in late Pleistocene times (ca. 30,000–12,000 B.C.) over what is now the low desert.

Late Paleolithic and Mesolithic peoples camped on the high ground along wadi channels and hunted, fished, and gathered plants in that rich habitat. When the Nile floods began to decline substantially after about

Table V. Average Concentrations of Several Elements Measured by INAA in the Sand-Sized Fraction of Sediments from Unit N at Neken and the Nekheb Formation at El-Kab

Element[a]	*Nine Samples from Nekhen*[b]	*Three Samples from El-Kab*[b]
La	25 ± 3	26 ± 3
Ce	18 ± 2	19 ± 3
Nd	11 ± 2	12 ± 1.5
Sm	7.6 ± 1	7.4 ± 0.6
Eu	4.4 ± 0.3	4.5 ± 0.4
Gd	6.3 ± 0.8	5.9 ± 0.7
Tb	5.7 ± 0.7	5.3 ± 0.6
Yb	3.8 ± 0.6	3.3 ± 0.2
Lu	3.6 ± 0.5	3.2 ± 0.6
Fe_2O_3 (%)	2.1 ± 0.3	3.5 ± 0.6
Co (ppm)	10.7 ± 3.7	11.2 ± 1.3
Sc (ppm)	3.3 ± 0.5	5.2 ± 0.6
Cr (ppm)	32 ± 6	39.6 ± 6.7
Th (ppm)	3.6 ± 0.6	3.2 ± 0.5
U (ppm)	0.5 ± 0.1	0.6 ± 0.15
Hf (ppm)	3.3 ± 0.6	3.56 ± 0.4

[a]REE are normalized to their concentration in chondritic meteorites. Fe concentration has been calculated as % Fe_2O_3. Other elements are reported as ppm (weight basis).
[b]Average of samples with standard deviation from the mean.

10,000 B.C., the rich late-Pleistocene silts were left high and dry, and this provided an ideal environment for early Predynastic agrarian peoples.

During the middle Holocene era, occasional summer rainy intervals (*see* ref. 22) carried desert sediments down the Great Wadi into the modern flood plain, building up a succession of alluvial fans under what would later be the site of the walled city of Nekhen (Kom el Ahmr). The level of Nile flooding in this area (marked by the Nekhen lithozone) reached to within about 100 m of the modern boundary between the cultivation and the low desert, at an elevation of approximately 80 m above sea level.

After about 3500–3400 B.C., rains ceased, the Great Wadi was temporarily deactivated; simultaneously, the level of the Nile floods dropped. Concurrently, the focus of the human occupation shifted from the low desert toward the flood plain, and the site of Nekhen experienced substantial growth (23). After the political unification of Egypt under the first pharaohs (ca. 3100 B.C.), Nekhen continued to flourish for centuries before going into slow decline after about 2500 B.C. Nekhen was finally abandoned (except for its temple) after 2230 B.C.

The political and cultural reasons for maintaining a nucleated population around Nekhen must have been strong indeed at this time, in light of the renewed light rainfall on the desert between about 3200 and 2500 B.C. that created the Unit C wadi deposits. In contrast to earlier Predynastic times (ca. 3800–3400 B.C.), no settlements sprouted in the desert. This last humid episode seems to have tapered off by the middle of the Pyramid Age (Old Kingdom, ca. 2500 B.C.), although sometime between 2500 B.C. and the Roman period, the Nile floods again ran high enough to reach to the desert's edge.

Although Nekhen had lost all but its religious importance after the political upheavals accompanying the end of the Old Kingdom (ca. 2230 B.C.), archaeological analysis of Unit A suggests that agriculture remained an important activity. Pottery recovered from the layer exhibited reverse stratigraphy that proved it to be a mixture of materials ranging from about 2500 B.C. to A.D. 100 and suggested that it was an anthropic soil (*24*). The trace element concentrations in Unit A suggest that it is a mixture of both older and younger Nile sediments. The archaeological and geochemical analysis, when combined with historical evidence for large-scale agricultural development in both Greek and Roman times (ca. 300 B.C.–A.D. 500), provides unusually consistent evidence for the impact of human activity on local flood plain morphology. It appears that sediments from nearer the Nile were moved to the area near Nekhen as part of a "land reclamation" program.

ARCHAEOLOGICAL ISSUE. A third issue clarified by geochemical research is archaeological and deals specifically with the changing technology of Predynastic pottery production. Over the last 90 years, the striking handmade pottery of the Predynastic era has been a major focus of archaeological and Egyptological inquiry. Archaeologists estimate that perhaps 50,000,000 broken pieces of Predynastic pottery litter the low desert at Hierakonpolis–a number and density far exceeding any known site in Egypt. In addition, there is extensive evidence for Predynastic pottery kilns in the settlements at Hierakonpolis. This extent is striking compared to only one possible example of a kiln found at any of the other sites in Egypt! Our ongoing studies of this pottery continue to rely upon our ability to understand the sediments (clays) used in the production of the pottery.

This research on pottery has three aspects. The ability to differentiate with trace elements different sedimentary units allowed us to conclude that pottery was produced from Nile deposits located in the immediate vicinity of the potter's workshop and kilns (*25*). The finer Plum Red Ware, most popular between ca. 4000 and 3400 B.C., and most common in cemeteries, was made from the same Nile sediments as the more common Straw Tempered Ware that dominated settlement ceramics. The Plum Red Ware was often fired in specific areas that may have behaved like natural wind tunnels (e.g., Locality 39 and 59 in Figure 1). Our more limited studies of Hard Orange Ware (a fine pottery most popular after about 3400–3200 B.C.) have shown that, although made from local Nile sediments, there was a significant addition of Ca salts to the paste as temper (*26*). This finding confirms morphological analysis that indicated the addition of crushed bone, shell, or calcite to the paste before firing. Chemical analysis reveals Ca concentrations of around 12% for Hard Orange Ware (Table IV) compared to 2–4% in the Nile sediments and an average of 3% for Plum Red Ware. Reaction of $CaCO_3$ by treatment of ground Hard Orange Ware sherds with dilute HCl and microscopic examination show clearly that Ca salts were deliberately added to the Nile sediments. Studies (*1, 4*) suggest that calcium carbonates lower the firing temperatures to below those necessary to produce the common Straw Tempered Ware, and allow the manufacture of harder, finer (Hard Orange) ceramics that rapidly replaced the earlier (Plum Red) fine ware that appeared to be somewhat more difficult to fire.

The replacement of one fine ware by another involved not only a technological change but correlated with the major shift of settlements from the desert toward the flood plain. The analysis of core samples from Kom El Ahmr also suggests a change in the patterns of Nile flooding at about this same time. In addition, all of these changes coincide with the emergence of the nation–state with the legendary earliest pharaohs from this locality. Only future research will reveal whether or how these events relate to each other.

A third research theme suggested by our geochemical studies of Predynastic ceramic technology at Hierakonpolis deals with the differences in methods of production between Plum Red and Straw Tempered wares. Initially, we suggested that the somewhat increased Ca content in the Plum Red Ware may have been due to the deliberate addition of Ca salts (*27*). The present studies of the wadi sediments and Unit C suggest, however, that the older Nile sediments contain an acid-soluble Ca phase deposited, much like the Fe, on the surface of particles. Finer grained particles with more surface area appear to contain most of this soluble Ca. This Ca probably originates from the Ca salt deposits in the older shale sediments that are redeposited from solution onto the Nile sediments.

The Plum Red Ware shows clear evidence that much of the coarser grained material was removed from the Nile sediment to produce the clay paste. Trace element concentrations (e.g., REE patterns) are generally higher (because of removal of the quartz dilutant), and microscopically the average grain size distribution is smaller. This finding reinforces earlier archaeological observations that the fine Plum Red Ware paste was probably cleaned by water separation. Thus, the grain size separation would have elevated the Ca to the levels in the paste, because as the finer grained particles have more surface area. This small increase in Ca content may not have substantially aided in reducing the melting or reaction point of the clay. However, the presence of fine-grained dispersions of Ca on the clay surfaces would have been more effective than large-grained Ca-containing particles present as mineral inclusions or added as temper.

Finally, as chemists and archaeologists, it is interesting for us to be able to demonstrate, with a variety of analytical techniques, that while the early Egyptians of Hierakonpolis were transforming their society into the world's first politically centralized nation–state, they were also recognizing the utility of natural materials to aid them in transforming humble Nile clays into new forms of economically, artistically, and symbolically important ceramics.

References

1. Edwards, W. I.; Segnit, E. R. *Archaeometry* **1984,** *26,* 69.
2. Tobia, S. K.; Sayre, E. V. In *Recent Advances in Science and Technology of Materials;* Bishay, A., Ed.; Plenum: New York, 1974; Vol. 3, pp 99–128.
3. Kaplan, M. F.; Harbottle, G.; Sayre, E. V. *Archaeometry* **1982,** *24,* 127.
4. Allen, R. O.; Rogers, M. S.; Mitchell, R. S.; Hoffman, M. A. *Archaeometry* **1982,** *24,* 199.
5. Allen, R. O.; Hamroush, H. A. In *Archaeological Chemistry III;* Lambert, J. B., Ed.; Advances in Chemistry Series No. 205; American Chemical Society: Washington, DC, 1984; pp 51–66.
6. Hoffman, M. A. *Egypt Before the Pharaohs;* Knopf: New York, 1979.
7. *The Predynastic of Hierakonpolis;* Hoffman, M. A., Ed.; Egyptian Studies Association: Alden, Oxford, 1982.
8. Hoffman, M. A. *Anthropology* **1980,** *4,* 51.
9. Butzer, K. *Science (Washington, DC)* **1960,** *132,* 1617.
10. Said, R. *The Geological Evolution of the River Nile;* Springer-Verlag: New York, 1981.
11. Hassan, F. A. *Science* **1981,** *212,* 1142.
12. Hamroush, H. A. In *The Predynastic of Hierakonpolis;* Hoffman, M. A., Ed.; Alden: Oxford, 1982; pp 93–100.
13. Wendorf, F.; Schild, R.; Issawi, B. *The Prehistory of the Nile Valley;* Academic: New York, 1976.
14. Vermeesch, P. M. *El Kab II*; University of Leuven: Leuven, 1978.
15. Vermeesch, P. M. *Chronique d'Egypte* **1970,** *45,* 89.
16. El Kammar, A. personal communication
17. Allen, R. O.; Pennell, S. E. In *Archaeological Chemistry-II;* Carter, G. F., Ed.; Advances in Chemistry Series No. 171; American Chemical Society: Washington, DC, 1978; pp 230–257.
18. Wendorf, F.; Hassan, F. A. In *The Sahara and the Nile. Quaternary Environments and Prehistoric Occupation in North Africa;* Williams, M. A. J.; Faure, H., Eds.; Academic: New York, 1980; pp 49–103.
19. Ions, V. *Egyptian Mythology;* Newnes Books: Middlesex England, 1982.
20. Hoffman, M. A.; Hamroush, H. A.; Allen, R. O. *J. Am. Res. Cent. in Egypt* (in press).
21. Shukri, N. M. *Quart. J. Geol. Soc.* **1950,** *106,* 466.
22. Ritchie, J. C.; Eyles, C. H.; Haynes, C. V. *Nature (London)* **1985,** *314,* 352.
23. Hoffman, M. A.; Hamroush, H. A.; Allen, R. O. *Geoarchaeology* **1987,** *2,* 1.
24. Allen, R. O.; Hamroush, H.; Hoffman, M. A. *Anal. Chem.* **1986,** *58,* 572A.
25. Jacquier, H.; Jacquier, J. personal communication
26. Allen, R. O.; Hamroush, H. *Chemtech* **1986,** *16,* 484.
27. Allen, R. O. *J. Chem. Ed.* **1985,** *62,* 37.

RECEIVED for review September 15, 1987. ACCEPTED revised manuscript July 11, 1988.

4

Compositional Data Analysis in Archaeology

Ronald L. Bishop and Hector Neff

Conservation Analytical Laboratory, Smithsonian Institution, Washington, DC 20560

As compositional analysis has become more routine in archaeological investigations, deficiencies in the numerical techniques used for data reduction and summary have become more apparent. A brief overview of techniques commonly used in the analysis of compositional data is presented as well as an example illustrating how data modeling (as opposed to data summary) can facilitate both the recognition of relevant data structure and inferences from data structure to underlying natural and cultural processes.

THE APPLICATION OF CHEMICAL ANALYTICAL TECHNIQUES to archaeological questions has a long history that extends back to the late 1700s (*1*). Many of the earlier investigations dealt with the compositional characterization of objects to elucidate aspects of their properties, such as color. Yet, as Harbottle (*2*) has noted, by the end of the 19th century, chemists like Damour and Helm viewed the chemical analysis of artifacts as a means of documenting long-distance traffic in particular materials. The basic approach of determining a chemical composition for an object and then comparing that profile to others similarly derived has been elaborated since that time. Today, the chemical characterization of artifacts constitutes a basic archaeological approach that can be used to address not only problems pertaining to long-distance exchange but to intraregional production and distribution (*3*), development of craft specialization (*4*), and typological refinement (*5*, *6*), among other issues.

Despite the volume of data generated and the variety of applications, development and testing of data-handling techniques have lagged. There is

0065–2393/89/0220–0057$08.50/0

not a "cookbook" approach to data analysis any more than there is some ideal group or number of elemental concentrations to determine for all applications. Complex natural and cultural interactions can account for much of the observed compositional variation, and one must be aware of these interactions to achieve a greater understanding of the data. In the discussion to follow, we will be concerned with aspects of multivariate data analysis that lead us toward the position that many of the questions being addressed in a compositional investigation require modeling rather than merely summarizing the data.

Background

The development of increasingly sophisticated analytical instrumentation that allows numerous elemental concentrations to be determined in a relatively short time and increased throughput of specimens has had a decided impact on archaeology. It has even been claimed (*7*) that the availability of analytical capability is partially responsible for concentrated archaeological attention to material exchange during the 1970s. With the interest among archaeologists and the technological advances has come a staggering amount of analytical data. Numerical summarization of these data, assisted by the increasing speed of the computer and availability of general-purpose statistical packages, is a vital link between the generation of data and its interpretation within the archaeological context.

Data analysis has not been neglected by archaeologists and "archaeometricians." Numerous papers have described various techniques applied to specific sets of data. Others have described how particular options of readily available commercial programs are used (*8–11*). At times, routine numerical procedures applied to well-determined data have been interpreted in a manner that fails to contribute to increased archaeological understanding. In many of these efforts there is an inappropriate use of statistics, failure to understand the component nature of the material being analyzed, or a failure to bridge from the analytical data to the archaeological context. In more general terms, the speed of data production and computation has outpaced the logic of the investigation.

Because we recognize that the merger of archaeological investigation with physicochemical analysis is still evolving, we will try to avoid reference to specific applications where we believe basic mistakes were made. Instead, we will discuss problems arising in the compositional analysis of archaeological materials in an abstract or generic manner. This approach may subdue the inclination some investigators may feel to engage in vitriolic rebuttal like that which followed Thomas's (*12*) general critique of statistical practice in archaeology (*13*).

Although many of the comments in this chapter are applicable to situations encountered during the analysis of data from diverse types of archaeological materials, we will illustrate our points in a later section with examples drawn from ceramic compositional systems. Our examples will incorporate very well-understood and artificial (or "dummy") data, as appropriate in a discussion of methodology.

Goals of Data Analysis

Compositional analysis of archaeological materials entails a series of nondiscrete steps of research design:

- problem formulation,
- sample selection,
- analytical approach,
- data analysis, and
- data integration.

The nature of the specified problem will suggest which samples and how many will be considered, whether raw source materials will be included in the investigation, the spatial and temporal extent of sampling, etc. Certainly, sampling of an interregional investigation will differ considerably from the more demanding requirements for an intraregional focus (*14*). Once the problem is formulated and samples are specified, selection of an appropriate analytical technique ideally depends upon the sensitivity and precision required to address the problem at hand. On a more practical level, one cannot dismiss considerations of instrumental availability and cost.

A data matrix produced by compositional analysis commonly contains 10 or more metric variables (elemental concentrations) determined for an even greater number of observations. The bridge between this multidimensional data matrix and the desired archaeological interpretation is multivariate analysis. The purposes of multivariate analysis are data exploration, hypothesis generation, hypothesis testing, and data reduction. Application of multivariate techniques to data for these purposes entails an assumption that some form of structure exists within the data matrix. The notion of structure is therefore fundamental to compositional investigations.

Structure within a compositional data set is the differential occurrence of data points in the n-space defined by elemental concentrations. One simple kind of structure consists of points grouped around two centroids, or centers of mass, in the elemental concentration space. Structure within a compositional data set is assumed, implicitly or explicitly, to reflect the underlying process responsible for the data. Thus, in the case of the two-centroid structure just mentioned, an underlying process, such as procurement of clay from two sources, is assumed.

Different operational levels may exist for the inference of process from structure. For example, principal-components analysis (a method described later in this chapter) permits compression of multivariate data into a few dimensions and yields a scatterplot of what appear to be two groups (Figure 1). The two groups are readily recognizable by using several different kinds of cluster analysis; the group separation is readily confirmed with discriminant analysis. In fact, one group was formed from the other by multiplying all elemental concentrations by 0.66. This example approximates the effect of a relatively pure temper (e.g., quartz sand) on the clay composition of ceramics (*16*). Particularly in ceramic production, an observed compositional profile might relate not only to the natural realm (source rocks, weathering, erosion, transportation, etc.), but also to the cultural realm (social and individual patterns of materials procurement and preparation).

The search for structure proceeds according to some mathematical model that can organize and represent the information in a data matrix. Particular kinds of associations between data entities or variables may be examined—but always relative to the particular model used (*17*). These models are at the same time structure-revealing and structuring. This concept is illustrated by three natural groups shown relative to concentrations of Fe and Sc in the scatterplot in Figure 2. Because of interelemental correlations, the groups form elongated ellipses, yet are fully separable at the 95% confidence interval. If a hierarchical cluster analysis based on Euclidean distances calculated from logged Fe and Sc values were carried out, the

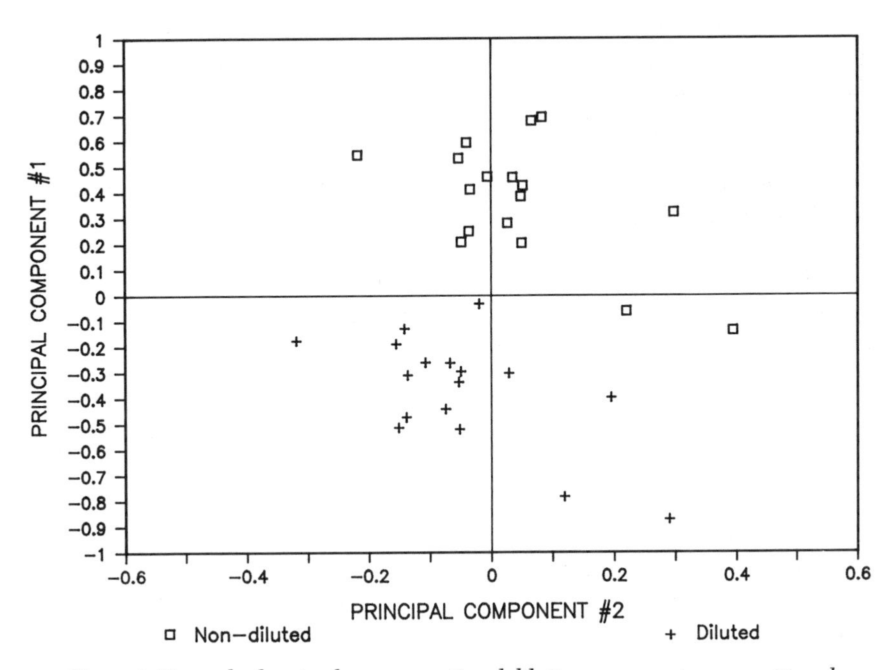

Figure 1. Example showing how proportional dilution may create compositional subgroups. "Diluted" specimens were created by multiplying 17 elemental concentrations in the nondiluted specimens by 0.66.

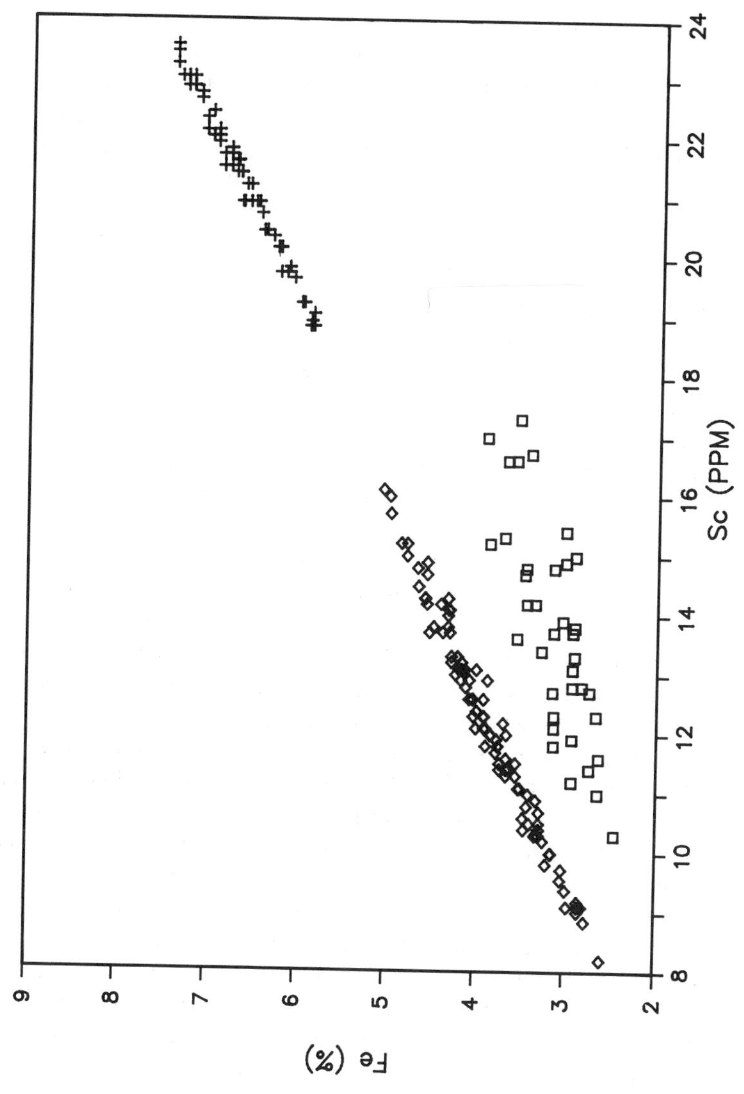

Figure 2. Scatterplot of Fe and Sc values for three distinct groups.

resulting spherical clusters would confound the group membership. If, however, a hierarchical cluster analysis were carried out by using Euclidean distances calculated from standardized principal-component scores, the resulting clusters would correspond to the groups evident in the scatterplot.

Without any prior knowledge of the structure in the data set in Figure 2, one might begin a search for structure with an analysis of the straight-line distances between points in the data set. Hierarchical cluster analysis of Euclidean distances implements a systematic approach to the analysis of straight-line distances. (*Euclidean distance* is the *n*-dimensional generalization of straight-line distance, and is discussed in greater detail later. *Cluster analysis* is a method of representing the Euclidean distances in two dimensions, and is also discussed later). However, the cluster analysis approach in this case fails to represent the known relationships among the known groups, although groups of closely similar samples are formed.

The problem lies in the model. The Euclidean distance calculation is inappropriate for use with correlated variables because it is based only on pairwise comparisons, without regard to the elongation of data point swarms along particular axes. In effect, Euclidean distance *imposes* a spherical constraint on the data set (*18*). When correlation has been removed from the data, (by derivation of standardized characteristic vectors) Euclidean distance and average-linkage cluster analysis return the three groups.

Most of the time, we do not have *absolute* a priori knowledge regarding the number of groups in a data set, or the relationships among the variates. Several rather distinct populations with differing patterns of interelemental correlations may be represented. In such cases, inspection of correlations whose pattern is pooled over all samples may not be informative (although, as in the example just presented, inspection of scatterplots alone may provide information on the number of groups likely to be found).

Assuming that a data set has some natural or optimum structure and that a given multivariate approach will be able to reveal it is a blind approach to data analysis. Because mathematical pattern-recognition techniques not only *reveal* structure but may *impose* structure as well, more informed application of multivariate techniques is needed. The choices among data analytical approaches must be made with reference to the stated research problem and an awareness of the requirements and assumptions of the various multivariate techniques. Different groups will be formed or different aspects of the data investigated depending upon specific problem formulation. From this perspective, one may reject outright naive notions of uniform methodology involving multivariate data analysis (*11*).

Multivariate Techniques and the Search for Structure

The literature dealing with the multivariate techniques of pattern recognition, numerical taxonomy, group evaluation, etc. is extensive (e.g., reference

19). This discussion provides only an outline of the techniques that have been used to search for structure in compositional data matrices generated by the analysis of archaeological materials. Before many of the techniques are used, however, some pretreatment of the data may be necessary.

Data Representation. Transformations can be applied to the data so that they will more closely follow the normal distribution that is required for certain procedures or for removing (or lessening) unwanted influences. Certainly for data analysis in which major, minor, and trace elemental concentrations are used, some form of scaling is necessary to keep the variables with larger concentrations from having excessive weight in the calculation of many coefficients of similarity.

Another form of scaling involves equalizing the extent of variation among the variables. In some instances, two transformations of the data are of interest (e.g., transforming the data to log-normal distributions and then ensuring equal weight through additional scaling). Using transformations that equalize the magnitude of the measurements and the amount of variation is in keeping with one of the premises of numerical taxonomy: that is, no individual variable should assume more weight than another in an analysis involving the calculation of resemblance (*19, 20*). However, when one is modeling, rather than merely summarizing the data, variable contributions may be adjusted according to different criteria. One example now under investigation is weighting chemical determinations as a function of their analytical errors.

Typical transformations include calculation of logarithms; standardization (mean of 0, standard deviation of 1); percent range; and percent of the maximum value. A different type of transformation has been used for summarizing chemical data from the analysis of steatite or soapstone. Arguing from principles of geochemistry, Allen and coworkers (*21–23*) normalized the rare earth concentrations relative to abundances in chondritic meteorites. Following Sayre (*24*) and Harbottle (*20*), we use base-10 logarithms in most of the examples discussed later in this chapter. A percent range transformation is also employed for one operation.

Ordination. Ordination procedures place a sample data point in a variable space to represent some trend or variation. No assumptions need to be made regarding the number of groups. A simple type of ordination would be to plot the coordinates of a sample relative to two variables as in Figure 2. For p-variables, higher dimensionality prohibits easy inspection, so most ordination techniques attempt to summarize the information within a data set and reduce the dimensionality (e.g., Figure 1).

Dimensionality reduction and ordination have had three main uses in compositional investigations. They have been used

1. to inspect the data to see if a general size component, one stemming from a proportional rather an absolute relationship, is present in the data (*16, 25*)
2. to project the data into a standardized space that offers a different, possibly more appropriate, perspective on interpoint distances (*26*)
3. to form a set of reference axes of reduced dimensionality for graphical display of grouped sample distributions determined by some other technique (*27*)

The most widely used ordination methods are based on extracting eigenvalues and eigenvectors (also called characteristic roots and characteristic vectors) from a minor product matrix, $\mathbf{X'X}$, or major product matrix, $\mathbf{XX'}$, of a data matrix, $\mathbf{X}$ (*28*). If the data matrix is centered by columns before calculating $\mathbf{X'X}$, the minor product matrix is a variance–covariance matrix. If the data matrix is first not only centered but standardized, the minor product matrix is a correlation matrix. Whether the starting point of the analysis is a variance–covariance matrix or a correlation matrix, the eigenvectors of $\mathbf{X'X}$ are usually called *principal components*. The eigenvectors of this matrix may be multiplied by their corresponding eigenvalues to produce factors that may be rotated to enhance their interpretability (in this case, the analysis is called a *factor analysis*).

In principal, neither centering nor standardization is necessary in eigenvector analyses, and each may be undertaken without the other. Orloci (*29*) and Noy-Meir (*30*) discuss the effects of centering in ecological applications. To our knowledge, there has been no careful consideration of the effects of centering and standardization on compositional data matrices. The popularity of standardized, centered eigenvector analyses has more to do with the availability of software than with the appropriateness of the assumptions.

The first principal component accounts for the direction of maximum variance through the data, with each successive component accounting for the maximum of the remaining variation. The length of each vector is determined by the square root of the associated eigenvalue. The derived components constitute a new set of reference axes that are linear combinations of the original measurement, but that now are orthogonal; that is, the variance of the original data is preserved but the covariance has been eliminated. Depending on how much of the variance one wishes to preserve in the analysis, the number of components may be truncated, but only at the loss of some information. Factor analysis includes an explicit statistical assumption that all meaningful variation in the data is accounted for by m underlying factors, with $m < p$, where p is the number of original variables (*28*). Variation remaining after extraction of the *m* factors is assumed to be purely stochastic. The derived factors may also be rotated to some simple structure. An example is varimax rotation, in which orthogonality is preserved but the factors are rotated so that the variance of the loadings on each factor is maximized. Rotation aids interpretation of the factors.

Q-mode factor analysis is based on a major product matrix, **XX′**. Whereas the R-mode analyses focus on interrelationships among variables, Q-mode analyses focus on interrelationships among objects. Accordingly, the major product matrix is usually a distance or similarity matrix. Formally, Q-mode and R-mode factor analyses are closely related because the nonzero eigenvalues of the major product matrix are identical to the eigenvalues of the minor product matrix, and the eigenvectors are easily derived from one another (*28*).

When applied to geological mixing problems, Q-mode factors are thought of as idealized end-members when the rotated factor matrix is normalized (*31–33*). When three such end members are extracted, the data points can be plotted on ternary diagrams, the apices of which can be thought of as representing the composition of components contributing to the mixtures in the data set. Mixtures derived from more than three components are more difficult to represent graphically.

If groups have been defined, then factor analysis can be used to determine to what extent elements are represented by the common portion of the factor. Not surprisingly, elements that are known to be mobile in several different environments (Na, K, Ba, Cs), or that are determined with less analytical precision, often have less of their variance represented in the factor solution.

Ordination has been useful for identifying regional patterning of compositional data on natural or cultural variables. For instance, in an effort to model neutron activation data from the analysis of pottery recovered along an escarpment of the Chiapas highlands, Bishop (*16*) utilized R-mode common-factor analysis. Nonchemical data were projected onto chemically based factor plots. After 13 years, and a certain diminishing of multivariate zeal, it can be questioned whether the same patterns could not have been derived from more straightforward principal components plots. The use of a common-factor model seemed warranted at the time as a means to partition analytical data into noise and signal components. In a similar effort, Bishop (*25*) projected nonchemical data onto principal-components plots and three-component Q-mode factor plots in an effort to identify patterning of natural and cultural variables on compositional data.

Aside from principal components, R-mode factor analysis, and Q-mode factor analysis, other techniques that have been used to reduce dimensionality in ungrouped compositional data include multidimensional scaling (*34*) and correspondence analysis (*35*).

Grouping. The most commonly employed techniques of data analysis in compositional investigations are those that seek to partition a data set into smaller groups that contain samples that are more similar to others in the group than to other samples in the data set. Cluster analysis, including both hierarchical and nonhierarchical variants, encompasses virtually the full range of grouping procedures used in compositional data analysis.

The manner in which sample-to-sample resemblance is defined is a key difference between the various hierarchical clustering techniques. Sample analyses may be similar to one another in a variety of ways and reflect interest in drawing attention to different underlying processes or properties. The selection of an appropriate measure of similarity is dependent, therefore, on the objectives of the research as set forth in the problem definition. Examples of different similarity measures or coefficients that have been used in compositional studies are average Euclidean distance, correlation, and cosine. Many others that could be applied are discussed in the literature dealing with cluster analysis (*15, 18, 19, 36, 37*).

Once a matrix of resemblance coefficients has been created, it can be summarized by partitions created to bring together the samples that share strong measures of resemblance. Most frequently, this partitioning is carried out by clustering the samples, linking together the pairs of most similar samples, and then stepping through the remaining samples until all are linked in a large treelike structure or dendrogram.

Hierarchical cluster analysis is a method for obtaining a quick approximation to grouping tendency in a set of data, but it is an insufficient representation of the data on which to base data analytic or archaeological inferences. In addition to the need to make an appropriate choice of resemblance measure and clustering algorithm, the reduction of the multivariate similarity matrix to a two-dimensional dendrogram can introduce considerable distortion. One measure of such distortion is the cophentic correlation (*38*). Although the resemblance relationships of the lower linkages of the dendrogram are usually well-represented, the relationship among the higher linkages is not. In using a hierarchical cluster model, no assumptions need to be made regarding the number of groups present in the data set. Inspection of the dendrogram does not inform directly on the best-levels linkage for group arrangement, although a method for testing the distinctness of clusters has been proposed by Sneath (*39*). A large change in cluster level of a dendrogram may be a *necessary* condition for cluster break, but it is not itself a *sufficient* condition (*40*). Additional problems arise in that once two samples are linked, that link can not be broken. Finally, hierarchical cluster procedures do not seek optimum partitions, and are usually found to perform poorly unless natural groups are well-separated.

A different approach to cluster analysis involves the use of some explicit criterion; the objective of that is to indicate the optimum number of groups in a data set. Several measures have been proposed that are based on the fundamental partition equation (*41*):

$$\mathbf{T} = \mathbf{B} + \mathbf{W} \tag{1}$$

where **T** is the matrix of the total variation in the data set, consisting of **B**, the variation between the groups, and **W**, the variation within the groups. A formal discussion of these matrix components is given in reference 42. By using these relationships, one can use minimum trace-**W**, determinant-**W**, or maximization of trace $\mathbf{W}^{-1}\mathbf{B}$, among others (*43, 44*). Like the resemblance measures mentioned for hierarchical clustering, these criteria have strengths and weaknesses. For example, the use of trace-**W** will result in spherical clusters. If one uses the determinant-**W** it is not necessary to assume that the natural groups exist in spherical distributions, but only that their shapes are roughly the same (*45, 46*).

The criteria just described are found in many of the k-means cluster programs that form disjoint partitions in to k-clusters. In use, the programs step from a minimum number of groups to some specified maximum number. Initial clusters are formed and samples are then moved in an interactive fashion until the criterion value is considered optimum (*43, 47*). Informal measures can be used to indicate which partition represents the number of natural groups in the data (*43*). k-Means procedures approach a global solution by deriving the best partitioning relative to a stated criterion value. Unfortunately, they have been found to be quite sensitive to outlying samples, the partioning of which can result in suboptimum solutions.

A basic question of whether hierarchical or nonhierarchical cluster analysis is used deals with the correct or best number of groups in a data set. The notion of "best" relates not only to a criterion value or large break in a dendrogram, but to the research objectives as well. We can not resist quoting from Everitt (*48*) what is probably the ultimate word regarding the number of groups:

> *In theory, of course, the problem is simple, to quote Dr. Idnozo Hcahscror-Tenib, that super galactian hypermetrician [sic] who appeared in Thorndike's 1953 Presidential address to the Psychometric society, "Is easy. Finite number of combinations. Only 563 billion billion billion. Try all. Keep best."*

Group Evaluative Procedures

Group evaluative procedures provide an assessment of the compactness of the groups resulting from the prior application of the ordination and grouping procedures just discussed. Although group evaluation logically should follow group formation, groups are sometimes formed by reference to the noncompositional data (e.g., type, form, provenance), with group evaluation procedures then used (as ad hoc group formation techniques) to include or exclude specimens on the basis of composition. For example, discriminant analysis may be applied to analyzed specimens in groups corresponding to several different sites to confirm a hypothesis of multilocus production. The problem is that discriminant analysis is very good at making groups out of whatever groups are assumed in the first place, particularly when the correlation or variance–covariance matrix is pooled over all groups.

The techniques used for group evaluation are based on multivariate generalizations of univariate statistics such as the centroid (a multivariate mean, or mean vector) and variance–covariance matrix (a multivariate generalization of the variance). Group evaluative procedures can be classified as single group procedures, in which specimens are evaluated as to their likelihood of membership in a single group, and multigroup procedures, such as discriminant analysis and canonical variates, in which several groups, or multivariate centers of mass are assumed to be represented in a data set.

Single-group evaluative procedures, because they are not implemented in most widely available statistical packages (SAS, SPSS, BMDP, etc.), are used less often in compositional data analysis than multiple-group evaluative procedures. ADCORR, a program developed at Brookhaven National Laboratory (*24*), performs single-group multivariate evaluation based on Mahalanobis (generalized) distance and the related Hotelling T^2 statistic (the latter is a multivariate generalization of the well-known Student's *t*). Mahalanobis distance, which is central to multiple-group as well as single-group evaluation, is defined for a group centroid and each individual member of the group as follows:

$$\mathbf{D}^2 = (\mathbf{X} - \mathbf{X}_{\mathrm{mean}})'\mathbf{S}^{-1}(\mathbf{X} - \mathbf{X}_{\mathrm{mean}}) \qquad (2)$$

where $(\mathbf{X} - \mathbf{X}_{\mathrm{mean}})$ is a vector of differences between the values for an observation and the mean values and $\mathbf{S}^{-1}$ is the inverse of the group variance–covariance matrix. Hotelling T^2 is essentially equivalent to Mahalanobis distance for individual points. The probabilities of membership in the core group for each specimen are then readily obtainable following transformation of the T^2 statistic to a related, F-distributed statistic (*28*).

Discriminant analysis evaluates the distance between individual points and *several* centroids hypothesized to exist in the hyperspace defined by elemental concentrations. Davis (*28*) provides a clear and concise description of the algebra involved in two-group and multiple-group discriminant analysis, showing that discriminant functions are equivalent to the eigenvectors of $\mathbf{W}^{-1}\mathbf{B}$, where $\mathbf{W}^{-1}$ is the inverse of the within-group sums of products matrix, and $\mathbf{B}$ is the between-group sums of products matrix. The Mahalanobis distances from an unknown point to each of the alternative centroids provide criteria for evaluating the relative probabilities of membership of the specimen in each of the groups.

Modeling vs. Summarizing Compositional Data: A Ceramic Example. Modeling is important; it involves informed interaction between the researcher, his or her objectives, analytical data, and multivariate data presentation. The potential for revealed data structures to vary depending upon the choice of numerical procedure is acknowledged explicitly. In addition, this approach requires the objectives of a particular investigation and the complexities of the compositional data base to be kept in mind during the stage of data analysis. The model provides the explicit rationale for reasoning from revealed data structure to inferences about human behavior. The remainder of this chapter illustrates some of the implications of this approach for the analysis of ceramic compositional data.

Any given analyzed sample of pottery is a small subset of a larger ceramic system. Pottery is formed from clays and nonplastic constituents according to shared customs of the local pottery-making group as well as idiosyncratic or stochastic effects. The compositional profile that is derived from the chemical analysis of a ceramic sample, therefore, is a weighted expression of both natural and cultural constraints.

One of the more obvious examples of this interaction involves the addition of temper to a clay matrix (*temper* may be another clay, but is more often a nonplastic material). The effect of tempering varies; a relatively pure material, such as quartz, may reduce elemental concentrations in a ceramic paste by a constant proportion (*49*). Addition of other kinds of temper or clay will result in a complex relationship of dilution and enrichment (*14*, *25*, *50*). Because elemental concentrations in sediments vary depending upon grain size (e.g., references 51–53), the size distributions of the added nonplastics also contribute to compositional complexity. If behavioral inferences are to be drawn, the culturally induced elemental variation arising from texture and temper differences among pottery produced from a single clay resource requires more than simple grouping and summary statistics.

Failure to consider the effect of temper during the analysis of compositional data might lead to spurious inferences in the following way: Suppose the object of an investigation is to identify local and nonlocal pottery at some site. If bimodal amounts of temper were added by local potters to ceramic paste (for example, depending on whether the potter planned to make a serving dish or a water storage jar), analyzed pottery from the site might fall into two compositional groups even though a single set of ceramic resources was used. If temper is left out of the model, the compositional data might suggest that one of the compositional groups is local and the other is nonlocal.

In the last example, the investigator's problem orientation might predispose him or her to accept the plausible but incorrect inference that local and nonlocal specimens are represented in the analyzed collection. Data analysis techniques that do not permit recognition of the source of group-separating variation can promote such erroneous interpretations of compositional patterning.

No data analysis technique is foolproof. Obtaining valid inferences from compositional analysis depends, above all, on considering alternative models of the processes responsible for data structure. To develop useful and plausible alternative models, nonchemical as well as chemical information must be considered. In the last hypothetical case, observation of a textural difference between serving vessels and water storage jars might suggest bimodal tempering of the same local clay, so that local production of both compositional groups could be suggested as an alternative to multilocus production. Our goal is to show how data analysis techniques, although they may not always lead to valid inferences, may at least facilitate the search for relevant data structure (*see* reference 54). The illustrations focus on the problem of tempering because it is an important consideration in most ceramic compositional investigations.

Data Used for Illustration. Our approach to the problem of how to fit data analysis techniques to expectations about structure in the data involves heuristic use of artificially structured data. In particular, we have relied on analyses of ceramics and ceramic raw materials from contemporary highland Guatemalan pottery-making towns. Analyzed clays, tempers, and ceramic sherds from three towns in the northern Valley of Guatemala (Durazno, Sacojito, and Chinautla) constitute a basic data set of 113 observations. With these data, a hypothetical data set is constructed, the structure of which we will examine by using some of the data analysis techniques just described. The analyzed specimens were furnished by Dean Arnold. Arnold's ethnographic observations (55–57) provide baseline expectations about patterning in the compositional data. Neutron activation analyses were carried out by using the standard procedures developed at Brookhaven National Laboratory (47).

Within the northern Valley of Guatemala whiteware tradition, there is a basic dichotomy between Sacojito–Chinautla whiteware and Durazno whiteware. Durazno potters exploit sources of whiteware clay different from the source used by Sacojito–Chinautla potters (55). Durazno whiteware pottery and raw materials therefore should be distinguishable from Chinautla–Sacojito whiteware.

The tempering material used in Durazno is volcanic ash from the ash blanket that covers the northern Valley of Guatemala. Chinautla–Sacojito potters exploit the same volcanic ash, albeit in different locations. Although potters in all three towns temper their clay, it is instructive to consider the potential effect on compositional patterning of variable tempering practices (e.g., a fine paste–medium paste dichotomy). Such a dichotomy characterizes many archaeological collections. If northern Valley of Guatemala potters made fine-paste pottery, its composition would resemble the composition of raw clay specimens included in the present data set; the analyzed sherds in the data set are taken to represent medium-paste pottery.

By artificially mixing clays and tempers, any number of hypothetical sherds can be added to the basic data set. In principal, the hypothetical mixtures weigh heavily in favor of finding suitable numerical techniques for discovering relevant structure in the data set. In the following example hypothetical fine-paste and medium-paste mixtures are added to the basic data set of clays and tempers. We feel that "stacking the deck" in this manner is not only appropriate but necessary in a study attempting to evaluate methodology.

Hypothetical ceramic paste mixtures are generated from clays and tempers of known concentrations by calculating elemental concentrations according to

$$S_i = \mathrm{PT}(T_i) + \mathrm{PC}(C_i) \quad (3)$$

where S_i represents the elemental concentrations in the ceramic, T_i represents the elemental concentrations in the temper, and C_i represents the elemental concentrations in the clay. Both T_i and C_i have been determined analytically. PT and PC are the proportions of temper and clay, respectively, mixed to make the ceramic paste.

The sum of PT and PC must equal one. Because any recipe for mixing components, no matter how strictly followed, will show some variation, the PTs and PCs are chosen randomly with mean, μ_P, as the proportion of clay or temper specified by the recipe and standard deviation, and σ_P indicating the relative standardization (i.e., how strictly the recipe was followed). In this example, the mean clay proportion in fine-paste mixtures was set at 0.85, and the mean proportion in medium-paste mixtures was set at 0.65. Relative homogeneity was assumed, and the standard deviation was set at 0.05 in both cases. The components (T_i and C_i) to be mixed in the randomly chosen proportions are chosen randomly from the analyzed data sets of clays and tempers. Unfortunately, only four samples of the clay used to make Durazno whiteware were analyzed, so the hypothetical fine-paste and medium-paste mixtures could exhibit some bias due to the small number of clays to choose from.

Analysis. Two separate patterns should be discernible in the artificial data set: sherds, clays, and hypothetical mixtures from Chinautla–Sacojito should be separable from raw materials and products from Durazno and fine- and medium-paste subgroups should be recognizable within the major groups. The first pattern involves *shape variation*, that is, differing mean vectors and differing variance–covariance structures that arise through natural processes of clay formation. The second pattern can be thought of as size variation, in which tempering introduces linear changes in the elemental concentrations. Dilution by a constant (e.g., by quartz sand) is an extreme example of size variation.

The elemental concentration data were first transformed to base-10 logarithms to counteract the implicit weighting due solely to variation in the abundance of elements in nature (cf. reference 20). Principal components were then calculated from the variance–covariance matrix of the log concentration data. A plot of the first two principal components (Figure 3) shows two major groups; one that is high on component No. 1 and one that is low on component No. 1. The groups represent the two clay sources; Chinautla–Sacojito-derived materials scored high and Durazno-derived materials scored low. The group separation is due largely to Cr, Rb, and Cs concentrations, that tend to be higher in the Chinautla–Sacojito materials. Table I contains coefficients of the original variables on the principal components along with the percentage of variance explained by each component.

Hierarchical and k-means cluster analysis distinguishes the two major source-specific groups evident on the scatter plot. The highest level division is between Chinautla–Sacojito material and Durazno material. Lower-level branches subdivide the Durazno material first, and the Chinautla–Sacojito texture subgroups remain in a single subdivision above a Euclidean distance of 0.11. k-Means cluster analysis allows the recovery of the major structure in the data, although the more subtle structure remains obscured, even when eight groups are formed. Discriminant analysis confirms the true structure if the true group membership is used as the classification criterion, or it confirms a six-group structure derived from k-means and hierarchical cluster analysis (if that structure is used as a basis for classification). The application of grouping procedures followed by discriminant analysis therefore would result in an erroneous interpretation of this data set.

If this data set were archaeological, a further problem of interpretation would involve the origin of separation between the two major groups. In other words, what would constitute the evidence that group separation on principal component No. 1 in Figure 3 represents a clay-source distinction rather than temper-related variation? The problem is analogous to the problem of separating size from shape in the field of multivariate morphometrics (*see* references 58–62). Theoretically, general size manifests itself as consistently positive coefficients of the original variables on the first principal component (59). However, because addition of a volcanic ash temper almost certainly will not produce uniform enrichment or dilution, the variation introduced by tempering cannot be considered analogous to general size. Because some elements may be diluted while others are enriched, the signs of the coefficients of the original variables on the first principal component will differ. Because the amount of enrichment or dilution is likely to vary, the magnitude of the coefficients also may differ widely.

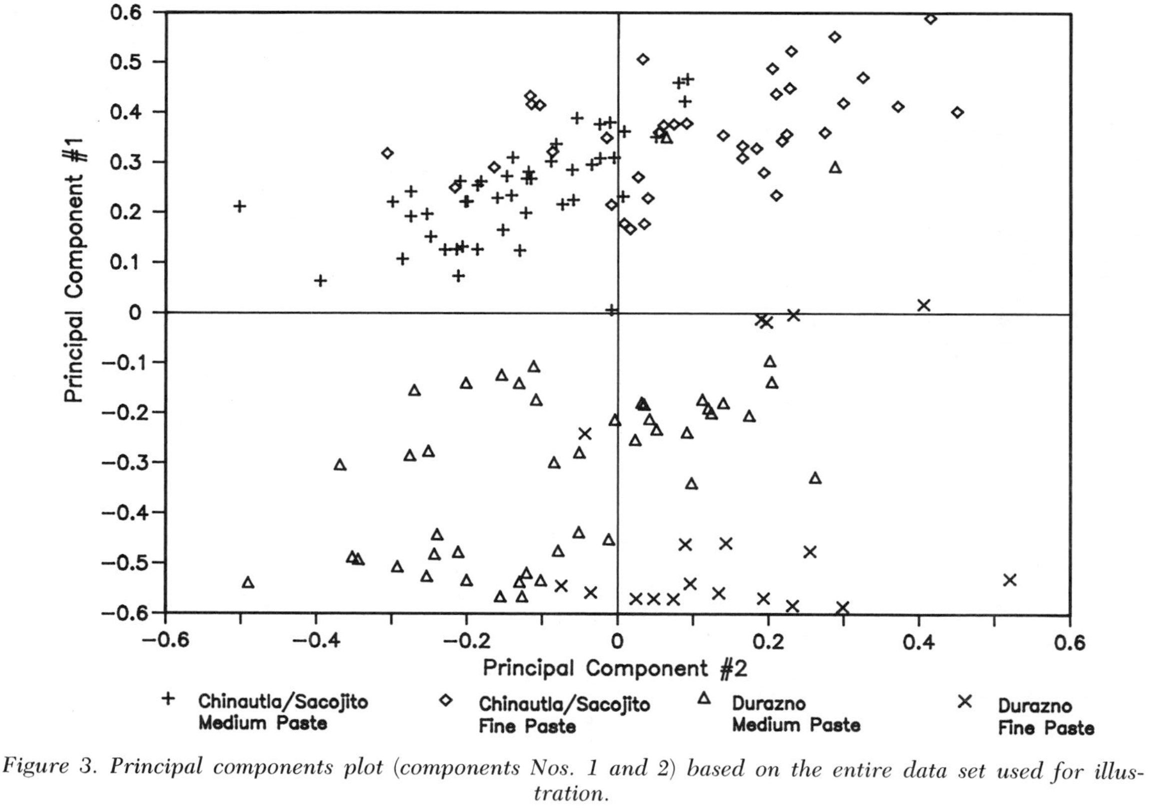

Figure 3. Principal components plot (components Nos. 1 and 2) based on the entire data set used for illustration.

Table I. Principal Component Coefficients for the Example Data Set

Element	*No. 1*	*No. 2*	*No. 3*
K	0.634	−2.421	2.242
Sc	0.030	1.492	−0.073
Cr	1.868	1.135	−1.038
Fe	0.144	1.263	0.240
Zn	0.618	1.040	0.022
Rb	1.223	−0.906	0.859
Cs	1.039	−0.175	−0.017
Ba	−0.310	−1.136	2.148
La	0.015	0.540	1.677
Ce	0.106	0.628	1.456
Sm	0.087	1.276	1.316
Eu	−0.109	1.561	1.347
Yb	−0.108	1.286	1.435
Lu	−0.159	1.307	1.347
Hf	−0.238	0.608	1.318
Ta	0.349	0.337	0.797
Th	0.293	−0.550	1.225
Percent of variance explained	54.817	17.327	14.420

More serious than the problems with the analogy between size and tempering, the fact that temper may interact differently with clays from the two distinct sources makes it unlikely that a single dimension analogous to allometric size will be recognizable in the data set. Nonetheless, if one possesses some information about the composition of likely tempering materials, that information can be used to help identify potential tempering dimensions. The volcanic ash temper used in the example data set is known to be enriched, relative to local clays, in K and Ba and diluted, relative to local clays, in most other elements in the analysis (Th tends to be enriched in temper, but the pattern is inconsistent). Although this information was obtained by neutron activation analysis of temper obtained from potters, it can be inferred from other sources, such as published studies of the composition of highland Guatemalan volcanic ash (*63*). In an archaeological situation, the information could have been obtained through compositional analysis of likely tempering materials collected during a carefully planned raw-materials sampling program or by separation and analysis of the nonclay component. Using this information, we expect the tempering components to be recognizable because K, Ba, and possibly Th should have high-magnitude coefficients of same sign.

The coefficients of the original variables on the principal components (Table I) do not contain a single dimension that is clearly related to temper. K and Ba, which are thought to be enriched in temper relative to clay, have opposite signs on the first component, so the first component is not a tempering dimension. The coefficients of K and Ba are fairly large and have the same sign on component No. 2, so component No. 2 may carry some temper-related variation. Similar reasoning suggests that principal component No. 3 also may reflect tempering.

Additional insight into the significance of group-separating variation can be obtained from further use of ordination. If the first three factors from a Q-mode factor analysis of data (expressed as a percentage of range) are normalized, data points are expressed as if they were mixtures of three components. The data points can then be plotted on ternary diagrams in which positions are determined by the percentage contribution from each factor. Figure 4 shows the data for the present example plotted against the first three varimax-rotated factors derived from Q-analysis. Two apices correspond to the two clay sources represented in the data. The third apex, which expresses primarily the amount of K, Ba, and Th in the mixture, probably corresponds with temper. Figure 5, which is derived from a Q-mode factor analysis of the original data set plus the tempers used to form the hypothetical mixtures, confirms the identification of the third apex as temper-related. These three-component plots identify the overall structure in the data set more clearly than any of the previous ordination and group-forming techniques.

Figure 5 illustrates why it may be useful to include potential tempering materials in a compositional study, even if exact sources probably have not been sampled. If tempering materials used in ceramics from a particular region can be assumed to be relatively uniform (like volcanic ash, which is a common temper in prehistoric ceramics from Mesoamerica), then inclusion of compositional data on the material identified as temper may, as in the present case, elucidate the interactions between temper and several different clays.

In summary, although division of the data set into two major groups attributable to clay sources can be accomplished by several methods (division is especially clear when ordination is combined with group formation procedures), it is more difficult to detect and interpret patterning within the major groups. This difficulty arises because processes operating on two hierarchical levels are being confused: the major division is produced by natural variation arising in the process of clay formation; whereas lower-level patterning is a result of two distinct cultural processes, Durazno potters mixing temper with their own unique clay and Chinautla–Sacojito potters mixing temper with their own unique clay. In the present example, a three-component ordination clarified both levels of structure, but this happy result in part reflects the artificial nature of the data and the fact that all raw materials had been analyzed. More complex data structures would have remained obscured in the Q-mode factor analysis.

A more generally useful approach is to incorporate into the assumption that multiple processes may create structure into the model, and to recognize that coarse-grained processes may obscure fine-grained processes. For in-

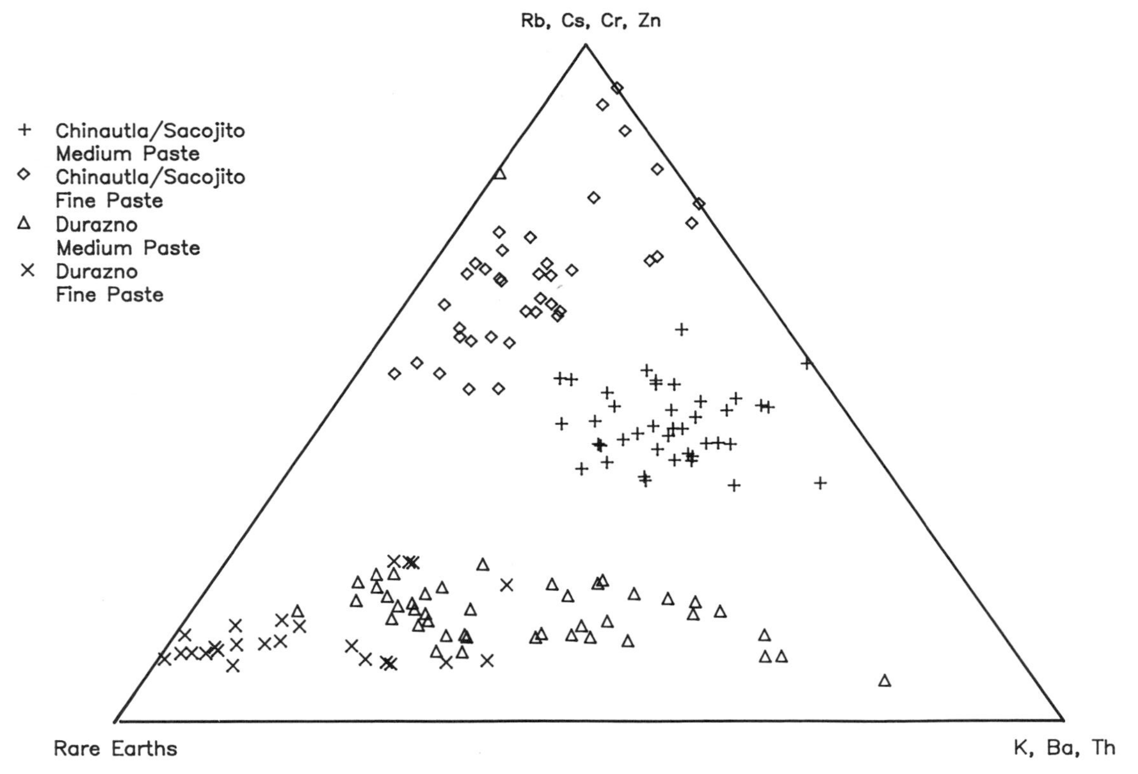

Figure 4. Plot of specimens in the example data set relative to three components defined by Q-mode factor analysis.

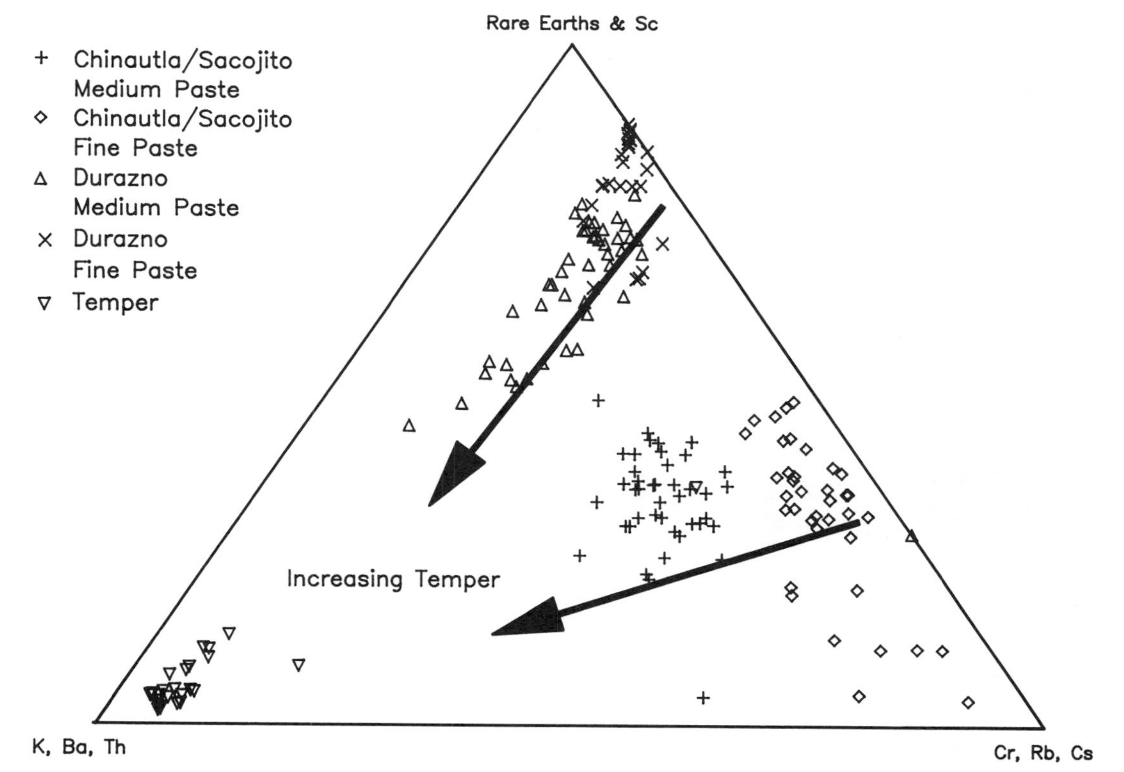

Figure 5. Plot of specimens in the example data set plus *northern Valley of Guatemala volcanic ash tempers relative to three components defined by Q-mode factor analysis.*

stance, having discovered two source-specific groups in the example data, a logical next step in the analysis would be to examine the subgroups individually. The following paragraphs describe this approach applied to the Chinautla–Sacojito subgroup.

Despite the complexities introduced by tempering with a noninert material, the analogy with allometric growth should hold when, as in the present case, the data set was generated by a group of potters working with a single set of ceramic resources. Tempering can be thought of as growth that involves change in shape or proportionality. Assuming that the data are all derived from a single, relatively homogeneous clay source (as in the present example) and, further, that the temper differs substantially from the compositional profile of the clay (as in the present example), all elemental concentrations should show linear relationships (positive or negative) with the proportion of temper in the mixture. The major dimension of variation (the first principal component) should carry most or all of the variation related to growth (tempering), particularly if temper-related variation is sufficient to create bipolar extremes recognizable as compositional groups.

Coefficients of the elemental concentrations on the first principal component (Table II) show the pattern expected for a temper-related dimension: the signs are positive for all variables except K and Ba, which are negative. Thus, the group separation on principal component No. 1 shown in Figure 6 can be interpreted as temper-related. As expected, the two major groups evident on the plot are (*1*) hypothetical fine-paste mixtures along with raw clay and (*2*) hypothetical medium-paste mixtures along with real sherds.

Table II. Principal Component Coefficients for the Chinautla–Sacojito Data Set

Element	*No. 1*	*No. 2*	*No. 3*
K	−0.669	2.231	4.865
Sc	1.374	−0.879	1.534
Cr	1.617	−1.271	5.321
Fe	1.127	−0.302	2.857
Zn	1.031	−1.253	−1.078
Rb	0.410	−0.059	2.541
Cs	0.665	−0.586	3.745
Ba	−0.528	3.314	1.254
La	0.999	1.451	−0.911
Ce	1.199	1.233	−0.223
Sm	1.585	0.742	−1.947
Eu	1.647	0.551	−2.672
Yb	1.439	0.892	−2.337
Lu	1.368	0.678	−2.158
Hf	0.602	0.882	0.768
Ta	0.721	0.378	2.338
Th	0.379	1.384	1.054
Percent of variance explained	44.419	31.947	8.047

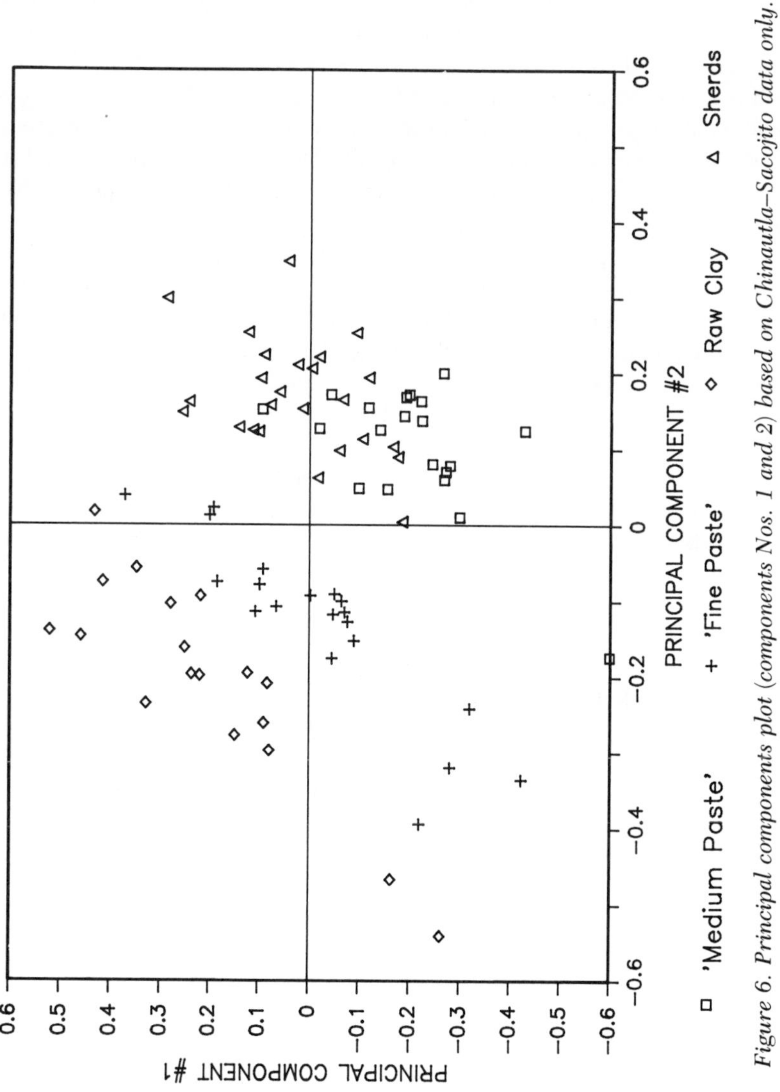

Figure 6. Principal components plot (components Nos. 1 and 2) based on Chinautla–Sacojito data only.

Both k-means and hierarchical cluster analysis yield two major groups attributable to temper-related variation (the medium-paste group included six fine-paste mixtures). Both techniques also extracted a small third group of outliers (the seven data points in the lower left corner of Figure 6). Discriminant analysis confirms the existence either of groups recovered by cluster analysis or groups defined previously as medium-paste or fine-paste. Single group evaluation with the program ADCORR also confirms the membership of the medium- and fine-paste core groups, although a few nonmembers show greater than 5% probability of membership in each group.

Figure 6 shows an unexpected result. The two major groups that are inferred to be temper-related are separated along principal component Nos. 2 and 1. This separation suggests that, contrary to expectations derived from analogy with the biological model, growth due to tempering is not confined to the major dimension in the data. The coefficients of the elemental concentrations of component No. 2 (Table II) further support this interpretation, a fact indicating that K and Ba, the two elements enriched by temper, are major determinants of variation on this second dimension. Temper-related variation (size), therefore, is expressed on both the first and second dimensions of variation in the second example data set.

Failure to isolate the effect of tempering on the first principal component appears to result from unexpected inhomogeneity in the data. The seven outliers identified by hierarchical and k-means cluster analysis cause the major axis of variation to veer away from the direction of temper-related size. When the outliers are removed and the principal components recalculated, component No. 1 (Table III and Figure 7) more closely approximates

Table III. Principal Component Coefficients Chinautla–Sacojito Data with Outliners Removed

Element	*No. 1*	*No. 2*	*No. 3*
K	−1.401	2.778	6.074
Sc	1.639	0.205	1.256
Cr	2.102	0.792	5.323
Fe	1.237	1.326	1.021
Zn	1.373	−1.457	2.029
Rb	0.436	0.572	3.564
Cs	0.917	0.563	3.685
Ba	−1.669	3.675	−0.930
La	0.375	2.001	−2.424
Ce	0.636	2.014	−1.398
Sm	1.145	1.215	−2.902
Eu	1.238	0.740	−2.928
Yb	0.934	1.168	−3.183
Lu	1.004	1.163	−4.114
Hf	0.245	1.537	−0.478
Ta	0.591	1.546	1.199
Th	−0.138	1.956	0.155
Percent of variance explained	50.529	23.255	7.281

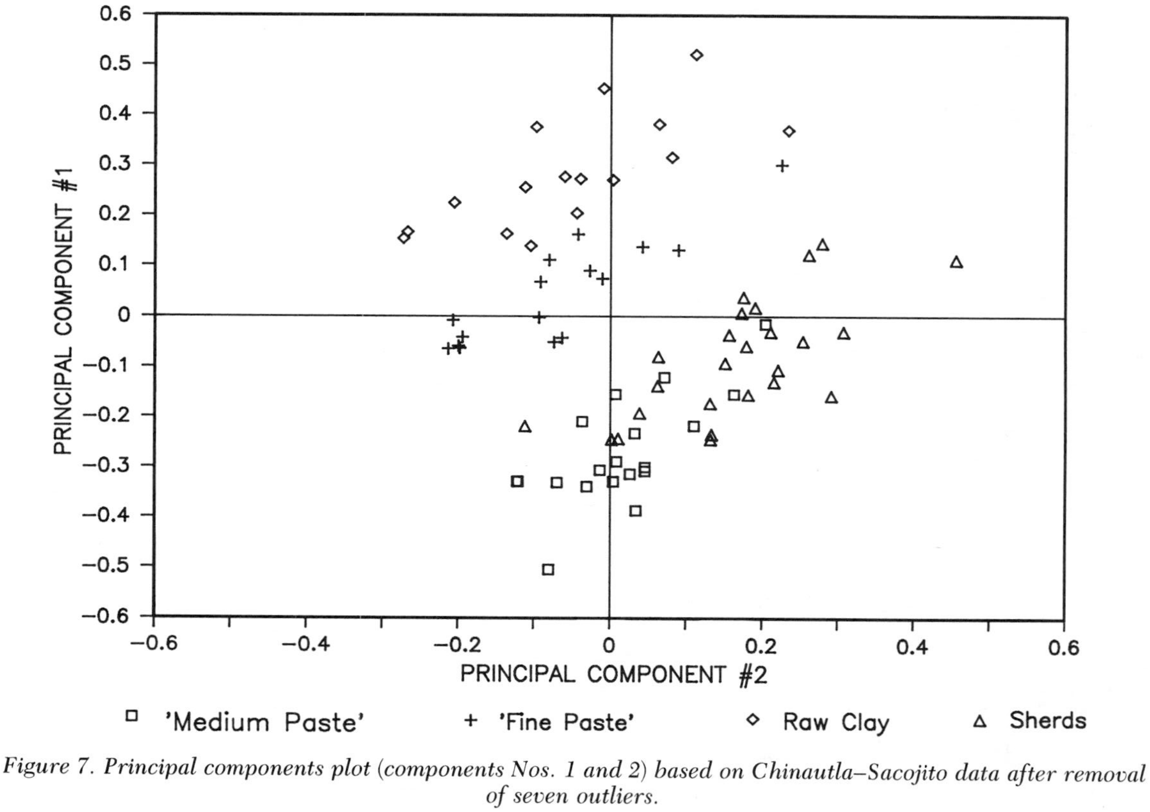

Figure 7. Principal components plot (components Nos. 1 and 2) based on Chinautla–Sacojito data after removal of seven outliers.

the expected allometric size dimension. The fine- and medium-paste groups that were separated on components Nos. 1 and 2 in the original data set were separated primarily on component No. 1 in the data set with outliers removed. (Judging from the component coefficients in Table III, some temper-related variation may still be on component No. 2.)

Once the effect of size has been recognized, it might be useful in some cases to remove its effects and reexamine the grouping tendencies in the data. Grouping tendencies may be examined independently of size-related variation by clustering the data by using principal component scores, exclusive of scores on the size-related component (principal component No. 1 in this case). Another means of accomplishing the same end is to regenerate the data from the principal components and their associated characteristic roots (eigenvalues), excluding the characteristic root of the temper-related dimension. Grouping procedures based on these two techniques yield identical results.

After removal of size from the data set (outliers removed previously), hierarchical and k-means cluster analyses fail to distinguish the two texture groups in the Chinautla–Sacojito data. Discriminant analysis with the size-corrected data yields nonsignificant test statistics for the effect of group (tempered vs. nontempered) and misclassifies 22 of 76 observations.

Complex enrichment–dilution, like simple proportional dilution, is expressed on the major axis of variation in the data, unless heterogeneity in the source clay overwhelms the effect of tempering. However, if heterogeneity in the clay overwhelms temper-related variation, one probably will not have identified subgroups anyway, and the problem is moot. If, as in the present example, two or more major groups are identified along with a smaller set of outliers, the outliers can be removed from the data set and the principal components recalculated to isolate the effect of temper. Although bimodal tempering may create two compositional groups in a data set of ceramics made from a single clay source, the effect of tempering can be identified analytically and removed in order to reveal the underlying homogeneity.

This example shows that, when a single process (potters interacting with a single clay and temper source) is involved, it is possible to isolate temper-related effects in compositional data. Such effects are more difficult or even impossible to identify when multiple cultural and natural processes are confounded within a compositional data set. As Read (*64*) has pointed out, the problem of multiple underlying processes also renders statistical theory inapplicable.

In general, one can expect only the most coarse-grained patterns in a compositional data set to be readily apparent and interpretable. In the original analysis, temper-related compositional patterning was largely obscured by major chemical differences between the clay sources and the fact that temper interacted differently with the two clays represented in the data, while second stage analysis of the Chinautla–Sacojito data readily revealed such patterning. The potential for a hierarchy of processes to create pattern in compositional data suggests that pattern-recognition techniques should be applied hierarchically. Such an approach to the present data set involves a second-stage search for pattern within each of the source-specific subgroups identified in the first stage of the analysis.

Conclusion

This chapter stresses the notion of modeling as it pertains to a structure or structures contained within a compositional data matrix and as revealed or imposed by choice of algorithmic approach. By using a generated example, the influence of such factors as outliers, transformations, interelemental correlation, choice of resemblance coefficients, grouping procedures, and group summary evaluation have been discussed. All of these factors are variable within the context of specific problem formulation.

Although the discussion has been restricted to ceramic systems, the basic theory holds for the summary of analytical data from stone (jade, turquoise, limestone); metals (native copper, iron); and glass. These materials are frequently complex mineralogical assemblages requiring that the multiple sources of variation be understood and modeled in relation to the problem being investigated. Even simple proportionality can be a pervasive influence in obsidian data, as the trace elemental concentrations may vary in relation to the silica content in a single flow, dependent upon the stage of discharge.

The concept of modeling tends to limit discussions of a best number or group of elements to use in compositional analysis (cf. 65, 66). What may be useful for one problem may not be useful for another. For example, at the global level of analysis, chromium concentration is an important regionally sensitive discriminator of ancient Maya pottery in the southwestern lowlands of Chiapas, Mexico. Yet, at the intraregional level of investigation, the range of variation frequently confuses rather than contributes to the local level of refinement.

Modeling by using the structure contained within a data set is a problem-oriented process, and as such it is fundamentally opposed to notions of a static mode of group formation. In the recent archaeometric literature, it was proposed that, as a convenient method of communication between researchers, the classification functions derived from discriminant analysis could be transmitted rather than the actual data (*11*)! Such a proposal seems to be insensitive to problem orientation, ceramic processes, or statistical influence.

Statistical modeling of compositional data, using both parametric and nonparametric aspects, must be considered along with problem definition, research design, sampling, etc., as these factors are part of the investigative program. Accordingly, what on first glance may be viewed as a misuse of some statistical approach may have less to do with some specific algorithm than it does with its use in isolation from the program's objectives. This isolation too frequently arises from the functional divisions of labor between the archaeologist and scientist. Part of a meaningful collaboration must entail consideration of how compositional data is to be modeled within the focus of a specific problem. Within such a context, archaeologically significant inferences will gradually replace object description as the main product of compositional investigations.

Acknowledgments

In this paper we have drawn on analyses carried out as part of the Maya Jade and Ceramics Project, a collaborative program of the Museum of Fine Arts, Boston and Brookhaven National Laboratory during 1977–1983. Work at Brookhaven was conducted under the auspices of the U.S. Department of Energy. Exploration into the interface between archaeological objectives, compositional variation and statistical modeling is an endeavor of the Smithsonian Archaeometric Research Collections and Records (SARCAR) facility located at the Smithsonian's Conservation Analytical Laboratory. Neff's participation in this research is made possible by a Smithsonian Institution Materials Analysis Postdoctoral Fellowship.

References

1. Caley, E. R. *J. Chem. Ed.* **1949,** *26*, 242–247.
2. Harbottle, G. In *Contexts for Prehistoric Exchange;* Earle, T. K.; Ericson, J. E., Eds.; Academic: New York, 1982; pp 13–51.
3. Rands, R. L.; Bishop, R. L. In *Models and Methods in Regional Exchange;* Fry, R. E., Ed.; SAA Papers No. 1; Society for American Archaeology: Washington, DC, 1980; pp 19–46.
4. Rice, P. In *The Ceramics of Kaminaljuyu, Guatemala;* R. K. Wetherington, Ed.; The Pennsylvania State University Monograph Series on Kaminaljuyu: University Park, 1978; pp 401–510
5. Neff, H. Ph.D. Thesis, University of California, Santa Barbara, 1984.
6. Bishop, R. L.; Lange, F. W.; Lange, P. C. In *The Art and Archaeology of Costa Rica;* Lange, F. W., Ed.; Phillips Exeter Academy through PDQ Printing: Urbana, 1987.
7. Earle, T.; Ericson, J. E. In *Exchange Systems in Prehistory;* Earle, T.; Ericson, J. E., Eds.; Academic: New York, 1977; pp 3–12.
8. Attas, M.; Fossey, J. M.; Yaffee, L. Paper presented at the 1984 Symposium on Archaeometry, Washington, DC.
9. Attas, M.; Fossey, J. M.; Yaffee, L. *Advances in Computer Archaeology* **1985,** *2*, 1–30.
10. Vitali, V.; Franklin, U. M. Paper presented at the Symposium on Archaeometry, Washington, DC, 1984.
11. Vitali, V.; Franklin, U. M. *J. Arch. Sci.* **1986,** *13*, 161–170.

12. Thomas, D. H. *American Antiquity* **1978**, *43*, 231–244.
13. Hole, B. *Annual Review of Anthropology* **1980**, *9*, 217–234.
14. Bishop, R. L.; Rands, R. L.; Holley, G. In *Advances in Archaeological Method and Theory*; Schiffer, M. B., Ed.; Academic: New York, 1982; Vol. 5; pp 275–330.
15. Bezdek, J. C. *Pattern Recognition with Fuzzy Objective Function Algorithms;* Plenum: New York, 1981.
16. Bishop, R. L. Ph.D. Thesis, Southern Illinois University, Carbondale, 1975.
17. Kemeny, J. G.; Snell, J. L. *Mathematical Models in the Social Sciences;* Ginn-Blaisdell: Waltham, 1962.
18. Everitt, B., In *Exploring Data Structures;* O'Muircheartaigh, C. A.; Payne, C.; Wiley and Sons: New York, 1977; Vol 1; pp 100–130.
19. Sneath, P. H. A.; Sokal, R. R. *Numerical Taxonomy;* W. H. Freeman: San Francisco, 1973.
20. Harbottle, G. In *Radiochemistry: A Specialist Periodical Report;* Newton, G. W. A., Ed.; The Chemical Society, Burlington House: London, 1976; pp 33–72.
21. Allen, R. O.; Luckenback, A. H.; Holland, C. G. *Archaeometry* **1975**, *17*, 69–83.
22. Allen, R. O.; Pennell, S. E. In *Archaeological Chemistry-II;* Carter, G. F., Ed.; Advances in Chemistry Series No. 171; American Chemical Society: Washington, DC, 1978; pp 230–257.
23. Moffat, D.; Buttler, S. J. *Archaeometry* **1986**, *28*, 101–15.
24. Sayre, E. V. *Brookhaven Procedures for Statistical Analysis of Multivariate Archaeometric Data;* Brookhaven National Laboratory: New York, 1975, unpublished.
25. Bishop, R. L. In *Models and Methods in Regional Exchange;* Fry, R. E., Ed.; SAA Papers No. 1; Society for American Archaeology: Washington, DC, 1980; pp 47–66.
26. Mertz, R.; Melson, W.; Levenbach, G. Proceedings of the 18th International Symposium on Archaeometry and Archaeological Prospection, Bonn, 14–17 March 1978, 1979, pp 580–596.
27. Bishop, R. L. *Mesoamerica* **1984**, *7*, 103–111.
28. Davis, J. C. *Statistics and Data Analysis in Geology, second edition;* Wiley and Sons: New York, 1986.
29. Orloci, L. J. *Ecology* **1966**, *54*, 193–215.
30. Noy-Meir, I. J. *Ecology* **1973**, *61*, 329–341.
31. Klovan, J. E. In *Concepts in Geostatistics;* McCammon, R. B., Ed.; Springer-Verlag: New York, 1975; pp 21–69.
32. Miesch, A. T. *Computers and Geosciences* **1976**, *1*, 147–159.
33. Full, W. E.; Ehrlich, R.; Klovan, J. E. *Computers and Geosciences* **1981**, *7*, 331–342.
34. Hammond, N.; Harbottle, G.; Gazard, T. *Archaeometry* **1976**, *18*, 147–168.
35. Underhill, L. G.; Jacobson, L.; Peisach, M. Paper presented at the 1984 Symposium on Archaeometry, Washington, DC, 1984
36. Anderberg, M. R. *Cluster Analysis for Applications;* Academic: New York, 1973.
37. Romesburg, H. C. *Cluster Analysis for Researchers;* Lifetime Learning: Belmont, California, 1984.
38. Sokal, R. R.; Rohlf, F. J. *Taxon* **1962**, *11*, 33–40.
39. Sneath, P. H. A. *Mathematical Geology* **1977**, *9*, 123–144.
40. Everitt, B. S. *Biometrics* **1979**, *35*, 169–181.
41. Wilks, S. S. *Mathematical Statistics;* Wiley and Sons: New York, 1962.
42. Cooley, W. W.; Lohnes, T. R. *Multivariate Data Analysis;* Wiley and Sons: New York, 1971.
43. Rubin, J.; Friedman, H. P. *A Cluster Analysis and Taxonomy System for Grouping and Classifying Data;* IBM, Scientific Center: New York, 1967.

44. MacRae, F. K. *Behavioural Science* **1971**, *16*, 423–424.
45. Marriot, F. H. C. *Biometrics* **1971**, *27*, 501–514.
46. Bishop, R. L.; Rands, R. L. In *Analysis of Fine Paste Ceramics;* Sabloff, J. A., Ed.; Memoirs of the Peabody Museum of Archaeology and Ethnology; Peabody Museum: Cambridge, 1982; Vol. 15, No. 2; pp 283–314.
47. Bishop, R. L.; Harbottle, G.; Sayre, E. V. In *Analysis of Fine Paste Ceramics;* Sabloff, J. A., Ed.; Memoirs of the Peabody Museum of Archaeology and Ethnology; Peabody Museum: Cambridge, 1982; Vol. 15, No. 2; pp 272–282.
48. Everitt, B. S. *Cluster Analysis;* Heinmann Educational: London, 1974.
49. Olin, J. S.; Sayre, E. V. In *Science and Archaeology;* Brill, R. H, Ed.; The MIT Press: Cambridge, 1971; pp 196–209.
50. Rice, P. In *The Ceramics of Kaminaljuyu, Guatemala;* R. K. Wetherington, Ed.; The Pennsylvania State University Monograph Series on Kaminaljuyu: University Park, 1978; pp 511–542.
51. Hirst, D. M. *Geochimica Cosmochimica Acta* **1962**, *26*, 1147–1187.
52. Slatt, R. M. *Can. J. Earth Sci.* **1975**, *12*, 1346–1361.
53. Slatt, R. M.; Sasseville, D. R. *Canadian Mineralogist* **1976**, *14*, 3–15.
54. Carr, C. In *For Concordance in Archaeological Analysis: Bridging Data Structure, Quantitative Technique, and Theory;* Carr, C., Ed.; Westport: Kansas City, 1986; pp 18–44.
55. Arnold, D. E.In *The Ceramics of Kaminaljuyu, Guatemala;* Wetherington, R. K., Ed.; The Pennsylvania State University Monograph series on Kaminaljuyu: University Park, 1978; pp 327–400.
56. Arnold, D. E. In *Spatial Organisation of Culture;* Hodder, I., Ed.; University of Pittsburgh: Pittsburgh, 1978; pp 39–60.
57. Arnold, D. E.; Rice, P. M.; Jester, W. A.; Deutsch, W. H.; Lee, B. K.; Kirch, P. V. In *The Ceramics of Kaminaljuyu, Guatemala;* Wetherington, R. K., Ed.; The Pennsylvania State University Monograph series on Kaminaljuyu: University Park, 1978; pp 543–546.
58. Reyment, R. A. In *Handbook of Statistics;* Krishnaiah, P. R.; Kanal, L. N., Eds.; North Holland: New York, 1982; Vol. 2; pp 721–745.
59. Reyment, R. A. *Mathematical Geology* **1985**, *17*, 591–609.
60. Reyment, R. A.; Blackith, R. E.; Campbell, N. A. *Multivariate Morphometrics;* Academic: New York, 1984; 2nd ed.
61. Humphries, J. M.; Bookstein, F. L.; Chernoff, B.; Smith, G. R.; Elder, R. L.; Poss, S. G. *Systematic Zoology* **1981**, *30*, 293–308.
62. Somers, K. M. *Systematic Zoology* **1986**, *35*, 359–368.
63. Drexler, J. W.; Rose, W. I., Jr.; Sparks, R. S. J.; Ledbetter, M. T. *Quaternary Research* **1980**, *13*, 327–345.
64. Read, D. In *For Concordance in Archaeological Analysis: Bridging Data Structure, Quantitative Technique, and Theory;* Carr, C., Ed.; Westport: Kansas City, 1986; pp 45–86.
65. Widemann, F.; Picon, M.; Asaro, F.; Michel, H. V.; Perlman, I. *Archaeometry* **1975,** *17*, 45–59.
66. Winther-Nielsen, M.; Conradsen, K.; Heydorn, K.; Mejdahl, V. In *Scientific Studies in Ancient Ceramics;* Hughes, M. J., Ed.; Occasional Paper No. 19, British Museum: London, 1981; pp 85–92.

RECEIVED for review June 11, 1987. ACCEPTED revised manuscript March 1, 1988.

5

Compositional Classification of Mexican Majolica Ceramics of the Spanish Colonial Period

Jacqueline S. Olin and M. James Blackman

Conservation Analytical Laboratory, Smithsonian Institution, Washington, DC 20560

This chapter is an attempt to refine the classification of Mexican majolica ceramics from Spanish Colonial sites by using chemical data obtained by neutron activation analysis. The ceramics examined came primarily from excavations in Mexico City and from the Santa Catalina de Guale Mission site, St. Catherines Island, GA. The majolica types from these sites are both Spanish and Mexican and date from the 16th to the late 17th century. A rationale for the chemical classification of Mexico City and Puebla production is proposed.

EXCAVATIONS AT THE SPANISH–INDIAN SITE OF FIG SPRINGS in Northern Florida in 1949 yielded a complex of majolica ceramics unlike those previously known from other Florida sites. The archaeologist, J. M. Goggin, carried out a detailed examination of the available majolica ceramics from late 15th- to 18th-century contexts and published an extensive typology for these ceramics (*1*). On typological and distributional evidence, Goggin recognized a Spanish and a Mexican production and made a tentative attempt to subdivide specific Mexican majolica types into productions at Mexico City, Puebla, or elsewhere in Mexico.

Lister and Lister (*2*) further refined the majolica typology for the 16th-century varieties. Their interpretation was based on the study of majolica ceramics found below a floor sealed in 1573 at the Metropolitan Cathedral excavation site along with other ceramics from less well-defined stratigraphic contexts at the Sagrario excavations in Mexico City. On the basis of this

0065–2393/89/0220–0087$07.50/0

study, they identified new majolica varieties, suggested revisions in the nomenclature of Goggin, and established a tentative chronology for the deposits beneath the Cathedral and Sagrario. The Listers attributed some majolica varieties to Puebla and argued that Mexico City was the origin for others.

In the initial chemical studies of majolica ceramics, Olin et al. (*3*) sought to establish, on a compositional basis, the Spanish and Mexican productions of majolica from excavations in the Caribbean, Venezuela, and Mexico. The research documented the chemical differences between the two production areas and showed that they could be readily differentiated on the basis of the cerium, thorium, and lanthanum concentrations. Maggetti et al. (*4*) further demonstrated that the Spanish and Mexican productions could also be distinguished by thin-section petrography. Maggetti et al. showed that although the Spanish production was characterized by inclusions of sedimentary origin (predominantly quartz), the Mexican production was characterized by inclusions of a volcanic origin.

Neutron activation and petrographic analysis of late medieval Spanish pottery from the major Spanish production centers of Seville, Granada, Paterna-Manises, Barcelona, and Talavera-Puente allowed progress to be made in uniquely characterizing these production centers (*5–7*). Efforts to identify different Mexican majolica productions petrographically have been unsuccessful, and an attempt at chemical characterization by directly coupled plasma–optical emission spectroscopy was later determined to have been flawed by problems encountered with the dissolution of the ceramic samples (*4*).

This chapter reports the results of research on chemical characterization of different production sites in Mexico by instrumental neutron activation analysis (INAA). The integration of the typological information of Goggin (*1*) and Lister and Lister (*3*) with chemical characterization data forms the basis for examining 16th-century ceramic samples from the Metropolitan Cathedral, Mexico City; 16th- and 17th-century ceramic samples from Santa Catalina de Guale Mission site on St. Catherines Island, GA; and modern majolica samples from Puebla, Mexico. We will show that the proposed Mexico City and Puebla productions can be distinguished chemically and, in addition, that temporal and spatial differences appear in the Puebla production.

History of Mexican Majolica Production

The production of majolica ceramics in Mexico during the 16th century is believed to have taken place in both Mexico City and in Puebla. Much of the evidence for Mexico City production consists of the ceramics from excavations at the Metropolitan Cathedral site.

Documentary information available from 16th-century archives concerning potters in Mexico City is exceedingly sparse and inferential. The early colonial roadways in Mexico City frequently bore the names of the particular crafts located along them, but there is no known 16th-century street name referring in any way to the pottery-making process. On the basis of documentary evidence from later centuries, the most probable location of the colonial potteries is believed to be west of the center of the city (*8*), although no colonial kilns or workshops have as yet been identified in this area.

Historical documents do, however, record dates for the construction of the Metropolitan Cathedral, the floor of which was laid in 1573, sealing underneath deposits of the first 50 years of colonial occupation and the earlier Pre-Colombian period. During the installation of support pylons at the cathedral in 1975 and 1976, 182 pits were sunk through the floor, allowing excavation and sampling of the early colonial deposits. Majolica ceramics from these excavations assigned to non-Spanish types are assumed to be from a Mexico City production. Despite the lack of more detailed information, the claim is made (*2*) that, as Mexico City was the most important city of New Spain for the entire 16th century, the earliest demands of the market for better dishes must have been met by local craftsmen.

The date of the beginning of majolica production at Puebla is uncertain. Puebla is well-known as the leading center for the manufacture of fine-grade majolica ceramics during the 17th and 18th centuries, and Lister and Lister (*2*) cite two references that describe potters in Puebla by the 1580s (*9, 10*). Actual production could, therefore, have begun during or somewhat before the 1580s. There are, however, few majolica sherds from controlled excavations in Puebla that are dated finely enough to establish the advent of production (*11, 12*).

More is known historically about the Spanish trade and supply in the early colonial period. Between 1564 and 1566, the Spanish established a pattern for shipping that involved two trading fleets. During this period, one fleet, the Tierra Firme flota, left Spain headed for Porto Bello, Panama, and the other, the New Spain flota, was destined for the port of Veracruz, Mexico. The New Spain flota dropped off ships headed for San Juan, Puerto Rico; Santo Domingo, Dominican Republic; various ports in Cuba; other Caribbean islands; and the Central American mainland (*13*). Supplies and trade goods from Spain most probably reached early missions on the North American mainland through this fleet, possibly via Cuba or Puerto Rico. Scattered records and widespread archaeological finds (mostly majolica) indicate that a flourishing trade existed in ceramics manufactured in Mexico.

The research for this chapter was undertaken in an effort to explore this early trade in Mexican-made majolica. The identification of the major production sites and the date that production began is critical to the study. It is, therefore, important to begin with the ceramic typology for 16th- and 17th-century Mexican majolica and to test its consistency and provenance attributions by chemical characterization. To this end, majolica from the sealed 16th-century context at the Metropolitan Cathedral in Mexico City and 16th- and 17th-century majolica from the mission context at Santa Catalina de Guale, St. Catherines Island, GA were examined.

Source and Typology of the Majolica Samples

Following the North American procedure in use at that time, the majolica types described by Goggin were given binomial designations. The first name is geographical, taken from an early find's locality or the assumed site of production, and the second name is descriptive. The typology, as developed by Goggin and revised by Lister and Lister, includes such characteristics as form, surface finish and decoration, and paste and glaze description.

The 16th-century majolica samples came from excavations at the Metropolitan Cathedral in Mexico City and can be assigned to a date before 1573. Other ceramics from excavations in Mexico City at the Sagrario and the Metro can not be so closely dated. The types of ceramics include Mexico City White, Fig Springs/San Juan Polychrome, Sevilla White, and Columbia Gun Metal.

Before 1982, Mexico City White, an undecorated, white glazed ware, was included with the similar type of Spanish ware called Columbia Plain. Subsequent reevaluation of the typology by Lister and Lister (*2*) resulted in the splitting of the type into the Spanish ware–Columbia Plain and Mexican ware–Mexico City White. Mexico City White, attributed by Lister and Lister to a Mexico City production (*2*), is the most common Mexican type from the Metropolitan Cathedral and Sagrario excavations. INAA analysis of 11 "Columbia Plain" sherds at Brookhaven National Laboratory (*3*) showed that nine were of Mexican origin and two did not match the composition of the Mexican sherds. These two sherds were later classified as Sevilla White (*2*), a third type of white glazed ware of 16th-century Spanish origin (*4*). Columbia Gun Metal is a variant of Columbia Plain with a darkened, rather than white, glaze that varies from dense iridescent black to speckled grey. This ware is attributed to a Spanish production.

Fig Springs/San Juan Polychrome majolica is painted in blue and yellow (and/or orange) on a cream to gray–white background and is the most abundant decorated fine ware in the excavations. It is attributed to a Mexico City production by Lister and Lister (*2*), although, Goggin (*1*) stated that it is possible these ceramics were manufactured in Puebla. The presence of Mexican production types among these excavated materials allowed the inclusion of unambiguous 16th-century majolica of Mexican production in this study.

To extend the sample of Mexican-attributed majolica into the 17th century, samples were obtained from sherds from the Spanish mission site of

Santa Catalina de Guale, on St. Catherines Island, GA. The site is believed to have been an important Guale Indian town by 1576 and to have been abandoned in the early 1680s (*14*). The Guale Indians of the southeastern coast were among the first indigenous peoples met by European explorers on the North American continent. In 1587, St. Catherines Island became the principal northern Spanish outpost on the Atlantic coast.

The sherds from St. Catherines Island include types attributed to both Spanish and Mexican productions. The Mexican types included Fig Springs/San Juan Polychrome, San Luis Blue-on-White, Aucilla Polychrome, Mt. Royal Polychrome, and Puebla Polychrome. The Fig Springs/San Juan Polychrome has already been described and sherds are shown in Figure 1. San Luis Blue-on-White is decorated in blue, and rarely with some yellow or black, on an off-white background. Goggin recognized two paste variants based on color: a reddish and a cream paste. He stated that the reddish paste variant is similar to Fig Springs/San Juan Polychrome paste and both may have been made in the same center. He proposed a Mexico City origin for the reddish paste variant and a Puebla origin for the cream paste variant. This proposal conflicts with the proposal made earlier by Goggin (*1*): that Fig Springs/San Juan Polychrome may have been produced in Puebla.

Figure 1. Fig Springs/San Juan Polychrome sherds from the Santa Catalina de Gaule Mission on St. Catherines Island, GA. Top row (left to right): SCI017, SCI037, SCI047; middle row: SCI009, SCI024, SCI018; bottom row: SCI031, SCI036.

The Listers considered all San Luis Blue-on-White to be of Mexico City manufacture. The San Luis Blue-on-White sherds from Santa Catalina de Guale are shown in Figure 2. Aucilla Polychrome is decorated in black, yellow (or orange), and green on a cream background. This type is thought to be a 17th-century production, attributed to an unknown place of manufacture in Mexico. Mt. Royal Polychrome is decorated with brown lines, yellow bands, and blue dots and designs on a cream background. Goggin assigned this type to the mid-17th century and to an unknown manufacturing site in Spain. The Aucilla and Mt. Royal Polychrome sherds are shown in Figure 3. Puebla Polychrome is decorated in blue and black (rarely with other colors) on a white background. It is assigned a late 17th-century date and attributed to Puebla.

Several known and attributed Spanish ware types were also examined from the St. Catherines Island ceramic corpus. These types included Columbia Plain, Yayal Blue-on-White, Santo Domingo Blue-on-White, Ich-

Figure 2. San Luis Blue-on-White sherds from the Santa Catalina de Guale Mission on St. Catherines Island, GA. Top row (left to right): SCI082, SCI027, SCI010, SCI076; second row: SCI069, SCI027, SCI041, SCI083; third row: SCI063, SCI048, SCI060, SCI073; fourth row: SCI059, SCI039; bottom row: SCI052, SCI075, SCI042.

Figure 3. Aucilla and Mt. Royal Polychrome sherds from the Santa Catalina de Gaule Mission on St. Catherines Island, GA. Top row (left to right): SCI038, SCI008, SCI080, SCI045; second row: SCI077, SCI055, SCI070, SCI057, SCI050; third row: SCI044, SCI058, SCI078, SCI072; bottom row: SCI074, SCI028, SCI030, SCI081.

tucknee Blue-on-White, Ichtucknee Blue-on-Blue, and Santa Elena Mottled Blue-on-White. Ichtucknee Blue-on-Blue has been reconsidered by the Listers, and an Italian and a Spanish variant have been proposed. They have chosen to name the proposed Spanish variant "Sevilla Blue-on-Blue" and to consider its manufacturing site to be in the Andalusian region of Spain. A single sherd, identified as Santa Elena Mottled Blue-on-White was included in the sample from St. Catherines Island. This type was described by South (*15*), who considered it to be a Spanish prototype for a similar red-paste ware from Mexico City.

The ceramic samples selected for discussion in this chapter are those of Mexican production. As reported by Maggetti et al. (*5*), the Mexican majolica production includes both calcareous and noncalcareous paste compositions. This chapter will focus on the calcareous Mexican majolica. The following ware types were selected to examine questions of the assignation to Mexico City or Puebla productions and to determine if chemical classification using neutron activation analysis data was possible:

- Mexico City White
- Fig Springs/San Juan Polychrome
- San Luis Blue-on-White
- Aucilla Polychrome
- Mt. Royal Polychrome

In addition to the archaeological material, majolica sherds were obtained from two modern factories in Puebla: La Trinidad and Santa Maria. Both of these potteries use a clay body blended from a mixture of equal amounts of a black volcanic clay and a white marl obtained from the immediate area around Puebla (*4*). Samples from these sherds were analyzed by neutron activation analysis and the data used to represent a Puebla composition.

Experimental Methods and Results

The data discussed in this chapter include the results of instrumental neutron activation analysis conducted at two different laboratories. The analyses of the Santa Catalina de Guale Mission samples, most of the Metropolitan Cathedral samples, and the modern Puebla samples were conducted at the National Bureau of Standards reactor by using procedures described by Blackman (*16*) and in Table I. The remainder of the samples were analyzed at Brookhaven National Laboratories (BNL), and reported in Olin et al. (*3*). Because different comparator standards were used in the two laboratories, all the BNL data were normalized to the Smithsonian Institution standard according to the procedure described by Blackman (*17*). The conversion factors are presented in Table I.

The data for all of the samples are presented in appendixes to this chapter. Appendix A includes samples from excavations in Mexico City at the Metropolitan Cathedral (SC 16, SC 20, SC 22, SC 29, SC 31, SC 37, and SC 38); the Sagrario (all other samples designated SC); and from the Metro excavations (designated SA). Appendix B includes all the samples from Santa Catalina de Guale (designated SCI). Appendix C contains the data for the modern majolica samples from Puebla (designated SD).

The data were initially screened to separate Spanish and Mexican productions. The ceramics typologically classified as Spanish had cerium, thorium, and lanthanum concentrations consistent with production in Spain (*3*). These types included Columbia Plain, Columbia Gun Metal, Sevilla White, Yayal Blue on White, Santo Domingo Blue-on-White, Ichtucknee Blue-on-White, Ichtucknee Blue-on-Blue, and Santa Elena Mottled Blue-on-White.

Table I. INAA Experimental Parameters

Element	Nuclide	Gamma Ray Energy (keV)	Conc. in Standard SRM 1633[a]	Count[b]	Analytical Precision SRM 679	BNL to SI Conversion Factors
Na	Na-24	1369	0.32	1	2.3%	1.037
K	K-42	1525	1.61	1	8.2%	
Ca	Ca-47	1297	4.70	1		
Sc	Sc-46	889	27.0	2	1.4%	0.893
Cr	Cr-51	320	131.0	2	3.1%	0.992
Fe	Fe-58	1099, 1292	6.20	2	2.9%	0.972
Co	Co-60	1173, 1333	41.5	2	1.5%	1.036
Zn	Zn-65	1115	213.0	2	3.5%	
As	As-76	559	61.0	1	6.0%	
Br	Br-82	554	8.6	1		
Rb	Rb-86	1077	125.0	2	9.1%	1.160
Sr	Sr-85	514	1700.0	2		
Zr	Zr-95	757	301.0	2		
Sb	Sb-122	564	6.9	1	9.9%	
Cs	Cs-134	796	8.6	2	2.7%	1.009
Ba	Ba-131	496	2700.0	1	13.2%	
La	La-140	1596	82.0	1	1.4%	1.095
Ce	Ce-141	145	146.0	2	1.8%	0.961
Nd	Nd-147	91	64.0	1		
Sm	Sm-153	103	12.9	1	1.6%	0.983
Eu	Eu-152	1408	2.5	2	2.2%	0.824
Tb	Tb-160	879	1.9	2	12.9%	
Yb	Yb-175	396	6.4	1	4.8%	1.023
Lu	Lu-177	208	1.0	1	6.7%	0.835
Hf	Hf-181	482	7.9	2	3.5%	1.213
Ta	Ta-182	1221	1.8	2	7.0%	0.821
Th	Pa-233	312	24.8	2	2.2%	0.992
U	Np-239	106	11.6	1	15.9%	
W	W-187	686	5.5	1		

[a]Na, K, Ca, and Fe results are given in percents; all others are in parts per million.
[b]Count 1: 1 h after a 5-day decay; count 2: 2 h after a 30-day decay.

Data for these types are presented in Appendixes A, B, and C. Three St. Catherines Island samples (SCI042, SCI052, and SCI075 in Appendix B), typologically classified with the San Luis Blue-on-White, a Mexican production, were chemically classified as Spanish. Close examination of these sherds (Figure 2) shows visible distinctions suggesting that these may be examples of a Spanish variant discussed by Lister and Lister (*18*).

To investigate the question of whether the majolica ceramics produced in Mexico during the 16th and 17th centuries could be assigned to a proposed Mexico City or Puebla production, the majolica with a Mexican composition and nine modern Puebla samples were subjected to cluster analysis. The hierarchical aggregative clustering program AGCLUS (*19*) used "average-link" clustering on a mean-Euclidean-distance matrix for the elements Sc, Cr, and Fe, and separated the ceramics from Mexico already discussed, into two large clusters (Figure 4). The upper cluster, Group A in Figure 4, contained all the late 16th- and 17th-century Santa Catalina de Guale samples, 11 of the 14 Fig Springs/San Juan Polychrome samples from the Mexico City excavations, and the nine modern Puebla samples. The lower cluster, Group B in Figure 4, contained only samples of 16th century context from the Metropolitan Cathedral and Sagrario excavations in Mexico City, including all the Mexico City White, the Mexican copies of Columbia Plain, and three Fig Springs/San Juan Polychrome samples.

The validity of these compositional groups was tested statistically by using Mahalanobis distance and Hotelling T^2 statistic. In the compositional group A, four Fig Springs/San Juan Polychrome samples (SCI009, SCI037, SCI047, and SC 44) and one San Luis Blue-on-White sample (SCI076) of the 54 samples were excluded at the 99% confidence interval. In the compositional group B, three samples of Mexico City White (SC 13, SC 20, and SC 24) were excluded at the 99% confidence level. These three samples were also outliers in the cluster analysis (Figure 4). With these samples eliminated, the reformed compositional groups were tested against each other. Individual members of each group displayed less than a 1% probability of membership in the other group. The two clusters, therefore, represent distinct and statistically verifiable chemical composition groups.

The means, standard deviations, and 95% confidence intervals for 22 elements in these two compositional groups and the modern Puebla samples are presented in Table II. The two compositional groups are virtually identical at the 95% confidence interval for all elements except Sc, Cr, and Fe. Primarily the elevated Cr values in the composition group A provide the discrimination. Figure 5 is a plot of Cr vs. Fe concentrations. The plot shows the separation of the two groups. The modern Puebla majolica samples (represented by the crosses) fall well within the 95% probability ellipse of the compositional group A.

On the basis of the association of modern Puebla majolica with the late 16th- to 17th-century majolica samples from the Guale Mission site in group A, we propose that this chemical compositional group represents the production of Puebla. The ceramic types that can be attributed to the proposed Puebla production include Aucilla Polychrome, Mt. Royal Polychrome, San Luis Blue-on-White, Fig Springs/San Juan Polychrome, Puebla Polychrome, and Puebla Blue-on-White. The association, in group B, of only samples from the Sagrario and the sealed 1573 Cathedral sites in Mexico City and Lister and Lister's attribution of the ceramic types (primarily Mexico City White and Mexico City copies of Columbia Plain) to Mexico City, provides a basis for proposing a Mexico City production for this compositional group. In contrast to the Puebla composition group, there is no independent evidence for a Mexico City origin and confirmation must await further research.

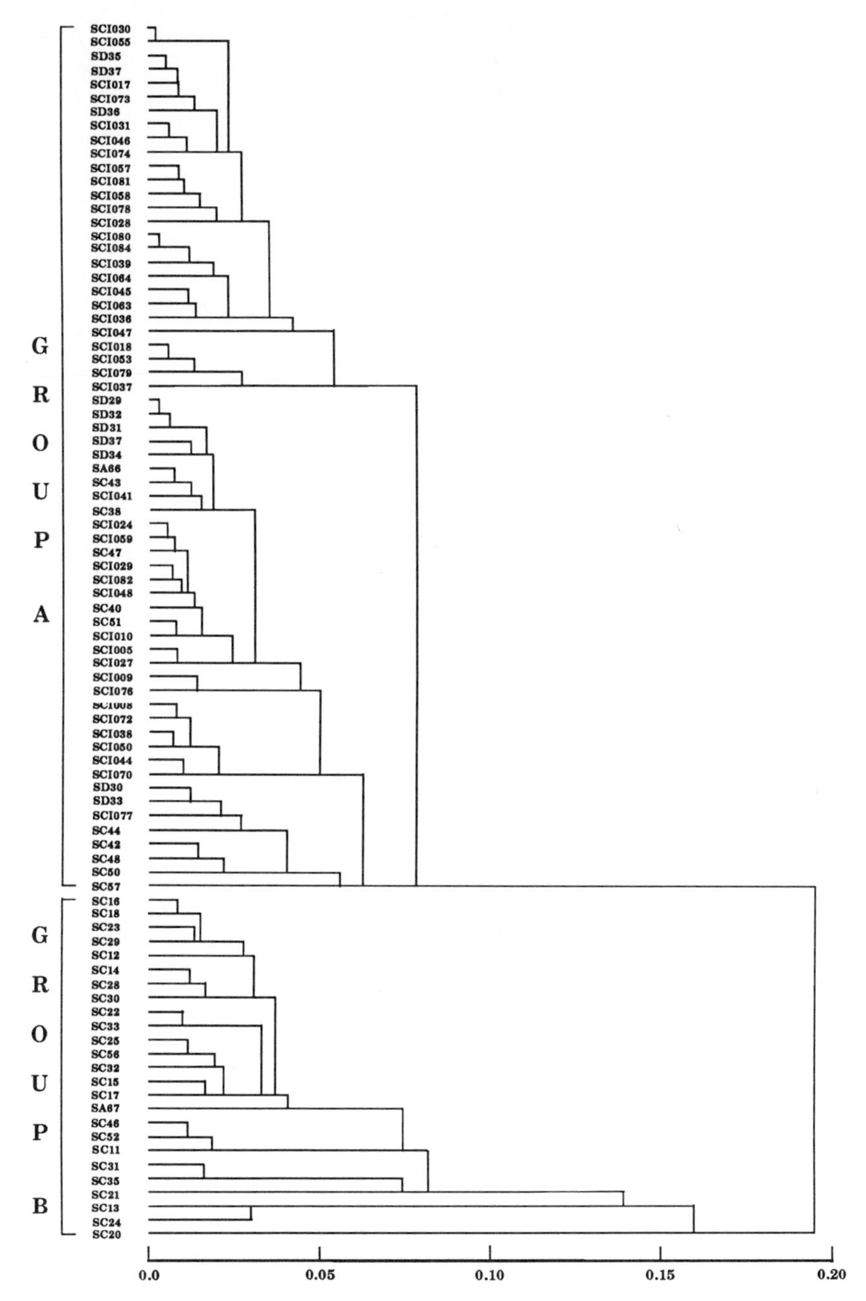

Figure 4. Dendrogram obtained by using the cluster analysis program, AGCLUS, for chromium, scandium, and iron values for majolica samples with Mexican chemical composition. Dendrogram shows two major groups, A and B, that are the proposed Puebla and Mexico City groups, respectively.

Table II. Chemical Data for the Mexico City and Puebla Production Samples and the Modern Puebla Samples

	Mexico City Production			Puebla Production			Modern Puebla Production		
Element	*Mean*	*Std Dev (%)*	*95% Confidence Interval*	*Mean*	*Std Dev (%)*	*95% Confidence Interval*	*Mean*	*Std Dev (%)*	*95% Confidence Interval*
Na	1.29	19.6	1.82 0.76	1.40	20.0	1.96 0.84	1.15	4.5	1.27 1.03
K	1.26	22.5	1.85 0.67	0.77	34.4	1.30 0.24	0.63	21.6	1.07 0.20
Ca	14.0[a]	25.2	21.4 6.5	9.8	35.7	16.8 2.8	12.7	6.5	14.6 10.7
Sc	9.5	10.1	11.4 7.5	12.7	5.8	14.2 11.2	12.0	8.1	14.2 9.7
Cr	54.3	10.8	66.6 42.1	98.3	14.0	126.0 70.6	94.5	11.5	120.0 69.4
Fe	2.71	9.4	3.24 2.17	3.57	8.7	4.19 2.94	3.45	9.8	4.22 2.67
Zn	75.6[a]	30.0	124.0 27.3	69.7	13.2	88.1 51.3	76.5	7.0	88.9 64.2
Rb	57.5	19.8	81.2 33.8	51.3	18.5	70.4 32.3	58.1	14.0	76.9 39.3
Sr	0.080[a]	12.6	0.102 0.059	0.055	24.5	0.083 0.028	0.028	8.3	0.033 0.023

Cs	3.9	22.2	5.7	4.5	35.0	7.6	6.8	11.5	8.6
			2.1			1.3			5.0
Ba	474.0	10.3	575.0	446.0	41.3	817.0	260.0	11.2	327.0
			372.0			76.0			193.0
La	22.4	18.5	31.0	21.0	16.1	27.8	20.1	6.2	23.3
			13.8			14.2			17.4
Ce	35.5	15.4	46.9	27.7	18.2	37.5	28.4	7.4	33.2
			24.2			17.4			23.6
Sm	4.13[a]	13.1	5.28	4.38	15.0	5.70	4.26	5.5	4.81
			2.98			3.07			3.70
Eu	0.94	13.9	1.21	1.11	13.7	1.42	0.96	7.4	1.12
			0.67			0.80			0.79
Yb	1.83[a]	12.3	2.32	1.88	11.7	2.32	2.16	5.1	2.42
			1.35			1.43			1.90
Lu	0.26[a]	11.5	0.33	0.28	17.2	0.38	0.33	12.3	0.43
			0.20			0.18			0.23
Hf	4.01	8.0	4.68	4.87	11.0	5.95	4.80	11.5	6.08
			3.34			3.79			3.53
Ta	0.51[a]	14.8	0.61	0.55	16.1	0.73	0.76	14.6	1.01
			0.41			0.37			0.50
Th	5.2	14.8	6.9	5.5	11.5	6.7	6.2	8.8	7.4
			3.6			4.2			4.9

NOTE: For Mexico City production $N = 22$; for Puebla production, $N = 51$; for Modern Puebla production, $N = 9$. Na, K, Ca, Fe and Sr results are given in percents; all others are in parts per million.

[a] $N = 16$.

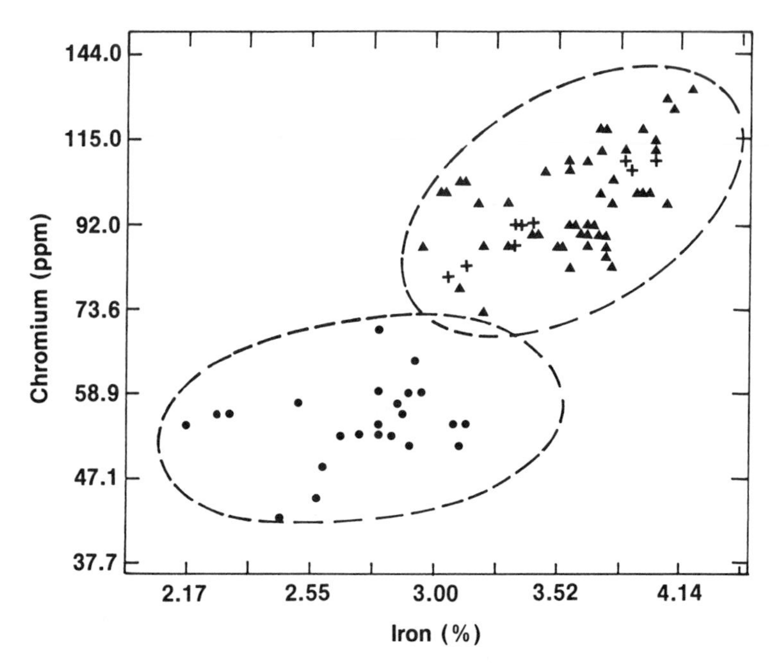

Figure 5. Logarithmic plot of the ratios of chromium to iron for the samples from sherds in group A and group B of the cluster dendrogram shown in Figure 4. Key: ●, archaeological samples attributed to Mexico City production; ▲, archaeological samples attributed to Puebla production; +, modern majolica produced in Puebla.

The ceramic types attributed to the proposed Mexico City production are Mexico City White, Mexico City copies of Columbia Plain, and Fig Springs/San Juan Polychrome.

Table III presents the proposed location of production on the basis of the chemical composition for the majolica types analyzed and the attributions made by Goggin (*1*) and Lister and Lister (*2*). All Mexico City White and Mexico City copies of Columbia Plain belong to the proposed Mexico City production as just discussed. All the San Luis Blue-on-White sherds reported here are from the excavations at Santa Catalina de Guale. With the exception of the three Spanish composition sherds discussed earlier, the remaining 17 sherds can be chemically classified as having been produced in Puebla. Both Goggin and the Listers propose a Mexico City production for San Luis Blue-on-White. Goggin's attribution to Mexico City is restricted to the reddish paste variant. We have no chemical confirmation of a Mexico City production for San Luis Blue-on-White.

Fig Springs/San Juan Polychrome is an interesting contrast to the San Luis Blue-on-White type. Fig Springs/San Juan Polychrome production has been attributed to Puebla by Goggin and to Mexico City by Lister and

Table III. Production Attributions

Majolica Type	*Chemical Attribution*	*Goggin*[a] *Attribution*	*Lister and Lister*[b] *Attribution*
Mexico City White	Mexico City	—	Mexico City
San Juan Polychrome	Puebla & Mexico City	Puebla	Mexico City
San Luis Blue-on-White	Puebla	Puebla & Mexico City	Mexico City
Aucilla Polychrome	Puebla	unknown Mexican	—
Mt. Royal Polychrome	Puebla	unknown Spanish	—
Puebla	Puebla	Puebla	—

[a]Ref. 1
[b]Ref. 2

Lister. The samples from Santa Catalina de Guale that can be attributed to a production site all appear to come from Puebla. The sherds from the Mexico City excavations are split between the proposed Puebla and Mexico City productions.

Two samples, SC 37 and SC 38, assigned to the proposed Puebla production group, in fact come from the sealed context at the Metropolitan Cathedral. The significance of these two sherds is that they provide evidence of Puebla production of majolica ceramics before 1573. Two of the Fig Springs/San Juan Polychrome sherds, SC 46 and SC 52, excavated at the Sagrario, have the proposed Mexico City composition. On this basis, we propose that there may be two varieties of Fig Springs/San Juan Polychrome, one from Puebla and one from Mexico City.

The later 17th-century types (Aucilla, Mt. Royal, and Puebla Polychrome) from Santa Catalina de Guale all group with the Puebla composition. Goggin attributed Aucilla to an unknown Mexican source and Mt. Royal Polychrome to an unknown Spanish production center. There is no evidence to date for sherds of the Mt. Royal Polychrome type having been produced in Spain, and we propose that both types are from a Puebla production.

Close inspection of the chemical data for the proposed Puebla production group shows that this group can be subdivided chemically into two typologically consistent subgroups. Figure 6 presents the means and standard deviations for selected alkali and rare earth elements in the two subgroups. Typologically, the two subgroups are Aucilla and Mt. Royal Polychrome and San Luis Blue-on-White. San Luis Blue-on-White is included with the 16th-century types (identified by the Listers); Aucilla and Mt. Royal Polychrome are not. Goggin, on the other hand, assigned San Luis Blue-on-White to

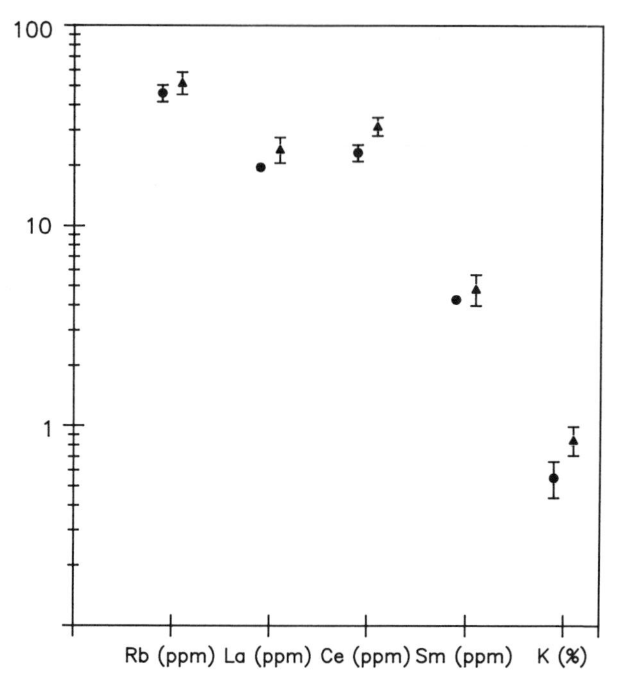

Figure 6. Log means and standard deviations for Rb, La, Ce, and K for San Luis Blue and White (▲) and Aucilla and Mt. Royal Polychrome (●) sherds from excavations at Santa Catalina de Guale.

the early 17th century and Aucilla and Mt. Royal Polychrome to the mid-to-late 17th century.

The concentration data for lanthanum and samarium, plotted in Figure 7, show distinctly different ratios with no overlap at the 95% confidence interval. Moreover, the modern Puebla majolica falls within the 95% confidence ellipse for the Aucilla and Mt. Royal Polychrome subgroup. These chemical differences strongly suggest a spatial or temporal distinction in clay resource use for the majolica production at Puebla. The association of the mid to late 17th-century majolica types with the modern Puebla production argues for a temporal continuity in resource procurement from the 17th century to the present. The sources of the modern clays used in majolica production at Puebla are described in Maggetti et al. (*4*).

Conclusions

The Mexican majolica production of the 16th and 17th centuries can be chemically classified into two very distinct groups based on chromium, iron, and scandium concentrations. One of these groups matches the composition of modern Puebla majolica and can be assigned to a Puebla production. This

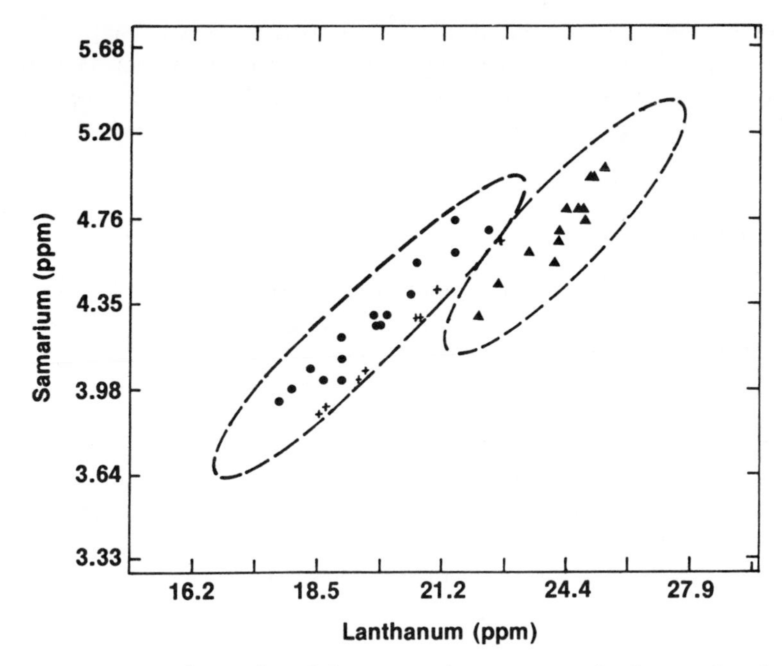

Figure 7. Logarithmic plot of the ratios of samarium to lanthanum for the samples from San Luis Blue-on-White (▲) and Aucilla and Mt. Royal Polychrome (●) with the modern Puebla samples included in the 95% confidence ellipse of the Aucilla and Mt. Royal Polychrome group.

Puebla production appears to have begun sometime before 1573, as evidenced by the presence of sherds of this chemical type beneath the floor of the Metropolitan Cathedral.

The Puebla production group can be further subdivided chemically into two subgroups that may arise from spatial or temporal shifts in resource procurement. The correspondence of the later production types with the modern Puebla majolica opens the possibilities for further investigations using ethnographic data.

The attribution of the second major chemical group to a Mexico City production is based, at present, on circumstantial evidence. Additional research using 16th-century majolica from the 1968–69 and 1975–76 excavations in Mexico City (*20*) will be needed to strengthen the attribution and to determine the range of types being manufactured.

On the basis of the evidence from the analyzed sherds from Santa Catalina de Guale, we find that only Spanish and Puebla production ceramics are present. No Mexico City production sherds are among those analyzed. The presence of Puebla production sherds from this 16th- and 17th-century site required transport from Mexico. To our knowledge, no specific Mexican sea route such as that for the Spanish flota (*21*), has been proposed for supply to the mission's sites in Spanish Florida. However, a flow of supplies from Mexico City through St. Augustine to St. Catherines Island has been proposed (22). Analysis of majolica sherds from carefully excavated shipwrecks in the Caribbean will be valuable evidence for establishing the routes taken by the supply ships.

Acknowledgments

We acknowledge the continued interest and participation of Florence and Robert Lister (University of Arizona) who provided the sherds from the excavations at the Metropolitan Cathedral in Mexico City that have been so important to our study. They developed the typology and selected the sherds that we have analyzed.

We also acknowledge Debra Peter and David Hurst Thomas for their help in providing sherds from the excavations being conducted by the American Museum of Natural History at the mission site of Santa Catalina de Guale on St. Catherines Island, GA. These excavations are supported by the Edward John Noble and St. Catherines Island Foundations. Debra Peter also assisted us by providing the typological classification of the ceramics from that site and by carrying out the sampling of the sherds that were selected and making arrangements for the photographic documentation.

We are most pleased to be able to use the modern majolica sherds collected by Marino Maggetti in Puebla, Mexico when he traveled there in 1981. These sherds have been of enormous value to us as a reference for the composition of majolica manufactured in Puebla.

Throughout this investigation we have been greatly aided in many ways by the efforts of Emlen Myers. We acknowledge the important support provided by the personnel of the Reactor Operations Division and the Nuclear Methods Group (Inorganic Analytical Research Division) of the National Bureau of Standards, who have made their facilities available to us both for this project and the numerous other projects that the Conservation Analytical Laboratory is carrying out with neutron activation analysis.

This research was supported by the Conservation Analytical Laboratory as a part of an ongoing project on the chemical classification of Spanish Colonial majolica.

Appendix A. Majolica from Excavations in Mexico City

Sample	*Na*	*K*	*Ca*	*Sc*	*Cr*	*Fe*	*Co*	*Zn*	*As*	*Rb*	*Sr*	*Cs*	*Ba*	*La*	*Ce*	*Sm*	*Eu*	*Yb*	*Lu*	*Hf*	*Ta*	*Th*
MEXICAN PRODUCTIONS																						
Mexico City White																						
SC 11	1.49	1.32	n.d.	10.5	55.0	2.2	7.1	n.d.	n.d.	63.0	n.d.	3.8	480.0	34.0	45.0	n.d.	1.20	n.d.	n.d.	4.4	n.d.	7.3
SC 12	1.35	1.31	n.d.	8.9	48.0	2.6	9.7	n.d.	n.d.	52.0	n.d.	5.0	520.0	25.0	33.0	n.d.	0.83	n.d.	n.d.	4.2	n.d.	5.4
SC 13	1.51	1.03	n.d.	6.6	42.0	2.1	6.0	n.d.	n.d.	47.0	n.d.	1.9	410.0	17.0	21.0	n.d.	0.68	n.d.	n.d.	3.5	n.d.	4.3
SC 14	1.76	1.65	n.d.	10.0	52.0	2.7	10.0	n.d.	n.d.	85.0	n.d.	5.2	520.0	27.0	35.0	n.d.	0.90	n.d.	n.d.	4.3	n.d.	5.6
SC 15	1.39	0.59	n.d.	9.3	59.0	2.7	10.0	n.d.	n.d.	53.0	n.d.	3.9	400.0	22.0	29.0	n.d.	0.86	n.d.	n.d.	4.0	n.d.	4.6
SC 17	1.02	0.92	14.3	8.91	57.8	2.82	9.89	69.0	8.34	46.0	760.0	3.27	515.0	18.3	33.0	3.64	0.859	1.67	0.244	3.75	0.460	4.22
SC 18	1.40	1.63	n.d.	9.2	52.0	2.6	10.9	n.d.	n.d.	77.0	n.d.	6.1	510.0	26.0	34.0	n.d.	0.95	n.d.	n.d.	4.4	n.d.	5.6
SC 20	2.40	1.43	n.d.	7.0	40.0	2.2	7.5	n.d.	n.d.	66.0	n.d.	2.8	520.0	21.0	27.0	n.d.	0.79	n.d.	n.d.	4.0	n.d.	5.3
SC 21	1.05	1.03	n.d.	7.1	49.0	2.9	11.1	n.d.	n.d.	71.0	n.d.	4.7	520.0	23.0	31.0	n.d.	0.77	n.d.	n.d.	4.4	n.d.	5.8
SC 22	1.16	1.53	15.5	9.31	54.6	3.03	9.82	73.3	6.35	67.0	847.0	4.76	500.0	21.9	39.3	4.39	0.979	1.96	0.279	4.04	0.535	5.48
SC 23	1.53	1.32	11.9	9.14	52.2	2.81	9.31	65.6	4.46	61.2	968.0	3.60	495.0	20.8	36.6	4.14	0.951	1.83	0.275	3.85	0.500	5.12
SC 24	1.02	1.40	16.4	6.21	39.0	2.08	5.73	43.8	3.56	41.5	1069.0	1.56	560.0	12.0	21.3	2.37	0.565	0.95	0.135	2.54	0.336	3.23
SC 25	1.61	n.d.	13.2	9.91	58.7	2.91	11.2	158.0	4.85	55.3	912.0	3.38	495.0	22.4	38.2	4.41	1.050	2.10	0.296	4.18	0.502	5.15
SC 28	1.68	1.63	11.6	10.0	51.6	2.87	10.5	71.6	5.92	64.4	836.0	4.11	525.0	22.0	39.5	4.33	0.964	1.90	0.261	4.18	0.548	5.51
SC 29	0.96	0.78	15.0	8.87	54.1	2.75	8.91	65.9	4.43	57.4	798.0	3.29	482.0	18.3	31.8	3.67	0.815	1.54	0.235	3.52	0.476	4.37
SC 30	1.31	1.33	12.3	10.1	55.3	2.84	10.7	72.6	5.57	63.4	885.0	3.97	561.0	22.9	40.6	4.54	1.009	1.98	0.281	4.18	0.577	5.58
SC 31	1.16	1.29	20.4	8.39	44.1	2.55	8.81	69.0	7.16	48.0	767.0	4.13	446.0	20.5	35.6	3.94	0.877	1.89	0.274	3.45	0.505	5.06
SC 33	1.14	1.12	11.3	9.04	53.8	3.10	8.53	55.0	7.05	39.4	817.0	2.64	465.0	17.0	28.1	3.37	0.759	1.33	0.191	3.92	0.536	4.56
SC 35	1.13	1.32	20.5	8.26	42.7	2.42	9.23	67.5	6.67	42.6	798.0	3.79	499.0	20.5	35.6	3.96	0.855	1.84	0.269	3.51	0.488	4.86
SC 56	1.47	1.34	11.4	9.51	58.9	2.88	11.1	69.3	10.00	56.2	641.0	4.29	427.0	23.0	39.6	4.58	1.059	2.28	0.309	4.21	0.578	5.40
SC 57	0.95	1.50	15.1	10.2	69.5	2.77	10.6	78.7	7.50	50.7	887.0	2.28	369.0	16.4	22.6	3.37	0.824	1.65	0.242	3.61	0.396	4.15
Mexico City Copies of Columbia Plain																						
SC 16	0.88	1.01	17.8	9.25	53.1	2.70	8.75	71.4	6.31	39.9	849.0	3.46	399.0	18.3	33.0	3.66	0.839	1.71	0.238	3.51	0.455	4.38
SC 32	1.07	1.25	15.2	9.77	63.2	2.90	9.59	69.2	5.18	56.9	817.0	3.96	465.0	19.1	34.6	3.88	0.910	1.71	0.251	4.06	0.519	4.66
San Juan Polychrome																						
SA 66	1.07	0.80	14.2	12.2	89.5	3.40	10.2	73.1	7.53	58.1	885.0	4.00	441.0	16.2	18.8	3.52	0.857	1.69	0.254	4.48	0.313	4.27
SA 67	1.07	1.36	n.d.	9.6	58.0	2.5	29.4	n.d.	n.d.	80.0	n.d.	3.7	470.0	17.0	20.0	n.d.	0.76	n.d.	n.d.	4.8	n.d.	4.4
SC 37	1.00	0.93	14.1	11.6	86.9	3.17	11.5	80.2	7.19	73.1	918.0	7.21	446.0	18.6	24.7	3.79	0.916	1.91	0.295	4.03	0.538	5.05
SC 38	1.02	1.45	13.3	12.3	85.9	3.27	10.5	88.9	7.10	68.4	731.0	3.82	427.0	19.7	25.2	4.01	0.923	1.92	0.281	4.25	0.519	4.97
SC 40	0.94	1.32	13.2	12.7	87.5	3.50	15.8	82.6	19.0	62.7	351.0	4.81	413.0	18.6	27.7	3.86	0.948	1.90	0.266	4.21	0.494	4.84
SC 42	1.08	1.12	14.7	11.9	71.4	3.14	10.1	78.3	10.4	59.2	425.0	3.44	398.0	18.7	29.5	3.82	0.910	1.83	0.254	4.24	0.473	4.91
SC 43	0.99	n.d.	14.2	12.5	90.2	3.38	9.6	81.5	7.13	60.1	340.0	3.66	383.0	19.7	28.5	3.79	0.908	1.77	0.279	4.33	0.593	5.05
SC 44	1.03	n.d.	14.9	10.8	82.4	3.33	16.3	116.0	8.36	46.9	851.0	4.50	413.0	15.0	23.3	3.05	0.752	1.40	0.211	3.57	0.499	4.52

Continued on next page.

Appendix A–*Continued*

Sample	*Na*	*K*	*Ca*	*Sc*	*Cr*	*Fe*	*Co*	*Zn*	*As*	*Rb*	*Sr*	*Cs*	*Ba*	*La*	*Ce*	*Sm*	*Eu*	*Yb*	*Lu*	*Hf*	*Ta*	*Th*
San Juan Polychrome—Continued																						
SC 46	1.48	n.d.	9.5	10.9	55.7	2.24	6.8	75.2	12.1	56.8	693.0	3.30	499.0	28.4	44.6	5.19	1.19	2.00	0.307	4.29	0.538	6.32
SC 47	1.19	n.d.	14.5	12.7	89.9	3.59	10.0	85.9	4.16	48.0	760.0	5.46	459.0	20.0	26.1	4.19	1.04	1.79	0.325	4.42	0.511	5.00
SC 48	0.88	1.40	16.7	12.4	74.3	3.17	14.9	87.7	19.7	84.9	690.0	6.08	535.0	22.1	28.5	4.27	0.946	2.13	0.348	4.06	0.508	5.52
SC 50	0.84	1.21	14.1	11.7	78.0	3.06	9.2	78.5	29.0	53.2	471.0	6.53	366.0	17.5	27.6	3.66	0.847	1.77	0.265	3.65	0.491	4.75
SC 51	0.96	1.15	10.9	13.4	90.8	3.54	11.6	87.5	17.9	49.8	432.0	4.82	340.0	19.0	29.4	3.96	0.944	2.02	0.299	4.06	0.553	4.93
SC 52	1.40	1.29	8.2	10.8	54.8	2.15	6.7	78.7	5.98	58.6	573.0	3.14	427.0	26.5	43.1	4.99	1.17	1.92	0.282	4.15	0.497	6.34
SPANISH PRODUCTIONS																						
Sevilla White																						
SC 10	0.69	2.10	n.d.	10.5	46.0	2.8	11.0	n.d.	n.d.	155.0	n.d.	10.4	430.0	54.0	82.0	n.d.	1.02	n.d.	n.d.	8.2	n.d.	19.5
SC 19	0.72	2.75	n.d.	10.7	48.0	2.9	10.4	n.d.	n.d.	178.0	n.d.	10.8	620.0	49.0	124.0	n.d.	1.01	n.d.	n.d.	6.8	n.d.	16.8
SC 26	0.51	2.35	13.6	11.7	50.4	3.43	14.7	83.2	16.7	167.0	351.0	10.1	511.0	42.0	77.1	7.14	1.07	3.23	0.457	6.03	1.51	15.6
SC 27	0.54	2.25	13.6	11.3	47.0	3.28	13.0	80.4	18.7	151.0	288.0	10.6	461.0	47.1	88.1	8.07	1.12	3.53	0.532	7.89	1.58	18.0
Columbia Gun Metal																						
SC 86	0.61	1.87	18.2	11.7	73.1	3.23	16.5	88.1	11.9	103.0	634.0	5.30	444.0	34.6	58.9	5.42	1.04	2.36	0.398	4.58	1.57	9.48
SC 87	0.55	0.83	19.7	10.0	67.6	2.82	10.4	74.8	11.6	74.5	545.0	4.51	325.0	32.1	55.7	4.99	0.998	2.49	0.356	5.26	0.964	9.06
SC 88	0.59	0.95	19.3	7.71	33.0	1.79	10.7	67.9	11.2	87.7	412.0	4.41	421.0	33.0	29.8	5.27	1.01	2.51	0.381	3.72	0.733	4.27
SC 89	0.73	1.48	14.7	10.3	71.3	3.01	10.0	78.2	11.9	88.5	450.0	5.21	389.0	33.9	59.3	5.38	1.05	2.28	0.339	5.64	0.929	9.40
SC 91	0.56	1.82	15.4	10.4	62.8	3.00	10.4	76.2	9.64	79.3	543.0	5.11	334.0	32.9	54.8	5.16	1.03	2.15	0.362	5.47	1.09	9.14

NOTE: Na, K, Ca, and Fe results are given in percents; all others are in parts per million.

Appendix B. Majolica Ceramics from Excavations at Santa Catalina de Guale Mission, St. Catherines Island, Georgia

Sample	Na	K	Ca	Sc	Cr	Fe	Co	Zn	As	Rb	Sr	Cs	Ba	La	Ce	Sm	Eu	Yb	Lu	Hf	Ta	Th
MEXICAN PRODUCTIONS																						
San Juan Polychrome																						
SCI009	1.89	0.964	8.4	11.6	83.8	3.92	15.0	66.5	6.14	44.2	634.0	4.83	385.0	22.0	34.8	4.31	1.08	1.66	0.217	4.74	0.429	5.74
SCI017	1.72	0.690	4.5	13.3	109.0	3.85	12.7	75.2	4.95	47.8	548.0	5.51	423.0	17.7	32.5	3.97	1.12	1.68	0.276	5.19	0.593	5.19
SCI018	1.36	0.614	4.5	13.8	126.0	4.05	10.1	70.6	3.80	56.8	581.0	6.22	311.0	18.1	29.9	3.81	1.02	1.51	0.283	5.21	0.627	6.00
SCI024	1.06	0.851	15.1	12.6	91.6	3.56	8.26	74.1	8.65	56.6	615.0	3.93	421.0	18.7	22.1	3.85	0.964	1.84	0.179	4.56	0.482	4.62
SCI031	1.79	0.745	4.5	13.7	115.0	3.96	14.3	70.0	4.76	45.1	698.0	5.90	313.0	19.1	34.8	4.16	1.13	1.95	0.228	5.98	0.578	5.31
SCI036	1.27	0.558	12.2	12.6	102.0	3.73	9.91	64.3	5.69	53.6	673.0	8.77	219.0	17.5	26.4	3.97	1.00	1.59	0.275	4.73	0.578	5.1
SCI037	1.29	0.920	6.7	15.1	122.0	4.20	11.8	71.4	12.6	68.2	417.0	5.35	337.0	20.8	32.0	4.24	1.07	2.09	0.348	5.36	0.710	6.05
SCI047	1.33	1.07	5.0	15.0	105.0	4.06	12.6	80.2	28.1	62.1	378.0	4.68	358.0	21.6	32.7	4.49	1.12	2.00	0.343	5.45	0.627	5.98
San Luis Blue-on-White																						
SCI010	1.08	1.02	10.8	13.2	91.2	3.62	13.4	79.3	9.10	68.5	356.0	5.42	495.0	21.9	36.0	4.29	1.02	1.98	0.280	4.65	0.467	5.38
SCI027	1.65	0.807	6.8	12.9	82.4	3.73	9.66	62.7	3.00	46.7	522.0	3.69	545.0	26.0	32.1	4.83	1.25	2.05	0.270	4.99	0.545	6.41
SCI029	1.67	0.863	10.5	12.5	88.5	3.64	9.27	75.9	3.70	47.2	530.0	4.17	406.0	23.9	29.2	4.70	1.19	1.86	0.216	5.20	0.537	5.96
SCI039	1.68	0.802	10.8	13.1	100.0	3.94	13.1	64.6	16.3	53.2	530.0	4.62	436.0	24.1	37.8	4.79	1.18	2.28	0.288	5.37	0.703	6.27
SCI041	1.63	0.836	9.5	12.0	87.7	3.49	9.10	64.4	6.04	45.5	644.0	3.74	506.0	23.7	30.1	4.54	1.15	1.79	0.290	4.92	0.617	6.10
SCI048	1.75	0.785	6.8	12.9	89.1	3.70	11.1	57.3	26.5	49.0	531.0	4.17	475.0	24.4	30.4	4.82	1.25	2.03	0.294	5.46	0.658	6.27
SCI053	1.40	0.684	9.0	13.8	128.0	4.01	9.68	58.7	6.65	49.3	871.0	9.27	336.0	19.9	27.0	4.26	1.08	1.75	0.285	5.14	0.585	5.60
SCI059	1.77	1.04	7.2	12.6	92.5	3.61	10.4	65.8	12.5	58.1	627.0	4.40	632.0	25.1	34.4	5.01	1.17	2.11	0.298	5.73	0.687	6.98
SCI060	1.77	0.957	7.5	12.7	85.9	3.72	10.9	76.2	7.14	50.9	521.0	4.48	574.0	24.8	35.1	4.95	1.22	2.14	0.293	6.07	0.628	6.79
SCI063	1.77	0.902	8.8	12.8	97.7	3.73	9.35	65.3	6.64	45.0	713.0	4.09	478.0	24.8	31.8	4.98	1.22	2.18	0.289	5.35	0.590	6.21
SCI064	1.17	1.01	8.5	14.4	97.1	4.01	10.9	77.4	22.1	61.0	518.0	4.14	n.d.	35.0	27.2	7.93	1.83	2.74	0.474	4.95	0.495	5.18
SCI069	1.67	0.798	9.1	12.2	82.2	3.52	9.04	72.8	2.56	50.4	565.0	4.06	533.0	24.0	29.6	4.68	1.11	1.97	0.324	5.35	0.557	6.14
SCI073	1.67	0.828	7.1	13.0	112.0	3.81	9.89	66.8	23.4	61.0	619.0	4.26	252.0	22.4	30.7	4.44	1.15	1.90	0.211	5.33	0.679	6.19
SCI076	1.96	1.099	8.3	12.0	81.8	4.06	13.6	69.5	7.73	52.0	500.0	5.02	501.0	23.2	35.6	4.58	1.15	1.98	0.297	4.85	0.515	5.65
SCI079	1.42	0.556	7.5	14.0	132.0	4.15	11.0	73.3	4.94	50.7	697.0	9.04	255.0	17.9	27.2	3.85	1.07	1.60	0.242	5.35	0.661	5.57
SCI082	1.67	0.687	9.8	12.5	88.7	3.72	9.53	68.9	2.49	47.2	528.0	3.95	537.0	24.5	29.6	4.82	1.20	1.99	0.301	5.30	0.395	6.14
SCI083	1.71	0.789	7.8	12.5	86.1	3.62	9.12	66.1	2.61	44.5	564.0	4.01	385.0	24.6	29.6	4.74	1.17	1.82	0.360	5.51	0.598	5.75
Aucilla Polychrome																						
SCI008	1.36	0.679	5.2	11.7	101.0	2.99	7.41	52.2	4.85	45.0	461.0	3.08	973.0	19.7	25.2	4.24	1.14	1.55	0.219	4.52	0.441	5.19
SCI038	1.37	0.682	3.7	11.7	103.0	3.06	8.17	53.3	5.98	52.2	537.0	3.24	1040.0	18.5	25.2	4.04	1.08	1.50	0.210	4.30	0.390	4.92
SCI044	1.39	0.555	11.7	11.9	96.8	3.26	8.75	69.8	3.24	44.1	445.0	2.88	314.0	20.4	22.5	4.25	1.12	1.85	0.364	5.25	0.604	5.19
SCI045	1.64	0.511	8.7	13.2	100.0	3.67	8.13	63.7	2.40	44.0	700.0	3.19	374.0	20.4	20.5	4.53	1.17	1.79	0.281	4.75	0.506	4.76
SCI050	1.41	0.608	5.4	12.0	103.0	3.09	8.24	65.2	3.81	45.9	573.0	3.71	869.0	19.6	25.0	4.27	1.15	1.75	0.308	4.55	0.553	5.12
SCI055	1.58	0.460	6.6	13.1	119.0	3.69	7.35	60.5	2.82	50.5	547.0	3.45	471.0	19.8	20.2	4.31	1.13	1.85	0.258	4.80	0.462	5.06

Continued on next page.

Appendix B.–*Continued*

Sample	*Na*	*K*	*Ca*	*Sc*	*Cr*	*Fe*	*Co*	*Zn*	*As*	*Rb*	*Sr*	*Cs*	*Ba*	*La*	*Ce*	*Sm*	*Eu*	*Yb*	*Lu*	*Hf*	*Ta*	*Th*
Aucilla Polychrome—Continued																						
SCI057	1.55	0.485	7.0	13.0	110.0	3.63	7.35	63.7	2.25	43.0	593.0	3.25	408.0	18.2	20.1	4.06	1.07	1.77	0.303	4.85	0.564	4.98
SCI058	1.48	0.479	5.9	12.8	106.0	3.52	9.44	71.8	2.88	43.1	362.0	2.64	433.0	22.1	26.6	4.69	1.21	2.00	0.299	4.82	0.550	5.65
SCI070	1.39	0.540	8.0	11.7	96.6	3.14	8.63	63.1	3.47	47.3	426.0	3.70	426.0	18.8	25.6	4.19	1.10	1.58	0.234	4.44	0.388	5.00
SCI072	1.33	0.729	9.5	11.4	101.0	3.02	7.55	58.2	2.99	48.6	574.0	3.36	975.0	18.8	23.8	4.05	1.12	1.73	0.262	4.31	0.416	4.65
SCI077	1.29	0.525	9.8	11.0	87.1	2.92	8.11	63.2	3.29	38.5	394.0	3.74	149.0	17.9	24.4	3.99	1.01	1.69	0.251	4.22	0.471	4.65
SCI078	1.53	0.521	8.2	12.6	113.0	3.69	7.78	66.1	2.58	49.1	490.0	3.39	482.0	17.6	20.4	3.92	1.11	1.67	0.236	5.21	0.561	4.92
SCI080	1.71	0.518	8.5	13.7	99.8	3.86	7.96	60.0	4.48	37.6	492.0	3.08	299.0	21.3	21.4	4.59	1.28	2.15	0.274	5.25	0.528	4.90
Mt. Royal Polychrome																						
SCI028	1.26	0.562	17.5	12.2	107.0	3.42	10.21	57.0	5.46	40.4	411.0	2.76	195.0	18.8	21.4	4.09	1.08	1.71	0.247	4.38	0.617	5.61
SCI030	1.33	0.256	11.8	13.1	118.0	3.68	9.82	73.3	6.82	48.9	442.0	3.37	248.0	20.3	23.8	4.38	1.18	1.91	0.304	5.00	0.603	5.51
SCI074	1.48	0.710	11.0	13.8	118.0	3.88	9.66	62.4	7.35	50.4	479.0	3.33	427.0	21.3	25.2	4.75	1.20	1.94	0.272	5.24	0.637	6.04
SCI081	1.29	0.538	11.6	12.7	110.0	3.55	8.51	55.1	7.28	51.6	426.0	2.96	262.0	19.5	23.7	4.33	1.17	1.85	0.229	4.58	0.610	5.37
Puebla Polychrome																						
SCI046	1.05	0.371	15.8	13.5	113.0	3.96	10.2	76.6	6.01	32.9	611.0	6.15	436.0	19.8	27.4	4.25	1.04	1.79	0.223	5.45	0.726	6.27
SCI084	1.47	0.624	7.1	13.7	98.9	3.89	8.99	71.0	5.13	38.4	403.0	3.37	363.0	24.3	25.0	5.07	1.28	2.04	0.313	5.22	0.548	5.75
Puebla Blue-on-White																						
SCI005	1.74	1.08	7.0	13.2	84.1	3.72	16.5	74.5	9.40	62.2	488.0	5.15	441.0	27.9	44.3	5.31	1.26	2.15	0.337	5.26	0.703	6.38
SPANISH PRODUCTIONS																						
Spanish Variant of San Luis Blue-on-White																						
SCI042	0.995	1.55	4.0	21.6	136.0	5.20	25.9	182.0	18.4	95.1	520.0	4.94	420.0	37.6	67.1	5.85	1.14	2.71	0.450	6.92	1.08	11.5
SCI052	1.03	1.24	5.7	21.9	135.0	5.30	24.9	192.0	15.5	100.0	800.0	5.07	384.0	38.8	70.8	5.94	1.24	2.87	0.505	8.32	1.08	11.9
SCI075	1.13	1.56	3.9	21.1	150.0	5.21	22.1	181.0	24.9	108.0	646.0	4.98	403.0	32.4	62.5	5.47	1.09	2.51	0.330	7.05	1.01	10.7
Columbia Plain																						
SCI012	0.891	0.547	10.3	12.9	90.4	3.66	15.6	68.4	13.3	83.4	285.0	6.87	361.0	40.1	69.3	6.31	1.21	2.81	0.433	7.05	1.04	11.3
SCI013	0.543	1.42	12.3	12.4	80.4	3.49	15.5	86.1	10.2	102.1	507.0	5.89	509.0	39.4	68.2	6.22	1.22	2.88	0.457	6.89	1.09	11.0
SCI014	0.614	1.09	11.0	13.2	87.9	3.68	14.2	88.7	15.7	81.1	369.0	5.19	421.0	43.3	75.3	6.87	1.32	3.10	0.492	7.21	1.17	11.8
SCI015	0.796	1.02	11.8	12.6	88.5	3.61	12.8	88.3	10.2	121.0	557.0	6.93	449.0	40.5	70.0	6.38	1.26	2.96	0.470	7.35	1.08	11.2
SCI016	0.755	1.39	9.8	13.6	86.5	3.79	14.1	97.9	10.0	135.0	432.0	6.71	402.0	42.7	76.9	6.76	1.30	3.00	0.447	6.79	1.06	11.9
SCI019	0.622	1.21	14.3	12.7	86.7	3.63	13.9	89.1	13.0	83.0	443.0	5.53	455.0	40.3	69.3	6.28	1.22	2.72	0.449	6.67	1.09	11.0
SCI020	0.809	0.820	12.8	12.2	84.1	3.50	12.1	83.6	10.7	117.0	472.0	6.71	443.0	40.8	70.3	6.41	1.21	3.15	0.383	6.53	0.995	11.6

Continued on next page.

Appendix B.–*Continued*

Sample	*Na*	*K*	*Ca*	*Sc*	*Cr*	*Fe*	*Co*	*Zn*	*As*	*Rb*	*Sr*	*Cs*	*Ba*	*La*	*Ce*	*Sm*	*Eu*	*Yb*	*Lu*	*Hf*	*Ta*	*Th*
Columbia Plain—Continued																						
SCI021	0.718	1.26	11.0	13.2	88.5	3.73	13.6	100.0	10.5	128.0	474.0	7.14	518.0	42.5	73.5	6.59	1.29	2.93	0.427	6.62	1.10	11.6
SCI022	0.766	0.998	13.6	13.5	93.5	3.70	13.5	89.9	12.6	103.0	472.0	6.44	428.0	41.0	71.0	6.44	1.28	3.19	0.451	6.03	1.172	11.9
SCI023	0.667	1.07	14.8	12.4	84.7	3.74	13.5	90.8	11.9	114.0	641.0	6.24	481.0	39.4	67.0	6.18	1.21	2.73	0.434	6.64	0.964	10.8
SCI025	0.615	1.13	9.2	14.2	95.9	4.05	16.3	156.0	17.5	102.0	434.0	7.46	472.0	43.3	71.3	6.75	1.35	3.07	0.452	7.48	1.12	11.7
SCI026	0.610	1.24	13.2	12.6	81.1	3.57	13.8	88.5	11.8	96.4	540.0	5.68	447.0	37.8	65.8	6.03	1.22	2.53	0.486	6.78	0.984	10.4
SCI034	0.614	1.30	13.2	13.1	88.7	3.66	14.1	89.5	14.8	96.2	540.0	5.71	491.0	40.9	71.4	6.75	1.30	3.18	0.495	6.52	1.15	11.7
SCI035	0.782	0.925	9.3	14.1	99.1	4.04	14.4	92.5	12.4	119.0	395.0	7.87	394.0	44.6	78.3	7.03	1.36	2.84	0.461	6.87	1.06	12.7
Yayal Blue-on-White																						
SCI001	0.385	2.04	9.3	13.6	81.7	3.86	14.2	90.4	9.08	122.0	474.0	6.70	412.0	42.9	74.0	6.73	1.26	2.82	0.421	6.73	1.07	12.2
SCI002	1.27	1.17	12.0	19.9	130.0	4.73	16.9	157.0	3.10	56.2	1521.0	3.66	265.0	27.2	49.7	4.55	0.989	2.26	0.286	4.06	0.652	8.20
SCI003	1.10	1.18	10.2	19.9	130.0	4.78	20.8	158.1	9.75	58.3	1750.0	2.80	348.0	27.9	50.7	4.73	1.00	2.03	0.386	4.06	0.671	8.51
SCI004	1.26	1.17	10.3	20.8	130.0	4.80	18.8	127.1	6.05	58.7	2070.0	2.82	326.0	28.1	51.1	4.83	1.03	2.56	0.362	4.54	0.724	8.71
SCI007	1.13	1.26	11.0	20.1	132.0	4.98	17.8	136.1	6.70	54.2	1849.0	2.92	389.0	28.2	51.5	4.73	0.993	2.17	0.284	4.25	0.63	9.25

NOTE: Na, K, Ca, and Fe results are given in percents; all others are in parts per million.

Appendix B.–*Continued*

Sample	*Na*	*K*	*Ca*	*Sc*	*Cr*	*Fe*	*Co*	*Zn*	*As*	*Rb*	*Sr*	*Cs*	*Ba*	*La*	*Ce*	*Sm*	*Eu*	*Yb*	*Lu*	*Hf*	*Ta*	*Th*
Yayal Blue-on-White																						
SCI001	0.385	2.04	9.3	13.6	81.7	3.86	14.2	90.4	9.08	122.0	474.0	6.70	412.0	42.9	74.0	6.73	1.26	2.82	0.421	6.73	1.07	12.2
SCI002	1.27	1.17	12.0	19.9	130.0	4.73	16.9	157.0	3.10	56.2	1521.0	3.66	265.0	27.2	49.7	4.55	0.989	2.26	0.286	4.06	0.652	8.20
SCI003	1.10	1.18	10.2	19.9	130.0	4.78	20.8	158.1	9.75	58.3	1750.0	2.80	348.0	27.9	50.7	4.73	1.00	2.03	0.386	4.06	0.671	8.51
SCI004	1.26	1.17	10.3	20.8	130.0	4.80	18.8	127.1	6.05	58.7	2070.0	2.82	326.0	28.1	51.1	4.83	1.03	2.56	0.362	4.54	0.724	8.71
SCI007	1.13	1.26	11.0	20.1	132.0	4.98	17.8	136.1	6.70	54.2	1849.0	2.92	389.0	28.2	51.5	4.73	0.993	2.17	0.284	4.25	0.63	9.25
Santo Domingo Blue-on-White																						
SCI006	0.538	1.61	12.9	11.9	77.4	3.40	13.2	84.7	7.94	103.0	481	6.43	484	38.8	67.0	6.04	1.15	2.83	0.395	6.67	0.986	11.0
SCI011	0.652	1.29	11.4	12.1	82.2	3.52	13.6	85.1	11.0	121.0	552	6.90	332	39.0	68.2	6.21	1.17	2.74	0.436	7.40	1.02	11.0
Ichtucknee Blue-on-Blue																						
SCI033	0.989	1.36	10.9	16.6	429.0	4.55	23.1	88.1	20.0	65.3	579	5.62	356	29.7	55.6	5.15	1.00	2.62	0.389	4.56	0.966	9.35
SCI065	0.499	0.89	10.6	16.9	105.0	4.65	17.9	87.7	34.5	60.7	463	6.55	524	48.1	83.6	7.71	1.49	3.53	0.492	6.64	1.36	13.5
Ichtucknee Blue-on-White																						
SCI049	0.785	1.28	18.2	12.0	99.8	3.40	12.4	84.7	12.1	109.0	443	6.67	332	39.8	72.6	6.55	1.29	2.36	0.358	6.37	1.61	11.1
SCI067	0.785	1.27	14.5	12.1	91.6	3.44	12.3	85.9	12.3	102.0	495	6.24	381	36.9	67.6	6.10	1.25	2.46	0.321	6.24	1.67	10.0
SCI071	0.716	1.11	12.1	13.5	106.0	3.78	13.6	90.8	12.9	88.3	317	6.04	394	39.4	71.9	6.49	1.32	2.31	0.371	6.40	1.75	10.9
Santa Elena Mottled Blue-on-White																						
SCI066	0.506	0.705	12.0	13.4	89.5	3.71	23.2	65.3	34.2	43.2	425	5.77	238	40.5	72.3	6.38	1.22	3.48	0.521	7.24	1.21	12.3
Unidentified Blue and Yellow																						
SCI061	0.646	0.796	13.3	16.1	376.0	4.45	23.3	130.0	11.2	44.8	592	3.51	222	31.0	55.2	5.30	1.04	2.35	0.418	4.46	0.916	9.95
Unidentified Blue-on-White																						
SCI040	0.750	0.940	9.6	12.9	90.8	3.84	13.0	95.3	14.1	118.0	352	6.92	435	43.2	75.0	6.65	1.31	3.10	0.420	7.45	1.11	12.5

NOTE: Na, K, Ca, and Fe results are given in percents; all others are in parts per million.

Appendix C. Modern Majolica Ceramics from Puebla, Mexico

Sample	*Na*	*K*	*Ca*	*Sc*	*Cr*	*Fe*	*Co*	*Zn*	*As*	*Rb*	*Sr*	*Cs*	*Ba*	*La*	*Ce*	*Sm*	*Eu*	*Yb*	*Lu*	*Hf*	*Ta*	*Th*
SD29	1.18	n.d.	13.4	11.8	92.0	3.33	25.4	73.1	3.53	65.0	244.0	6.34	236.0	20.6	27.8	4.30	0.957	2.12	0.340	4.67	0.794	6.00
SD30	1.20	0.77	11.4	10.7	79.8	3.01	7.74	68.9	3.99	61.2	282.0	6.34	252.0	18.7	25.8	3.95	0.822	2.11	0.365	4.26	0.766	5.60
SD31	1.08	n.d.	13.2	11.7	93.1	3.37	8.36	71.3	3.71	47.6	265.0	6.04	258.0	20.3	27.3	4.22	0.975	2.28	0.311	4.62	0.673	5.92
SD32	1.06	0.59	12.8	11.7	91.2	3.32	10.7	76.9	4.16	50.4	294.0	6.58	228.0	20.7	28.6	4.30	0.944	2.15	0.324	4.53	0.617	5.96
SD33	1.19	0.72	11.5	10.9	83.0	3.08	8.69	73.5	3.17	51.1	274.0	6.44	250.0	18.6	26.1	3.94	0.879	2.03	0.248	4.18	0.762	5.81
SD34	1.15	n.d.	13.3	11.4	87.9	3.30	10.9	77.6	4.58	54.3	304.0	6.10	236.0	20.4	27.3	4.25	0.973	2.04	0.331	4.46	0.587	5.77
SD35	1.15	0.47	12.4	12.9	109.0	3.80	15.0	81.5	4.38	58.5	292.0	7.78	272.0	21.1	31.2	4.46	1.01	2.30	0.343	5.35	0.867	6.64
SD36	1.16	n.d.	13.4	13.5	108.0	3.95	16.9	85.3	n.d.	73.1	312.0	7.82	294.0	22.5	31.5	4.65	1.06	2.28	0.385	5.62	0.863	7.10
SD37	n.d.	n.d.	n.d.	13.0	107.0	3.84	12.7	80.9	n.d.	61.9	251.0	7.93	316.0	n.d.	30.1	n.d.	0.995	n.d.	n.d.	5.55	0.889	6.90

NOTE: Na, K, Ca, and Fe results are given in percents; all others are in parts per million.

References

1. Goggin, J. S. *Spanish Majolica in the New World, Types of the Sixteenth to Eighteenth Centuries*; Yale Univ. Publications in Anthropology No. 72, Yale University Press: New Haven, CT, 1968.
2. Lister, F. C.; Lister, R. H. *Sixteenth Century Majolica Pottery in the Valley of Mexico*; Anthropological Papers of the University of Arizona No. 39: University of Arizona Press: Tucson, AZ, 1982.
3. Olin, J. S.; Harbottle, G.; Sayre, E. V. In *Archaeological Chemistry II*; Carter, G. F., Ed.; Advances in Chemistry Series No. 171; American Chemical Society: Washington, DC, 1978; pp 200-229.
4. Maggetti, M.; Westley, H.; Olin, J. S. In *Archaeological Chemistry III*; Lambert, J. B., Ed.; Advances in Chemistry Series No. 205; American Chemical Society Washington, DC, 1984; pp 151-191.
5. Jornet, A.; Blackman, M. J.; Olin, J. S. In *Ceramics and Civilization*; Kingery, D., Ed.; American Ceramic Society: Columbus, OH, 1985; pp 235-255.
6. Hughes, M. J.; Vince, A.G. In *Proceedings of the 24th International Archaeometry Symposium*; Olin, J. S.; Blackman, M. J., Eds.; Smithsonian Institution: Washington, DC 1986; pp 353-367.
7. Jornet, A.; Blackman, M. J.; Westley, H.; Olin, J. S. In *Abstracts of the 1984 Symposium on Archaeometry*; Olin, J. S.; Blackman, M. J., Organizers; Smithsonian Institution: Washington, DC, 1984; p 73.
8. op. cit. Lister, F. C.; Lister, R. H., 1982, pp 90-91.
9. Cervantes, E. *Loza Blanca y Azulejo de Puebla*, 2 vols. Mexico, 1939.
10. Leicht, H. *Las Calles de Puebla*, Estudio Historico, Mijares, Puebla, 1934.
11. op. cit. Goggin, J., p 53-56.
12. Lister, F. C.; Lister, R. H.; *Historical Archaeology* **1984,** *18*, 87-102.
13. op. cit. Goggin, J. p 214.
14. Thomas, D. H. *Anthropological Papers of the American Museum of Natural History* **1987,** *63, Part 2*, p 57.
15. South, S. *Research Manuscript Series 184* Institute of Archaeology and Anthropology, University of South Carolina: Columbia, SC, 1985; pp 45-46.
16. Blackman, M. J. In *Archaeological Chemistry III*; Lambert, J. B., Ed.; Advances in Chemistry Series No. 205; American Chemical Society: Washington, DC, 1984; pp 19-50.
17. Blackman, M. J., Olin, J. S., Jornet, A. In *Abstracts of the 1984 Symposium on Archaeometry*; Olin, J. S.; Blackman, M. J., Organizers; Smithsonian Institution, Washington, DC, 1984, p 12.
18. op. cit Lister, F. C.; Lister, R. H., 1982, p 18.
19. Bieber, A. W. Jr.; Brooks, D. W.; Harbottle, G.; Sayre, E. V. *Archaeometry* **1976,** *18(1)*, 59–74.
20. Vega Sosa, C. *El Recinto Sagrado de Mexico-Tenochtitlan: Excavaciones 1968-69 y 1975-76*; INAH, 1979.
21. Martin, C. J. M. *The International Journal of Nautical Archaeology and Underwater Exploration* **1978,** *8(4), 279-302.*
22. Deagan, K. personal communication, Dec. 2, 1987.

RECEIVED for review July 22, 1987. ACCEPTED revised manuscript February 11, 1988.

6

Analysis of Neolithic Iranian Ceramics

R. G. V. Hancock[1], S. J. Fleming[2], and W. D. Glanzman[2]

[1]SLOWPOKE Reactor Facility and the Department of Chemical Engineering, and Applied Chemistry, University of Toronto, Toronto, Canada, M5S 1A4

[2]MASCA, The University Museum, University of Pennsylvania, Philadelphia, PA 19104

Neolithic pottery (207 samples) from Hajji Firuz, Dalma Tepe, and Pisdeli Tepe in the Solduz Valley of northwestern Iran and from Tepe Gawra and Tell 'Ubaid in northern and southern Iraq, has been analyzed for 14 elements by neutron activation analysis. The three Iranian sites are type sites for three important archaeological periods in Iran. Small temporal differences were found in the Hajji Firuz sherd chemistry, but the predominant cause of variations of elemental concentrations was determined to be a calcium-rich phase-based dilution. Small, but probably distinct, chemical differences were found between the Solduz Valley sherds and small samplings of sherds from the major trading center of Tepe Gawra and the more distant Tell 'Ubaid. Archaeologically anomalous sherds from Dalma Tepe appear to have been made from local materials.

THE 5TH MILLENNIUM B.C. was a period of significant social change in West Asia. A general population growth encouraged migration to new settlement areas, and established village communities became more complex (*1*). All of these communities interacted to varying degrees at different times, and traded raw materials and finished products. The nature and extent of these interactions is of primary archaeological interest. These interactions reveal information about economic organization within these early societies and their degree of reliance on one another. Analyses of macroscopic features of surviving ceramics, such as vessel form, surface decoration, and fabric texture are the most common investigative tools.

0065–2393/89/0220–0113$06.00/0

Because geochemically different clay sources may have been used by potters to produce ceramics for both domestic and trade purposes, neutron activation analysis (NAA) has been used as an independent means of ceramic characterization. Because of the relatively good analytical precision possible with NAA, statistical patterning of NAA data for major, minor, and trace element concentrations may be used as a powerful provenancing tool.

To assess the degree of spatial resolution that such a scientific approach might achieve, analytical data from ceramics from five early settlement sites are presented and evaluated. Three of the sites, Hajji Firuz, Dalma Tepe, and Pisdeli Tepe, are within 9 km of each other in the Solduz Valley of Azerbaijan (northwestern Iran). Tepe Gawra, in northern Iraq, is relatively distant from the Solduz Valley (200 km west-southwest, over part of the Zagros Mountains). It is believed that the Tepe Gawra had significant trade links with the settlements in the Solduz Valley. Tell 'Ubaid, in southern Iraq, is more than 1000 km south of Tepe Gawra (*see* Figure 1). All of the material from the Solduz Valley was excavated as part of the Hasanlu Project of the University Museum, directed by R. H. Dyson, Jr. The pottery used for this study was recovered from limited test excavations at the sites of Hajji Firuz (*2*), Dalma (*3*), and Pisdeli Tepes by T. C. Young, Jr. in 1961, and from a horizontal clearance at Hajji Firuz Tepe by M. M. Voigt in 1968.

Experimental Details

Sampling. Pottery (207 pieces) was analyzed in this project. Samples varied in size from 25 to 500 mg. Individual samples were cut with a diamond saw or were broken from their parent sherd. The outer surfaces of each sample were removed with a carborundum bit. After cleaning, the samples were weighed and stored in polyethylene vials.

Chemical Analysis. The sherd samples were analyzed by instrumental neutron activation analysis (INAA) at the SLOWPOKE Reactor Facility of the University of Toronto (*4*). The first part of the analysis determined the concentrations of elements that produce short-lived radionuclides, including U, Dy, Ba, Ti, Mg, Na, Al, Mn, Cl, and Ca. This determination was achieved by irradiating the larger samples (>200 mg) for 1 min at 1.0×10^{11} neutrons cm^{-2} s^{-1}. After a delay time of approximately 19 min (to allow the ^{28}Al to decay to reasonable levels), each sample was assayed for 5 min with either a germanium–lithium or a hyperpure germanium detector-based γ-ray spectrometer. Samples were irradiated sequentially at 6.5-min intervals. Smaller samples were irradiated with higher neutron doses so that they produced as much radioactivity as the larger samples. Chemical concentrations were calculated by using the comparator method, established over the past decade at Toronto, with the aid of an assortment of international multielement rock standards and in-house chemical standards.

The second part of the analysis, to determine the concentrations of elements that produce radioisotopes with longer half-lives, including Sm, Eu, Na, and K, was

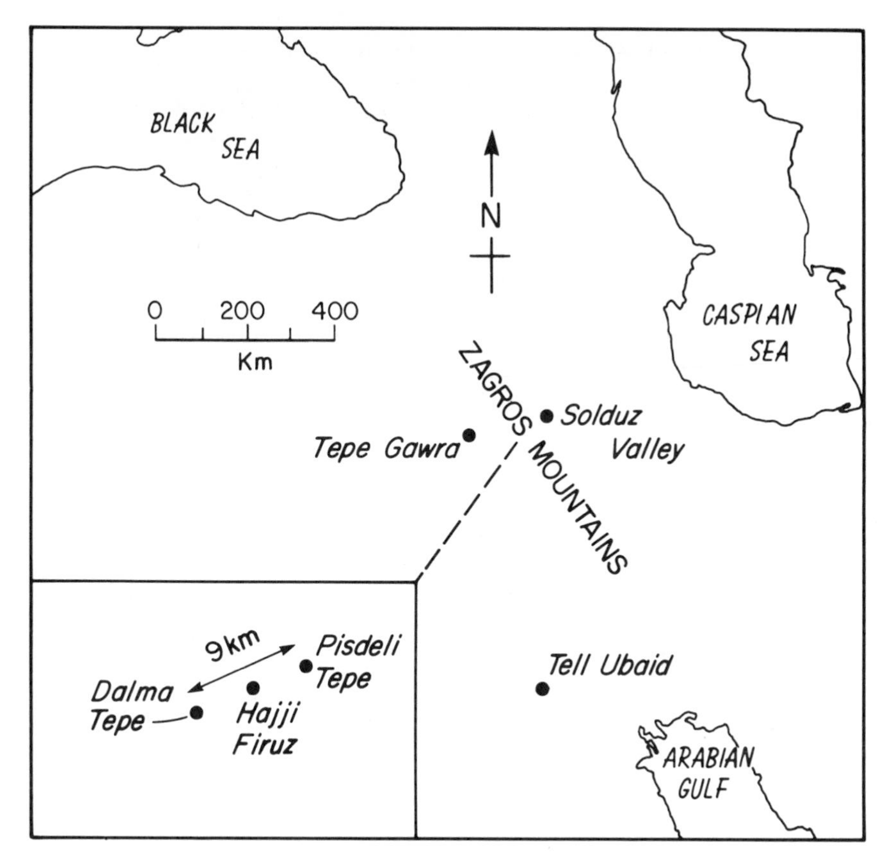

Figure 1. Map of site locations in Iran and Iraq.

performed by irradiating the samples in batches with neutron doses of 10 to 20 times the original dose. The samples were then left to decay overnight and were assayed sequentially with the same γ-ray spectrometers used previously. Chemical concentrations were calculated as before.

Results and Discussion

The results of the analyses of pottery from the Solduz Valley (51 samples from Hajji Firuz, 91 samples from Dalma Tepe, and 45 samples from Pisdeli Tepe) are presented in Table I as group means and standard deviations for the elements Na, Al, K, Ca, Ti, V, Mn, Sm, Eu, and Dy. Uranium, magnesium, and barium have not been included because their values were invariably at or close to their INAA detection limit. Also, Cl has been ignored because it is accumulated from the environment during pot or sherd burial.

Table I. Analytical Data for Sherds from the Solduz Valley Sites

Element	*Hajji Firuz (51 samples)*	*Dalma Tepe (91 samples)*	*Pisdeli Tepe (45 samples)*
Na (%)	0.96 ± 0.23	1.11 ± 0.18	1.08 ± 0.23
Al (%)	6.6 ± 0.9	7.1 ± 1.0	7.6 ± 0.3
K (%)	2.1 ± 0.4	2.2 ± 0.3	2.3 ± 0.3
Ca (%)	8.3 ± 2.7	7.9 ± 3.0	6.1 ± 2.7
Ti (%)	0.47 ± 0.11	0.52 ± 0.13	0.53 ± 0.11
V (ppm)	120 ± 20	140 ± 30	150 ± 30
Mn (ppm)	750 ± 200	990 ± 210	1000 ± 250
Sm (ppm)	4.5 ± 0.8	4.7 ± 0.8	5.0 ± 0.8
Eu (ppm)	1.2 ± 0.2	1.3 ± 0.2	1.4 ± 0.2
Dy (ppm)	3.8 ± 0.6	4.2 ± 0.8	4.4 ± 0.7

NOTE: All values are given as mean ± standard deviation.

Two points are apparent from this data. The first is that, apart from Ca, the trends in mean chemical concentrations, although very small, are consistently low to high, from Hajji Firuz sherds through Dalma Tepe sherds to Pisdeli Tepe sherds. This trend indicates the probability of dilution by a Ca-rich phase in this series of ceramics. The second point is that, for each element, the data are not particularly tightly clustered, so that all the data sets overlap broadly. With single standard deviations of 15 to 20%, the three data sets are relatively dispersed and again suggest a Ca-rich phase dilution effect (5). This effect can be seen in Figure 2, which is a Ca–Al scattergram for all the Solduz data, in which the Al content is taken as a marker for the clay fabric that the Ca dilutes. The positive correlation of Al and Ti (*see* Figure 3) confirms this conclusion.

It is presumptuous to claim site-specific separations of the three sets of ceramics from the intermingling of the data in these scattergrams without looking at the intergroup elemental ratios (*see* Table II). These ratios show that, apart from Ca, the Pisdeli Tepe to Dalma Tepe group ratios are relatively consistent at 1.04 ± 0.04 and indicate strong geochemical similarities. However, in addition to the expected Ca anomalies, the Dalma Tepe to Hajji Firuz and the Pisdeli Tepe to Hajji Firuz group ratios show abnormal behavior for Mn, with ratios of 1.41 and 1.43 compared with the ratios of the other elements of 1.11 ± 0.05 and 1.15 ± 0.05, respectively. This abnormal behavior may reflect a slightly different clay source at Hajji Firuz relative to Dalma Tepe and Pisdeli Tepe (*see* Table III and the relevant discussion).

To establish the relative internal consistency of the data, the large group of sherds from Dalma Tepe was somewhat arbitrarily divided unequally into two groups. The data for the first 25 archaeologically chosen sherds (representing a minimum statistically reliable number of sherds and the sherds

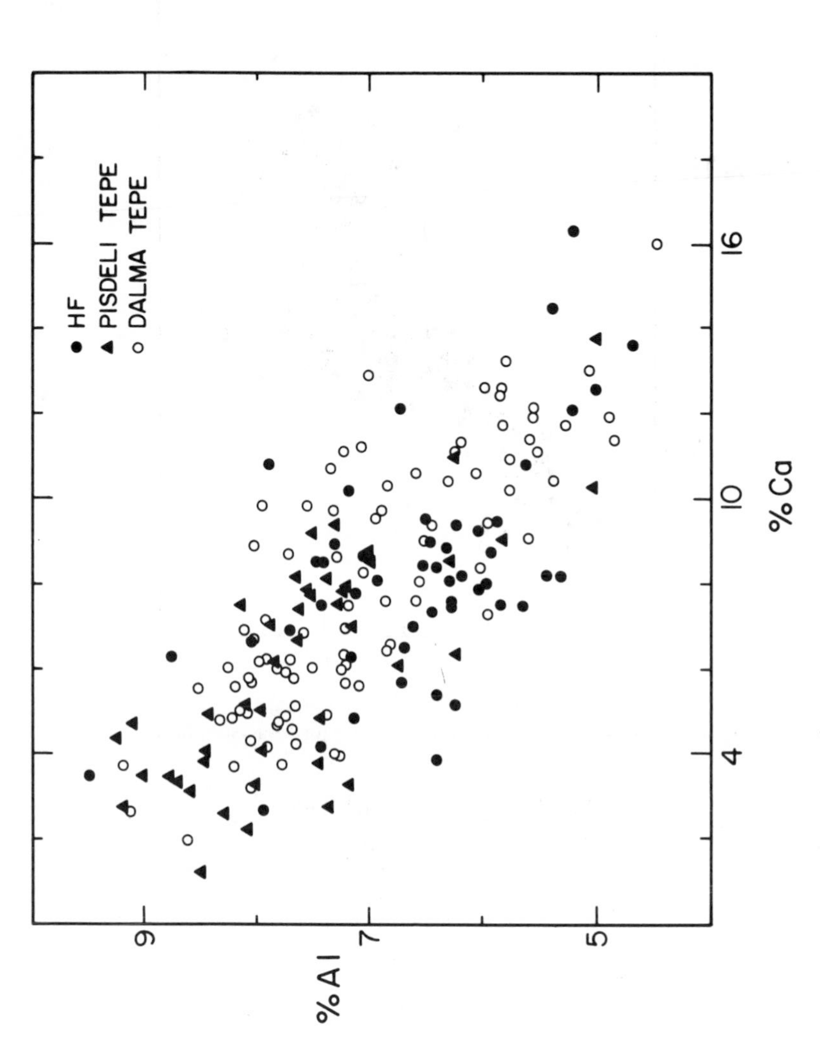

Figure 2. Al–Ca scattergram showing a negative correlation associated with Ca-rich phase dilutions.

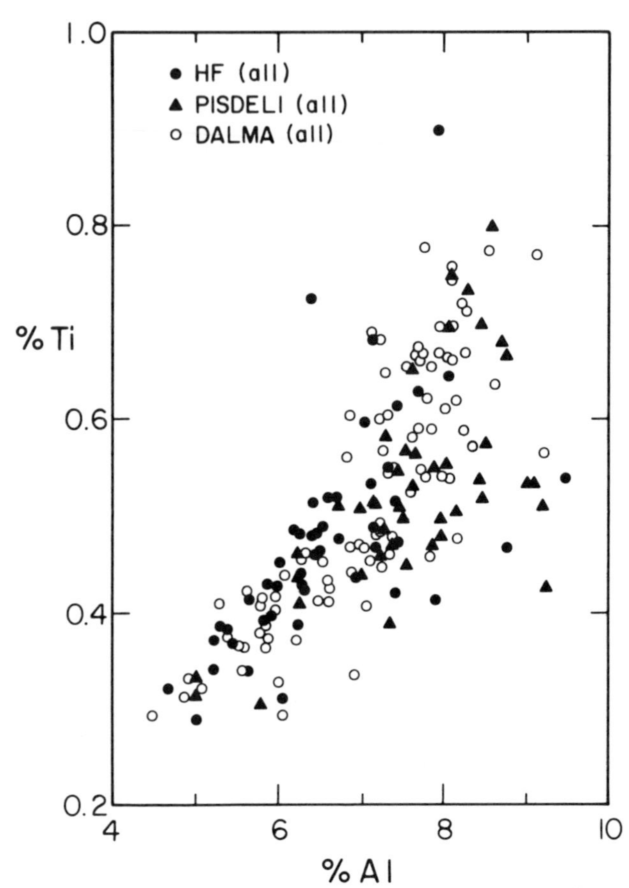

Figure 3. Ti–Al scattergram illustrating a positive correlation.

Table II. Group Elemental Ratios

Element	*Pisdeli/Dalma*	*Dalma/Hajji Firuz*	*Pisdeli/Hajji Firuz*
Na	0.97	1.16	1.10
Al	1.07	1.08	1.15
K	1.05	1.05	1.10
Ca	0.77	0.95	0.73
Ti	1.02	1.11	1.13
V	1.07	1.17	1.25
Mn	1.01	1.41	1.43
Sm	1.06	1.04	1.11
Eu	1.08	1.08	1.17
Dy	1.05	1.11	1.17
Average	1.04 ± 0.04[a]	1.11 ± 0.05[b]	1.15 ± 0.05[b]

[a]Without Ca.
[b]Without Ca and Mn.

Table III. Check of the Internal Consistency of the Data from Dalma Tepe

Element	*First 25 Samples*	*Remaining 66 Samples*
Na (%)	1.08 ± 0.29	1.12 ± 0.21
Al (%)	7.0 ± 1.1	7.2 ± 1.0
K (%)	2.3 ± 0.4	2.2 ± 0.3
Ca (%)	8.1 ± 3.6	7.8 ± 2.8
Ti (%)	0.51 ± 0.14	0.53 ± 0.13
V (ppm)	130 ± 40	140 ± 30
Mn (ppm)	950 ± 270	1000 ± 190
Sm (ppm)	4.9 ± 0.9	4.6 ± 0.8
Eu (ppm)	1.3 ± 0.2	1.3 ± 0.2
Dy (ppm)	4.3 ± 0.9	4.1 ± 0.8

NOTE: All values are given as mean ± standard deviation.

thought to be most representative of the archaeological group) were compared with the data for the remaining 66 sherds in Table III. This comparison reveals remarkable agreement between the two data sets. The relative sizes of the group mean precisions are consistent with the more variable Ca content in the small group relative to the larger group and confirm the intragroup consistency.

Temporal Differences. The Solduz Valley pottery studied here covers the period from ca. 5500 B.C. to ca. 3550 B.C. The appreciable scatter in the NAA data presented in Table I raises the question of whether some of the scatter is the result of a temporal change in the chemistry of the clay sources used by potters at any or all of the three Solduz Valley sites.

Complete cross comparison between the sites is not feasible because the Hajji Firuz phase (ca. 5500 to 4750 B.C.), the Dalma phase (ca. 4750 to 4350 B.C.), and the Pisdeli phase (ca. 4350 to 3550 B.C.) (6) are not represented at every site. All three phases occur at Hajji Firuz, though material from the Pisdeli phase is quite limited. The Dalma phase occurs at Dalma Tepe, with a very limited amount of material attributed to the Hajji Firuz/Dalma phase transition. Only the Dalma and Pisdeli phases occur at Pisdeli Tepe. Accordingly, temporal comparisons may be made at Hajji Firuz and Pisdeli Tepe, that yield quite different kinds of information.

Data showing the temporal differences for the selection of sherds from Hajji Firuz are presented in Table IV and Figure 4. Considering the small number of Dalma-phase and Pisdeli-phase samples, and the enhanced possibilities of their group mean concentrations deviating from "normal", these two groups tend to be close in average composition to that of the Dalma Tepe and Pisdeli Tepe ceramics listed in Tables I and III. This similarity is most pronounced for Al and Mn. On the other hand, the Hajji Firuz-phase

Table IV. Test for Temporal Differences in Sherd Compositions at Hajji Firuz

Element	*Hajji Firuz Phase* *38 samples*	*Dalma Phase* *5 samples*	*Pisdeli Phase* *6 samples*
Na (%)	0.95 ± 0.21	1.20 ± 0.24	0.88 ± 0.23
Al (%)	6.3 ± 0.7	7.1 ± 0.7	7.7 ± 1.5
K (%)	2.1 ± 0.4	2.2 ± 0.3	2.4 ± 0.3
Ca (%)	8.4 ± 2.3	5.9 ± 2.1	8.6 ± 4.4
Ti (%)	0.45 ± 0.08	0.65 ± 0.17	0.49 ± 0.10
V (ppm)	120 ± 20	150 ± 30	140 ± 30
Mn (ppm)	680 ± 150	1100 ± 170	890 ± 150
Sm (ppm)	4.3 ± 0.8	5.2 ± 0.4	4.9 ± 1.1
Eu (ppm)	1.2 ± 0.2	1.2 ± 0.1	1.3 ± 0.2
Dy (ppm)	3.6 ± 0.5	4.6 ± 0.4	4.3 ± 0.5

NOTE: All values are given as mean ± standard deviation.

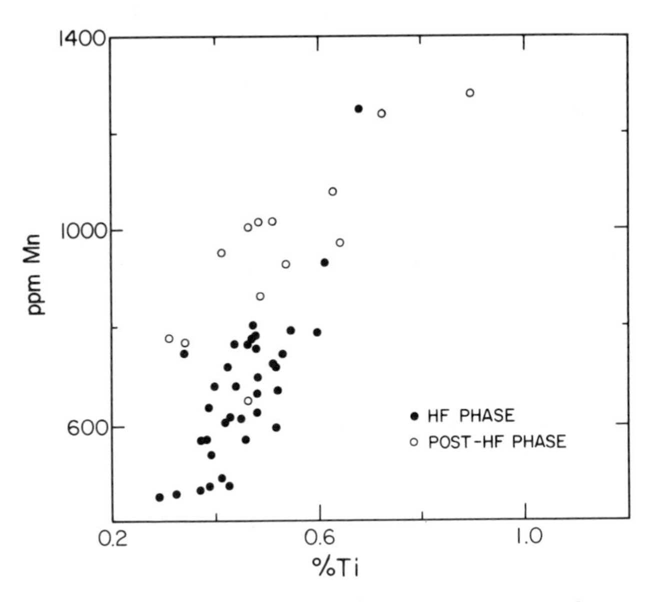

Figure 4. Mn–Ti scattergram for sherds from Hajji Firuz indicating probable geochemical differences between archaeological periods.

sherds exhibit significantly lower average levels of both Al and Mn than was found in the sherds from Dalma Tepe or Pisdeli Tepe. With the removal of the 11 later phase sherds, the Hajji Firuz-phase Al and Mn group mean concentrations decrease in magnitude and become more precise relative to the data displayed in Table I for all sherds from Hajji Firuz; this decrease sets them further apart chemically from the sherds of the other two Solduz Valley sites.

The composition profiles for the two phases of sherds found at Pisdeli Tepe are displayed in Table V and Figure 5, and do not indicate temporal differences.

Comparison of Sherds from Tell 'Ubaid, Tepe Gawra, and the Solduz Valley. Ten sherds from Tell 'Ubaid and 10 sherds from Tepe Gawra were analyzed in the same way as those from the Solduz Valley. The data for the sherds from these diverse regions are presented as group means and standard deviations in Table VI. Large differences are not apparent in the geochemistry of the sherds from the three regions, with the following exceptions: Relative to the Solduz Valley sherds, the Tell 'Ubaid sherds are low in K, and the Tepe Gawra sherds are low in both Na and in Mn. In fact, the Tepe Gawra sherds have a composition profile that is close to that of the Hajji Firuz phase of the Hajji Firuz sherds as displayed in Table IV, with the exception of the low level of Na found in sherds from Tepe Gawra. The Tell 'Ubaid and Tepe Gawra sherds are also more calcareous than the majority of the Solduz sherds, but, because only a small sampling of Tell 'Ubaid and Tepe Gawra sherds has been made to date, this observation may or may not be significant.

Anomalous Sherds from Dalma Tepe. A preliminary investigation has been made to determine whether the INAA data for Dalma Tepe was somewhat biased by the possible presence of imports from Tepe Gawra, which, throughout the period from ca. 5100 to 2200 B.C., was one of the primary trading settlements in northern Mesopotamia. Eight sherds from the Dalma Tepe corpus were identified typologically as anomalous. They did not fit into the Dalma phase typological mainstream, but were strongly affiliated in painting style and fabric to wares from Tepe Gawra. Data for these anomalous sherds are presented in Table VII.

Table V. Test for Temporal Differences in Sherd Compositions at Pisdeli Tepe

Element	*Pisdeli Phase 31 samples*	*Dalma Phase 14 samples*
Na (%)	1.12 ± 0.24	1.00 ± 0.20
Al (%)	7.7 ± 0.9	7.6 ± 1.1
K (%)	2.3 ± 0.4	2.2 ± 0.3
Ca (%)	6.2 ± 2.7	5.8 ± 2.7
Ti (%)	0.51 ± 0.09	0.57 ± 0.13
V (ppm)	140 ± 30	160 ± 30
Mn (ppm)	960 ± 270	1040 ± 300
Sm (ppm)	5.0 ± 0.8	5.1 ± 0.8
Eu (ppm)	1.4 ± 0.2	1.4 ± 0.2
Dy (ppm)	4.4 ± 0.7	4.2 ± 0.9

NOTE: All values are given as mean ± standard deviation.

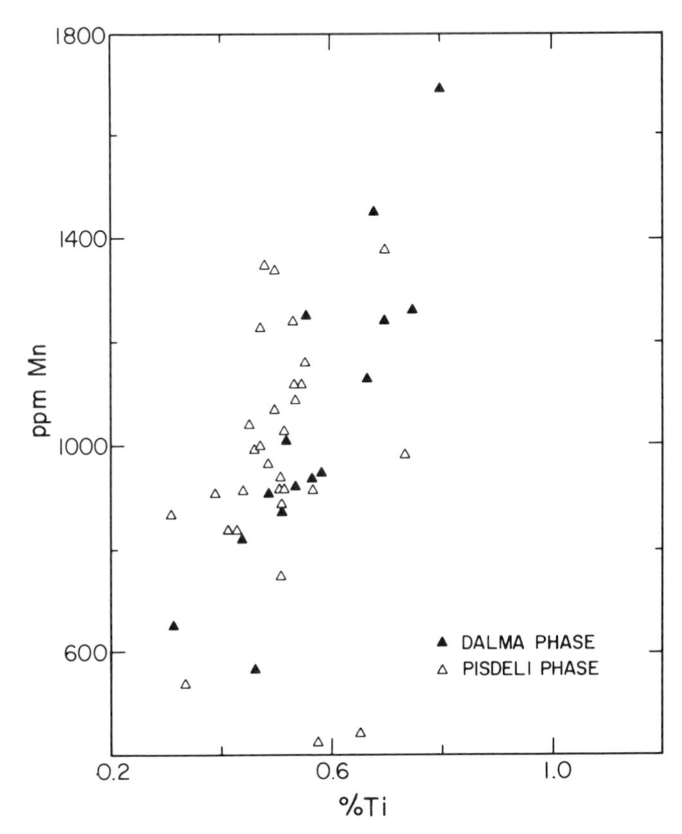

Figure 5. Mn–Ti scattergram for sherds from Pisdeli Tepe showing similarities between sherds of two periods.

Table VI. Analytical Results for Sherds from the Solduz Valley

Element	*Tell 'Ubaid* *10 samples*	*Tepe Gawra* *10 samples*	*Solduz Valley* *187 samples*
Na (%)	1.10 ± 0.16	0.65 ± 0.11	1.06 ± 0.24
Al (%)	6.9 ± 0.4	6.6 ± 0.6	7.7 ± 1.1[a]
K (%)	1.3 ± 0.3	2.0 ± 0.4	2.2 ± 0.3
Ca (%)	10.2 ± 2.3	11.3 ± 1.3	7.6 ± 3.0
Ti (%)	0.47 ± 0.05	0.46 ± 0.04	0.51 ± 0.12
V (ppm)	140 ± 20	140 ± 30	140 ± 30
Mn (ppm)	1100 ± 100	630 ± 110	920 ± 250[a]
Sm (ppm)	4.3 ± 0.5	4.1 ± 0.6	4.7 ± 0.8
Eu (ppm)	1.1 ± 0.2	1.2 ± 0.1	1.3 ± 0.2
Dy (ppm)	3.6 ± 0.4	3.9 ± 0.2	4.1 ± 0.8

NOTE: All values are given as mean ± standard deviation.
[a]Includes the Al and Mn depleted Hajji Firuz phase sherds.

Table VII. Anomalous Wares Found at Dalma Tepe

Element	*M52*	*M53*	*M55*	*M57*	*M69*	*R142*	*R143*	*R144*
Na (%)	0.63	0.69	0.94	0.85	1.19	0.80	1.06	0.84
Al (%)	5.8	6.9	7.2	7.0	7.7	5.9	7.4	8.0
K (%)	1.7	1.9	2.0	1.7	2.2	1.9	1.7	2.2
Ca (%)	10.9	9.7	11.1	12.9	4.2	12.4	10.7	9.8
Ti (%)	0.38	0.34	0.45	0.47	0.59	0.37	0.48	0.54
V (ppm)	120	140	130	120	150	100	130	170
Mn (ppm)	980	960	940	980	920	640	970	1100
Sm (ppm)	3.8	4.2	5.0	3.9	6.0	3.0	4.4	4.5
Eu (ppm)	1.0	1.1	1.2	1.0	1.5	1.1	1.2	1.2
Dy (ppm)	3.4	3.9	4.1	3.5	4.9	3.1	4.3	4.7

With the exception of one sample (No. M69), all atypical sherds have the high Ca content typical of the 10 Tepe Gawra sherds analyzed to date. If the 10 Tepe Gawra sherds represent the total corpus of Tepe Gawra ceramics over time, then No. M69 is probably a Dalma Tepe copy of the northern 'Ubaid ceramics manufactured at Tepe Gawra. Only one sherd (No. R142), with a low Al and Mn content and a high Ca content, has a significant chance of having been made in Tepe Gawra.

When considering the remaining six sherds, it is tempting to ignore the K values and claim that Nos. M52, M53, M57, and R143 could have originated from Tell 'Ubaid. If, however, the K values are not ignored, then it is probable that, with the exception of No. R142, all the anomalous sherds are chemically compatible with the mainstream sherds from Dalma Tepe.

Conclusions

- Of the ceramics analyzed from the Solduz Valley in northwestern Iran, only those manufactured during the Hajji Firuz phase at Hajji Firuz appear to be chemically distinguishable from ceramics made at later times at Hajji Firuz, Dalma Tepe, and Pisdeli Tepe.
- Small samplings of sherds from Tepe Gawra and Tell 'Ubaid indicate small, but probably significant, chemical differences between the clay sources at each site and between them and the ceramics of the Solduz Valley.
- The main intrasite source of elemental compositional variation is either a natural or man-made Ca-rich phase dilution of the parent clay, and this effect must be addressed before intersite chemical variations may be assessed.

Acknowledgments

The authors acknowledge the generous cooperation of M. Voigt and M. de Schauensee in providing the material analyzed in this study. The analytical work was made possible by an infrastructure grant to the SLOWPOKE Reactor Facility of the University of Toronto from the Natural Sciences and Engineering Research Council of Canada.

References

1. Voigt, M. M. *Relative and Absolute Chronologies for Iran Between 6500 and 3500 BC, in Chronologies in the Near East*; Aurenche, O.; Evin, J.; Hours, F., Eds; B.A.R. International Series 379; B.A.R. Oxford, England, 1987; pp 615–646.
2. Voigt, M. M. *Hajji Firuz Tepe, Iran: The Neolithic Settlement*; University Museum Monograph No. 50; The University Museum of the University of Philadelphia: Philadelphia, 1983.
3. Hamlin, C. *Iran*, **1975,** *13,* 111–127.
4. Hancock, R. G. V. *Archaeometry* **1984,** *26,* 210–217.
5. Hancock, R. G. V.; Millett, N. B.; Mills, A. J. *J. Archaeol. Sci.* **1986,** *13,* 107–117.
6. Henrickson, E. F. *Iran* **1985,** *XXIII,* 63–108.

RECEIVED for review June 11, 1987. ACCEPTED revised manuscript March 14, 1988.

7

Trace Element Analysis of Pueblo II Kayenta Anasazi Sherds

Richard D. Foust, Jr., J. Richard Ambler, and Larry D. Turner

Northern Arizona University, Ralph M. Bilby Research Center, Flagstaff, AZ 86011

Thirty-two sherds representing five different examples of Kayenta Anasazi Pueblo II pottery (Tusayan Corrugated [TC], Medicine Black-on-Red [MB], Tusayan Black-on-Red [TB], Dogoszhi Black-on-White [DB], and Sosi Black-on-White [SB]) have been analyzed for the elements As, Ba, Co, Cr, Cu, Fe, Mn, Ni, Pb, Se, V, and Zn by using the techniques of flame atomic absorption spectroscopy (FAA) and electrothermal atomic absorption spectroscopy (ETAA). Analytical procedures for the chemical analysis of ceramics afford accuracy and sensitivity and require only a modest capital investment for instrumentation. The sherd samples were collected at two sites, one in southern Utah (Navajo Mountain [NM]) and the second in northern Arizona (Klethla Valley [KV]). These sites are approximately 60 km apart. Statistical treatment of the data shows that only three clay types were used in the 32 sherds analyzed, and that only three elements (Fe, Pb, and Ni) are necessary to account for 100% of the dispersion observed within this sample set.

MANY INSIGHTS INTO EARLY CIVILIZATIONS have been provided by the study of ceramics. Various physical and chemical methods are considered standard techniques for modern archaeology (*1*, *2*). The macroanalysis, microanalysis, and trace chemical analysis of artifacts have added greatly to the understanding of prehistoric civilizations (*3*–*6*). The chemical techniques used for ceramic analysis have included spark source mass spectrometry (*7*);

0065–2393/89/0220–0125$06.00/0

X-ray fluorescence emission (*8, 9*); atomic absorption (*10–12*), Mössbauer (*13*), and scanning Auger (*14*) spectroscopic methods; and neutron activation analysis (NAA) (*15–17*). In addition, several exotic physical methods such as X-ray xeroradiography (*18, 19*) and photoacoustic analysis (*20*) have been applied to the study of ceramics. Of the analytical methods listed, neutron activation has been the most frequently used technique.

Although nondestructive analytical techniques are preferred for archaeometry, these methods are not without their faults. Surface analytical methods are susceptible to the glazing and firing practices of individual potters and do not provide a representative analysis of the entire sample. Neutron activation analysis can be nondestructive, but standard practice for ceramics involves taking a 50–100-mg sample for analysis from an inconspicuous place. Atomic absorption and plasma emission spectroscopy require larger sample sizes than NAA and require a lengthy digestion process. The advantage of atomic absorption spectrometry is that the analytical techniques are easily learned by a chemical technician and the capital investment for instrumentation (compared to the cost of a nuclear reactor) is small. Meschel (*21*) has given a complete discussion of the analytical methods used in archaeological chemistry and provides a comparison of the advantages and disadvantages of each method.

Traditional flame atomization methods have been preferred for atomic absorption analysis of ceramics. Very little mention has been made of electrothermal atomization (ETAA) in the literature. Although ETAA offers increased sensitivity with lower detection limits and smaller sample sizes, the problems of matrix interference (*22–25*) have resulted in the development of sample-specific methods for ETAA.

Each element has unique chemical and physical properties that require individual determination of the optimum times and temperatures for drying, charring, and atomizing in the graphite furnace (*26*). In addition, the use of "matrix modifiers" has been found to improve the analytical quality of the ETAA analysis for some elements (*27*). Modern atomic absorption instruments are capable of reading both the peak shape and peak area of the vaporized element, and it is possible to improve the analysis by selecting the proper signal reading mode (*28*). Improved vaporization characteristics are often obtained with a stabilized temperature platform (STP or L'vov platform, named after its inventor) used in the graphite tube (*29, 30*), and a background corrector (either deuterium lamp or Zeeman effect) greatly improves ETAA accuracy (*31, 32*).

Experimental determination of the optimum values for these instrumental and procedural variables makes ETAA a tedious and unattractive method when compared with faster multielement procedures. However, once the procedures and optimum conditions have been determined, samples can be analyzed in about 10 min per element per sample. The procedures and instrumental conditions reported in this chapter permit the analysis of ceramic samples accurately, economically, and efficiently by similarly equipped laboratories.

Several early studies of southwestern pottery have been reported (*33, 34*), but Shepard (*1*) is credited with the first systematic approach to ceramic analysis. By using the physical properties of pottery (color, hardness, texture, luster, porosity, and strength) combined with an analysis of manufacturing methods and decorative techniques, Shepard was able to make statements about the distribution and exchange of ceramics in the Rio Grand drainage area of New Mexico. Deutchmann (*35, 36*) used neutron activation in a study of 200 sherds to measure the concentrations of 22 chemical elements in an attempt to define locations of ceramic manufacture and regions of exchange for two types of Kayenta Anasazi Pueblo II ceramics (Sosi Black-on-White and Dogoszhi Black-on-White). Deutchmann analyzed sherds from 10 sites located on or near Black Mesa in northeastern Arizona. Through a sophisticated statistical analysis of the data, some remarkable compositional patterns were revealed that were taken as evidence for exchange of ceramics between the sites studied.

Geib and Callahan (*37*) have conducted the most recent study of ceramic exchange in the Kayenta Anasazi region. They concluded that the Klethla Valley region of northern Arizona was the initial production site of ash-tempered Tusayan White Ware. This conclusion was based on an analysis, by traditional methods, of ceramics from approximately 300 sites. Geib and Callahan ended their paper with a call to more definitively establish similarities and differences among the ceramics of the Kayenta Anasazi region through chemical means.

The purpose of our project was twofold. First, developments in atomic absorption spectrophotometry (electrothermal atomization and background correction techniques) have made this method of analysis more applicable to archaeological samples by requiring significantly smaller amounts of material (about 250 mg) and by lowering the detection limits for many elements to the parts-per-billion range. Development of the laboratory procedures and instrumental conditions necessary to analyze archaeological samples would, therefore, be of benefit to many archaeological laboratories. This point is particularly significant because of atomic absorption's large economic advantage over neutron activation techniques.

The second goal of this project was to learn something about the manufacture and exchange of pottery in this region through the application of flame atomic absorption spectroscopy (FAA) and ETAA to the study of ceramics found at two sites in southern Utah and northern Arizona.

Experimental Details

Chemical Methods. Sherds were prepared for analysis by removing surface glazing and designs with a fine-grit silicon carbide abrasive paper.

A 250-mg sample was taken for analysis, weighed to 0.1 mg, and digested by the following procedure.

The sherd sample was placed into a tared, acid-washed polyethylene bottle (125-mL). Aqua regia solution (0.50 mL) was added to the sample, followed by the addition of 5.0 mL of hydrofluoric acid (48%). The aqua regia solution (stored in an acid-washed polyethylene bottle) was prepared by combining 12.5 mL of HNO_3 (70%) with 37.5 mL of HCl (37%). The polyethylene bottle containing the sample, aqua regia, and HF was capped and shaken on a horizontal shaker for 12–17 h. The sample bottle was then removed from the shaker and placed in a water bath at 90 °C for 1 h. The White precipitate (metal fluorides) observed at this stage of the digestion was dissolved upon the addition of 75.0 mL of a saturated boric acid solution. The final sample volume was adjusted to a mass of 100.0 g by the addition of deionized, double-distilled water. Aliquots of this solution were used without further treatment for both ETAA and FAA analysis.

All chemicals and reagents used in this work were ACS reagent grade, all volumetric glassware was class-A, and all beakers, bottles, and volumetric glassware were washed with aqua regia and rinsed with deionized, double-distilled water before use.

Instrumental Conditions. FAA analyses for Ba, Fe, and Mn were done on a Perkin Elmer model 560 atomic absorption spectrophotometer with an air–acetylene flame for Fe and Mn and a nitrous oxide–acetylene flame for Ba (*38*, *39*). The remaining elements were determined by ETAA with the conditions listed in the appendix (Table A-1) with a Perkin Elmer model 5000 atomic absorption spectrophotometer equipped with a model HGA-400 graphite furnace and a deuterium background corrector. The precision of FAA was ±1.0% and the precision of ETAA was ±10%. Absorbance readings for FAA were converted directly to concentrations by comparison with absorbance readings for standards that had been previously stored in the instrument's computer.

Absorbance readings for ETAA were converted to concentrations by applying a modified method of standard additions (*39*). Two solutions of the element of interest were prepared (along with a blank) within the concentration range expected for the samples. These solutions were used to prepare a standard curve that was used to calculate concentrations from the absorbance data of unknown solutions. The percent recovery for the unknown was verified by spiking the sample with a known quantity of the element of interest, and repeating the ETAA analysis. The readings were taken as valid if the spiked sample produced the expected absorbance increase. Accuracy of the analytical procedures was verified by analysis of National Bureau of Standards Reference Material No. 1645, river sediments. The results of this comparison are shown in Table I.

Table I. Comparison of Analytical Methods against NBS Standard Reference Material, No. 1645, River Sediments

Element	*NBS*[a]	*BRC*[b]
Arsenic	66	68.8
Barium	—	182
Chromium	2.96 ± 0.28%	3.05%
Cobalt	8.0	8.0
Copper	109 ± 19	92.6
Iron	11.3 ± 1.2%	9.2%
Lead	714 ± 28	715
Manganese	785 ± 97	763
Nickel	45.8 ± 2.9	43.1
Vanadium	23.5 ± 6.9	23.0
Zinc	1720 ± 169	1670

NOTE: All concentrations are reported as micrograms per gram (ppm) except chromium and iron, which are reported as weight percent.
[a]National Bureau of Standards.
[b]Bilby Research Center.

Experimental Design. To simplify the interpretation of results, sherd samples were selected for analysis from the Pueblo II era (ca. 900–1150 A.D.). Samples from only two sites that were known to have been occupied by the Kayenta Anasazi were examined (*40*). Figure 1 shows the area of the Kayenta Anasazi and identifies the two sites (Navajo Mountain and Klethla Valley) where sherds were collected.

Three common wares of Pueblo II ceramics are found at the two sites selected for this study (Tusayan Gray Ware, Tsegi Orange Ware, and Tusayan White Ware), and they are easily distinguished by their physical appearance. Tsegi Orange Ware has a red or orange color and dramatic black designs painted on its surface. In contrast, the Tusayan Gray Ware and Tusayan White Ware are light gray. Tusayan Gray Ware has a textured, woven appearance with no painted designs. Tusayan White Ware has a smooth surface that does contain black painted designs. As shown in Figure 2, Tsegi Orange Ware and Tusayan White Ware are classified further by their painted surface designs. All three ceramic wares were found at excavations in the Kayenta Anasazi region, but the percentages of each changed with location. These differences in distribution have been the basis for many claims of ceramic exchange within the Kayenta Anasazi region.

Results and Discussion

The results of the chemical analysis of the sherds are listed in Table II (Klethla Valley samples) and Table III (Navajo Mountain samples). Each number shown in Tables II and III is the average value of at least three separate

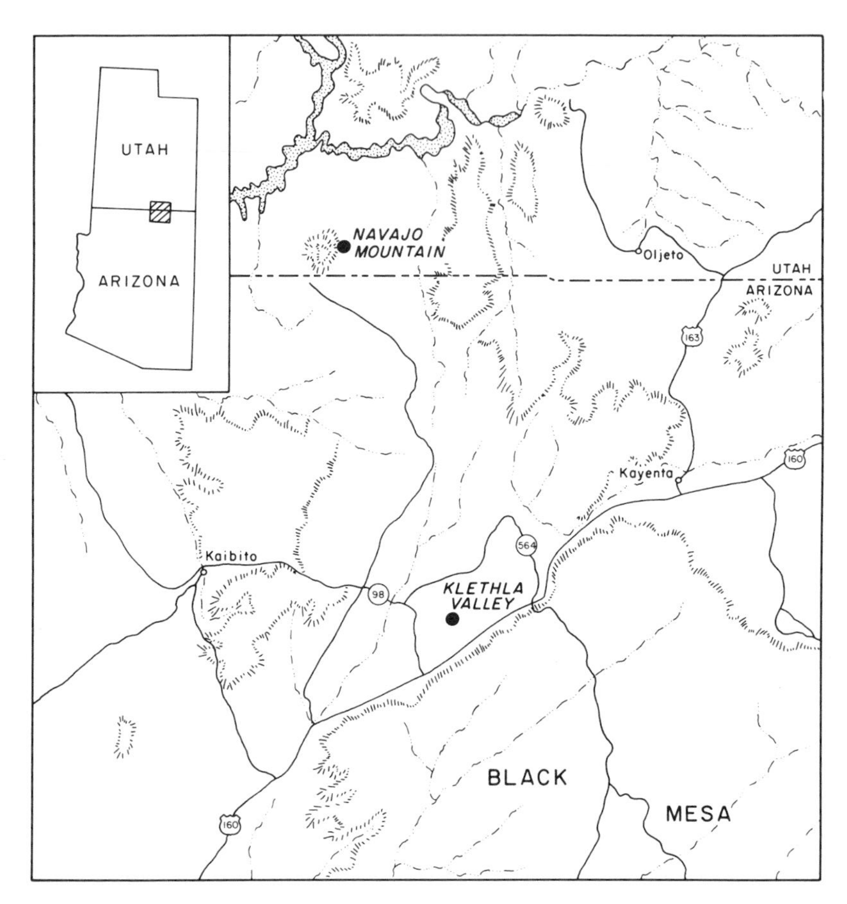

Figure 1. Map of the Kayenta Anasazi region of northern Arizona and southern Utah showing locations of archaeological sites.

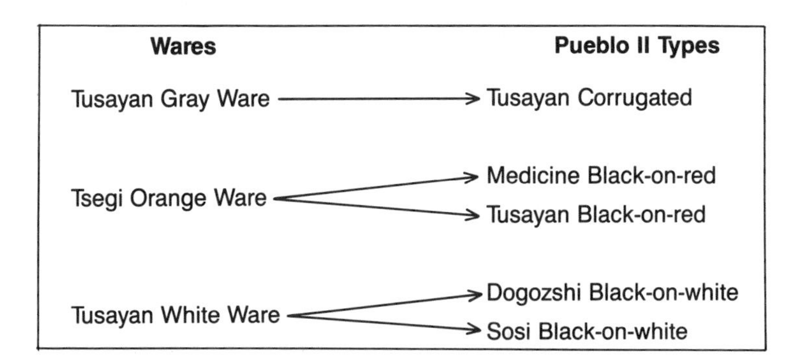

Figure 2. Classification of Pueblo II Kayenta Anasazi pottery.

Table II. Elemental Concentrations (ppm) of Pueblo II Kayenta Anasazi Sherds from Klethla Valley

Sherd Description	*As*	*Ba*	*Co*	*Cr*	*Cu*	*Fe*	*Mn*	*Ni*	*Pb*	*Se*	*V*	*Zn*
Tusayan Corrugated	96.9	5350	5.0	65.6	44.4	12,500	43	24.8	35.5	0.290	154	37.6
Tusayan Corrugated	6.7	2250	5.5	55.9	74.7	15,400	74	13.1	38.0	0.003	158	44.1
Tusayan Corrugated	11.8	1250	2.0	80.5	25.7	24,100	106	11.1	39.3	0.020	126	12.3
Medicine Black-on-Red	67.3	1690	4.5	71.9	19.8	21,900	242	29.5	28.4	0.003	96.0	53.5
Medicine Black-on-Red	6.8	1590	6.9	67.0	20.9	19,800	299	25.4	29.9	0.075	168	43.8
Tusayan Black-on-Red	5.0	1580	6.2	46.6	21.9	28,000	918	28.8	30.5	0.010	114	69.2
Tusayan Black-on-Red	7.8	1700	6.3	62.8	18.4	26,500	296	28.3	26.0	0.085	109	58.1
Dogozshi Black-on-White	6.6	1670	4.1	43.3	12.7	13,100	83	18.5	29.8	0.065	86	28.2
Dogozshi Black-on-White	106	1580	2.8	50.8	13.3	7,500	99	10.2	24.8	0.003	80	23.0
Dogozshi Black-on-White	6.1	2430	5.0	43.5	37.0	18,800	144	13.1	30.6	0.150	123	25.9
Dogozshi Black-on-White	17.5	2250	10	52.2	40.6	13,700	160	35.7	35.4	0.080	92	66.8
Sosi Black-on-White	20.7	2070	3.1	43.0	35.2	13,500	126	20.0	31.3	0.045	136	53.9
Sosi Black-on-White	31.2	2400	4.3	40.8	25.5	15,100	189	14.5	29.1	0.003	80	34.9
Sosi Black-on-White	119	2560	1.9	55.7	48.6	14,800	101	18.1	29.1	0.003	197	30.6
Sosi Black-on-White	0.2	2170	9.1	52.3	24.3	15,900	76	24.4	38.2	0.020	148	64.3
Sosi Black-on-White	4.8	2040	2.4	44.8	20.4	14,300	85	8.6	32.2	0.030	79	35.0

Table III. Elemental Concentrations (ppm) of Pueblo II Kayenta Anasazi Sherds from Navajo Mountain

Sherd Description	*As*	*Ba*	*Co*	*Cr*	*Cu*	*Fe*	*Mn*	*Ni*	*Pb*	*Se*	*V*	*Zn*
Tusayan Corrugated	39.8	2320	2.6	40.6	26.4	20,100	100	9.9	37.4	0.003	64	37.5
Tusayan Corrugated	2.4	2450	3.7	59.5	67.4	15,700	89	14.7	35.3	0.170	126	22.2
Medicine Black-on-Red	12.8	1810	6.1	59.7	20.0	23,800	298	27.5	27.6	0.070	119	54.0
Medicine Black-on-Red	23.9	1580	5.6	70.5	21.3	24,200	290	26.2	23.3	0.003	132	52.4
Tusayan Black-on-Red	24.2	1910	7.6	71.6	16.3	23,400	250	32.1	25.9	0.290	110	63.8
Tusayan Black-on-Red	3.4	2090	6.1	60.8	17.8	24,400	283	26.5	24.0	0.003	151	72.0
Dogozshi Black-on-White	4.3	2420	3.1	39.8	23.3	13,600	83.0	10.9	32.4	0.025	122	44.9
Dogozshi Black-on-White	3.2	3080	3.3	41.4	31.8	13,500	140	18.8	32.2	0.080	82	24.0
Dogozshi Black-on-White	2.6	2870	3.3	42.0	22.5	12,100	82.0	15.3	30.1	0.003	124	38.2
Dogozshi Black-on-White	26.3	2660	2.0	72.6	27.0	14,600	102	11.9	34.9	0.003	122	21.2
Sosi Black-on-White	15.9	1380	4.7	55.5	20.6	21,400	84.0	11.5	32.5	0.003	110	30.7
Sosi Black-on-White	1.9	1070	4.2	53.5	27.4	12,600	52.0	23.9	25.9	0.095	109	25.3
Sosi Black-on-White	65.1	1130	5.2	62.0	30.0	12,900	157	16.4	28.4	0.003	131	24.2
Sosi Black-on-White	4.1	1600	3.5	46.1	42.2	16,500	129	14.2	32.4	0.003	343	27.1
Sosi Black-on-White	0.6	2600	3.1	41.4	15.8	16,200	135	12.7	23.1	0.003	69	60.8
Sosi Black-on-White	130	1890	5.6	54.1	17.7	11,700	211	16.4	21.5	0.015	71	16.0

analyses. Three runs were performed on each sherd sample for each element. When an agreement of ±10% was achieved between runs, the mean value was taken as the elemental concentration for that sherd. The 10% agreement requirement was usually satisfied with three runs, but occasionally five runs were required. Each number shown in Tables II and III has an uncertainty of ±10%.

The chemical concentration data from the 32 sherds (Tables II and III) were then analyzed by a series of two-way Hotelling T^2 tests (*41*). The Hotelling test was chosen because of its applicability to multivariate analysis (*42, 43*) and was used to test for statistically significant differences between locations and between ware types. A statistically significant difference is identified when the calculated T^2 value exceeds T^2 for the specified level of confidence (95% in this case). The results of the Hotelling tests are shown in Table IV.

The results of the Hotelling T^2 test show that there is no significant chemical difference between the Tusayan Corrugated, Medicine Black on Red, or Tusayan Black-on-Red sherds, even though the comparison is between samples collected at sites that are 60 km apart. By contrast, the Dogoszhi Black-on-White and the Sosi Black-on-White sherds are significantly different when compared between sites. Furthermore, comparison of the pooled Medicine Black-on-Red sherds to the pooled Tusayan Black-on-Red sherds gives a T^2 value well in excess of the critical value. This result indicates that these two types are chemically different.

The chemical concentrations from Tables II and III were then combined to give mean element concentrations for each grouping as identified from the Hotelling T^2 test (Tusayan Corrugated [TC], Medicine Black-on-Red [MB], Tusayan Black-on-Red [TB], Navajo Mountain Dogoszhi Black-on-White [NMDB], Klethla Valley Dogoszhi Black-on-White [KVDB], Navajo Mountain Sosi Black on White [NMSB], and Klethla Valley Sosi Black-on-White [KVSB]). These data are listed in Table V along with the experimental 95% confidence limits shown for each mean concentration.

Table IV. Results of the Two-Way Hotelling T^2 Test When Applied to the Original Atomic Absorption Data for Element Concentrations

Interaction Tested	*Degrees of Freedom*	T^2_{calc}	$T^2_{0.95}$
NMTC vs. KVTC	12	20.9	39.8
NMMB vs. KVMB	9	2.5	83.2
NMTB vs. KVTB	9	58.6	83.2
NMDB vs. KVDB	21	12,005.0	33.1
NMSB vs. KVSB	30	40.2	17.9
MB vs. TB	21	861.4	33.1

Abbreviations used: Dogozshi Black-on-White (DB), Klethla Valley (KV), Medicine Black-on-Red (MB), Navajo Mountain (NM), Sosi Black-on-White (SB), Tusayan Black-on-Red (TB), and Tusayan Corrugated (TC).

A one-way analysis of variance (ANOVA) (*44, 45*) was then performed on the data listed in Table V for each element. The ANOVA was followed with a Student–Newman–Keuls test (*46*) to determine the number of subgroupings that resulted from differences in the mean metal concentrations. Six elements (Cu, Fe, Mn, Ni, Pb, and Zn) were shown to be at significantly different concentrations when compared between the seven groupings.

The original data for these six elements were then subjected to a discriminate function analysis with the 7M program of BMDP Statistical Software, Inc. (*47*). Discriminate analysis is a technique used to classify individual objects into one of two or more alternative groups that are known to be distinct (*48, 49*). An advantage of discriminate function analysis is that the variables that contribute to making the classification are identified. By identification of those variables (element concentrations) that contribute the most to the discriminate function, it is possible to reduce the number of analyses that are required for future sherd classifications. The discriminate function analysis resulted in only three elements (Fe, Pb, and Ni) being required to account for 100% of the dispersion between the wares and types within the seven groups. Even more surprising is the fact that two elements (Fe and Pb) account for 99.1% of the dispersion.

The discriminate function analysis also yields classification functions for each variable (Fe, Pb, or Ni) within each group (TC, MB, TB, NMDB, KVDB, NMSB, and KVSB) and a constant for each group. Once known, the classification functions can be used to classify each of the original sherds into one of the seven possible groups. The classification matrix, obtained by treating data from the 32 original sherds with the classification functions, is given as Table VI.

The predicted classifications are correct 100% of the time at the ware level for Tsegi Orange Ware, but at the type level, one Medicine Black-on-Red sherd is incorrectly classified as Tusayan Black-on-Red. One Tusayan Corrugated sherd is incorrectly classified as being a Klethla Valley Sosi Black-on-White sherd, but the four remaining Tusayan Corrugated sherds are correctly classified for a success rate of 80%. No sherds from other types were incorrectly classified as being Tusayan Corrugated. The ability to predict between the four remaining groups drops dramatically, with 33% correct for Navajo Mountain Sosi Black-on-White, and 25% correct each for Navajo Mountain and Klethla Valley Dogoszhi Black-on-White sherds. There were no correct predictions for the Klethla Valley Sosi Black-on-White samples. The failure to correctly classify Tusayan White Ware sherds into one of the four possible Tusayan White Ware groups means it is not possible to chemically distinguish between these groups by this technique.

Canonical correlation analysis is a technique that can be used to test whether two or more groups, thought to be different, are indeed distinct (*50, 51*). This analysis yields a set of variables that are combined in linear

Table V. Mean Elemental Concentrations (ppm) of Pueblo II Kayenta Anasazi Sherds for the Wares and Types Used for Discriminate Analysis and Canonical Correlations

Sherd Description	*Tusayan Corrugated*	*Medicine Black-on-Red*	*Tusayan Black-on-Red*	*Navajo Mountain Dogozshi Black-on-White*	*Klethla Valley Dogozshi Black-on-White*	*Navajo Mountain Sosi Black-on-White*	*Klethla Valley Sosi Black-on-White*
n	5	4	4	4	4	6	5
As	31.5 ± 48.8	27.7 ± 43.5	10.1 ± 15.2	9.1 ± 18.3	34 ± 76.8	36.3 ± 54.6	35.2 ± 60.2
Ba	2724 ± 1916	1668 ± 170	1820 ± 359	2758 ± 450	1982 ± 670	1612 ± 600	2248 ± 278
Co	3.76 ± 1.86	5.78 ± 1.59	6.55 ± 1.12	2.92 ± 0.99	5.48 ± 5.01	4.38 ± 1.02	4.16 ± 3.60
Cr	60.4 ± 18	67.3 ± 8.68	60.4 ± 16.5	49 ± 25.1	47.4 ± 7.50	52.1 ± 7.66	47.3 ± 7.91
Cu	47.7 ± 28.2	20.5 ± 1.14	18.6 ± 3.79	26.1 ± 6.77	25.9 ± 23.9	25.6 ± 10.3	30.8 ± 14.1
Fe	17,560 ± 5651	22,425 ± 3217	25,575 ± 3300	13,450 ± 1639	13,275 ± 7370	15,217 ± 3798	14,720 ± 1111
Mn	85.4 ± 40.1	282 ± 43.3	437 ± 512	102 ± 43.2	122 ± 58	128 ± 58.4	115 ± 56.2
Ni	14.7 ± 7.36	27.2 ± 2.85	28.9 ± 3.72	14.2 ± 5.72	19.4 ± 18.2	15.8 ± 4.62	17.1 ± 7.39
Pb	37.1 ± 2.11	27.3 ± 4.52	26.6 ± 4.40	32.4 ± 3.13	30.1 ± 6.92	27.3 ± 4.87	32 ± 4.63
Se	0.097 ± 0.160	0.038 ± 0.064	0.097 ± 0.213	0.028 ± 0.0578	0.074 ± 0.096	0.020 ± 0.059	0.020 ± 0.029
V	126 ± 46.6	129 ± 48	121 ± 32.1	112 ± 32.4	95.2 ± 30.5	139 ± 107	128 ± 61.8
Zn	30.7 ± 16.2	50.9 ± 7.65	65.8 ± 9.80	32.1 ± 18.1	36 ± 32.9	30.7 ± 16.3	43.7 ± 18.1

NOTE: The numbers reported are ± 95% confidence limits.

Table VI. Classification Matrix Obtained from the Original Sherd Data as Predicted from the Classifications Functions

Group	*Percent Correct*	*TC*	*MB*	*TB*	*NMDB*	*KVDB*	*NMSB*	*KVSB*
TC	80.0	4	0	0	0	0	0	1
MB	75.0	0	3	1	0	0	0	0
TB	100.0	0	0	4	0	0	0	0
NMDB	75.0	0	0	0	3	0	0	0
KVDB	50.0	0	0	0	1	2	1	0
NMSB	66.7	0	0	0	0	1	4	1
KVSB	.0	1	0	0	1	1	2	0
Total	62.5	5	3	5	5	4	7	3

NOTE: Data presented as number of cases classified into group.

combinations to yield canonical variables. Canonical variables, once obtained, can be used to examine the relationships between groups (canonical correlations). A canonical correlation analysis was preformed on the original sherd data using only iron, lead, and nickel concentrations as independent variables with the 7M program of BMDP Statistical Software, Inc. (*41*). The most effective way to display canonical correlations is to plot the two canonical variables against each other.

Figure 3 shows the results of the canonical correlation analysis and vividly illustrates the separation of the sherd data into two distinct groups. The open symbols in Figure 3 represent individual sherds, and the filled symbols represent group means. The data clustered on the left half of Figure 3 represent the two types of Tsegi Orange Ware studied (Medicine Black-on-Red and Tusayan Black-on-Red). The data clustered on the right are from the Tusayan Corrugated and Tusayan White Ware samples. Although there is considerable overlap between the Tusayan White Ware and Tusayan Corrugated samples, they are clearly resolved from the Tsegi Orange Ware sherds. Tusayan Corrugated samples appear higher on the plot than the Tusayan White Ware samples, but overlap the White Ware sherds sufficiently to preclude resolution as a separate group by canonical correlation.

The results of this study are consistent with the interpretation that only three clay sources were used to manufacture the three ceramic wares. An alternative explanation, that the same clay source was used to manufacture both Tusayan Gray Ware and Tusayan White Ware (with minor chemical differences being the result of firing and glazing practices), is possible. However, the analysis was done on whole sherds after all surface markings had been removed. Therefore, the small chemical differences observed between Tusayan Gray Ware and Tusayan White Ware were probably not due to surface designs or firing procedures. Finally, evidently the three ceramic wares were exchanged freely over a distance of 60 km.

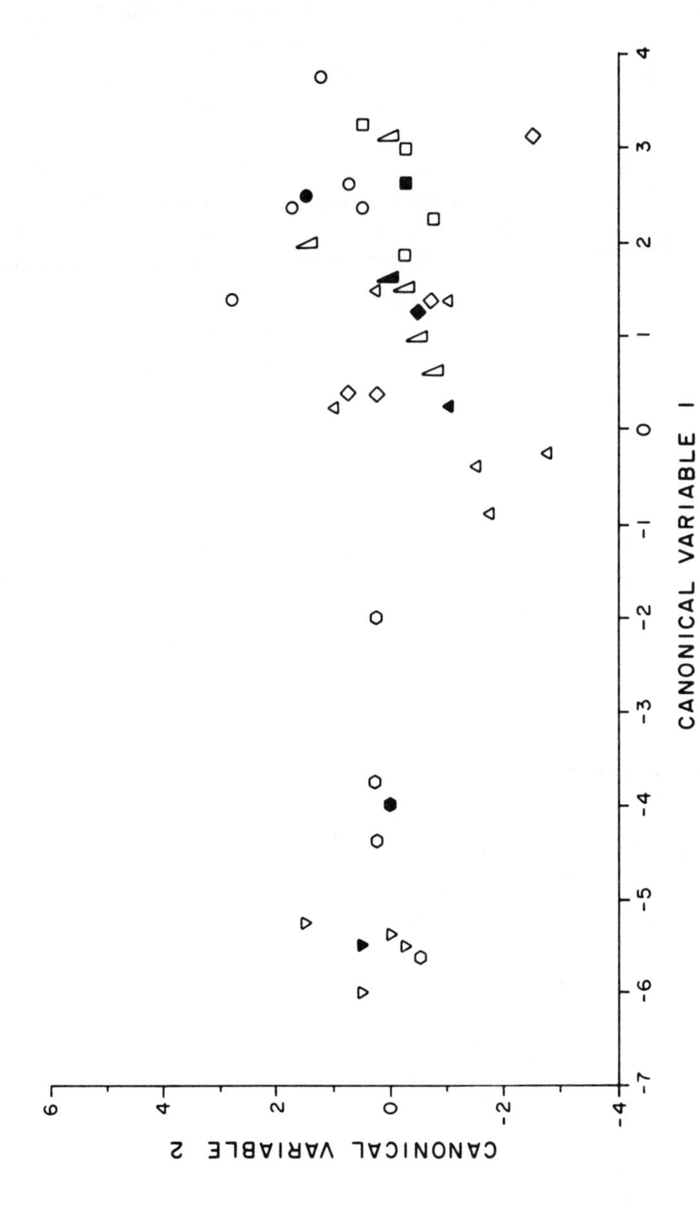

Figure 3. Canonical correlation plot of 32 Pueblo II Kayenta Anasazi sherds. Symbol key: ○, *TC;* ▽, *TB;* ⬡, *MB;* □, *NMDB;* ◇, *KVOB;* ◺, *NMSB;* △, *KVSB (filled symbols represent group means).*

Summary

Methods have been developed for the chemical analysis of prehistoric ceramics by flame atomic absorption and electrothermal atomic absorption spectrophotometry, and these techniques have been applied to the analysis of 32 Kayenta Anasazi Pueblo II sherds. Differences in the chemical compositions of three ceramic wares (Tusayan White Ware, Tusayan Gray Ware, and Tsegi Orange Ware) are clearly evident and have been used to group the sherds into classifications long accepted by archaeologists for these ceramics. In addition, the chemical similarity of sherds located at two sites, separated by a distance of 60 km, demonstrates ceramic exchange among the Pueblo II Kayenta Anasazi. Finally, only three elements (Fe, Pb, and Ni) are needed to account for 100% of the dispersion of the ceramics studied between seven possible groups. For future studies, then, measuring only these three elements affords good discriminating power and greatly reduces the time and cost involved.

Acknowledgments

The authors thank Phil Geib, Research Associate at Northern Arizona University, for many helpful comments throughout this project and for reviewing the final manuscript, and Richard Turek, of the Northern Arizona University mathematics department, for many helpful suggestions for the statistical treatment of the data.

List of Abbreviations

BRC	Bilby Research Center
DB	Dogoszhi Black-on-White
ETAA	electrothermal atomic absorption spectroscopy
FAA	flame atomic absorption spectroscopy
KV	Klethla Valley
MB	Medicine Black-on-Red
NAA	neutron activation analysis
NBS	National Bureau of Standards
NM	Navajo Mountain
ppb	parts per billion
ppm	parts per million
SB	Sosi Black-on-White
STP	stabilized temperature platform
TB	Tusayan Black-on-Red
TC	Tusayan Corrugated

Appendix. Instrumentatal Conditions for Electrothermal Atomization Atomic Absorption Spectroscopy of Sherd Samples

Table A-1. Arsenic

•Wavelength: 193.7 nM
•Background correction: yes
•L'vov platform: yes

•Argon flow rate: 80 mL/min
•Lamp: EDL
•Slit: 0.7nM

Process	*Temperature, °C*	*Ramp Time, s*	*Hold Time, s*
Drying	110	15	15
Charring	300	10	15
Atomization	2300	0	3
Burn-out	2500	1	4

NOTES: Baseline correction: 20 s. Read: peak area, 3.5 s. 10-μL injections, followed with the addition of 50 μL of a 1000-ppm $NiNO_3$ solution pipetted into the graphite tube before the analysis.

Table A-2. Copper

•Wavelength: 324.8 nM
•Background correction: no
•L'vov platform: yes

•Argon flow rate: 200 mL/min
•Lamp: HCL
•Slit: 0.7 nM

Process	*Temperature, °C*	*Ramp Time, s*	*Hold Time, s*
Drying	110	15	15
Charring	500	5	15
Atomization	2600	0	3
Burn-out	2500	1	3

NOTES: Baseline correction: 19 s. Read: peak area, 3.5 s. 10-μL injections.

Table A-3. Chromium

•Wavelength: 357.9 nM
•Background correction: no
•L'vov platform: yes

•Argon flow rate: 200 mL/min
•Lamp: HCL
•Slit: 0.7 nM

Process	*Temperature, °C*	*Ramp Time, s*	*Hold Time, s*
Drying	110	15	15
Charring	250	5	20
Atomization	2500	0	3
Burn-out	2500	1	4

NOTES: Baseline correction: 24 s. Read: peak area, 3.5 s. 10-μL injections.

Table A-4. Cobalt

•Wavelength: 240.7 nm		•Argon flow rate: 50 mL/min	
•Background correction: yes		•Lamp: HCL	
•L'vov platform: yes		•Slit: 0.2 nM	
Process	*Temperature, °C*	*Ramp Time, s*	*Hold Time, s*
Drying	110	15	15
Charring	1400	20	10
Atomization	2500	0	3
Burn-out	2500	1	4

NOTES: Baseline correction: 29 s. Read: peak area, 3.5 s. 10-μL injections.

Table A-5. Nickel

•Wavelength: 232.0 nM		•Argon flow rate: 60 mL/min	
•Background correction: yes		•Lamp: HCL	
•L'vov platform: yes		•Slit: 0.2 nM	
Process	*Temperature, °C*	*Ramp Time, s*	*Hold Time, s*
Drying	110	15	15
Charring	800	10	20
Atomization	2650	0	3
Burn-out	2600	1	3

NOTES: Baseline correction: 28 s. Read: peak area, 3.5 s. 10-μL injections.

Table A-6. Lead

•Wavelength: 283.3 nM		•Argon flow rate: 100 mL/min	
•Background correction: yes		•Lamp: EDL	
•L'vov platform: yes		•Slit: 0.7 nM	
Process	*Temperature, °C*	*Ramp Time, s*	*Hold Time, s*
Drying	110	15	15
Charring	300	10	15
Atomization	1800	0	3
Burn-out	2500	1	4

NOTES: Baseline correction: 24 s. Read: peak height, 3.5 s. 10-μL injections.

Table A-7. Selenium

•Wavelength: 196.0 nM		•Argon flow rate: 50 mL/min	
•Background correction: yes		•Lamp: EDL	
•L'vov platform: yes		•Slit: 0.7 nM	
Process	*Temperature, °C*	*Ramp Time, s*	*Hold Time, s*
Drying	110	10	30
Charring	200	10	40
Atomization	2700	0	3
Burn-out	2600	1	4

NOTES: Baseline correction: 45 s. Read: peak height, 3.5 s. 20-μL injections.

Table A-8. Vanadium

•Wavelength: 318.4 nM
•Background correction: no
•L'vov platform: no
•Argon flow rate: 80 mL/min
•Lamp: HCL
•Slit: 0.7 nM

Process	*Temperature, °C*	*Ramp Time, s*	*Hold Time, s*
Drying	110	15	15
Charring	800	10	20
Atomization	2650	0	3
Burn-out	2800	1	4

NOTES: Baseline correction: 28 s. Read: peak area, 3.5 s. 10-μL injections. It was necessary to run a distilled water blank and re-zero the instrument after every three runs with these conditions.

Table A-9. Zinc

•Wavelength: 213.9 nM
•Background correction: yes
•L'vov platform: yes
•Argon flow rate: 200 mL/min
•Lamp: EDL
•Slit: 0.7 nM

Process	*Temperature, °C*	*Ramp Time, s*	*Hold Time, s*
Drying	110	15	15
Charring	250	5	10
Atomization	2400	1	2
Burn-out	2500	1	5

NOTES: Baseline correction: 14 s. Read: peak area, 3.5 s. 10-μL injections. Sample solution diluted 1:2 with deionized double-distilled water before analysis.

References

1. Shepard, A. O. *Ceramics for the Archaeologist*; Carnegie Institution of Washington: Washington, DC, 1956.
2. *Archaeological Ceramics*; Olin, J. S.; Franklin, A. D., Eds; Smithsonian Institution: Washington, DC, 1982.
3. *Archaeological Chemistry*; Beck, C. W., Ed.; Advances in Chemistry 138; American Chemical Society: Washington, DC, 1974.
4. *Archaeological Chemistry II*; Carter, G. F., Ed.; Advances in Chemistry 171; American Chemical Society: Washington, DC, 1978.
5. *Archaeological Chemistry III*; Lambert, J. B., Ed.; Advances in Chemistry 205; American Chemical Society: Washington, DC, 1984.
6. Goffer, Z. *Archaeological Chemistry: A Sourcebook on the Applications of Chemistry to Archaeology*; Chemical Analysis Series 55; Wiley: New York, 1980.
7. Friedman, A. M.; Lerner, J. In *Archaeological Chemistry II*; Carter, G. F., Ed.; Advances in Chemistry 171; American Chemical Society: Washington, DC, 1978; pp 70–78.
8. Carter, G. F. In *Archaeological Chemistry III*; Lambert, J. B., Ed.; Advances in Chemistry 205; American Chemical Society: Washington, DC, 1984; pp 311–329.
9. Stos-Fertner, Z.; Hedges, R. E. M.; Evely, R. D. G. *Archaeometry* **1979,** *21*, 187.

10. Gritton, V.; Magalousis, N. M. In *Archaeological Chemistry II*; Carter, G. F., Ed.; Advances in Chemistry 171; American Chemical Society: Washington, DC, 1978; pp 258–270.
11. Tennent, N. H.; McKenna, P.; Lo, K. K. N.; McLean, G.; Ottoway, J. M. In *Archaeological Chemistry III*; Lambert, J. B., Ed.; Advances in Chemistry 205; American Chemical Society: Washington, DC, 1984; pp 133–150.
12. Torres, L. M.; Arie, A. W.; Sandoval B. In *Archaeological Chemistry III*; Lambert, J. B., Ed.; Advances in Chemistry 205; American Chemical Society: Washington, DC, 1984; pp 193–213.
13. Maniatis, Y.; Simopoulos, A.; Kostikos, A. In *Archaeological Ceramics*; Olin, J. S.; Franklin, A. D., Eds.; Smithsonian Institution: Washington, DC, 1982; pp 97–108.
14. McNeil, R. J. In *Archaeological Chemistry III*; Lambert, J. B., Ed.; Advances in Chemistry 205; American Chemical Society: Washington, DC, 1984; pp 255–269.
15. Harbottle, G. In *Archaeological Ceramics*; Olin, J. S.; Franklin, A. D., Eds.; Smithsonian Institution: Washington, DC, 1982; pp 69–77.
16. DeAtley, S. P.; Blackman, M. J.; Olin J. S. In *Archaeological Ceramics*; Olin, J. S.; Franklin, A. D., Eds.; Smithsonian Institution: Washington, DC, 1982; pp 79–87.
17. Perlman, I. In *Archaeological Chemistry III*; Lambert, J. B., Ed.; Advances in Chemistry 205; American Chemical Society: Washington, DC, 1984; pp 117–132.
18. Meyers, P. In *Archaeological Chemistry II*; Carter, G. F., Ed.; Advances in Chemistry 171; American Chemical Society: Washington, DC, 1978; pp 79–96.
19. Alexander, R. E.; Johnston, R. H. In *Archaeological Ceramics*; Olin, J. S.; Franklin, A. D., Eds.; Smithsonian Institution: Washington, DC, 1982; pp 145–154.
20. Frost, A. D. In *Archaeological Ceramics*; Olin, J. S.; Franklin, A. D., Eds.; Smithsonian Institution: Washington, DC, 1982; pp 155–163.
21. Meschel, S. V. In *Archaeological Chemistry II*; Carter, G. F., Ed.; Advances in Chemistry 171; American Chemical Society: Washington, DC, 1978; pp 1–24.
22. Baudin, G.; Chaput, M.; Feve, L. *Spectrochim. Acta* **1971,** *268*, 425.
23. Warren, J.; Harrison, M. P. *Proc. Anal. Div. Chem. Soc.* **1976,** *13*, 287.
24. Czobik, E. J.; Matousek, J. P. *Anal. Chem.* **1978,** *50*, 2.
25. Beaty, R. D.; Cooksey, M. M. *At. Absorpt. Newsl.* **1978,** *17*, 53.
26. Sturgeon, R. E. *Anal. Chem.* **1977,** *1255A*, 49.
27. Ediger, R. D. *At. Absorpt. Newsl.* **1975,** *14*, 127.
28. Issaq, H. J.; Young, R. M. *Appl. Spectrosc.* **1977,** *31*, 171.
29. L'vov, B. V. *Spectrochim. Acta* **1978,** *33b*, 153.
30. Carnrick, G. R.; Slavin, W.; Manning, D. C. *Anal. Chem.* **1981,** *53*, 1866.
31. Newstead, R. A.; Whiteside, P. J. *Eur. Spectrosc. News* **1977,** *3*, 8.
32. Slavin, W.; Carnrick, G. R. *At. Spectrosc.* **1985,** *6*, 157.
33. Colton, H. S.; Hargrave, L. L. *Handbook of Northern Arizona Pottery Wares*; Museum of Northern Arizona Bulletin II; Flagstaff, AZ, 1937.
34. Colton, H. S. *Pottery Types of the Southwest: Wares 8A, 8B, 9A, 9B*; Museum of Northern Arizona Ceramic Series, No. 3A; Flagstaff, AZ, 1955.
35. Deutchman, H. L. In *Models and Methods in Regional Exchange*; Fry, R. E., Ed.; Society for American Archaeology Papers, No. 1; Washington, DC, 1980; pp 119–133.
36. Deutchman, H. L. Ph.D. Dissertation, Southern Illinois University, 1979.
37. Geib, P. R.; Callahan, M. M. *Kiva* **1987,** 52, 95.
38. *Methods for Chemical Analysis of Water and Waste*; U.S. Environmental Protection Agency: Cincinnati, OH, 1976.

39. *Analytical Methods for Atomic Absorption Spectrophotometry*; Perkin-Elmer: Norwalk, CT, 1976.
40. Ambler, J. R. *The Anasazi: Prehistoric People of the Four Corners Region*; Museum of Northern Arizona: Flagstaff, AZ, 1977.
41. *BMDP Statistical Software*; Dixon, W. J., Ed., University of California: Los Angeles, CA, 1983; pp 98–100.
42. Zar, J. H. *Biostatistical Analysis*, 2nd ed.; Prentice-Hall: Englewood Cliffs, NJ, 1984; pp 455–459.
43. Davis, J. C. *Statistics and Data Analysis in Geology*, 2nd ed.; Wiley: New York, 1986; pp 493–497.
44. *SPSS-X User's Guide*, 2nd ed.; SPSS: Chicago, 1986; pp 451–474.
45. Zar, J. H. *Biostatistical Analysis*, 2nd ed.; Prentice-Hall: Englewood Cliffs, NJ, 1984; pp 168–178.
46. Zar, J. H. *Biostatistical Analysis*, 2nd ed.; Prentice-Hall: Englewood Cliffs, NJ, 1984; p 190.
47. *BMDP Statistical Software*; Dixon, W. J., Ed.; University of California: Los Angeles, CA, 1983; pp 519–535.
48. Davis, J. C. *Statistics and Data Analysis in Geology*, 2nd ed.; Wiley: New York, 1986; pp 478–491.
49. Afifi, A. A.; Clark, V. *Computer-Aided Multivariate Analysis*: Lifetime Learning: Belmont, CA, 1984; pp 246–275.
50. Afifi, A. A.; Clark, V. *Computer-Aided Multivariate Analysis*: Lifetime Learning: Belmont, CA, 1984; pp 361–373.
51. Davis, J. C. *Statistics and Data Analysis in Geology*, 2nd ed.; Wiley: New York, 1986; pp 607–615.

RECEIVED for review June 11, 1987. ACCEPTED revised manuscript March 15, 1988.

8

Heated Mineral Mixtures Related to Ancient Ceramic Pastes

X-ray Diffraction Study

Richard S. Mitchell† and Steven C. Hart[1]

Department of Environmental Sciences, University of Virginia, Charlottesville, VA 22903

The temperatures at which ceramic pastes were fired and data indicating the compositions of original clay materials can be determined by using X-ray diffraction powder techniques. Until now, very little systematic work has been done on the effects of admixed nonclay minerals on the final paste mineralogy. The following mixtures of 95% clay mineral and 5% carbonate mineral were incrementally heated at 100 °C intervals from 300 through 1100 °C: kaolinite–calcite, kaolinite–dolomite, montmorillonite–calcite, and montmorillonite–dolomite. At elevated temperatures new phases form. These phases are anorthite, cristobalite, dehydroxylated montmorillonite, gehlenite, lime, metakaolinite (dehydroxylated kaolinite), mullite, periclase, spinel, and magnesium aluminum silicates with stuffed high-quartz derivative structures. These phases individually provide clues about the original paste composition and the firing temperature.

ALTHOUGH X-RAY DIFFRACTION HAS BEEN USED FOR MANY YEARS in the identification of crystalline materials, its special application to the study of ceramic temper and paste minerals is relatively recent. Because many tem-

† Deceased.
[1] Current address: Department of Geology, Texas A&M University, College Station, TX 77843

0065–2393/89/0220–0145$06.00/0

per fragments are essentially unaltered rock or mineral fragments, they usually can be easily identified by the standard X-ray diffraction powder techniques. On the other hand, the use of this method for the identification of paste materials is more problematic because the clay materials used for ceramics are always changed by the high temperatures of kiln firing. However, a knowledge of the changes in clays with progressive heating allows the possibility of gathering information about firing that would not otherwise be available (*1*, *2*). Data indicating the temperatures at which the ceramics were fired and the compositions of the original clay materials can often be deduced.

Several studies of the effects of heating on pure clays have been reported in archaeological literature, but very little systematic work has been done on the effects of admixed mineral impurities upon the clays that constitute ceramic paste. The purpose of this chapter is to study the controlled firing of four measured mixtures of the clays, kaolinite and montmorillonite, with the common carbonates, calcite and dolomite.

Isphording (*3*) explored the usefulness of X-ray diffraction in the identification of pure clay pastes. Various clay minerals, when fired, become amorphous or form new mineral phases. Isophording explored some of these changes for pure samples of kaolinite, palygorskite, and montmorillonite and concluded that, in many instances, it should be possible to identify the original paste clay from which a ceramic was derived, although the clay itself may have been destroyed in the initial kiln-firing process. Heimann (*4*) pointed out that a knowledge of the temperature-dependent properties of a particular clay also allows one to deduce, from the state of transformation reached in a resulting fired paste, the maximum firing temperature for that ceramic. He also pointed out that if an ancient object is refired, its paste mineralogy remains fairly constant until the original firing temperature has been exceeded.

Although there is a vast literature on the dehydration and decomposition of pure clay minerals upon heating, impure clays, and especially clays related to ceramic pastes, have come under investigation relatively recently. For example, in 1962 Périnet (*5*) studied mineral phases that formed when calcareous kaolinite was heated at various temperatures. Kupfer and Maggetti (*6*) explored the stability ranges for certain mineral phases that form during the firing of calcareous illitic clay pastes. Maggetti (*7*) summarized data for the phases derived from kaolinitic clay pastes during controlled firing as well as mineralogical phases formed during the firing of noncalcareous illitic and calcareous illitic clays. Maniatis et al. (*8*) briefly considered the X-ray data for Iraqi sherds apparently derived from calcareous clays.

A factor that complicates the analysis of a clay paste, as noted by some of the previously mentioned workers, is the presence of additional constituents in the clay, like calcite, dolomite, iron oxides, gypsum, opaline silica, and others. These constituents, or their decomposition products, often enter into the chemical reactions to form new crystalline phases. Although their study did not focus on ceramic pastes, Mitchell and Gluskoter (*9*) considered the mineralogical changes that occur during the sintering at various temperatures of impure clay samples obtained from low-temperature coal ash. These pastes were mixtures containing silicates (clays), sulfides, sulfates, carbonates, etc. Allen et al. (*10*) and Hamroush (*11*), who observed that these impure clays had compositions similar to clay-bearing sediments such as those found along the Nile, found the results useful in interpreting the temperatures of firing and the original sediment compositions of Predynastic Egyptian ceramics. The phases observed in these ceramics included metakaolinite, calcite, anorthite, gehlenite, hematite, magnetite, and cristobalite.

The desirability of determining the role played by impurities in clay in the formation of high-temperature pastes has also come into focus during the following X-ray studies made in our laboratory: lamp sherds from the Eastern Mediterranean containing metakaolinite, calcite, anorthite, magnetite, and clinopyroxene (Rydin, S.M., University of Virginia, unpublished data, 1982); and American Indian pot sherds from the Donnaha site, North Carolina, containing metakaolinite, dolomite, anorthite, hematite, and clinopyroxene (?) (Davis, J.D., University of Virginia, unpublished data, 1986).

Experimental Procedure

This study is concerned with four different mixtures, including kaolinite and calcite (kc); kaolinite and dolomite (kd); montmorillonite and calcite (mc); and montmorillonite and dolomite (md). All the mixtures, by weight, were 95% clay and 5% carbonate mineral. The minerals were first ground to a fine powder and thoroughly mixed by hand before heating in a muffle furnace (temperature controlled to within ±20 °C). Each mixture was heated for 1 h, air quenched at room temperature, and analyzed by X-ray diffraction. X-ray films were made in cameras of 11.46-cm diameter with filtered copper radiation and exposure times of 6 h. Several wet mixtures that simulated ceramic paste before firing were heated and studied in like manner, but they showed no differences from the dry mixtures.

The initial minerals used were

1. white, well-crystallized pure kaolinite (locality uncertain): $Al_4(Si_4O_{10})(OH)_8$
2. montmorillonite from the Polkville mine, Polkville, MS (API reference clay No. 19): $(Al_{1.49}Mg_{0.53}Fe_{0.08})$ $(Al_{0.10}Si_{3.90})$-$O_{10}(OH)_2(Na_{0.02}Ca/2_{0.37})$ (*12*)
3. pure transparent calcite (locality unknown): $CaCO_3$
4. ferroan dolomite from the Old Dominion quarry, Albermarle County, VA: $Ca(Mg_{0.92}Fe_{0.08})(CO_3)_2$ (Fordham, O.M., Jr., Virginia Division of Mineral Resources, personal communication, 1986).

Minor quartz was detected in some of the montmorillonite, but, for the temperature range considered, this did not influence the results of this study.

The mineralogical phases formed by heating the clay–carbonate mixtures are summarized in Tables I and II. Solid-state chemical reactions at temperatures below fusion are rather common.

Table I. Results of Heating Mixtures of Kaolinite (k) or Montmorillonite (m) with Calcite (c) or Dolomite (d)

Temp. (°C)	*kc*	*kd*	*mc*	*md*
1100	Mullite Anorthite	Mullite Anorthite Periclase(?)	Crist. Spinel Anorthite Indialite(?)	Crist. Spinel Anorthite Indialite(?)
1000	Mullite Anorthite Gehlenite Lime	Mullite Anorthite Lime (tr.) Periclase(?)	Crist. Anorthite Mg–Al–Si No.2 Spinel	Crist. Mg–Al–Si No.2 Spinel Anorthite Cordierite(?) Diopside(?)
900	Lime Metakaol. Gehlenite Mullite Anorthite(?)	Metakaol. Mullite Lime Periclase Gehlenite	Mg–Al–Si No.1 Lime Anorthite	Mg–Al–Si No.1 Anorthite Crist.(?) Lime (tr.) Periclase(tr.)
800	Metakaol. Lime	Metakaol. Lime Periclase	Lime Anorthite	Lime Periclase Mg–Al–Si No.1(?)
700	Metakaol. Lime	Metakaol. Lime Periclase	Dehy. Mont. Lime	Dehy. Mont. Lime Periclase
600	Metakaol. Calcite Lime	Metakaol. Lime Periclase	Dehy. Mont. Lime Calcite (tr.)	Dehy. Mont. Lime Periclase
500	Metakaol. Kaolinite Calcite	Metakaol. Dolomite Kaolinite	Dehy. Mont. Calcite	Mont. Dehy. Mont. Dolomite
400	Kaolinite Calcite	Kaolinite Dolomite	Mont. Calcite	Mont. Dolomite
300	Kaolinite Calcite	Kaolinite Dolomite	Mont. Calcite	Mont. Dolomite

Abbreviations used: cristobalite (crist.) dehydroxylated montmorillonite (dehy. mont.), matakaolinite (metakaol.), montmorillonite (mont.), and trace (tr.).

NOTE: Clay 95%, carbonate 5%, by weight; major phases are at the top of each list.

Table II. Mineral Phases Observed in Heated Mixtures of Kaolinite–Calcite (kc), Kaolinite–Dolomite (kd), Montmorillonite–Calcite (mc), and Montmorillonite–Dolomite (md)

Phase Observed	*Temperature Range (°C)*	*Clay Mixture*
Anorthite	900–1100	kc, md
	800–1100	mc
	1000–1100	kd
Calcite	room–600	kc, mc
Cristobalite	900–1100	md
	1000–1100	mc
Dolomite	room–500	kd, md
Gehlenite	900–1000	kc
	900	kd
Indialite (?)	1100	mc, md
Kaolinite	room–500	kc, kd
Kaolinite, dehydroxylated (metakaolinite)	500–900	kc, kd
Lime (if high humidity, portlandite)	600–1000	kc, kd
Mg–Al–Si No.1 (like high quartz)	800–900	md
	900	mc
Mg–Al–Si No.2	1000	mc, md
Montmorillonite	room–400	mc
	room–500	md
Montmorillonite, dehydroxylated	500–700	mc, md
Mullite	900–1100	kc, kd
Periclase	600–1100	kd
	600–900	md
Spinel	1000–1100	mc, md

Results

Kaolinite–Calcite (kc) Mixture. Table I shows that there were no appreciable changes for the kaolinite or calcite at 300 or 400 °C. Kaolinite began to disappear between 400 and 500 °C, and it lost its structure by 600 °C. The breakdown of kaolinite upon heating has been studied extensively (numerous papers summarized by Grim, *13*) and is caused by the loss of the hydroxyl ions (OH^-) upon heating. This loss causes the structural layers to collapse and, hence, the disappearance of the normal kaolinite X-ray diffraction reflections. With the loss of kaolinite reflections, the mineral is considered essentially amorphous, although often there are few very weak reflections from an ill-defined phase commonly referred to as metakaolinite (or dehydroxylated kaolinite). Metakaolinite is considered to be a structural remnant of the collapsed kaolinite. According to Grim (*13*), if the original kaolinite was well-crystallized it often retains some degree of order when dehydrated by heating.

Published X-ray diffraction data for metakaolinite are sparse; however, Hill (*14*) and Mitchell and Gluskoter (*9*) have each reported two weak diffuse reflections: Hill, 4.43 and 3.52 Å; Mitchell and Gluskoter, 4.48 and 2.59 Å. For the present study, a reference metakaolinite pattern was prepared by heating the pure kaolinite at 800 °C for 1 h. Reflections at 3.51, 4.51, and 1.89 Å were observed and used to detect the presence of metakaolinite. In these kaolinite–calcite mixtures, metakaolinite was present from about 500 through 900 °C. This stability range generally agrees with the observations of Mitchell and Gluskoter (*9*) and those workers cited by Grim (*13*).

At elevated temperatures, calcite decomposed to form lime (CaO) and carbon dioxide. In a laboratory environment with a high relative humidity, after the sample cools, the lime easily reacts with moisture in the air to form portlandite ($Ca(OH)_2$). In these experiments, lime and portlandite were considered equal phases. Lime began to form at 600 °C and persisted through 1000 °C, although it began to decrease in quantity at 900 °C, at which chemical reactions between it and metakaolinite began to yield anorthite ($CaAl_2Si_2O_8$) and gehlenite ($Ca_2Al(AlSi)O_7$). Also at 900 °C, mullite ($Al_6Si_2O_{13}$) began to form from the excess metakaolinite in the system. At 1100 °C, the only phases remaining in the sample were anorthite and mullite.

Several other mixtures of kaolinite–calcite were studied for preliminary data, and in an 80% kaolinite–20% calcite mixture, both lime and gehlenite also occurred in addition to anorthite and mullite at 1100 °C. Except for relative mineral quantities, the temperature and phase relationships observed in this calcium-rich mixture were identical to those mixtures containing less calcite.

Kaolinite–Dolomite (kd) Mixture. Table I shows that the heating history of the kd mixture is similar to that for kc, except for the formation of periclase (MgO) from the decomposition of the dolomite ($CaMg(CO_3)_2$). Again, kaolinite altered to metakaolinite between 400 and 500 °C, and this persisted until 900 °C. Dolomite decomposed between 500 and 600 °C to form lime (CaO) and periclase. Lime began to decrease at 900 °C with the formation of gehlenite and anorthite, and periclase persisted as other high-temperature crystalline compounds increased. Mullite, which began to form at 900 °C, and anorthite were the major components at 1100 °C. Gehlenite and periclase could have also been present at the highest temperatures, but because their relative amounts were so small in comparison to the mullite and anorthite, they were not easily detected.

The decomposition of the ferroan dolomite in this experiment occurred at a temperature below that for chemically pure dolomite, which is near 807 °C. Smykatz-Kloss (*15*) has shown that ferroan dolomite decomposes at lower temperatures; in fact, he has shown how the temperature of decomposition from differential thermal analyses can be used to determine the weight percentage of FeO in the mineral.

Montmorillonite–Calcite (mc) Mixture. Heated mixtures of montmorillonite and calcite yielded the phases given in Table I. Although the montmorillonite structure persisted through 400 °C, it underwent dehydroxylation between 400 and 500 °C. Grim and Bradley (*16*) have shown that the general layered structure is able to survive the elimination of the (OH) water with moderate readjustments. This structure produces an X-ray diffraction pattern like that given in Table III. Table III represents data close to those observed in this study. This phase is called dehydroxylated montmorillonite in Table I. This phase disappeared between 700 and 800 °C as a result of the complete destruction of the montmorillonite crystal structure. Calcite decomposed between 500 and 600 °C to form lime that was present through 900 °C.

Table III. X-Ray Powder Diffraction Data for Dehydroxylated Montmorillonite

d *in* Å	*Relative Intensity*
9.7	m
4.85	m
4.48	vs
3.22	s
2.59	m
2.5	m (diff.)
2.23	m
2.1	w (diff.)
1.93	w
1.72	w
1.67	m (diff.)
1.51	m
1.49	w (diff.)
1.31	w
1.25	w

NOTE: Abbreviations used: medium (m), strong (s), very strong (vs), weak (w), and diffuse (diff.).
SOURCE: Modified from Bradley and Grim (*19*).

Anorthite began to crystallize at 800 °C and was observed at all temperatures through 1100 °C. Anorthite apparently formed by chemical reactions between lime and the dehydroxylated montmorillonite.

At 900 °C, the dominant phase was a magnesium aluminum silicate (here named Mg–Al–Si No. 1) that has X-ray diffraction data very close to high (beta)-quartz (Table III). Schreyer and Schairer (*17*) and Schulz et al. (*18*) have studied several compounds in which Mg^{2+} and Al^{3+} can substitute for Si^{4+} in the high quartz structure to form "stuffed high quartz derivatives". These solid solution phases can form over a wide range of compositions (e.g., $MgO{\cdot}Al_2O_3xSiO_2$, where $x = 2, 3, 4$, or 6) in the temperature range of about 800 to 950 °C. Although the compositions vary considerably, the X-ray diffraction data remain consistently close to high quartz (with slight variations in the unit cell sizes). The following cards in the Powder Diffraction File illustrate this feature: 14-249, 25-511, 27-716. In their earlier study of the heat treatment of montmorillonite, Bradley and Grim (*19*) mistakenly referred to these phases as high quartz. Others have named them μ-cordierite (*20*) or silica O (*21*). Also accompanying this phase at 900 °C were anorthite and lime.

At 1000 °C, the magnesium aluminum silicate stuffed high quartz derivative (Mg–Al–Si No. 1) apparently altered to give slightly different diffraction data. This phase is referred to here as Mg–Al–Si No. 2 (Table III). Schreyer and Schairer (*17*) showed that all of these stuffed high quartz derivatives are metastable and are easily converted to other compounds or show superstructure reflections with increased temperatures or prolonged heating. At 1000 °C, anorthite was also present and was accompanied by spinel ($MgAl_2O_4$) and cristobalite (SiO_2). Both minerals apparently formed from the breakdown of Mg–Al–Si No. 1.

At the highest temperature studied (1100 °C), the major phase was cristobalite, followed by spinel and anorthite. There was slight evidence for the continued presence of minor magnesium aluminum silicate. The strongest reflection for this minor component was at 8.48 Å, a diagnostic reflection for indialite ([$Mg_2Al_4Si_5O_{18}$], Powder Diffraction File card 13-293), but here again, the exact nature of this phase is uncertain.

Montmorillonite–Dolomite (md) Mixture The crystalline phases formed by heating the md mixture at various temperatures (Table I) were similar to those found in the mc mixture, with some variations. Montmorillonite dehydroxylated between 400 and 500 °C to form the phase shown in Table III and finally broke down completely between 700 and 800 °C. The dolomite decomposed to yield lime and periclase between 500 and 600 °C. These phases diminished by 900 °C and were not detected at higher temperatures, although theoretically they are stable at these temperatures. Anorthite formed at 900 °C and above.

Table IV. Comparison of X-Ray Data for Magnesium Aluminum Silicates Found in this Study with High (beta) Quartz

High quartz[a]		*Mg–Al–Si No.1*		*Mg–Al–Si No.2*	
d (Å)	*Intensity*	d (Å)	*Intensity*	d (Å)	*Intensity*
4.339	16	4.44	m	4.37	w
3.399	100	3.42	vs	3.40	s
2.505	4	2.56	m–		
2.314	1	2.31	w+	2.29	w
2.169	3	2.22	m	2.21	w
2.017	4	2.05	m–		
1.8473	18	1.86	s–		
1.6996	1	1.71	w–	1.76	w+
1.6809	1				
1.5708	9	1.60	m	1.63	m
1.4463	1	1.43	m–	1.45	m
1.4065	5	1.40	m		
1.3958	9			1.39	w
1.3043	1	1.30	w		
1.2785	3	1.28	w		
1.2525	2				
1.2193	3	1.23	w		
1.2034	1	1.20	w		
1.2003	3	1.19	w		
1.1753	4				
1.1015	2	1.11	w–		

[a]High quartz data is adapted from ref. 22.
NOTE: Abbreviations used: m, medium; s, strong; w, weak; vs, very strong. + or – indicate high or low end of the indicated intensity.

Traces of the Mg–Al–Si No. 1 phase appeared at 800 °C and became major at 900 °C. By 1000 °C, the Mg–Al–Si No. 2 phase took its place, and at 1100 °C, the major evidence for a magnesium aluminum silicate was the indialite reflection mentioned earlier.

Also at 1000 °C, spinel formed and was a major phase at 1100 °C. Cristobalite began forming at about 900 °C and became a major phase by 1100 °C. Minor, weak X-ray reflections at 1000 °C also suggested cordierite and diopside (clinopyroxene), but both of these were uncertain and disappeared by 1100 °C.

Discussion and Conclusions

The mineral phases observed in this study are summarized in Table II. For each phase, the mixtures that yielded them and their observed temperature stability ranges are given. When these phases are encountered in X-ray diffraction studies of ceramic pastes, it should be possible to determine the maximum temperature at which the ceramic was heated, providing it did not exceed 1100 °C and contained a mixture close to the clay–carbonate mixtures used in this study. Precise information on the length of time needed to achieve the results observed by heating has not been determined. The heating times used here were shorter than those used with a kiln, but Maniatis and Tite (*23*) showed that longer heating times would have made little difference. In the present study, the samples were heated for 1 h.

Furthermore, most of the phases show a close affinity to the original clay mineral, either kaolinite or montmorillonite. The most notable exception is anorthite, which resulted from all the mixtures, irrespective of the clay, at various temperatures between 800 and 1100 °C. Periclase also formed from mixtures involving both clay types, but in each instance it was derived from the dolomite mixed with them.

These results are based entirely on mixtures of two components. Natural clays from which ceramics are made often have several additional minerals; that is, they may contain two or more clay types, as well as anhydrite, feldspars, gypsum, iron oxides, manganese oxides, opaline silica, pyrite, etc. The presence of these additional components would not only affect the temperatures at which reactions occur, but would strongly influence the nature and chemistry of the final minerals formed in the systems. A possible exception is quartz, which is stable over the temperature range used in this study (*9*), but opaline silica (e.g., from sediments containing siliceous diatoms, radiolaria or sponge spicules) would form additional phases in this range. Another group of minerals that is stable over a broad temperature range and that can cause difficulty in interpreting the results is the feldspars. If plagioclase feldspar (e.g., anorthite) was in the original paste, it could survive the kiln firing of the ceramic. Therefore, in some instances, X-ray diffraction analysis may overestimate the firing temperature of an ancient ceramic if the paste from which it was made already included high-temperature minerals such as anorthite. This problem was encountered by Allen et al. (*10*) in their study of some early Egyptian ceramics.

Ancient ceramic materials are often susceptible to mineralogical changes after burial. For firing temperatures below 700 °C, the loss of the water on firing for some minerals can be reversed during subsequent burial (*23*, *24*). Also, some calcite or lime may be leached out by acid water if the burial is shallow. On the other hand, if calcite is observed it might be of secondary origin, having been deposited later in pores and voids in the ceramic. Periclase, originally present in a ceramic ware, might eventually be hydrated to form brucite ($Mg(OH)_2$). Maggetti (*7*) has summarized some of the other possible changes in mineralogy that could take place during a long period of burial.

References

1. Perinet, G. *Seventh Int. Ceram. Congr.* (*London*), 1960, pp 371–376.
2. Shepard, A. O. *Ceramics for the Archaeologist*; Carnegie Inst. Washington Publ. 609: Washington, DC, 1954.

3. Isphording, W. C. *Am. Antiq.* **1974,** *39,* 477–483.
4. Heimann, R. B. In *Archaeological Ceramics*; Olin, J. S.; Franklin, A. D., Eds., Smithsonian Institution Press: Washington, DC, 1982; pp 89–96.
5. Perinet, G. *Bull. Soc. Fr. Mineral. Cristallogr.* **1962,** *85,* 120–122.
6. Kupfer, T.; Maggetti, M. *Schweiz, Mineral. Petrogr. Mitt.* **1978,** *58,* 189–212.
7. Maggetti, M. In *Archaeological Ceramics*; Olin, J. S.; Franklin, A. D., Eds.; Smithsonian Institution Press: Washington, DC, 1982; pp 121–133.
8. Maniatis, Y.; Simopoulos, A.; Kostikas, A. In *Archaeological Ceramics*; Olin, J. S.; Franklin, A. D., Eds.; Smithsonian Institution Press: Washington, DC, 1982; pp 97–108.
9. Mitchell, R. S.; Gluskoter, H. J. *Fuel* **1976,** *55,* 90–96.
10. Allen, R. O.; Rogers, M. S.; Mitchell, R. S.; Hoffman, M. A. *Archaeometry* **1982,** *24,* 199–212.
11. Hamroush, H. A. Ph. D. dissertation, University of Virginia, Charlottesville, 1985.
12. Kerr, P. F.; Hamilton, P. K.; Pill, R. J.; Wheeler, G. V.; Lewis, D. R.; Burkhardt, W.; Reno, D.; Taylor, G. L.; Mielenz, R. C.; King, M. E.; Schieltz, N. C. Analytical Data on Reference Clay Materials, Report 7; Am. Pet. Inst., Project 49, Clay Mineral Standards, Columbia University: New York, 1950; 160 pp.
13. Grim, R. E. *Clay Mineralogy*; McGraw–Hill: New York, 1968; 2nd ed.; 596 pp.
14. Hill, R. D. *Trans. Br. Ceram. Soc.* **1956,** *55,* 441–456.
15. Smykatz-Kloss, W. *Differential Thermal Analysis: Application and Results in Mineralogy*; Springer–Verlag; New York, 1974; 185 pp.
16. Grim, R. E.: Bradley, W. F. *J. Am. Ceram. Soc.* **1940,** *23,* 242–248.
17. Schreyer, W.; Schairer, J. F. *Z. Kristallogr.* **1961,** *116,* 60–82.
18. Schulz, H.; Muchow, G. M.; Hoffman, W.; Bayer, G. Z. *Z. Kristallogr.* **1971,** *133,* 91–109.
19. Bradley, W. F.; Grim, R. E. *Am. Mineral.* **1951,** *36,* 182–201.
20. Rankin, G. A.; Merwin, H. E. *Am. J. Sci.* **1918,** *45,*301–325.
21. Roy, R. *Z. Kristallogr.* **1959,** *111,* 185–189.
22. Borg, I. Y.; Smith, D. K. *Mem. Geol. Soc. Am.* **1969,** *122,* 743–744.
23. Maniatis, Y.; Tite, M. S. *J. Arch. Sci.* **1981,** *8,* 59–74.
24. Tite, M. S. *Methods of Physical Examination in Archaeology*; Seminar Press: London and New York, 1972.

RECEIVED for review June 11, 1987. ACCEPTED revised manuscript February 29, 1988.

METALS

9

Bronze Age Archaeometallurgy of the Mediterranean: The Impact of Lead Isotope Studies

Noël H. Gale and Zofia A. Stos-Gale

Department of Earth Sciences, University of Oxford, 1 Parks Road, Oxford OX1 3PR, England

For more than 50 years, it has been a goal to use scientific methods to establish which ore deposits were the ultimate sources of the metals from which Bronze Age metal objects were made. Solution of this problem would allow ancient trade routes and cultural contacts to be established. Approaches based solely on trace element analyses have largely failed, and, in many cases, have resulted in archaeological confusion. Success necessitates an approach that takes into account metallurgy; ore deposit geology; and isotope geochemistry, especially lead isotope studies. The methodological background and the success that we have attained in solving this problem are discussed against the background of our archaeometallurgical investigations into the sources of silver, lead, and copper in the Bronze Age Mediterranean.

THE AEGEAN REGION, the area occupied by modern Greece, Crete, and western Turkey, saw a number of remarkable and original changes in the years roughly between 3500 and 1500 B.C. At the beginning of this period, life throughout the Aegean may be described as "neolithic." Life was a matter of simple farming and fishing, and the population lived in small villages and was occupied with the crafts of pottery, woodwork, and perhaps weaving. Tools were almost exclusively of stone or bone.

By the end of the third millennium, the first true civilization of Europe was born in Crete. The great Minoan palaces there, such as Knossos, Phaistos, and Mallia, functioned as production and distribution centers with a

0065-2393/89/0220-0159$11.00/0

written script for record keeping. Along with the rich site of Akrotiri on Thera, the Minoan palaces were centers of high artistic achievement. Throughout the Aegean, there had developed at different times, other small, protourban, communities, each acting as a center for its region, and as a focus of production and trade. Moreover, society had become stratified, both hierarchically and functionally into a number of craft specializations, among them, the production and working of metals.

By 1200 B.C., the procurement of metals had become a very important matter for Aegean societies. This situation is reflected in the Late Bronze Age copper oxhide* ingots that were distributed throughout the Mediterranean from Sardinia in the West, through Greece, Crete, Bulgaria, and Cyprus to Syria, Anatolia and Egypt in the East (*1–3*). In the Late Bronze Age, tin bronze became the dominant alloy in use. Tin deposits do not occur in the Aegean region and, therefore, must have been acquired somewhere else.

Our thesis is that the adoption of the smelting and working of metals was a decisive step, perhaps one of the more important steps that led directly to the emergence of civilization in the Aegean. Metallurgy created a new kind of wealth that led to increasing social stratification. The practice of metallurgy created new classes of craftsmen and their products: tools, which transformed carpentry and shipbuilding, and weapons, which revolutionized war. The procurement of metals led to the early development of underground mining and gradually led to the development of trade.

The production of metals from their ores, which involved the development of smelting, furnaces, and eventually alloying technologies, was a crucial step in the metallurgical revolution, because only when these technologies had been developed could copper be produced on a scale large enough to have a significant impact on society. The earliest copper objects were probably fashioned from native copper, which was extremely scarce everywhere in the Aegean but common in parts of Turkey, or from copper produced by small-scale crucible smelting of oxidized ores.

Two-thousand years elapsed from the time the first metal objects were made in the Aegean to the supreme accomplishments of the Shaft Graves in Mycenae. Metallurgy was adopted slowly because of the time required to understand that metals could be produced from ores, and then the time needed to master the slagging and furnace technology required to produce metals in sufficient quantity. Even with modern knowledge of the physics and chemistry of the processes involved, successful replication of ancient methods of copper smelting is a long and tedious business, as the work of Tylecote and others has shown (*4*).

* An oxhide ingot is a copper ingot that weighs approximately 30 kg. It has the approximate shape of the flayed skin of an ox. Oxhide ingots form an important and characteristic component of Late Bronze Age trade in copper in Cyprus, Greece, and Egypt.

Apart from the Minoans and, later, the Mycenaeans, the Early Cycladic people played an important part in the development of Mediterranean metallurgy. In fact, the Cycladic islands of the central Aegean came into extraordinary prominence in the third millennium B.C. The islands were the home of a flourishing culture with prominent settlements, a rather abundant population, well-developed pottery, and striking achievements in marble sculpture. Production of silver, lead (*5*), and copper from their ores was developed early, along with a rather vigorous trade.

Although they did not develop any indigenous use of tin bronze (*6*), the Cycladic peoples were preeminent in their use of arsenical copper. The Cycladic peoples were the dominant seafarers of the early third millennium Aegean. Their cemeteries, as described by Renfrew (*7*) and Doumas (*8*), give evidence of a gradually developing social stratification that was associated with their development of metallurgy.

Trade in Metals

Studies of the development of ancient metallurgy and of the trade in metals that eventually ensued are central to Late Chalcolithic and Bronze Age archaeology. Real progress in these studies can be achieved only by using an integrated approach that involves archaeology, metallurgy, geology and ore mineralogy, mining studies, the analyses of stratified metal objects, and the analysis and dating of the remains of ancient mining and metal smelting.

A central problem in archaeometallurgical studies is the determination of the provenance of copper, lead, silver, tin, etc. The development, in recent years, of the lead isotope technique for analyses of metal artifacts (*9, 10*) has provided, for the first time in the history of archaeometallurgy, a direct analytical method for linking metal artifacts to ore deposits. Lead isotope analysis must be used with intelligence and caution against the background of metallurgy and geology. The study of the metal resources used by different cultures and of metal trading networks is so important that the impact of the lead isotope method for metal provenancing has been compared with the development of C-14 dating. Recent work using this technique has, radically changed the archaeological theories of the past 40 years concerning the sources of copper, lead, and silver in the Bronze Age Mediterranean.

Foundations and Development of Lead Isotope Archaeology

Recent reviews (*11–12*) make it unnecessary to do more than draw attention to some salient points of lead isotope archaeology. For many years, comparative lead isotope studies of ancient metals and ores from the appropriate ore deposits have been in the forefront of metal provenance studies in archaeology (*13–15*). The earliest lead isotope studies by Brill (*17–18*) and

Grögler et al. (*19*) were hampered by the low accuracy of isotope composition measurement that was then attainable (the absolute 2σ errors in the isotope ratios were about 1%), by the lack of a sufficient data base for ore samples, and by the lack of a coherent archaeological strategy. Nevertheless, this early work provided a base for the subject of lead isotope archaeology.

In contrast, chemical analyses of metal artifacts have not solved the problem of determining the sources of metal (*20–22*). Indeed, a comparison of analyses of minor elements of metal artifacts and ore deposits may never pinpoint the sources of ore used. Not only do the minor elements usually vary widely in content through a given ore body, but in the smelting of an ore to yield a metal, differences between of the minor elements in the ore and the metal come from the added flux, the fuel, and from variable partitioning between metal and slag (*4*). In addition, some elements will be lost by volatilization, depending on the oxidizing or reducing environment encountered.

Lead isotope analysis presents an alternative to chemical analysis in the determination of metal sources. Lead isotope analysis is a physical method that depends on the variability of the isotopic composition of lead (present either as the chief or as a subsidiary element) in metal ore sources and on comparative analyses of the lead isotope composition in metal artifacts and ore sources. Lead isotope analysis is free from most of the limitations of chemical analysis. The isotopic composition of lead usually varies only within narrow limits in a given ore body. The lead isotopic composition of the ore passes unchanged through the smelting, refining, working, casting, or corrosion processes into the metal (*23*). Though corrosion does not alter lead isotope compositions by fractionation (*24*), corroded samples of copper objects of low lead content can be dangerous for lead isotope analysis. Corrosion may alter the true lead isotope composition of an artifact by exchanging its lead with lead of different isotopic composition from other objects in the burial environment or from the soil itself (*25*).

Naturally occurring lead is made up of four isotopes in varying proportions. According to their atomic mass, measured in atomic mass units, they are designated ^{204}Pb, ^{206}Pb, ^{207}Pb, and ^{208}Pb. ^{204}Pb is nonradiogenic in origin. The other three isotopes of lead derive, in part, from the radioactive decay of uranium and thorium. For a particular ore body, the present-day lead isotope ratio is determined by the integrated effects of all associations with uranium and thorium between some initial time and a later time, when the ore was formed. After the ore is formed, the lead is usually no longer associated with uranium and thorium.

So far, the only available technique capable of sufficiently accurate measurement of lead isotope compositions, both for isotope geochemistry and for archaeological applications, is thermal ionization mass spectrometry. Any well-equipped laboratory can routinely measure lead isotope ratios with absolute 2σ errors of 0.1% on samples of 100 ng of lead. The accuracy of determination of the isotopic composition of lead in a particular laboratory can be assessed by making replicate measurements of the lead isotopic composition of standards issued by the United States National Bureau of Standards (NBS). For archaeological purposes, the usual way of presenting the lead isotope measurements is to relate the atomic ratios of $^{208}Pb/^{206}Pb$ and $^{206}Pb/^{204}Pb$ to the ratio of $^{207}Pb/^{206}Pb$. Our measurements of NBS standards demonstrate reproducibility of better than one part per thousand (1/1000) for each of these three ratios.

Archaeometrists often ask whether the isotopic composition of lead in an ore deposit is uniform throughout. Doubt has also been expressed as to whether the lead in different minerals from the same ore deposits, especially from the primary sulfides (as opposed to oxidized minerals) will have the same isotopic composition. To answer these doubts, a number of investigations have been carried out by isotope geochemists. For two deposits in Australia, Gulson (*26*) has shown that the complete range of lead isotope compositions in the primary sulfide ores varies less than 0.3%, and that the oxidized ores in the overlying gossan have lead isotope compositions indistinguishable from the deep primary sulfide ores. Similar results were obtained in Oxford for a range of ores from Cyprus and Laurion (*27*).

The question of lead isotope uniformity in an ore body has also been investigated in several other laboratories. Many ore deposits, especially the so-called strata-bound or comformable ore deposits, have a lead isotope composition that varies very little throughout the whole deposit, often by less than 0.3%. Such ore deposits commonly have lead isotope compositions that lie close to a single-stage model evolution curve (*28*). Such a simple model cannot describe the composition in any real ore body.

Several, more realistic, models have recently been constructed (*29*). The importance of these models is that they allow calculations of the so-called "model ages" from the observed lead isotope composition of a particular ore body. These calculations should give the geological age of the time of emplacement of the ore body. The importance of this information for isotope archaeology is that, once the lead isotope composition of a metal artifact has been measured, a model age can be calculated that will give a guide to the rough geological age of the ore deposit that supplied its metal. This calculation indicates which ore deposits should be investigated in further provenance studies.

In contrast with conformable ore deposits, some vein-type ore deposits have variable lead isotope compositions that plot as linear arrays of data points in one or both lead isotope diagrams. Such anomalous, or multistage, ore deposits can sometimes cause difficulties for archaeological provenance studies (*30–31*). Fortunately, deposits of this type are not common in the region of the Mediterranean; indeed, none have yet been found. Only in

Fenan, Timna, and central Anatolia, have ore sources been found that retain sufficient uranium or thorium to cause variable present-day lead isotope compositions in a single ore deposit.

Methodology and Problems Encountered

Seven years ago we started the first systematic research program on the application of the lead isotope techniques to provenance studies in archaeology. Particular stress was placed on the sources of metals in the Mediterranean Bronze Age. For the first 2 years we worked mostly on the sources of lead and silver in Bronze Age Greece, Cyprus, and Egypt (*32–36*). In 1982, we pioneered the application of the lead isotope method for provenancing copper-based artifacts (*15, 37–38*).

Considering the quantity of metal from Bronze Age archaeological sites in the Mediterranean, and the quite insufficient geological information about the ore deposits in this region that may have supplied the metal, the work was immense. During the past century, archaeological sites of a technological nature, such as smelting sites, slag heaps, and ancient mining galleries, have been largely ignored by archaeologists. There is , therefore, a necessity for extensive field work based on geological and historical information in parallel with laboratory studies.

Apart from direct studies of ancient mining or metallurgical remains, extensive fieldwork is needed also to establish the lead isotope characteristics of copper and lead–silver deposits that could have, or are known to have been, worked in the Bronze Age. In the past 8 years we have explored ore deposits in Greece, Cyprus, Sardinia, and the Sinai.

Our fieldwork has two main objectives:

1. to collect ores and slags for chemical, mineralogical, and lead isotope analyses, and
2. to establish whether any archaeological technique (like C-14 or thermoluminescence dating) or other evidence (like pottery, tools, tool marks, or character of mining) points to the Bronze Age exploitation of the deposit.

Lead isotope and chemical analyses of byproducts of copper smelting from archaeological sites are important for the interpretation of the technology that might have influenced the lead isotope ratios of the metal produced on a given site. Copper ores usually contain unwanted gangue that has to be separated from the metal in the smelting process by the addition of a flux. In principle, lead of an isotopic composition different from that in the metal ore might be introduced with the flux or the fuel ash. Siliceous fluxes generally have a very low lead content. Ferrugineous fluxes have usually been taken from the nearest source, which will usually be the gossan of the ore source itself, and have the same lead isotope composition as the ore source.

Fuel, chiefly charcoal, usually has a low lead content and, in many cases, the trees from which it came were growing near the ore deposit and tend to have incorporated lead of the same isotopic composition as the ore deposit. In most cases, the isotope composition of lead in the smelted copper has not been perturbed away from that of lead in the copper ore. Further isotopic comparisons of copper ores and associated Bronze Age copper slags are needed to examine this question more extensively. Comparisons that have already been made bear out the hypotheses just advanced.

To establish the range of lead isotope compositions, the lead isotope field characteristic for a given ore deposit, of about 50 samples of ores, or ores and slags from a given ore deposit should be analyzed. Figure 1 demonstrates the construction of lead isotope fields characteristic of three different copper deposits in the Eastern Mediterranean. The final result of the determination of the lead isotope composition characteristic of a given ore deposit consists of a set of three lead isotope ratios: $^{208}Pb/^{206}Pb$, $^{207}Pb/^{206}Pb$, and $^{206}Pb/^{204}Pb$. These ratios form a three-dimensional, approximately egg-shaped, distribution in space.

Two-dimensional representation of the lead isotope data requires two separate diagrams that present the ratios $^{208}Pb/^{206}Pb$ versus $^{207}Pb/^{206}Pb$ and $^{206}Pb/^{204}Pb$ versus $^{207}Pb/^{206}Pb$. Figure 2 shows the alternative ($^{206}Pb/^{204}Pb$ versus $^{207}Pb/^{206}Pb$) diagram for the same copper deposits as in Figure 1. Sometimes, only by using all three measured lead isotope ratios is it possible to decide if particular characteristic fields overlap. For example, the Cypriot and Kythnian fields overlap on Figure 1, but are well-separated on Figure 2.

A better representation of the three-dimensional lead isotope fingerprint can be made by using a multivariate discriminant analysis. Figure 3 shows the characteristic lead isotope composition of the same three ore deposits that was prepared by using multivariate discriminant analysis (Pollard, M., University of Cardiff, personal communication in 1986). It shows clearly, as the two-dimensional diagrams do not, that these fields may be resolved by using all three isotopic ratios.

Certain ore deposits may partially overlap in their lead isotope composition. In such a case, it might not be possible to decide on the basis of the lead isotope data alone which of the ore deposits in question provided the ore for metal artifacts that have lead isotope compositions that fall into the overlapping space. Sometimes, one of these ore deposits can be ruled out by using trace element data, particularly from gold and silver analyses (*6*). Lead isotope analyses by themselves can make a negative statement with absolute certainty in a way which chemical analysis can never hope to do. If the lead isotope composition of an artifact falls well outside the lead isotope field characteristic of a particular ore deposit then it is certain that the metal

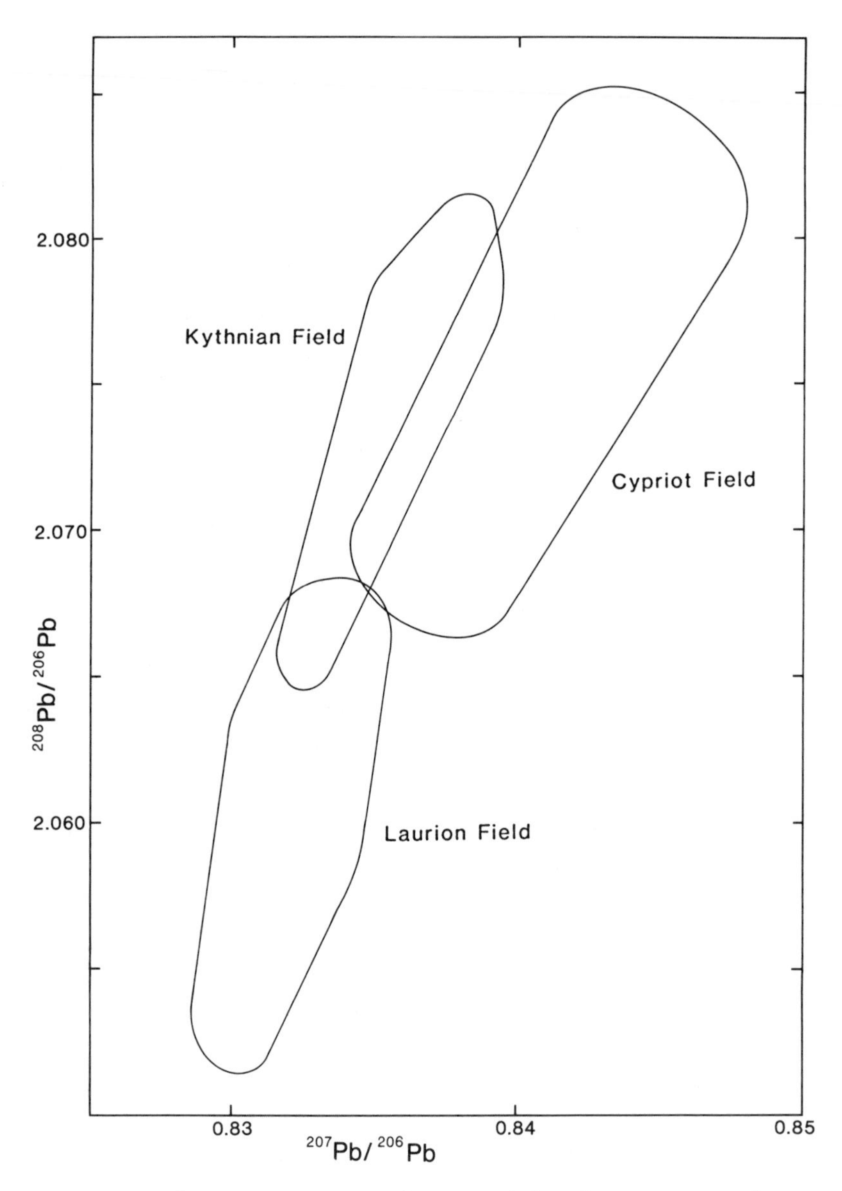

Figure 1. Lead isotope compositions for ores and slags from Cyprus, Kythnos, and Laurion. The fields overlap in this two-dimensional diagram but can be separated by using all three available lead isotope ratios.

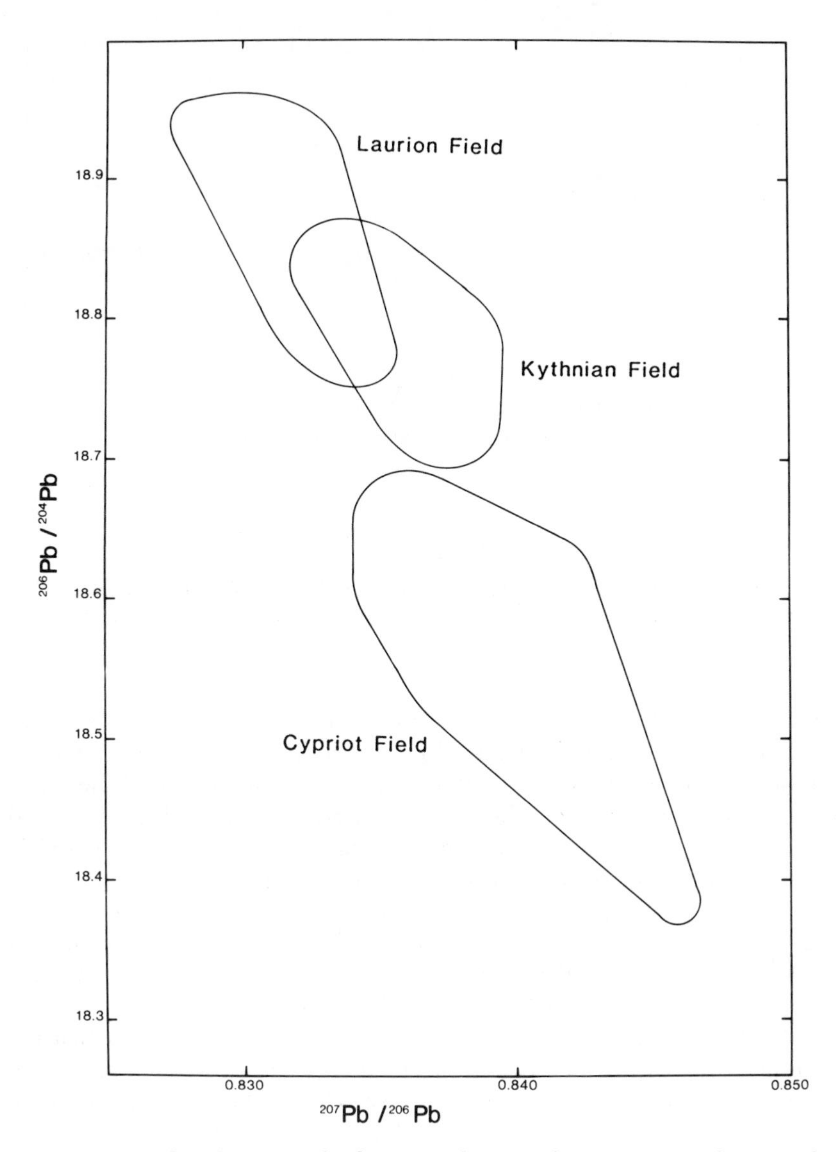

Figure 2. The alternative lead isotope diagram for Cyprus, Kythnos, and Laurion fields.

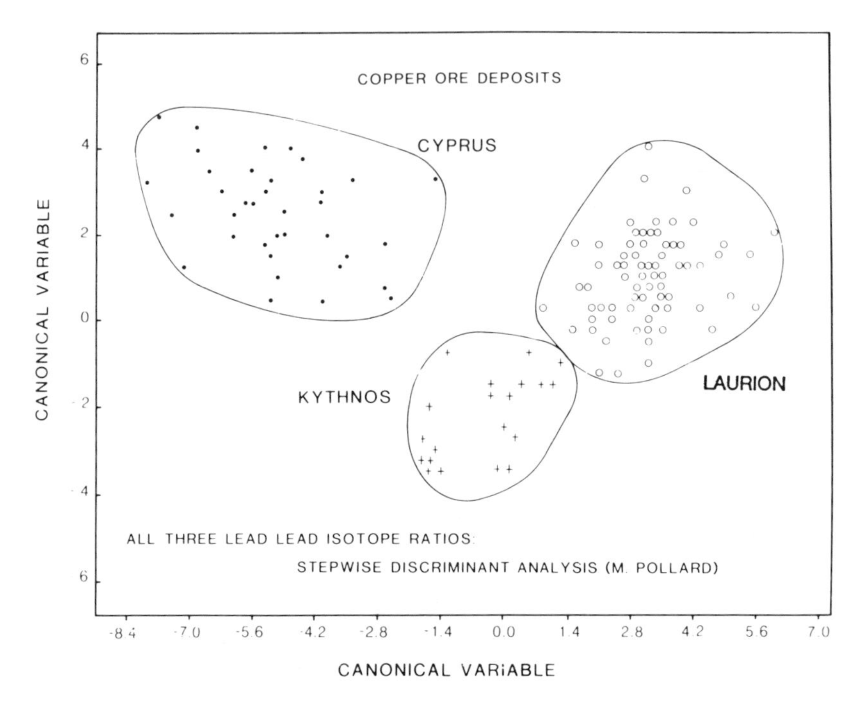

Figure 3. Stepwise discriminant analysis by M. Pollard of all three measured lead isotope ratios for ore samples from Cyprus, Kythnos, and Laurion ore deposits.

for that artifact did not come from that ore source. This statement requires revision only if there is evidence of perturbation of the lead isotope composition of the metal away from the lead isotope composition of the metal ore. This perturbation could be caused by lead from the fuel and flux used in smelting. In the few cases examined so far by workers in Mainz and in Oxford, evidence of such perturbation has not been found.

For each ore deposit under investigation, the ore for lead isotope analyses was collected from various parts of the deposit to check the uniformity of the characteristic lead isotope composition within the deposit. Various types of minerals were analyzed. If available, stratified archaeological material representing byproducts of ancient metal production from that ore deposit (e.g., slags and litharge) were used for construction of a set of lead isotope ratios characterizing the ore source. The results of this approach for the Cypriot copper ores are shown in Figure 4, in which the lead isotope ratios for each of the individual Cypriot ore deposits clusters tightly together, independently of the mineralogical type and spatial origin within the deposit of the sample analyzed.

In addition to the lead isotope analyses, it is necessary to take into account isotope geochemistry, metallurgy, and ore deposit geology. In seek-

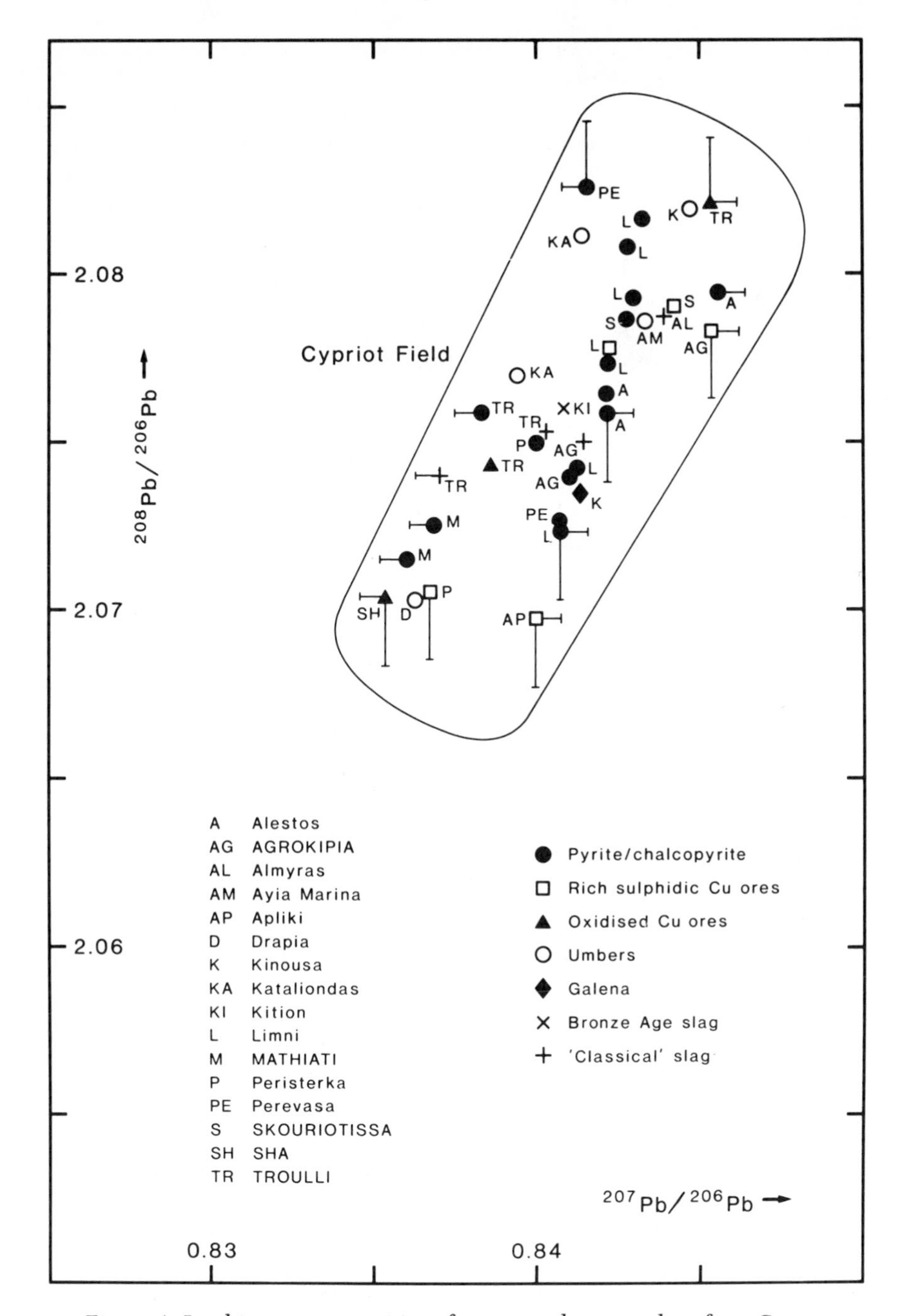

Figure 4. Lead isotope compositions for ores and copper slags from Cyprus. Copper ores, umbers, galena and copper slags have lead isotopic compositions falling into the same field. Error bars at the 95% level for the lead isotope ratios are shown on this and subsequent figures. (Reproduced with permission from ref. 3. Copyright 1986 University of Birmingham.)

ing the provenance of copper it is necessary to compare only artifacts with the lead isotope compositions of copper ore deposits. Pernicka et al. (*42*) compared copper artifacts with pyrite–blende–galena ore deposits that contained no copper. In some cases, both copper and lead ores occur in the same ore deposit (e.g., Funtana Raminosa in Sardinia). In such cases, the lead isotope compositions of the copper ores will usually (in base metal ore deposits free of uranium and thorium) be the same as those of the lead ores.

A knowledge of the geology is important in order to determine the proportions of lead and copper and whether the copper and lead ores may have been so finely intergrown that it was impossible for Bronze Age peoples to separate or even recognize them. Again, careful investigations of isotopic homogeneity are necessary in the case of vein deposits of copper, in which variable contamination by country rocks may have occurred. Caution is again indicated if high U/Pb ratios in a given copper deposit are suspected. It is very helpful if mining or metallurgical (slags) activities associated with a given copper ore deposit can be dated and if it can be established that nearby archaeological sites have used the ores at a known period.

The application of lead isotope analyses to provenancing almost pure copper artifacts seems relatively straightforward in most instances, but what of arsenical copper and tin bronze alloys? In principle, the arsenic in arsenical copper may have been derived from high-arsenic minerals (such as tennantite or basic copper arsenates) containing lead of a different isotopic composition from that of the copper ore used to produce the copper. However, absolutely no archaeological or other evidence of proves that this was ever done in the Bronze Age Aegean. On the contrary, at Kythnos definite evidence shows that arsenical copper was produced in EBII times from arsenical copper ores of variable (sometimes zero) arsenic content but of uniform lead isotope composition.

Tin bronze is another matter because, apart from rare copper ore deposits also containing tin (e.g., some of those in Cornwall), tin from a quite different source than the copper has almost always been added to copper to produce bronze. The only tin mineral that is likely to contain lead is stannite, which is not common except as a mineralogical curiosity. Tin has always been obtained in very large quantities from cassiterite, which almost never contains even a trace of lead. The tin ingots found underwater off the coast of Israel contain no lead. Our work on Cypriot Late Bronze Age bronzes indicates no perturbation away from the characteristic Cypriot lead isotope composition, even for bronzes containing 18% tin.

The most serious potential problem for the use of lead isotope analyses for provenancing copper in copper-based alloys is the possible remelting together of metal derived from different ore sources. The effect of melting together, in various proportions, two copper artifacts of different lead isotope compositions is to generate a number of objects having isotopic compositions falling, in the usual lead isotope diagrams, on a straight line joining the two end-member compositions. Melting together in random ways a number of objects, each of which has a different lead isotope composition, would generate even more smeared-out isotope compositions.

If remelting together of scrap from diverse copper ore sources were common, then one would commonly observe "smeared-out" lead isotope patterns that covered larger areas than are characteristic of ore deposits. So far, we have not seen these lead isotope patterns in our studies; this finding argues that remelting may not have been a common practice in the Aegean Bronze Age. Some archaeological evidence that suggests that remelting may not have been a very common practice at any time in the Aegean, where it was, at all times, common to bury metal objects with the dead (*7*).

For Middle Minoan Crete, Jennifer Moody (*39*) has shown that metal objects are predominantly found not in settlements or palaces, but in graves where they were largely safe from reuse. Further, Branigan (*40*) has shown that the reriveting of daggers was a common practice in Early and Middle Minoan Crete, and some 15% of reriveted daggers show clear evidence of having been reriveted on two or more separate occasions. Branigan (*40*, p. 46) comments that

> *all the rerivetings are perhaps indicative of the trouble to which Minoans of the Early Bronze Age went in order to avoid scrapping their weapons, or even having them melted down and recast.*

In fact, it was always easier to repair riveting or rework a damaged edge than to melt down and recast a tool or a weapon. Even in the Late Bronze Age, the Aegean was unusual in comparison to other parts of Europe in that the quantity of hoarded metal was very small in absolute terms and also in comparison to the amount of metal deposited in graves (*22*). The quantities of metal deposited in Mycenean and Minoan graves were much greater than in most of Europe. In many cases, hoards of metal in the Aegean were not metalworkers' hoards but clearly metal possessions hidden away in time of danger that the owner hoped to recover in happier times.

Even if a limited amount of melting and mixing of metal from discarded objects did take place, it may not have caused much confusion in the lead isotope pattern. If the various discarded objects in a founder's hoard were all made from copper that derived from the same ore source, then the mixing together of their metal to make new objects would not alter the lead isotope composition characteristic of that ore source. Strong economic and technical arguments suggest that, at a given time, a particular site or area will tend to receive copper predominantly from one, or perhaps two, ore sources. For the Late Minoan II founder's hoard found in the Unexplored Mansion in Crete, this seems to be the case. Eighty-five percent of the objects were made of copper from the Laurion ore source, and 15% from other copper ore sources (*41*).

Pernicka and his colleagues (*42*) have suggested a very valuable method to prove whether a given body of data exhibits mixing. The method depends on having both lead isotope and trace element data for the metal artifacts in question. Collinearity between three objects in a particular lead isotope–trace element diagram admits the possibility that the middle object may have been made of metal mixed from the other two. This hypothesis is negated if, in another such diagram, for different trace elements, these three objects do not lie in a straight line. The method is laborious because all possible object combinations must be examined, but this could readily done by using a computer. We have applied this technique to investigating whether the analyzed objects from Troy and the "Troas" (*6*, *42*) were made from mixtures of remelted objects: they were not.

The question of the "remelt effect" is still stressed by archaeologists who have not presented any properly discussed archaeological evidence that it was a common phenomenon in the East Mediterranean Bronze Age.

Ore Sources for the Bronze Age Mediterranean

The three copper ore deposits mentioned (Kythnos, Laurion, and Cyprus) played important roles as copper sources in the Eastern Mediterranean at different stages of the Bronze Age. Among these deposits, only Cyprus had the reputation as a copper source in the archaeological literature at the outset of our research. Theories about the metal sources exploited in Bronze Age Greece varied from the use of small sources local to any of the cultural centers throughout Greece, (*see* references 43 and 44), to the theory of centralized copper production on Cyprus (*1*) and the sudden introduction of metals and metallurgy into the Aegean from the East [the "metallschock" theory introduced by Schachermeyr (*45*)] in Early Bronze Age II times.

Early Bronze Age. Our work on the Early Bronze Age metal sources started as an investigation of the sources of lead and silver for the Early Cycladic people (*5*). The result of this work, as shown in Figure 5, was proof that the dominant source of silver and lead for the Early Bronze Age Cyclades was the Cycladic island of Siphnos, and that in the later part of the Early Bronze Age the Laurion lead–silver deposits also provided these metals (*16*).

Since 1982, we have extended our research to the possible sources of copper in Greece and have surveyed many of the known copper deposits for any evidence of Bronze Age exploitation. In this research, perhaps the most rewarding site was discovered on Kythnos.

Kythnos, a Cycladic island in the neighborhood of Kea and Seriphos, came to our attention when surveying the Cycladic ore deposits. Fiedler, a German emissary of King Otto of Greece, mentioned in his report (*46*) an ancient iron slag heap on Kythnos (*46*). After our field work on Kythnos in 1982, it became clear that the slag heap near Ayios Yoannis Bay contained not iron but copper slag and that most probably it was the residue of Early Bronze Age copper production (*47*). On the basis of the lead isotope analyses of the slags and ores found within the slag, as well as other ores nearby, we have established the lead isotope field characteristic of the Kythnos copper ore deposit. Dating the site was made possible by fragments of Early Cycladic II pottery found within the slag heap. Kythnos is, therefore, the earliest proven copper smelting site known in Greece. This archaeological evidence was confirmed by the results of some lead isotope analyses of Early Cycladic copper artifacts.

Figures 6 and 7 show the lead isotope data for the Kythnos Hoard, a hoard of Early Cycladic tools now held in the British Museum (*48*), and Early Cycladic daggers from the Ashmoleum Museum collection. All tools from the Kythnos Hoard and 12 out of 17 daggers from Amorgos were consistent with the lead isotope ratio characteristic for Kythnian slag and ore, as seen by comparing Figures 6 and 7.

Kythnian copper was distributed to other islands in the Aegean. Several Early Bronze Age copper objects of Kythnian origin were found on the island of Kea (Figure 8), and preliminary work with K. Davaras shows that the majority of metal objects from the Early Minoan site of Ayia Photia in Eastern Crete originate from Kythnian copper. The latter is not too surprising because the general archaeological affinities of the Ayia Photia copper, ceramic, and marble finds are Cycladic.

Chemical analyses of slags and ores from the Ayios Yoannis slag heap on Kythnos yielded some interesting technological information. Fragments of ore found among the discarded slag at this site prove that the chief copper ore smelted there was malachite plus azurite in a ferruginous matrix. Analyses show that the ore sometimes contained a few percent of arsenic. Analyses of copper prills contained within different fragments of the smelting slag show arsenic levels from zero to 4.5%. The arsenic levels prove that in the third millennium B.C. arsenical copper was sometimes accidentally produced in the smelting of copper from its ores. Most likely, the superior properties of some batches of this accidentally produced arsenical copper were recognized, and these batches of copper were selected for making weapons and tools.

Accidental smelting of arsenical copper was likely wherever copper was smelted and the ores used happened to contain some arsenic. The widespread early use of arsenical copper could possibly be the result of accidental independent "invention" in many different copper-producing centers. So far, there is no evidence that copper was produced on Kythnos before or after Early Cycladic II times.

Turning to another important Cycladic site, the settlement of Kastri on Syros, first excavated by Tsountas (*49*) and systematically excavated later by Bossert (*50*), occupies the uppermost part of a barren tongue-shaped hill, surrounded by deep ravines. Across a very steep ravine from Kastri lies the

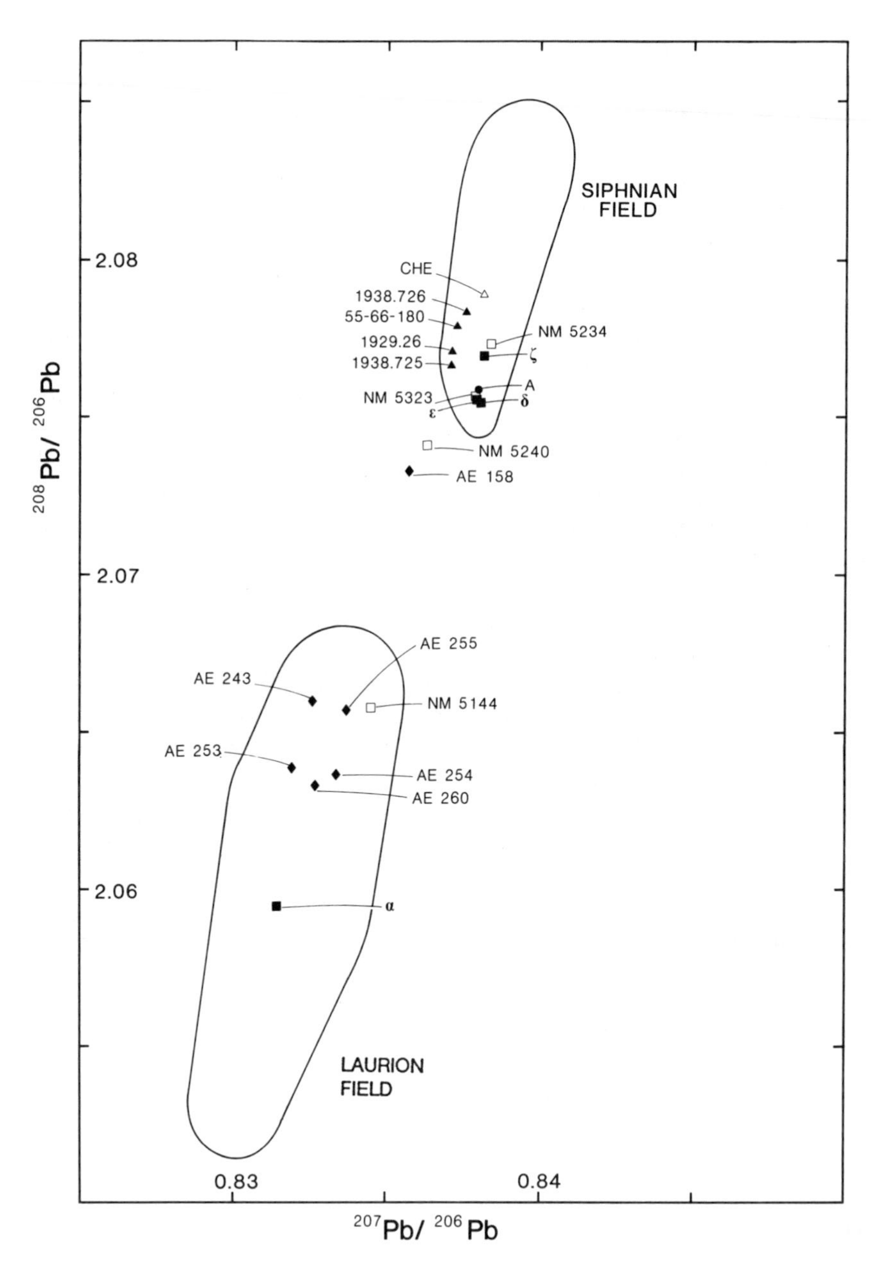

Figure 5. Lead isotope compositions of Early Cycladic lead and silver artifacts.

◆ AE 158 Amorgos, Kapros, Ag Bowl

◆ AE 253 Amorgos, Ag Bracelet

◆ AE 254 Amorgos, Ag Bracelet

◆ AE 255 Amorgos, Ag Bracelet

◆ AE 260 Amorgos, Ag Bracelet

◆ AE 243 Amorgos, Arkesine, Pb object

▲ 1929.26 Naxos, Pb Boat Model

▲ 1938.725 Naxos, Pb Boat Model

▲ 1938.726 Naxos, Pb Boat Model

▲ 55-66-180 Naxos, Pb Boat Model

□ NM 5234 Syros, Kastri, Ag Diadem

□ NM 5144 Syros, Chalandriani, Ag Pin

□ NM 5323 Syros, Kastri, Pb Weight

□ NM 5240 Syros, Kastri, Pb Rivet

△ CHE Cheiromylos, Pb fragment

● A Antiparos, Pb Figurine, BM 84, 12-13, 20

■ NM 11.310 Naxos, Pb Rivets

Figure 5—Continued

large necropolis of Chalandriani, also excavated by Tsountas. Modern archaeological studies show (*51*) that, whereas Kastri was occupied only for a short time in ECIIIA times, the cemetery at Chalandriani was in use both before and during the occupation of the settlement of Kastri.

Many authors have referred to the occurrence of Early Cycladic tin bronze artifacts (e.g. *48*, *52*), but a closer examination shows that the only tin bronze objects from proven Early Cycladic contexts are those from Kastri on Syros (*6*, pp. 41–42). Tin bronze was not found at all in our analyses of 31 Early Cycladic objects from Amorgos, Paros, Kythnos, and Chalandriani on Syros (*53*).

In other ways, the nature of the Kastri settlement is not purely Cycladic. Bossert (*50*, pp. 70–75) remarked that although it was difficult to find parallels between the finds at Kastri and those in the graves of Chalandriani, the differences were very apparent. Although sauce boats found at Chalandriani pointed to connections with the Greek mainland, they were completely absent at Kastri, where the distinctive pottery forms have clear Anatolian comparisons at Troy II. Notable features of the finds at Kastri are the metal objects and the evidence that metallurgy was practiced there. Tsountas (*49*)

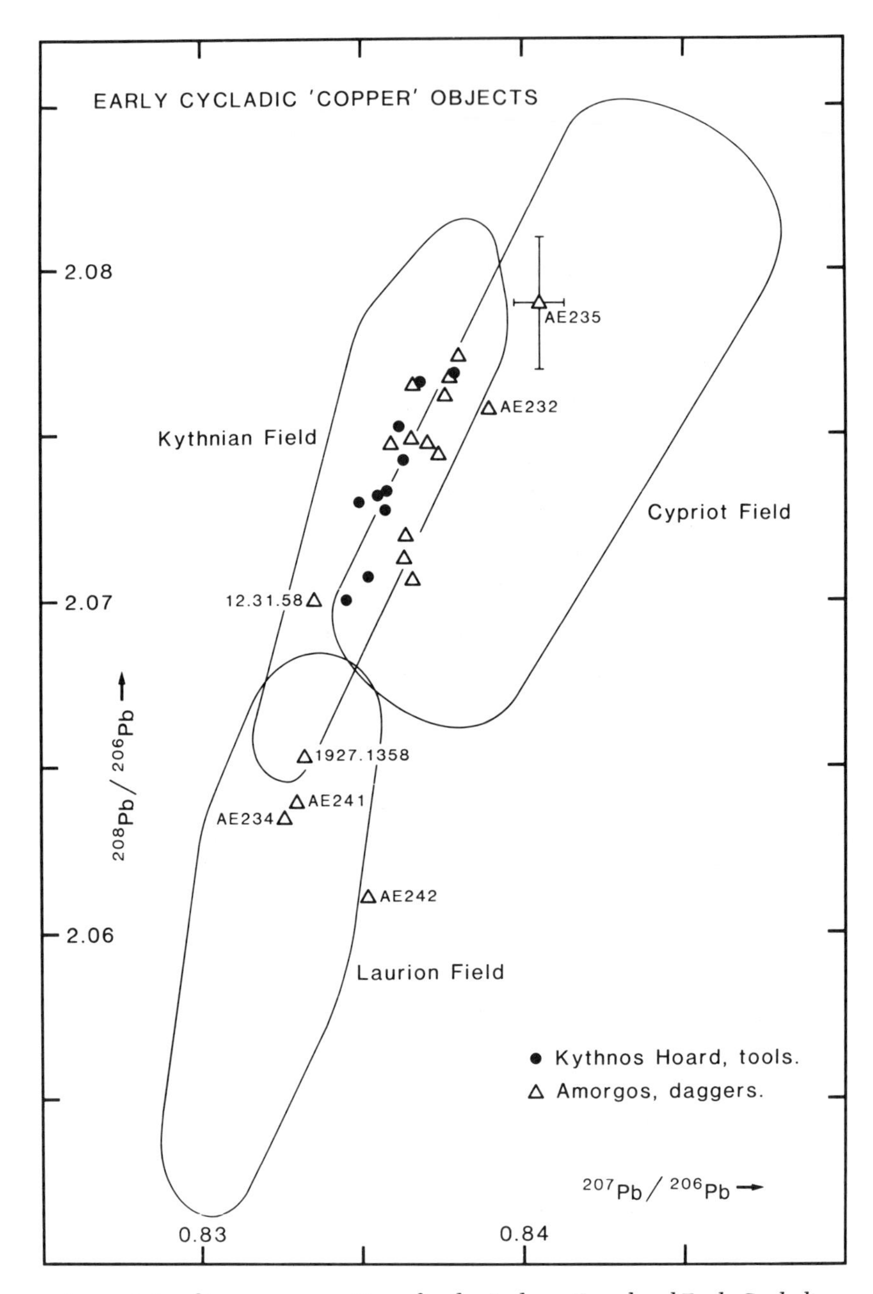

Figure 6. Lead isotope compositions for the Kythnos Hoard and Early Cycladic copper-based alloy artifacts from Amorgos.

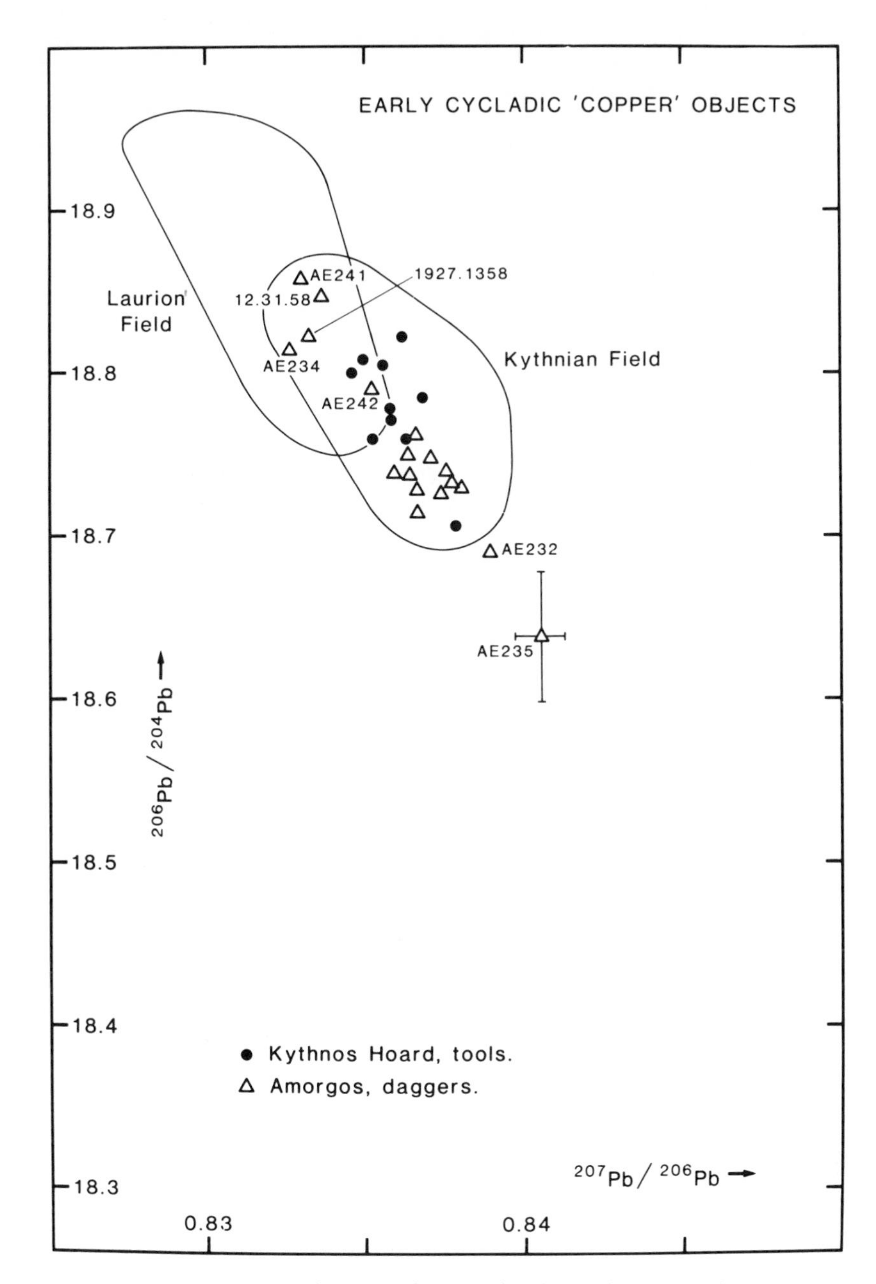

Figure 7. Alternative lead isotope diagram for the Kythnos Hoard and Early Cycladic copper alloy artifacts from Amorgos.

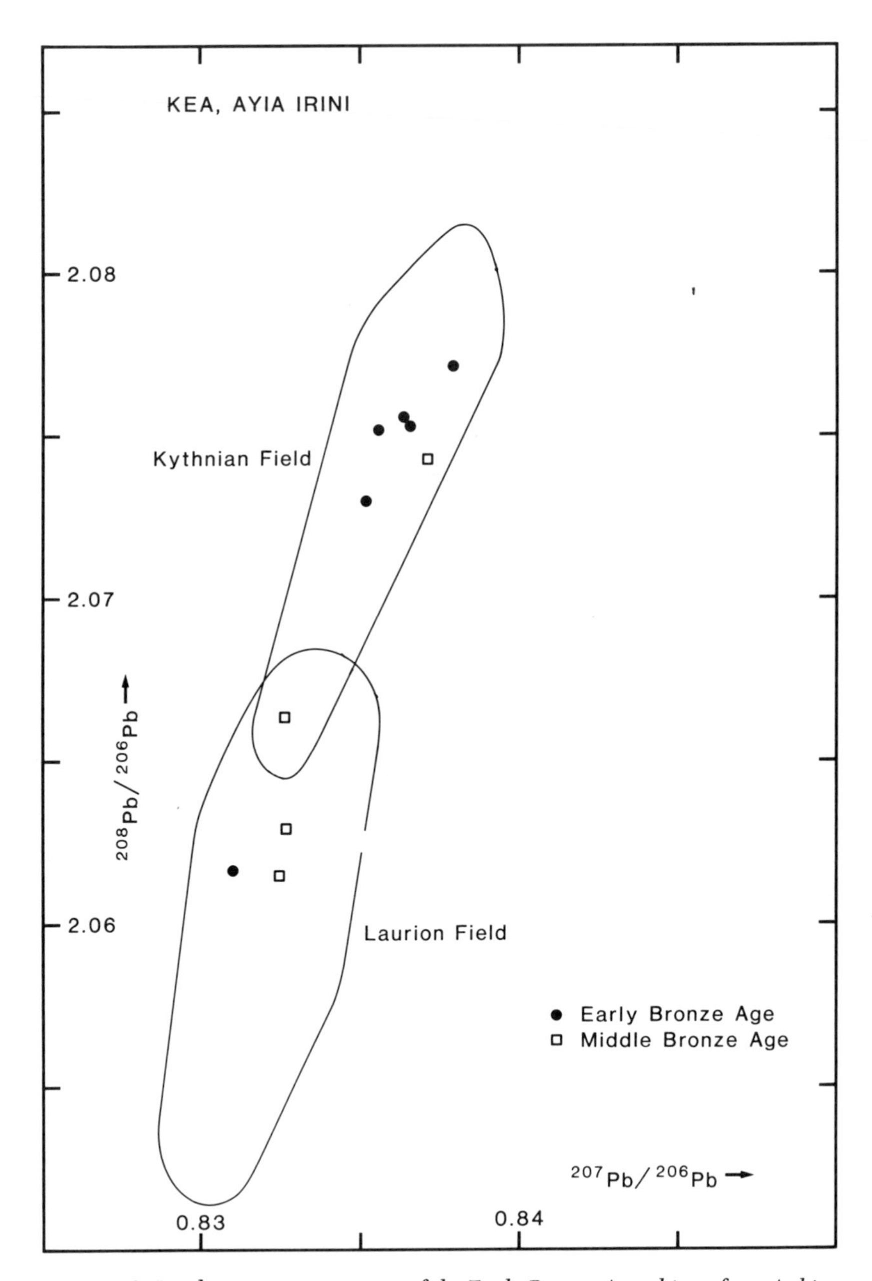

Figure 8. Lead isotope composition of the Early Bronze Age objects from Aghia Irini on Kea.

and Bossert (*50*) found crucibles containing traces of bronze and lead; molds for casting flat axes, arrowheads, swords, spearheads and chisels (*54*); a silver diadem and some lead objects (*5*); a few objects of arsenical copper; and many objects of tin bronze.

Although rich copper ores, particularly azurite, malachite, or cuprite, can be smelted in crucibles, the so-called "slag" from Kastri reported by Tsountas (*6*, pp. 31, 42) is, in fact, a lump of arsenical copper metal. We concluded that there is no positive evidence of smelting at Kastri. The archaeological evidence is for melting and casting metals only.

How then the metal artifacts from Troy II compare with those from Kastri on Syros? Chemical analyses of 19 objects from Troy II (*see also* reference 42) revealed that 14 are high tin bronzes (*6*, pp. 39–40). Tin bronze was not common at this time in Anatolia, but it seems to have been quite abundant in Troy and in Thermi on Lesbos (*43*). The lead isotope analyses of copper-based objects from Troy show that Troy obtained copper from a surprisingly large number of sources (*see also* reference 42).

Figure 9 shows the results of the lead isotope analyses of these objects compared with the lead isotope composition of the objects from Kastri on Syros and Chalandriani. On the basis of our measurements of the amount of variation in lead isotopic composition found for other copper ore deposits, we have estimated approximate bounds for at least five different copper ore sources from which Trojan copper must have been derived. As yet, there is not enough lead isotope data on the Anatolian copper ore deposits to be able to directly link the estimated lead isotope fields with particular copper occurrences. However, because the lead isotope composition is, to a first approximation, controlled by the geological age of the ore deposit, these five different ore sources, of widely different lead isotope compositions, must have been formed at quite different times.

The rough geological age of ore formation corresponding to lead isotopic compositions that fall into various parts of a diagram, such as Figure 9, can be calculated from models of the evolution with time of ore lead isotope compositions (*55*). On this basis, one can deduce the approximate geological model ages of the copper ore deposits associated with the five groups of Trojan artifacts as follows:

Group A	Pliocene to Oligocene (2–34 million years ago)
Group B	Eocene to Triassic (34–200 million years ago)
Group C	Triassic to Devonian (200–360 million years ago)
Group D	Silurian to Precambrian (400–700 million years ago)
Group E	Precambrian (700–900 million years ago)

The Turkish Mineral Research and Exploration Institute has classified Turkish copper ores according to geological age. Figure 10 shows eight different regions in Anatolia and one in Bulgaria where copper deposits exist

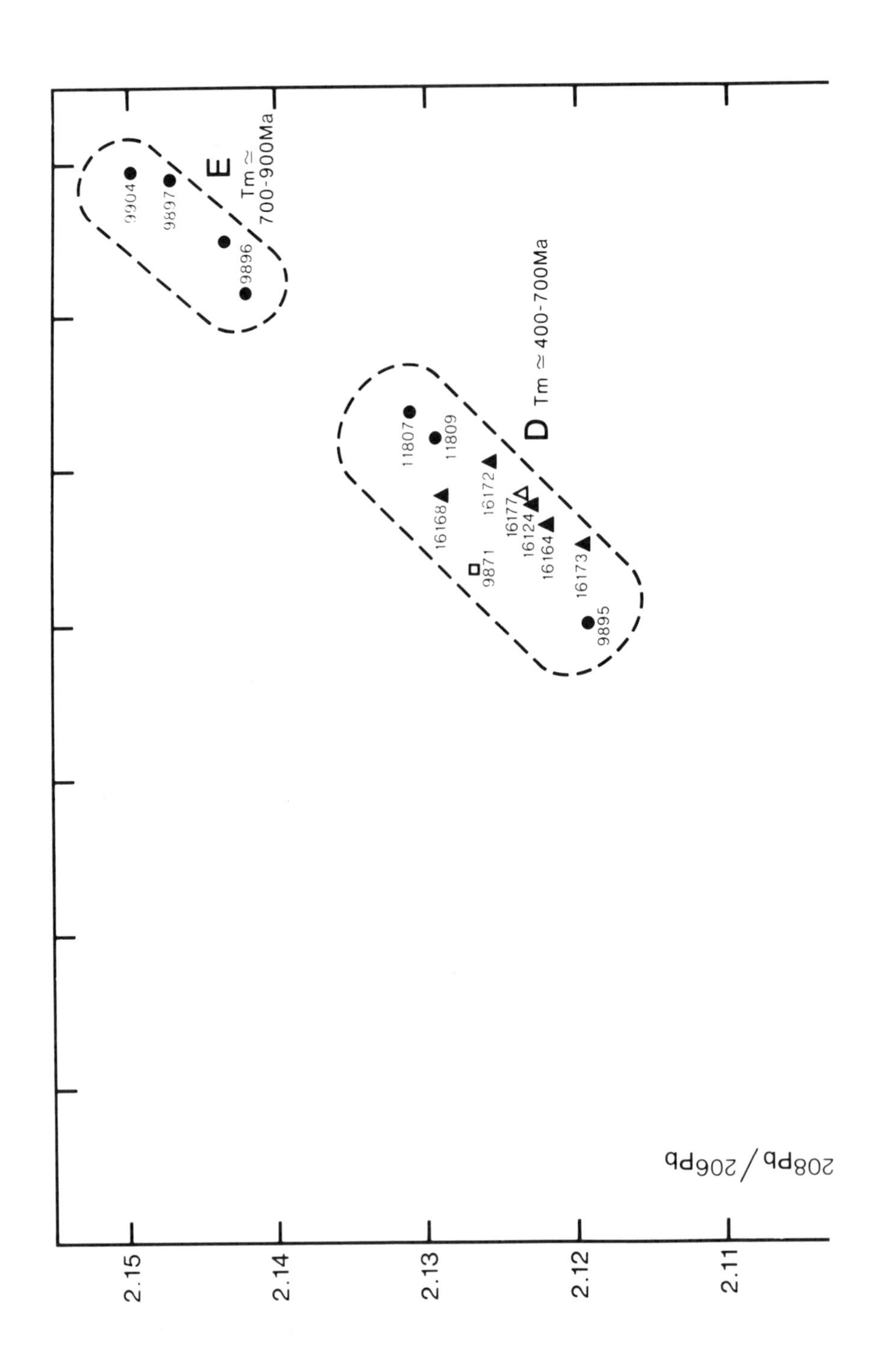

E
Tm ≃ 700-900Ma
9904
9897
9896
D
Tm ≃ 400-700Ma
11807
11809
16172
16168
16177
16124
16164
9871
16173
9895
2.15
2.14
2.13
2.12
2.11
$^{208}Pb/^{206}Pb$

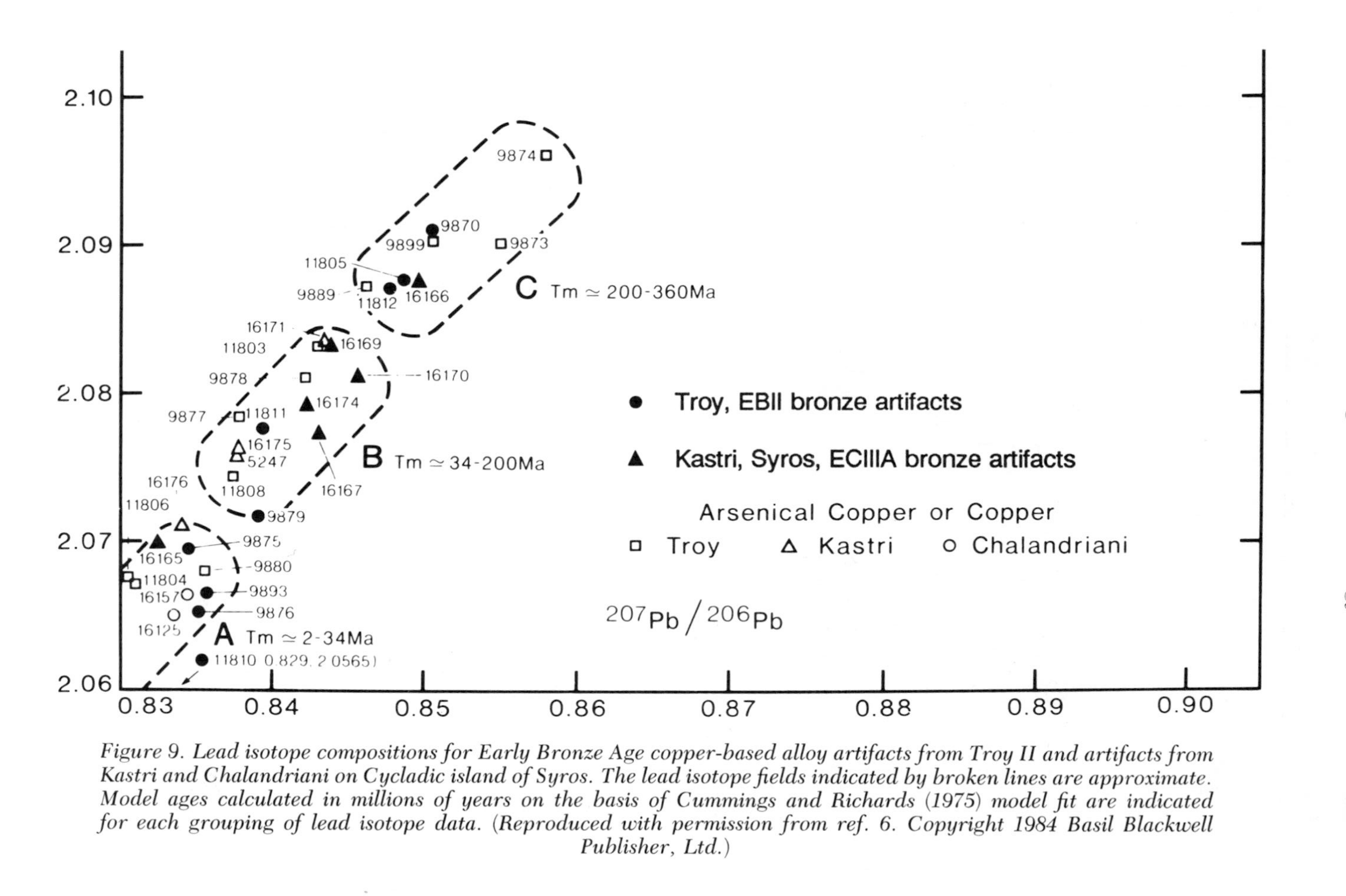

Figure 9. Lead isotope compositions for Early Bronze Age copper-based alloy artifacts from Troy II and artifacts from Kastri and Chalandriani on Cycladic island of Syros. The lead isotope fields indicated by broken lines are approximate. Model ages calculated in millions of years on the basis of Cummings and Richards (1975) model fit are indicated for each grouping of lead isotope data. (Reproduced with permission from ref. 6. Copyright 1984 Basil Blackwell Publisher, Ltd.)

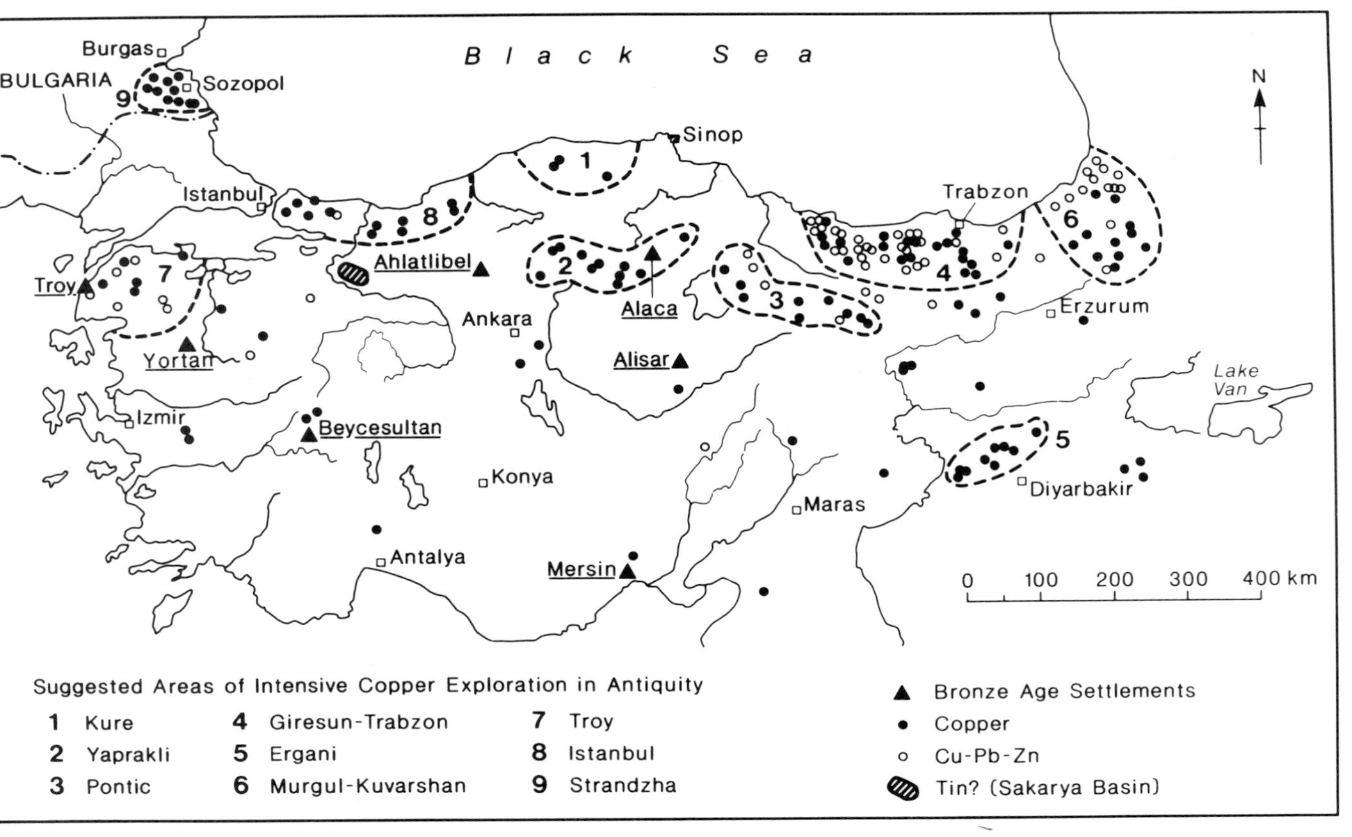

Figure 10. Suggested areas of intensive copper exploitation in Antiquity, adapted from de Jesus (1980, 397). (Reproduced with permission from ref. 6. Copyright 1984 Basil Blackwell Publisher, Ltd.)

that may have been worked intensively for copper in ancient times. Six of these regions were identified by de Jesus (*56*). The Strandzha region in Bulgaria was discussed extensively by Chernykh (*57*). There are a number of copper ore deposits near Troy. Some of the regions of Turkey from which the copper ores may have come to produce the five lead isotope Trojan artifacts groups can now be predicted. Relying on the geological classification, the associations are as follows:

Group A	areas 7, 5, 2, and southern part of 6
Group B	areas 7, southern part of 3, 4, 5, 6, and 9 (9 on the basis of ref. 58.)
Group C	areas 1, 7, and 8
Group D	areas 1, 7, and 8
Group E	no known possible sources in Anatolia or Bulgaria

These associations are, at present, only tentative and assume a roughly single-stage lead isotope evolution, with negligible contamination of the copper ore deposits by lead derived from older country rocks. Nevertheless, this approach should be used as a basis for further field exploration of the areas in question. The mines and ore deposits in north and northwestern Turkey have recently been explored by a team from the Max Planck Institute for Nuclear Physics in Heidelberg (*42*, *59*), and we hope that there will soon be more information available for the lead isotope "fingerprints" of Anatolian ore sources. Information available so far proves only that Troy did not get copper from its hinterland and does not identify its source.

The lead isotope analyses of the copper-based objects from Kastri on Syros, represented in Figure 9, show that none of these objects is made of copper from Kythnos, Laurion, or any other source typical of indigenous Early Cycladic copper objects. Six of the Kastri objects cluster in the Trojan lead isotope field D, eight in field B and two in field A. The isotopic evidence suggests strongly that the objects from Kastri were made with copper from ore sources that were also exploited in Troy in EBII times. Together with the high tin content in all of the Kastri objects the evidence seems compelling that Trojan metal sources were used to manufacture the tools and weapons found at Kastri.

Some of the objects from Kastri falling in Trojan field B need further discussion. Four of these objects (16167, 16169, 16171, and 16174) have lead isotope compositions that resemble the fingerprint of Cypriot copper ores; however, we have rejected the hypothesis that these metal objects at Kastri are Cypriot in origin by comparing their trace elemental composition, particularly gold and silver content, with some Cypriot copper artifacts excavated on Cyprus. The results of the neutron activation analyses of these objects are presented in Figure 11.

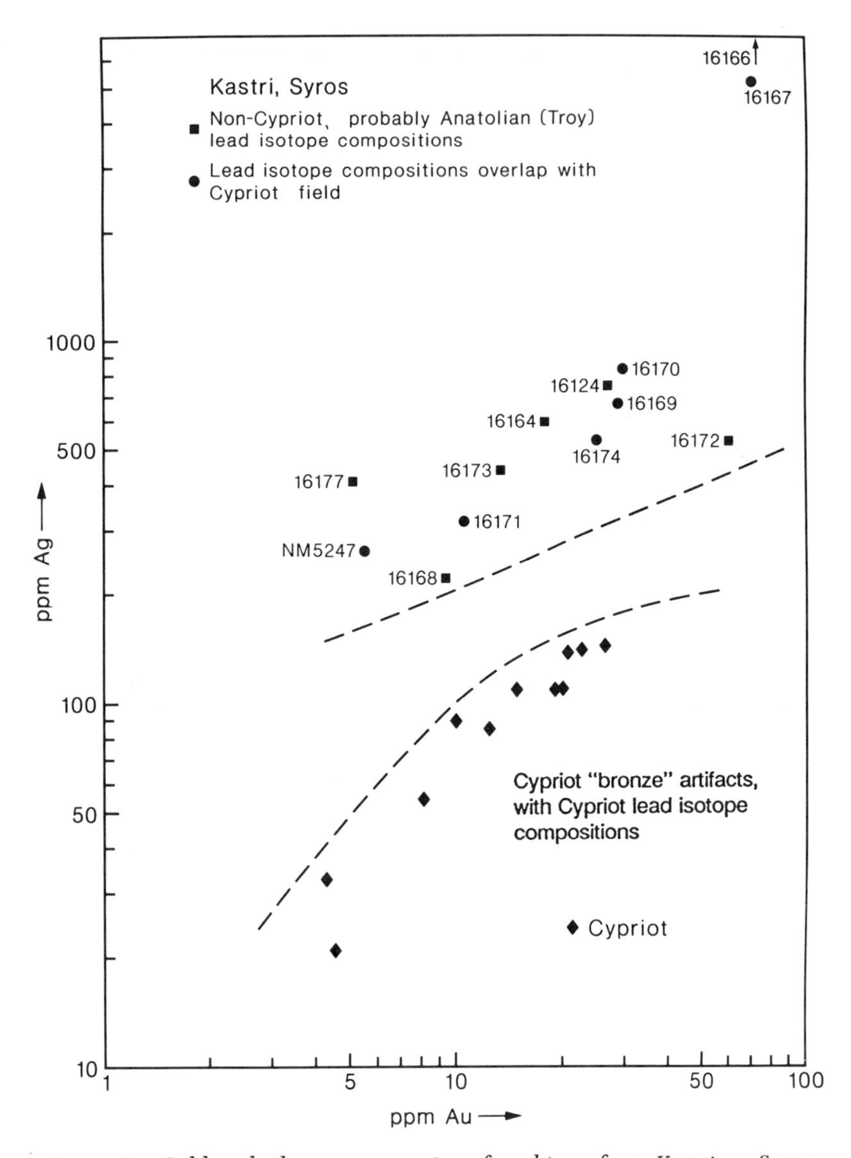

Figure 11. Gold and silver concentrations for objects from Kastri on Syros with Cypriot lead isotope composition compared with the gold and silver content of Cypriot artifacts. (Reproduced with permission from ref. 6. Copyright 1984 Basil Blackwell Publisher, Ltd.)

All of the objects from Kastri whose lead isotope compositions were confused with the Cypriot field fall in the upper part (silver-rich) of the diagram, along with those Kastri objects that fall in Trojan isotope field B, which is quite different isotopically from the Cypriot isotope field. It seems clear that the 16 analyzed objects from Kastri are Trojan both in alloy type and in provenance of the copper; moreover, many of these objects are also of Trojan typology (*50*). The objects might have been made in Troy and brought to Kastri as finished objects or perhaps, in view of the metallurgy practiced in Kastri, the metal was brought from Troy to Kastri and made into objects of Trojan type there. In either case, the objects are quite unlike the two analyzed from Chalandriani, both in alloy type and copper source. The combined archaeological and analytical evidence suggests very strongly that Kastri was a settlement of Anatolians, most probably from Troy.

Late Bronze Age. It seems very likely that, at least in the Late Bronze Age, the Minoans were involved in a substantial metal trade, if for no other reason than that metal deposits on Crete are rare, but Late Minoan bronze objects are plentiful. The nature and organization of the copper trade in the Late Bronze Age is particularly intriguing.

Because it is known both that Cyprus had large copper deposits and that copper metallurgy was actively practiced on Cyprus at these times, it has been most often assumed that the Myceneans and Minoans engaged in trade with Cyprus to procure copper.

Further evidence caused Aegean archaeologists generally to accept the idea that Cyprus was a very important source of copper for the Minoans and the Myceneans. That evidence is the appearance in Cyprus of Mycenean and Minoan pottery. Most of this pottery was dated by Strom and others to the 14th and much of the 13th century B.C. This pottery and its contents were taken to represent the Cypriot end of a trade route that was motivated by the Mycenean and Minoan need of copper. So far, however, there has been no direct proof that the copper or bronze objects in Minoan Crete were indeed made of Cypriot copper. Neither was there any direct proof that Cypriot copper ever left Bronze Age Cyprus in any quantity.

Some have adduced that the widespread distribution of copper oxhide ingots is evidence of a vigorous external Late Bronze Age trade in Cypriot copper. They assume, however, that all such ingots were made of Cypriot copper, a fact not hitherto either proved or disproved. What is the position revealed by the pottery finds? If we equate finds of Cypriot pottery overseas, as reviewed by Catling (*60*), with Cypriot trade, then in Late Cypriot I times the Cypriots continued to develop their trade with the Levant and Egypt while their relations with the West remained insignificant.

Cypriot pottery of this period is found in considerable amounts in Ugarit and Alalakh in North Syria, Gaza, Tell el Ajjul in Palestine, and in Egypt. Cypriot pottery was hardly found at all in Cilicia or in the Aegean. Very few

Cypriot objects are known in the Bronze Age Aegean. Cypriot pottery abroad suggests strong Cypriot trade with its coastal Eastern neighbors and very little trade with the Aegean.

Strom and others recorded (1979) large quantities of MycIIIa and MycIIIb pottery found in Enkomi, Kition, Hala Sultan Tekke, and Maroni. These pottery finds have been interpreted as evidence of a great Mycenean trading expansion and have been linked with the fall of Knossos at about 1380 B.C. In fact, Minoan pottery also came to Cyprus later in the 14th and 13th centuries B.C. The pottery came both as storage stirrup jars and as as fine LMIIIA2 and LMIIIB vases, from Kouklia in south–west Cyprus, near Larnaka and Enkomi in the south–east and at Akanthou on the north coast, and near Morphou Bay at Toumba tou Skourou (*60*).

Neopalatial hoards show that in Late Minoan Ia times (1550–1500 B.C.) Crete was very rich in bronze; most of the copper oxhide ingots found on Crete, in particular those of Kato Zakro and Aghia Triadha, are from this period. Catling (*60*) suggested that, on the basis of some LMIA Minoan pottery excavated in the Morphou Bay area in northern Cyprus, one could expect that the nearby Skouriotissa mines were the source of copper for neopalatial Crete in general and the LMIA oxhide ingots in particular.

Oxhide ingots have a special connection with Cyprus. Apart from the oxhide ingots and ingot fragments found on Cyprus, two Cypriot openwork bronze stands contain offering bearers carrying oxhide ingots. Moreover, there is the bronze statuette of a female divinity standing on an oxhide ingot and the horned, armed god standing on an oxhide ingot excavated by Schaeffer at Enkomi (both of these date to the 12th century B.C.). Oxhide ingots have a wide distribution in the Mediterranean. They are found *in corpore* from Sardinia and Sicily in the West, through Crete and mainland Greece, to Bulgaria, Cyprus, Syria, and Anatolia in the East.

The importance of oxhide ingots in reconstructing ancient trade in the Mediterranean was shown by the discovery of the Cape Gelidonya Bronze Age shipwreck (*61*) off the coast of Turkey that had on board 39 such ingots. The importance of this find has recently been highlighted by the spectacular discovery of the Kas shipwreck that carried another cargo of oxhide ingots. All oxhide ingots appear to belong to the second millennium B.C. and most to its second half. They clearly represent a form in which raw copper was transported in the Late Bronze Age.

So far, no oxhide ingots have been found in western Crete and comparatively few on mainland Greece. A few fragments of oxhide ingots were excavated in Aghia Irini on the Cycladic island of Kea, one whole ingot and several fragments in Mycenae, and 19 were found in the sea off the coast of Euboea.

Samples from four different oxhide ingot fragments have been analyzed out of the total of 11 fragments found in the Mathiati hoard, as well as samples from four different fragments found at Skouriotissa together with tuyere and crucible fragments. All this material is in the Nicosia Museum on Cyprus. The lead isotope data for all these ingots is consistent with the lead isotope composition characteristic for Cypriot copper ores (Figure 12). It would have been surprising if any other result had been obtained. Nineteen oxhide ingots were found together in the Italian excavations at Aghia Triadha. They represent, according to Buchholtz, an early type and date approximately to late MMI or early LMIA (*1*); therefore, they were made a century earlier than the oxhide ingots from Cyprus. At present, we have the results of lead isotope analyses of nine of the Aghia Triadha ingots.

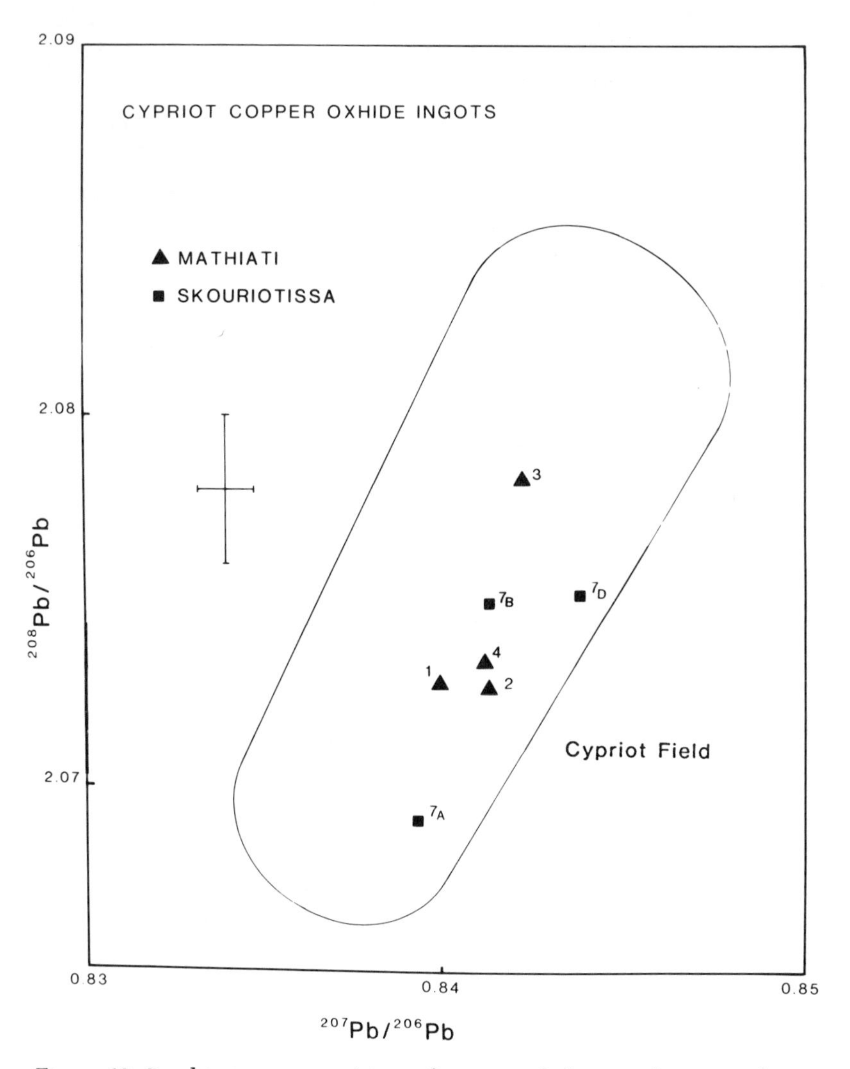

Figure 12. Lead isotope compositions of copper oxhide ingot fragments from Mathiati and Skouriotissa, Cyprus. (Reproduced with permission from ref. 3. Copyright 1986 University of Birmingham.)

Figure 13 shows that these ingots do not originate from Cypriot copper ores or Laurion, nor are they made of copper from the Ergani Maden mines in Turkey, which were frequently suggested as an important Bronze Age source of copper in Anatolia. The symbol Tm (Figure 13) gives lead isotope model ages calculated theoretically. The model age for the Ergani Maden

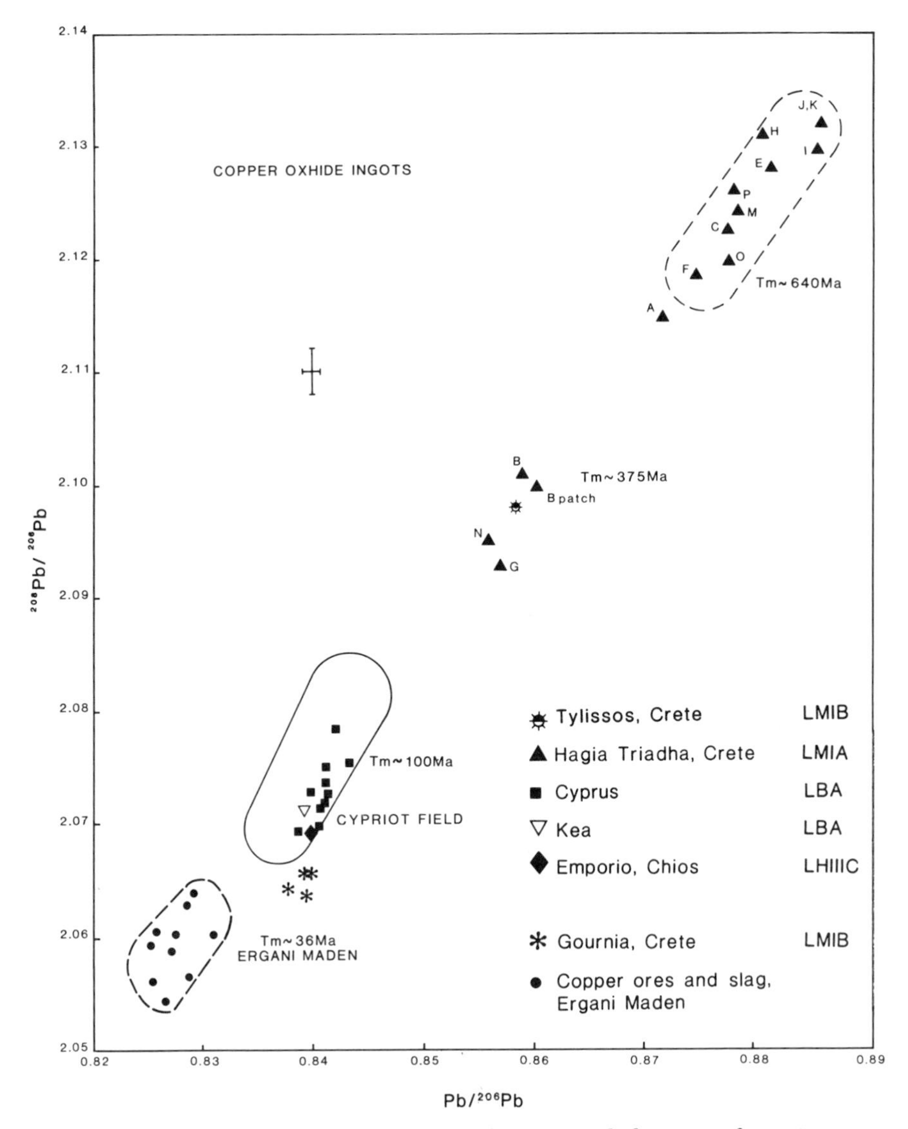

Figure 13. Lead isotope compositions of copper oxhide ingots from Ayia Triadha, Crete, in relation to the lead isotope fields for ores from Cyprus and Ergani Maden, Anatolia. The fields outlined by hatched lines are not yet firmly established.

mine is in the range of about 31.5 million years. The Cypriot copper ore deposit was dated to about 80 to 100 million years. In contrast, the model age calculations suggest that eight of the Aghia Triadha ingots were made of copper that came from a Precambrian ore deposit of about 640-Ma age and the ninth ingot came from a different ore deposit with a model age of about 370 Ma. These ingots are certainly not made from Cypriot copper.

The lead isotope composition of Lavrion mines in Attica is again quite different, so the possibility that the Aghia Triadha ingots originate from Lavrion can be rejected. Is it possible that the ingots are made from Cretan copper? We have sampled the scanty copper ores of Sklavopoulou and Chrysostomos and were given some copper ores from Miamou, Lasaia, Panagia, and Lebena on the southern coast of Crete. The lead isotope data for these Cretan ores can be plotted in the general region of the Cypriot copper ores (Figure 14); consequently, the Aghia Triadha ingots are not made from any known Cretan ores.

We cannot yet identify the copper sources used to make the Aghia Triadha ingots, but they may have been in Anatolia. If the lead isotope data of some of the Anatolian bronzes of an earlier date is compared with with those of Aghia Triadha ingots, there are clear similarities; however, the trace element compositions of these Minoan ingots are considerably different from those of the Anatolian objects.

The reason for this difference might be explained by the use of a different type of copper ore or different fluxes for smelting of the ore over the 500-year time span that separates the ingots from the Anatolian objects. We have also analyzed two other groups of copper-based Late Minoan objects. The lead isotope analyses included 23 bronze artifacts from the Late Minoan II destruction level in the Unexplored Mansion at Knossos and objects from the LMIIIB site of Armenoi in West Crete. The lead isotope compositions of the objects from the Unexplored Mansion are given on Figure 15. The results of lead isotope analyses indicate that most of the objects from this site originate from Lavrion mines; however, there are also three objects consistent with the Cypriot lead isotope fingerprint and three objects of a quite different lead isotope composition from an unknown ore deposit.

From western Crete we analyzed objects from the LMI site of Nerou Kourou. Figure 16 shows that all these objects are consistent with an origin from Lavrion copper ores. Other objects from western Crete that have been analyzed originated from the LMIII sites of Samonas and Armenoi. All copper-based objects from these sites are consistent with the Lavrion lead isotope composition and were certainly not made from Cypriot or Cretan copper. These results are not surprising. In contrast with Cyprus, Lavrion is not often mentioned in the geological literature as a copper source, but copper deposits in this area are well-known to economic geologists, even if in modern times they are not economically important (*62*, *63*).

A body of evidence has accumulated (primarily on the basis of lead isotope and chemical analyses) that copper from Lavrion was exploited at

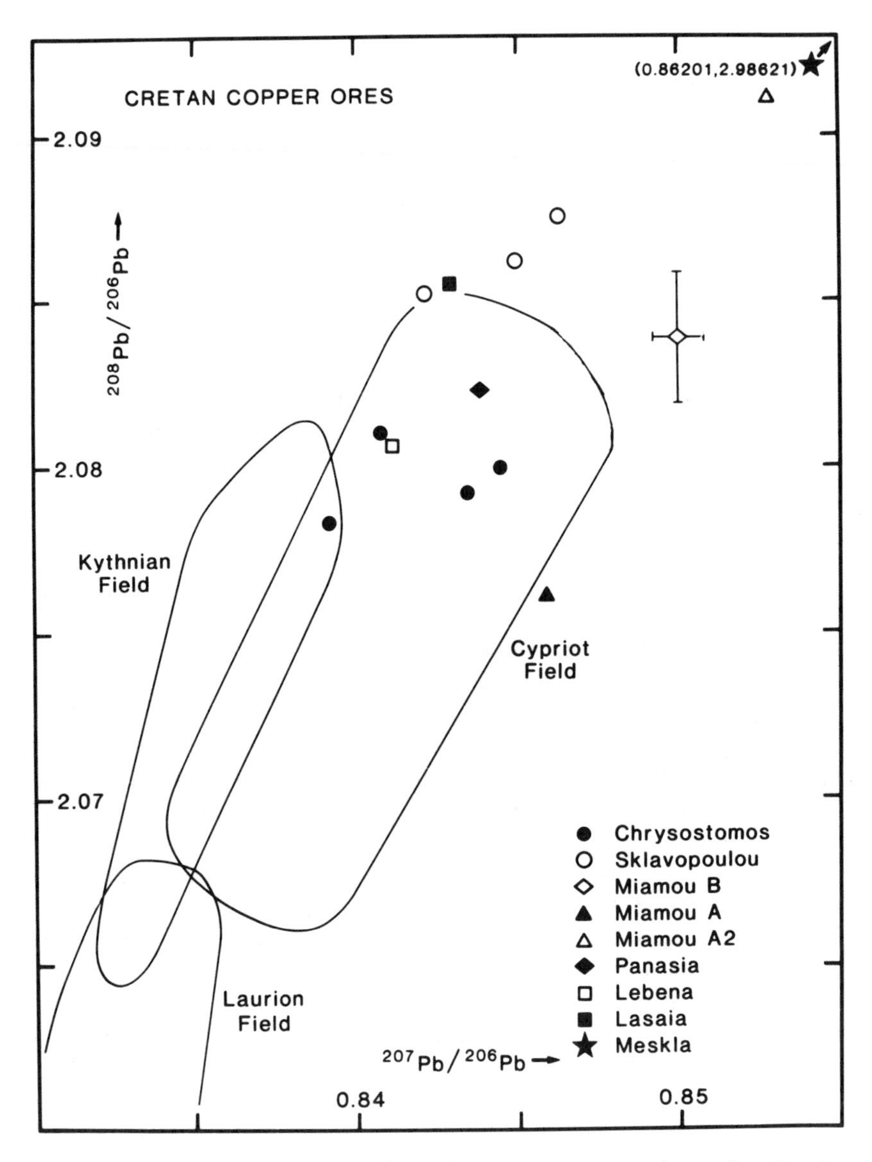

Figure 14. Lead isotope compositions of Cretan copper ores. (Reproduced with permission from ref. 3. Copyright 1986 University of Birmingham.)

least since Middle Minoan times. Copper slag of this date was excavated on the site of Aghia Irini on the Cycladic island of Kea; lead isotope analyses of these slag pieces, as well as later slags and bronzes, are fully consistent with the lead isotope composition characteristic of copper and silver ores from Lavrion (Figure 17). On this diagram, the only copper object analyzed

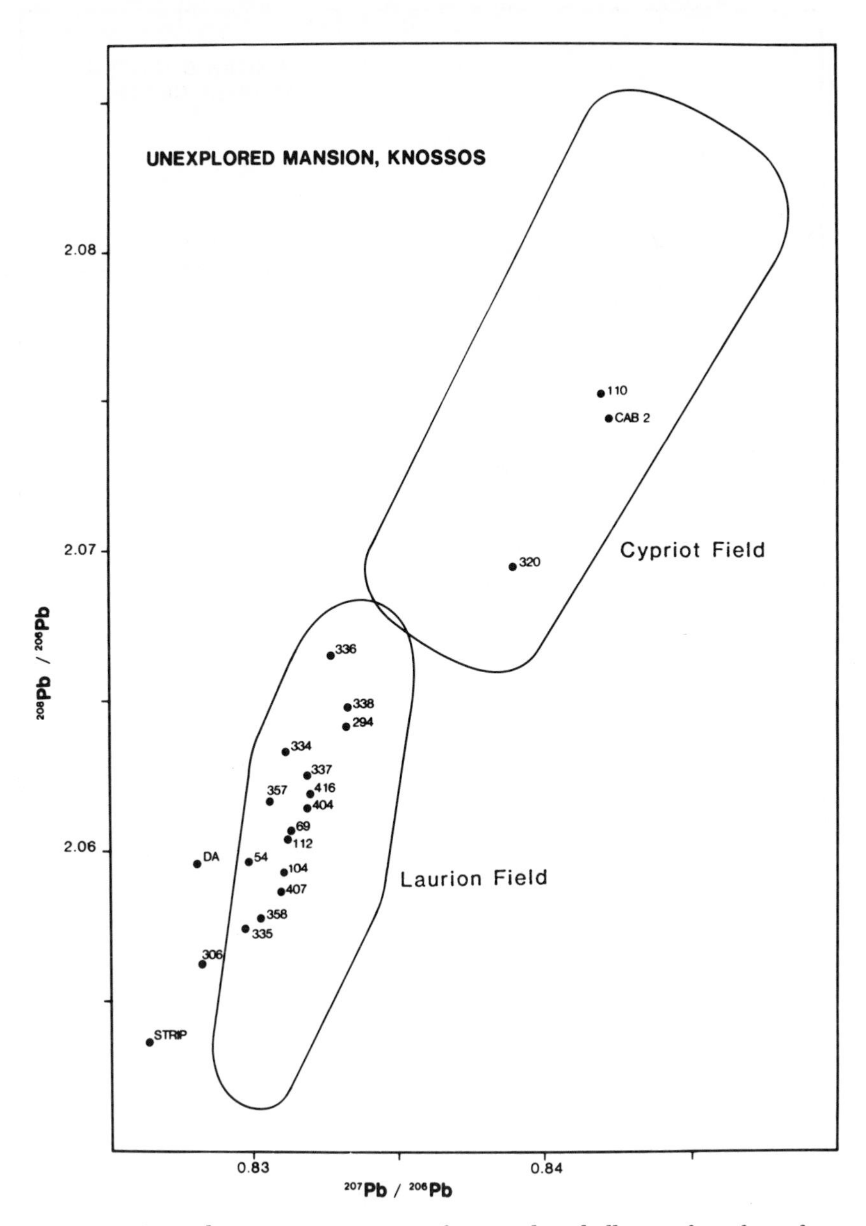

Figure 15. Lead isotope compositions of copper-based alloy artifacts from the Unexplored Mansion, Knossos.

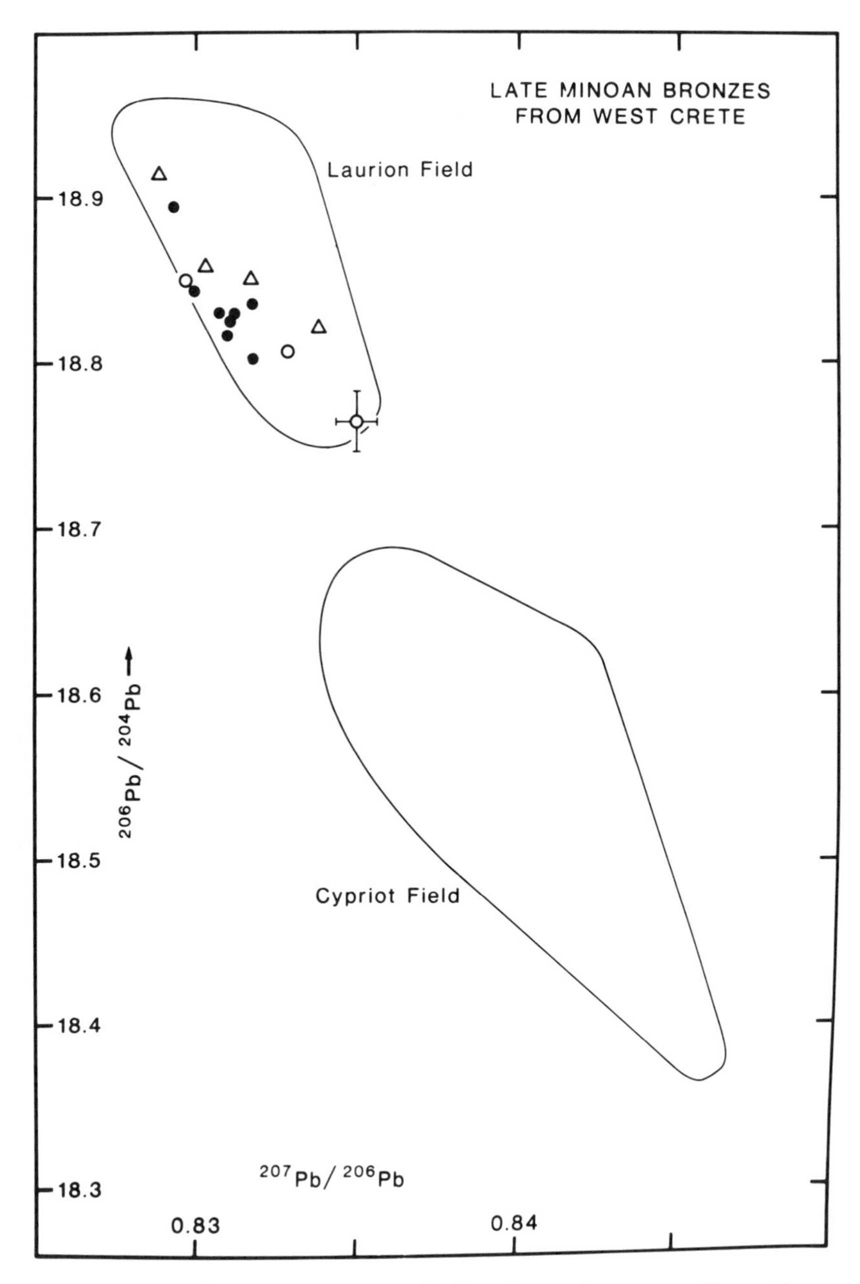

Figure 16. Lead isotope compositions for Late Bronze Age copper alloy artifacts from Armenoi, Samonas, and Nerou Kourou, West Crete.

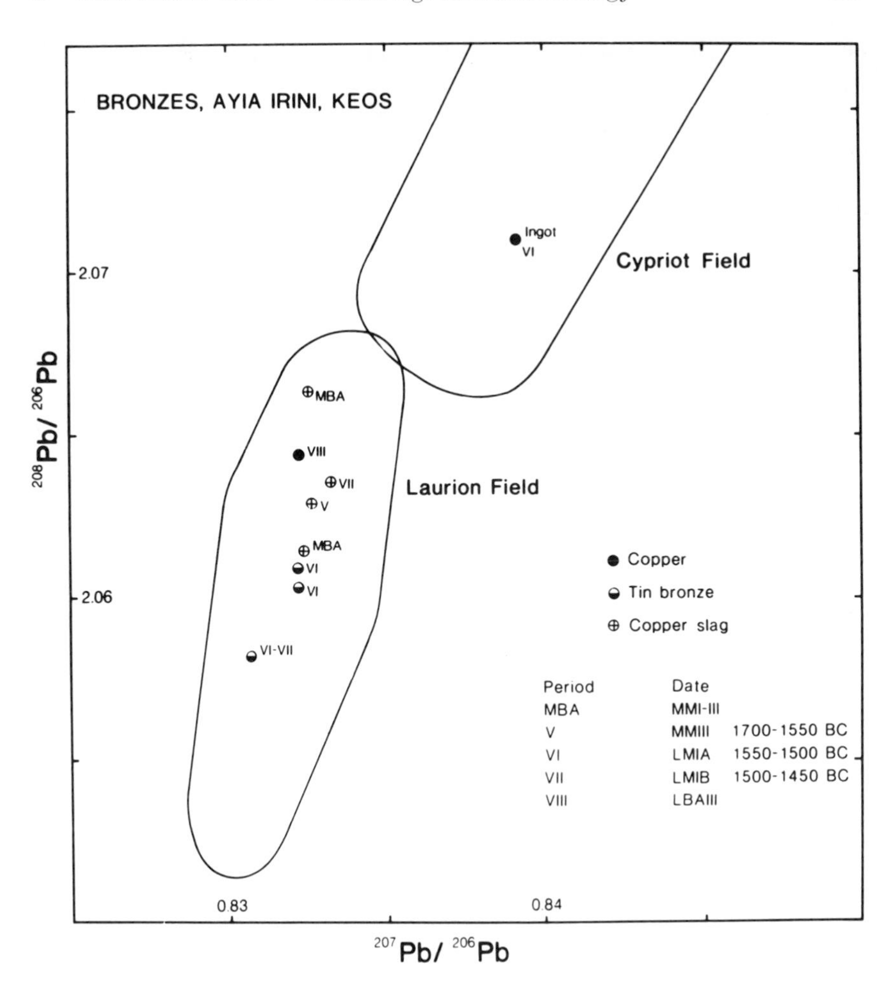

Figure 17. Lead isotope compositions for Late Bronze Age copper alloy objects from Aghia Irini on the Cycladic island of Keos.

from Kea with an isotopic composition consistent with Cypriot copper ores is a fragment of an oxhide ingot that dates to LMIA times.

It is still too early to draw sweeping conclusions from lead isotope evidence, but so far we have not analyzed an oxhide ingot that would be consistent with the Lavrion lead isotope composition. Also, so far, we have no evidence that Lavrion copper traveled much outside the borders of present day Greece. If, in the Late Bronze age, copper from Lavrion was used chiefly to supply Mycenean and Minoan markets, then perhaps there was no need to cast the metal smelted from Lavrion ores into an internationally recognized shape. It would be expected that in the Late Bronze Age Cypriot copper was in strong competition with any other copper source in the Aegean. In fact, the lead isotope analyses of 6 LMIA objects from Thera (Figure

18) revealed that the copper metal at that time came from both Lavrion and Cyprus.

In light of these results, the lead isotope composition of the objects from LMI and LMIII West Crete (Figure 16) is surprising in its uniformity. It is much too early to say that no copper sources other than Lavrion were used in western Crete. Statistically, however, it may be significant that not even one of the 15 objects analyzed so far falls out of the Lavrion field.

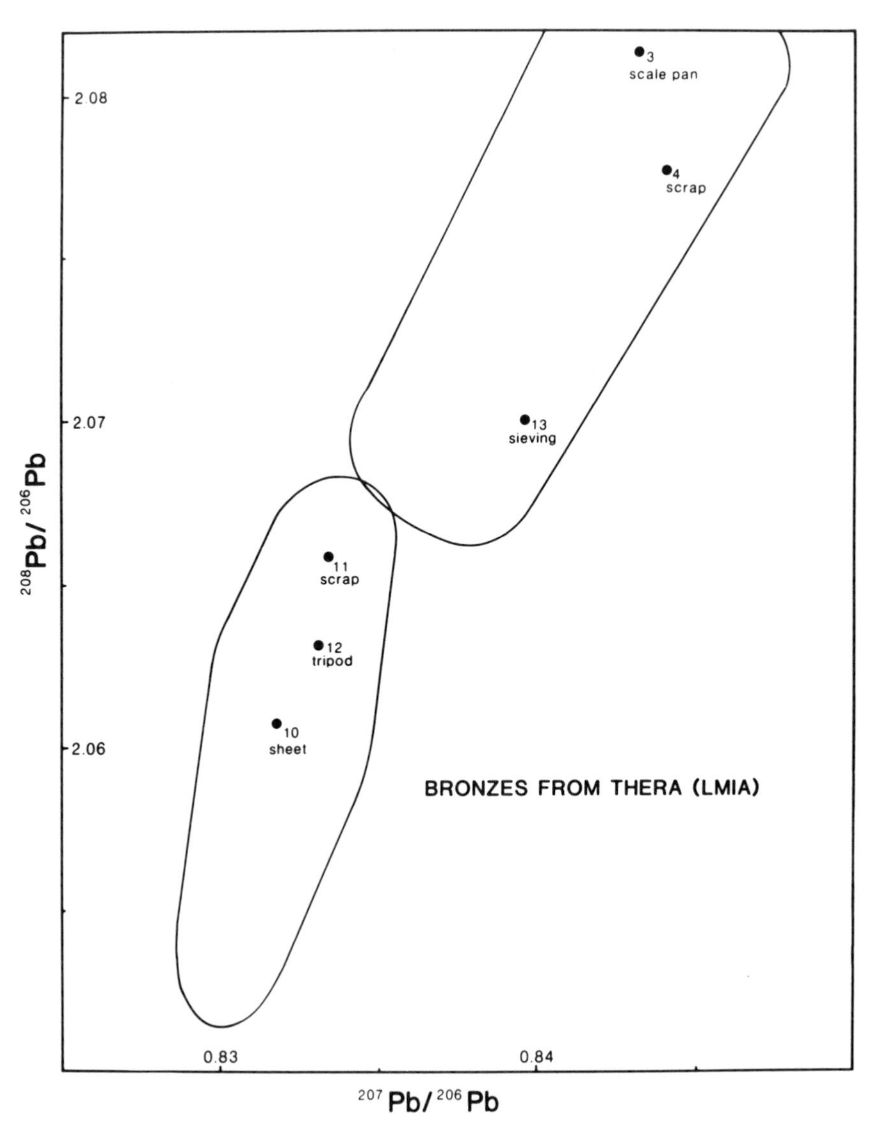

Figure 18. Lead isotope composition of the Late Bronze Age objects from the Cycladic island of Thera.

Summary

Cypriot copper does not account for the majority of the metal in Late Bronze Age copper-based objects from central and western Crete, Thera and Kea. With the exception of western Crete, several sources of copper reached these sites, but by far the major source in the later phases of the LBA was Lavrion in Attica. All Cretan oxhide ingots analyzed so far and a fragment of an oxhide ingot from Kea originated from different copper deposits, and not Lavrion. These ingots are earlier, however, and date to LMIA.

It is not surprising that copper from Lavrion was found in Crete. According to Schofield (*64, 65*) and Davis (*66*), a great deal of evidence shows that the Cycladic islands of the western string (including Thera, Milos, and Kea) formed part of a regular exchange network and that this was a major route by which Cretan goods and ideas were transmitted to mainland Greece. Lead isotope data (*16, 33, 41*) shows that in the Late Bronze Age lead and silver came from Lavrion to Crete via Kea, Milos, and Thera. It is not surprising that copper should have come from the same source.

Another route is possible by which metal supplies could have come to Crete. Western Crete is near the southern coast of the Peloponesus and the island of Kythera, where there is archaeological evidence suggesting considerable trade with Crete. The connections between Kythera and western Crete were noted by Coldstream and Huxley (*67*) in the similarities of pottery as early as EMII. This evidence suggests that in the relations between the islands and the various Cretan centers, there were certain spheres of interest corresponding to the natural sailing routes.

The suggestion that Laurion copper came to western Crete via Kythera would get some archaeological backing if it was possible to decide unequivocally that the LMIB pottery of the Alternating Style originated in western Crete. Elisabeth Schofield (*64*) concluded that

> *Kastri on Kythera and Ayia Irini on Kea are two of the sites at which [Alternating Style pottery] figures most prominently, and at both it is considered by the excavators to be an import. The suggestion that it may have originated in western Crete receives some backing from the finding of considerable quantities at Chania; where the excavators regard it as a local product. . . .*

The concentration of this apparently West Cretan pottery in Kythera and Kea is significant because it is now accepted that Aghia Irini on Kea was a distribution center throughout the Bronze Age It would not be surprising if Aghia Irini distributed silver and copper from Lavrion It is likely, therefore, that the metal supplies for West Crete were coming directly through this route and that there was no particular demand for more copper from elsewhere.

Cyprus was not a major source of copper for the LM objects found in West Crete The majority of these objects are consistent with having been made of copper from the mines of Laurium. It seems that Cypriot copper, even in the Late Bronze Age, did not "flood" the Aegean. More lead isotope analyses of archaeological material from the Mycenaean world, from the Eastern Coast of the Mediterranean, Syria and Anatolia, and also from the Cape Gelidonya and Kas wrecks, are needed to determine the distribution of Cypriot copper in the Mediterranean. It is already clear that lead isotope provenance studies have a major role to play in deciphering the metal trade in the Bronze Age Mediterranean.

Abbreviations

EBII	Early Bronze
EC	Early Cycladic
ECIIIA	Early Cycladic IIIA
EMII	Early Minoan II
LBA	Late Bronze Age
LM	Late Minoan
LMIA	Late Minoan IA
LMI	Late Minoan I
LMIB	Late Minoan IB
LMIII	Late Minoan III
LMIIIA2	Late Minoan IIIA2
LMIIIB	Late Minoan IIIB
MMI	Middle Minoan I
MycIIIa	Mycenaean IIIA
MycIIIb	Mycenaean IIIB

Acknowledgments

This work was supported by the Leverhulme Trust, the Science and Engineering Research Council and the North Atlantic Treaty Organization (Grant 574/83). We are greatly indebted to the archaeological authorities in Greece and Cyprus, to numerous archaeologists in both countries and to the British School at Athens for its unfailing support.

References

1. Buchholz, H. G. *Praehistorische Z.* **1959,** *37*, 1–40.
2. Muhly, J. D. In *Acts of the International Archaeological Symposium on Early Metallurgy in Cyprus (4000–500 B.C.), Nicosia*; Muhly, J. D.; Maddin, R.; Karageorghis, V., Eds.; Pierides Foundation: Nicosia, Cyprus, 1982; pp 251–270.
3. Gale, N. H.; Stos-Gale, Z. A. *Annu. Br. School Athens,* **1986,** *81*, 81–100
4. Tylecote, R. F.; Ghaznavi, H. A.; Boydell, P. J. *J. Archaeol. Sci.* **1977,** *4*, 305–33.
5. Gale, N. H.; Stos-Gale, Z. A. *Annu. Br. School Athens,* **1981.** *76*, 169–224.
6. Stos-Gale, Z. A., Gale, N. H.; Gilmore, G. R. *Oxford J. Archaeol.* **1984,** *3*(3), 23–45.
7. Renfrew, A. C. *The Emergence of Civilisation; Methuer: London, 1972.*
8. Doumas, C. *Early Bronze Age Burial Habits in the Cyclades* (*SIMA XLVII*); Paul Åströms: Göteborg, 1972.
9. Brill, R. H.; Wampler, J. M. *Am. J. Archaeol.* **1967,** *71*, 63–77.
10. Gale, N. H. In *Thera and the Aegean World*; Doumas, C., Ed.; Förlag: London, 1978; Vol. 1, pp 529–545.
11. Farquhar, R. M.; Vitali, V. *MASCA J.* **1987,** 1–8.
12. Gale, N. H. In *Chemical Analyses of Art and Archaeological Artifacts*; Hughes, M. J., Ed.; Wiley: New York, in press.

13. Brill, R. H. In *The Impact of the Natural Sciences on Archaeology*; Allibone, T. E., Eds.; The British Academy: London, 1970; pp 143–164.
14. Stos-Gale, Z. A.; Gale, N. H. *Rev. Archeometrie* **1980,** *111*(*5*), 285–295.
15. Gale, N. H.; Stos-Gale, Z. A. *Science* **1982,** *216*, 11–19.
16. Gale, N. H.; Stos-Gale, Z. A.; Davis, J. L. *Hesperia* **1984,** *53*(*4*), 389–406.
17. Brill, R. H.; Wampler, J. M. *Am. J. Archaeol.* **1965,** *69*, 165-166.
18. Brill, R. H.; Shields, W. R. *Lead Isotopes in Ancient Coins*; Special Publication No. 8, Royal Numismatic Society; Oxford University: Oxford, 1972; pp 279–303.
19. Grögler, N.; Geiss, J.; Grurenfelder, M.; Houtermans, F.G. *Z. Naturforsch.* **1966,** *21*, 1167-1172.
20. Muhly, J. D. In *Acts of the International Archaeological Symposium on the Relations Between Cyprus and Crete ca. 200–500 B.C., Nicosia*; Karageorghis, V., Ed.; Society for Cypriot: Nicosia, Cyprius, 1979; 87–100.
21. Coles, J. M. In *Advances in World Archaeology*; Wendorf, F.; Close, A. E., Eds.; MacMillan: New York, 1982; pp 265–321.
22. Harding, A. F. *The Mycenaeans and Europe*; London, 1984.
23. Russell, R. D.; Farquhar, R. M. *Lead Isotopes in Geology*; Interscience: New York, 1960; pp 1–2.
24. Barnes, I. L.; Gramlich, J. W.; Diaz, M. G.; Brill, R. H. In *Archaeological Chemistry—II*; Carter, G. F., Ed.; Advances in Chemistry 171; American Chemical Society: Washington, DC, 1978; pp 273–277.
25. Gale, N. H. In *Proceedings of the 25th Symposium on Archaeometry, Athens, 1986*; Maniatis, Y.; Grimanis, A. P., Eds.; in press.
26. Gulson, B. L.; Mizon, K. J. *J. Geochem. Explor.* **1979,** *2*, 299–320.
27. Stos-Gale, Z. A.; Gale, N. H.; Zwicker, U. "The Copper Trade in the South East Mediterranean Region: Preliminary Scientific Evidence"; Reports of the Department of Antiquities, Cyprus, 1986; pp 122–144.
28. Faure, G. *Principles of Isotope Geology*; Wiley: New York, 1977; p 227.
29. Amov, B. G. *Earth Planet. Sci. Lett.* **1983,** *65*, 61–74.
30. Gale, N. H.; Mussett, A. E. *Rev. Geophys. Space Phys.* **1973,** *2*, 37–86.
31. Köppel, V.; Grurenfelder, M. In *Lectures in Isotope Geology*; Jager, E.; Hunziker, J. C., Eds.; Springer: Berlin, 1979; pp 134–153.
32. Stos-Fertner, Z.; Gale, N. H. *Archeo-Physika* (*Bonn*) **1978,** *10*, 299–314.
33. Gale, N. H.; Stos-Gale, Z. A. *Sci. Am.* **1981,** *244*(*6*), 176–192.
34. Gale, N. H.; Stos-Gale, Z. A. In *Acts of the Second International Congress of Cypriot Studies, Nicosia, 1982*; Leventi, A. G., Ed.; Society for Cypriot: Nicosia, Cyprus, 1985; pp 51–66.
35. Stos-Gale, Z. A. In *Acts of the Second International Congress of Cypriot Studies, Nicosia, 1982*; Leventi, A. G., Ed.; Society for Cypriot: Nicosia, Cyprus, 1985; pp 67–72.
36. Stos-Gale, Z. A.; Gale, N. H. *J. Field Archaeol.* **1982,** *9*, 467–485.
37. Gale, N. H.; Stos-Gale, Z. A. "Lead Isotope Analysis and Alashiya: 3"; Reports of the Department of Antiquities, Cyprus, 1985; pp 83–99.
38. Gale, N. H.; Stos-Gale, Z. A.; Gilmore, G. R. *Anatolian Studies* **1985,** *35*, 143–173.
39. Moody, J. A.; Lukermann, F. E. In *Contributions to Aegean Archaeology*; Wilkie, N. C.; Coulson, W. D. E., Eds.; University of Minnesota: Minnesota, 1985; pp 61–89.
40. Branigan, K. *Copper and Bronze Working in Early Bronze Age Crete* (*SIMA XIX*); Lund, 1968.
41. Stos-Gale, Z. A.; Gale, N. H. *Skr. Utgivna av Svenska Instituteti Athen, Ser. 4* **1984,** *32*, 59–64.

42. Pernicka, E.; Seeliger, T. C.; Wagner, G. A.; Begemann, F.; Schmitt-Strecker, S.; Eiber, C.; Oztunali, O; Baranyi, I. *Jahrb.* **1984,** 533–599.
43. Branigan, K. *Aegean Metalwork of the Early and Middle Bronze Age*; Oxford University: Oxford, 1974; pp 59–63.
44. Faure, P. In *Proceedings of the IVth Cretological Congress, Athens*; Athens, 1980; Vol. 1, pp 150–168.
45. Schachermeyr, F. In *Realencyclopdie der Klassischen Altertumswissenschaften*; Pauly-Wissowe, Ed.; 1954; Vol. 22, p 2.
46. Fiedler, K. G. *Reise durch alle Theile des Konigreiches Griechenland*; F. Fleischer: Leipzig, 1841.
47. Gale, N. H.; Stos-Gale, Z. A.; Papastamataki, A.; Leonis, K. In *Proceedings of the Symposium on Early Furnace Technology, British Museum, London, 1982*; Craddock, P. T.; Hughes, M. J., Eds.; British Museum occasional paper 48; British Museum: London, 1985; pp 81–102.
48. Renfrew, A. C. *Am. J. Archaeol.* **1967,** *71,* 1–20.
49. Tsountas, C. *Archaiologike Ephemeris,* **1889,** 73–134.
50. Bossert, K.-M. *Archaiologikon Deltion* **1967,** *22,* 53–76.
51. Barber, R. L. N.; MacGillivray, J. A. *Am. J. Archaeol.* **1980,** *84,* 141 and 157.
52. de Jesus, P. S. *In The Search for Ancient Tin*; Franklin, A. D.; Olin, J. S.; Wertime, T. A., Eds.; Smithsonian Institution: Washington, DC, 1979; pp 33–9.
53. Gale, N. H.; Stos-Gale; Z. A. In *Proceedings of the Cycladic Workshop, Institute of Archaeology, London, 1983*; MacGillivray, J. A.; Barber, R. L. N., Eds.; Department of Classicla Archaeology: Edinburgh, 1984; pp 255–276.
54. Papathanassopoulous, G. *Neolithic and Cycladic Civilisation*; Melissa: Athens, 1981; pp 124–125.
55. Cumming, G. L.; Richards, J. R. *Earth Planet. Sci. Lett.* **1975,** *28,* 155–171.
56. de Jesus, P. S. *The Development of Prehistoric Mining and Metallurgy in Anatolia;* BAR International Series, **1980,** *74.*
57. Chernykh, E. N. *Gornoie Dielo i Metallurgiya v Drevneyshey Bolgarii*; Bulgarian Academy of Science, Sofia, 1978.
58. Petrascheck, W. E. in *Mineral Deposits of Europe, Vol. II;* Dunning, F. W.; Mykura, W.; Slater, D., Eds.; London, 1982; pp 1–12.
59. Seeliger, T. C.; Pernicka, E.; Wagner, G. A.; Begemann, F. Schmitt-Strecker, S.; Eibner, C.; Oztunali, O.; Baranyi, I. *Jahrb. Romisch–Germanischen Zentralmuseums* **1985,** 597–659.
60. Catling, H. "Cyprus and the West: 1600–1050 B.C."; Ian Sanders Memorial Lecture, Sheffield, 1980; 35pp.
61. Bass, G. F. *Trans. Am. Philos. Soc., New Series* **1967,** *57*(*8*).
62. Phillips, T. A. *A Treatise on Ore Deposits*; Macmillan: London, 1984; 485 and 490.
63. Marinos, G. P.; Petrescheck, W. E. *Lavrion, Geological and Geophysical Research* **1954,** *4*(*1*), 184.
64. Schofield, E. In *Temple University Aegean Symposium 7*; Betancourt, P., Ed.; Temple University: Philadelphia, 1982, pp 9–14.
65. Schofield, E. *Oxford J. Archaeol.* **1982,** *1,* 9–25.
66. Davis, J. L. In *Papers in Cycladic Prehistory*; Davis, J. L.; Cherry, J. F., Eds.; University of California: Los Angeles, 1979; pp 143–154.
67. Coldsteam, T. N.; Huxley, G. L. Eds. *Kythera: Excavations and Studies*; British School at Athens: London, 1972.

RECEIVED for review June 11, 1987. ACCEPTED revised manuscript May 13, 1988.

10

Instrumental Neutron Activation Analyses of Metal Residues Excavated at Tel Dan, Israel

Mirela C. Manea-Krichten[1], Nancy Heidebrecht[2], and George E. Miller[1]

[1]Department of Chemistry, University of California, Irvine, CA 92717
[2]Department of Ancient Near Eastern Languages and Cultures, University of California, Los Angeles, CA 90024

Instrumental neutron activation analysis was used to determine concentrations of several major and trace elements in samples of heavily corroded residues found in crucible fragments excavated at Tel Dan, Israel. The residues were mostly hard, metallic phases admixed with nonmetallic inclusions that appeared to be ceramic material from the loose porous interior of the crucible itself. The objective was to identify the metals that had been melted in these crucibles. A method is described that attempts to separate nonmetallic and metallic phase data. In comparison to previous reports on analyses of source materials thought to have been used at Dan in this period (Late Bronze II Age–Early Iron I Age: 1400–1000 B.C.), high gold concentrations were found. These appear to be correlated to arsenic and antimony concentrations. This finding is discussed in relation to possible changes in the source of tin at this period.

NEUTRON ACTIVATION ANALYSIS IS A VERY SENSITIVE TECHNIQUE for trace element determinations in various samples. If there are no elements that mutually interfere, the purely instrumental version of this method is often chosen for its established advantages such as accuracy, speed, sensitivity, simultaneous multielement determination, and sample preservation (*1*). For these reasons, instrumental neutron activation analysis (INAA) was applied to samples taken from a series of metal-working residues excavated at Tel Dan, Israel, from 1985 to 1986.

0065–2393/89/0220–0199$06.00/0

The city of Tel Dan, located at the border of what are now Israel and Lebanon, has been the center of much archaeological interest. It is located on an established Near Eastern tin trade route that originated somewhere near Susa during the Middle Bronze Age (*2*). The same trade route supplied tin for Crete. The source of copper throughout this time was postulated to be Cyprus. Tel Dan's inclusion in the Mari texts attests to the significance of metal working at Dan (or Layish as it was then known) (*3–5*). Full details of the archeological finds and context are reported elsewhere (*6*).

At Tel Dan, two metal-working areas have been located just inside the upper city gate system on the southern perimeter of the mound. A very large workshop (Area A–B) has been dated to 1100–1000 B.C. Smaller subsidiary working areas found slightly to the north (Area B-1) indicate considerable continuity in working metal at this site from Early Iron I Age (ca. 1100 B.C.) back to the Late Bronze IIa Age (ca. 1400 B.C.). The workshop contents discovered in Areas B-1 and A–B during 1984 and 1985 include steep-sided thimble-shaped crucibles with residues; ceramic tuyeres; and stone, bone, and horn tools. In Area B-1, the crucible working area was scattered on tamped earth surfaces. In Area A–B there was an extensive cobbled area with crucible fragments in situ among the cobbles (*7*).

Most of the samples were taken from corroded, dense metallic residues found in fragments of thick, flat-base porous ceramic crucibles, very similar to those described by Tylecote (*see* ref. 8, p. 20) as type B4 excavated at sites near Timna and Tel Qasile (*9, 10*). The residues may be what Tylecote called "internal slag" (*8, 11, 12*). As no evidence of smelting activity was noted, these residues were presumed to be from bronze alloying and purification activities for artifact production at Dan, even though no artifacts were located in the vicinity. Extensive alloying was needed before fashioning artifacts, as there was little benefit to the smelter in carrying out extensive purification (*13*). The ore smelting site, be it at Dan or elsewhere, is not established. The form of the samples, best described as a hardened "froth" of metal with ceramic inclusions, was somewhat unusual and posed several challenges.

Sample Identification

- Samples 806, 807, 808, 809, 1006, 1008, 1009, 1010, 1011, and 1014 are from workshop area A–B that was excavated in the 1985 season.
- Samples 801, 802, 803, 804, 805, and 1015 are from area B-1 (dated to Late Bronze Age I/Late Bronze IIa Age transition period) that was excavated in the 1985 season.
- Sample 1012 is from a small pin-shaped metal fragment that was found in connection with the Late Bronze Age crucibles in area B-1.

- Sample 1013 is similar to sample 1012 and was also found in B-1, but in an earlier Middle Bronze II Age/Late Bronze I Age destruction layer

For comparison, similar types of materials from another site were sampled. Samples 1003 and 1004 are from Tel Akko on the Mediterranean coast and are attributed by the excavator to the Sherden, a group of coast-dwelling peoples (1400–1200 B.C.). Sample 1003 is from Akko VI, and 1004 is from Akko VII.

Background

The original interest in these excavated samples stemmed from the important role of bronze and the historical implications of the extensive tin trade routes in the Middle Bronze Age (2200–1600 B.C.). As bronze metallurgy developed from the earlier use of copper and arsenical copper, tin sources became vital (*14*). By the Middle Bronze Age, the limited distribution of tin deposits throughout the world and the ample supply of copper were well-known, and extensive tin trade routes had been established to supply the metallurgical centers of the world with this then-considered "precious" metal (*15–16*). The Ugaritic tablets from Ras Shamra, however, indicate a severe drop in the value of tin around the Late Bronze Age (1400–1200 B.C.). It has been postulated (*17*) that this drop was related to the discovery of more accessible sources of tin.

General Considerations

The primary objective of our research was the analysis of the mixtures of pottery (crucible wall) and corroded metal residue (slags) by INAA in an attempt to provide information on the metals used at Dan. In addition, two small metal fragments, with little or no ceramic adhesion were sampled. All materials were highly corroded and showed various green, grey, and brown encrustations, and a complete absence of obvious metallic surface.

As mentioned earlier, the samples were very heterogeneous. Either the melting process was incomplete and the higher melting solids that remained closest to the crucible walls were unevenly mixed and not representative of the completely melted portion that was decanted from the crucible, or selective corrosion or fractionation of certain elements occurred.

Thus, rather than undertaking a detailed metallurgical study, it was decided to attempt to provide as representative a sample as possible from each specimen. The heterogeneous mass was drilled with a carbide burr from as many facets of the fragment as possible. Hardness inhibited drilling beyond approximately 5 mm from the surface.

In one case, (sample 1010) sufficient identifiable crucible was adhering to the metal to allow the sampling of what could be representative ceramic.

The data from this sample were later used to estimate the amount of ceramic in each "metal" sample (Table I). This estimation was done by assuming a constant composition for the ceramic component (mostly Si, Al, K, Mg, and Na) and by assuming that all of the silicon in the composite was there by virtue of ceramic contamination. Thus, the total "ceramic" could be subtracted, and the remainder could be assumed to be "metal". This process is not precise, but it did seem to account for the bulk of the Al, Na, K, and Mg in the samples.

All elemental concentrations in this study were measured by using the comparator form of INAA, in which standards and samples are subjected to the same conditions of irradiation, decay, and counting. This method minimizes the errors due to geometry, flux gradients, detector efficiency, etc. This approach was validated by employing standards similar in physical and chemical composition to the samples. National Bureau of Standards (NBS) standard reference materials (SRMs) (37E, sheet brass; 872, phosphor bronze; 158A, silicon bronze; and 1633, coal fly ash) were used in duplicate. Two irradiations were made at different flux, decay, and counting times. Very small polyethylene vials (0.1-mL) were used to maintain good geometric reproducibility.

To address a concern that there might be substantial quantities of unobserved elements, the total accountable mass of each sample was computed. Only one sample (804) had an accountable mass close to 100%. This mass was calculated without corrections for the possible corrosion of copper and iron that contributes oxygen. The corroded appearance of the sample would

Table I. Elemental Analysis of Sample 1010.

Element	*1010 Metal & Ceramic (0.214 g)*	*Ceramic (0.083 g)*	*1010 Metal[a] (0.198 g)*
Al (%)	1.9 ± 0.2	8.4 ± 0.07	0.6 ± 0.07
As (ppm)	710 ± 70	15 ± 4	770 ± 70
Au (ppm)	2.0 ± 0.04		2.2 ± 0.04
Cu (%)	26 ± 2		29 ± 2
Fe (%)	6.8 ± 0.7	7.0 ± 1	7.3 ± 0.8
K (%)	1.3 ± 0.2	1.5 ± 0.2	
La (ppm)	14 ± 2	54 ± 4	15 ± 2
Mg (%)		1.3 ± 0.4	
Mn (%)	0.09 ± 0.01	0.2 ± 0.02	0.1 ± 0.01
Na (%)	0.26 ± 0.02	0.20 ± 0.02	
Sb (ppm)	82 ± 10		90 ± 9
Sc (ppm)	3.7 ± 0.5	16 ± 2	4.0 ± 0.5
Si (%)	2.9 ± 0.4	12 ± 1	
Sn (%)	12 ± 2		13 ± 2
V (ppm)	66 ± 8	150 ± 20	71 ± 9
Zr (ppm)	66 ± 9	250 ± 20	71 ± 10

NOTE: All values are ± 1σ. (Errors based on a single analysis.)
[a]Corrected eliminating "ceramic" contribution.

warrant such a correction. Perhaps the anomaly was caused by over-correcting for ceramic contamination. In all the other cases, this summation failed to total 100%, so an INAA prediction program developed at the University of California–Irvine (UCI) (*18*) was used to to help eliminate possible additional components. The analytical methods could not exclude the presence of substantial lead, nickel, or sulfur, but would exclude silver, calcium, cobalt, chromium, titanium, or zinc, as shown in Table II.

Experimental Procedure

Material (50–500 mg) was removed from each specimen by drilling small holes (5 mm deep) at various locations on each sample. Carbide drills were used to to minimize contamination. The sample drillings were collected on weighing paper.

Polyethylene vials (2.66 mm high × 9.4 mm o.d., Type-H container, Biologisch Laboratorium, Vrije Universiteit, Amsterdam) were cleaned with nitric acid followed by three deionized water rinses and a warm air dry. Samples were transferred into the preweighed clean vials with a Teflon [poly(tetrafluoroethylene)]-coated spatula. The vials were subsequently handled only with forceps or plastic-gloved hands. An iron-coated tipped soldering iron was used to heat-seal the vials. The drill bit and the spatula were rinsed after each sample to prevent sample-to-sample contamination. An empty vial (cleaned, labeled, and sealed as if it contained sample) was processed in parallel. Portions (100–400-mg) of several different NBS SRMs were similarly encapsulated and analyzed with each group of samples. A small section of gold-doped aluminum wire (Reactor Experiments, Inc., San Carlos, CA), certified as 0.11% gold, was used as a gold standard.

Two series of irradiations were made with the UCI 250-kW TRIGA Mark I reactor. The first run consisted of irradiation for 30 s at a thermal neutron flux of 4.8×10^9 neutrons cm^{-2} s^{-1} using the facility's pneumatic transfer system. Samples were permitted to decay for 1 min and then counted for 2 min on a Ge–Li detector (21% efficiency) at a distance of 1.3 cm from the crystal and a gain setting of 0.8 keV per channel. Elements determined in this run were Al, Mg, Mn, Si, Sn, V, and Zr.

Table II. Lower Limits of Detection (Sample 1006, 0.167 g)

Element	*Energy (KeV)*	$t_{1/2}$[a]	*Concentration* (μg)	*Proportion* (%)	*Possible Major Component*
Ag	633	2.42 min	16	0.010	no
Ca	3084.4	8.72 min	6900	4.2	no
Cl	2166.8	37.2 min	650	0.390	no
Co	1173.2	5.27 years	0.7	4.2×10^{-4}	no
Cr	320	27.7 days	24	0.014	no
Ni	1481.7	2.5 h	2.6×10^{16}	100	yes
Pb	569.7	0.7 s	3.3×10^{24}	100	yes
Rb	1076.6	18.6 days	9.4	5.6×10^{-3}	no
S	3102.4	5.05 min	3.0×10^{5}	100	yes
Sr	514	64.8 days	170	0.10	no
Ti	320.1	5.76 min	300	0.18	no
Zn	1115.4	243.8 days	22	0.013	no

[a]Half-life.

The second run consisted of a 10-min irradiation at full power (250 kW) in a thermal neutron flux of 1.2×10^{12} cm^{-2} s^{-1} with the samples loaded in the facility's rotary specimen rack. Samples were allowed to decay for approximately 6 days. (A shorter decay was tried, but the observed dead time was prohibitive). Counts were made for 30 min each with a different Ge–Li detector (11% efficiency), at the same crystal-to-sample distance and gain setting previously described. The concentrations of the remainder of the elements were determined as a result of this irradiation.

Discussion of Results

The elemental concentrations determined in the "metallic phase" are given in Table III. The crudeness of the "ceramic correction" method just described adds additional uncertainty to some of these values (especially aluminum, manganese, and vanadium). Most values, however, are quite meaningful.

The appreciable iron content is expected if substantial flux residue remains in the slag, as has been observed elsewhere (*19*). The high gold concentrations seen in many samples were the most unexpected. Because the samples are dated to the time of the postulated change in tin trade routes, and therefore tin source, it is tempting to suggest that the gold concentration values provide a valuable clue. The "accepted" trade route in the Middle to Late Bronze Age serving this part of the world originated in Cornwall, England (*20–21*). Table IV compares our data with data reported by Rapp (*22*) from analysis of smelts of Cornwall tin. The discrepancy between these data suggests either a different source of tin or a link between gold and an element other than tin. However, Rapp obtained his data from Cornish ores smelted in the present and not in ancient time.

Evidence presented by Gale (*23*) using lead isotope data indicates a change in Crete's source of copper by the Late Bronze Age. The proximity of Dan to Crete, as well as the established parallels in their trade routes, may be significant. Gale postulates a switch for the source of copper from Cyprus to the Laurion in Greece. If a change in tin source is the cause of this unusual elemental pattern (gold in particular), it would be important to investigate the recently located silver and gold mine in the Taurus mountains in Turkey, which has been reported to contain some tin (*24*). Also meriting further inquiry are the elemental concentrations in metals dating to the Bronze Age and found on the shipwreck off Cape Galidonya (*25*).

Although absolute interpretation is risky, further examination of this data indicates a number of rough correlations. The first is the parallel between arsenic and gold concentrations (Figure 1). Table V lists all of the samples analyzed along with their ratios of gold to other elements. Examination of the arsenic-to-gold ratios indicates that use of this ratio as a criterion would group many of these samples together, with most falling in the range of 250–400:1 (As:Au). The fact that not all of the samples fall within this range, is not surprising because there is no reason to expect that all of the excavated samples had the same source. Some samples may even have originated from remelting older cast objects, a common practice.

Table III. Elements Observed in 15 Samples of Residues

Element	*1006* *0.167 (g)*	*1010* *0.198 (g)*	*1011* *0.074 (g)*	*1014* *0.295 (g)*	*1015* *0.052 (g)*	*802* *0.475 (g)*	*804* *0.061 (g)*	*806* *0.274 (g)*
Al (%)	0.5 ± 0.004	0.6 ± 0.07	0.2 ± 0.002	0.9 ± 0.01	0.2 ± 0.003	–	0.2 ± 0.002	–
As (ppm)	1700 ± 200	770 ± 70	1700 ± 200	690 ± 70	1900 ± 200	2200 ± 200	2200 ± 200	620 ± 60
Au (ppm)	6.9± 0.1	2.2 ± 0.04	4.8 ± 0.1	0.4 ± 0.01	14 ± 0.2	8.8 ± 0.1	7.5 ± 0.1	1.8 ± 0.1
Cu (%)	28 ± 2	29 ± 2	33 ± 4	12 ± 1	35 ± 4	54 ± 2	64 ± 4	42 ± 2
Fe (%)	20 ± 2	7.3 ± 1	16 ± 2	22 ± 2	–	0.6 ± 0.1	5.8 ± 1	4.3 ± 0.6
La (ppm)	21 ± 3	15 ± 2	64 ± 5	21 ± 2	–	–	44 ± 5	7 ± 2
Mn (%)	0.2 ± 0.02	0.1 ± 0.01	0.2 ± 0.02	0.5 ± 0.05	0.04 ± 0.01	–	0.2 ± 0.02	0.03 ± 0.01
Sb (ppm)	180 ± 20	90 ± 9	130 ± 10	38 ± 4	340 ± 30	270 ± 30	290 ± 30	54 ± 5
Sc (ppm)	7.2 ± 0.8	4.0 ± 0.5	18 ± 2	7.7 ± 0.8	7.9 ± 1	1.1 ± 0.2	13 ± 2	2.3 ± 0.4
Sn (%)	5.5 ± 0.9	13 ± 2	7 ± 1	1.5 ± 0.4	7.9 ± 2	2.1 ± 0.4	32 ± 2	4.1 ± 0.7
V (ppm)	110 ± 10	71 ± 9	200 ± 30	97 ± 10	110 ± 20	39 ± 5	220 ± 30	110 ± 10
Zr (ppm)	64 ± 10	71 ± 10	240 ± 20	75 ± 7	50 ± 20	–	150 ± 20	19 ± 7
O (%)	16	10	15	13	9	14	–	13
Total (%)	77 ± 5	60 ± 5	72 ± 7	50 ± 5	52 ± 5	70.5 ± 2	102 ± 7	63 ± 3

Continued on next page.

Table III. Continued

Element	*807* *0.163 (g)*	*808* *0.193 (g)*	*809* *0.111 (g)*	*1003* *0.117 (g)*	*1004* *0.169 (g)*	*1012*[b] *0.301 (g)* *(metal)*	*1013*[b] *0.200 (g)* *(metal)*
Al (%)	0.5 ± 0.003	0.6 ± 0.004	0.3 ± 0.003	0.4 ± 0.01	0.5 ± 0.004	0.4 ± 0.01	0.6 ± 0.01
As (ppm)	1700 ± 200	750 ± 70	1700 ± 200	1600 ± 200	1300 ± 100	2800 ± 300	12000 ± 1000
Au (ppm)	1.8 ± 0.1	0.3 ± 0.1	0.8 ± 0.1	12 ± 0.1	3.1 ± 0.1	60 ± 0.3	14 ± 0.1
Cu (%)	25 ± 2	21 ± 2	36 ± 3	68 ± 4	16 ± 2	62 ± 3	71 ± 3
Fe (%)	20 ± 2	20 ± 2	22 ± 3	–	6.0 ± 0.9	0.6 ± 0.2	–
La (ppm)	28 ± 2	25 ± 2	17 ± 3	–	23 ± 2	–	–
Mn (%)	1.0 ± 0.1	0.05 ± 0.01	0.5 ± 0.1	–	0.1 ± 0.001	–	–
Sb (ppm)	130 ± 10	120 ± 10	85 ± 8	110 ± 10	95 ± 9	210 ± 20	480 ± 50
Sc (ppm)	6.1 ± 0.8	7.8 ± 0.8	4.4 ± 0.9	–	7.9 ± 0.8	1.3 ± 0.2	–
Sn (%)	4.5 ± 0.8	14 ± 0.9	7.2 ± 1	5.4 ± 1	12 ± 2	3.2 ± 0.6	–
V (ppm)	140 ± 20	99 ± 10	160 ± 20	82 ± 10	170 ± 10	30 ± 5	39 ± 6
Zr (ppm)	93 ± 10	82 ± 10	72 ± 10	21 ± 10	83 ± 9	–	–
O (%)	15	14	19	17	7	16	18
Total (%)	66 ± 5	69 ± 5	85 ± 6	91 ± 5	41 ± 5	82 ± 3	91 ± 3

NOTE: Column headings are sample numbers. All values are ± 1σ (errors based on a single analysis). Values not detected indicated by –.
[a]Assuming pure CuO and Fe_2O_3.
[b] Metal fragment samples, no ceramic correction applied.

Table IV. Trace Element Concentrations in Cassiterite (SnO_2) Smelts from Cornwall, England, Deposits and Mean Values Obtained in Residues from Tel Dan, Israel

Element	*Cornwall (ppm)*	*Tel Dan (ppm)*
Ag	9	–
Au	0.11	9.2
Cr	66	–
Fe	1000	121,000
Hg	9	–
Ni	670	–
Ru	20	–
Se	39	–
Ta	2	–
W	15	–

SOURCE: Adapted from reference 22.
NOTE: – indicates not detected.

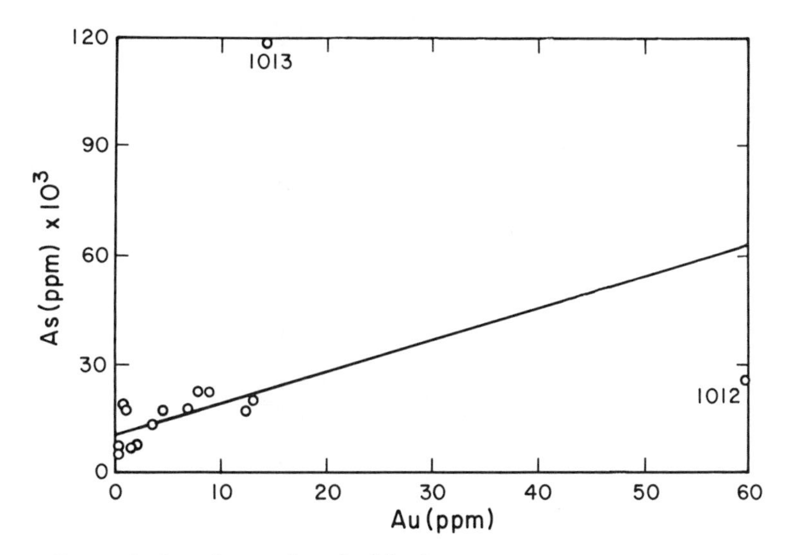

Figure 1. Correlation plot of gold relative to arsenic concentration.

Correlation of the tin-to-gold and copper-to-gold concentrations is not as good (Figures 2 and 3). This result may indicate that the arsenic, rather than tin or copper, was the source of the gold in the residue (assuming that high arsenical copper ore was not used in this period). This idea seems to be supported by the degree of correlation found between the antimony and gold values (Figure 4). Ratios of 25–40:1 (Sb:Au) are seen for the correlated group. Most likely, antimony would enter as an impurity with added arsenic along with much of the gold. However, during high-temperature working, much arsenic is commonly lost because of the low sublimation temperature of the metal and its compounds relative to antimony. Perhaps at the tem-

Table V. Ratios of Elements Determined in Samples Analyzed

Ratio	*1006*	*1010*	*1011*	*1014*	*1015*	*802*	*804*	*806*	*807*	*808*	*809*	*1003*	*1004*	*1012*	*1013*
As : Au (ppm:ppm)	250	360	360	1900	140	260	300	340	950	2900	2040	130	420	50	850
Sb : Au (%:ppm)	26	41	26	95	25	30	38	30	72	410	110	9	30	4	34
Sn : Au (%:ppm)	0.8	5	1.4	3.3	0.6	0.2	5	2.5	2.5	2.5	50	10	0.4	3.3	0.1
Cu: Au (%:ppm)	4.1	13	7	32	2.5	6.1	8.6	23	14	80	44	6	5	1	5

NOTE: Column headings are sample numbers. All samples were residues except 1012 and 1013, which were metals. All samples were from Tel Dan except 1003 and 1004, which were from Tel Akko.

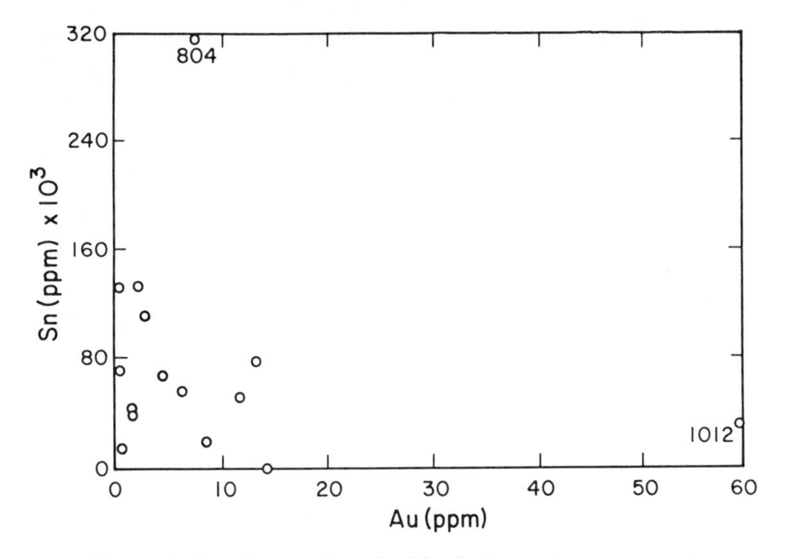

Figure 2. Correlation plot of gold relative to tin concentration.

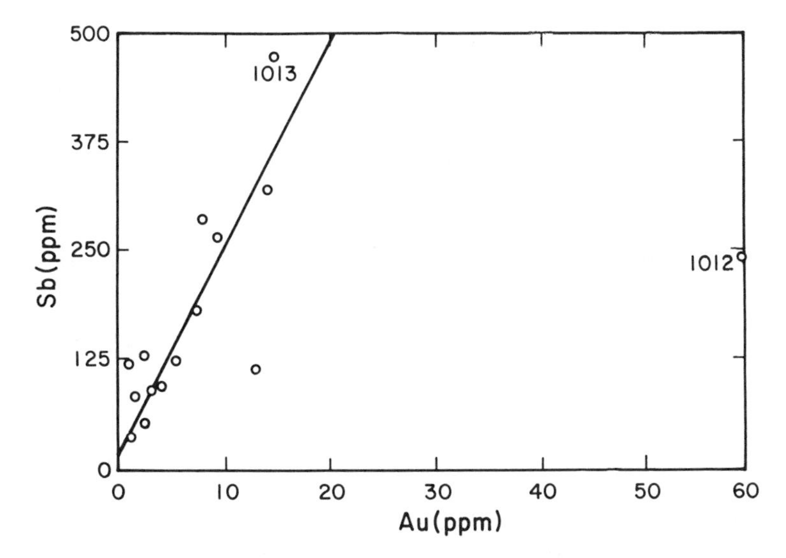

Figure 3. Correlation plot of gold relative to copper concentration.

peratures reached when these materials were worked an equilibrium was possible that led to a relatively stable arsenic concentration in the residues.

Sample 1013 is a corroded piece of metal that showed no ceramic contamination (macroscopically or chemically) and contained no detectable tin. If this were a copper ingot, or pin resulting from the melting of a plain

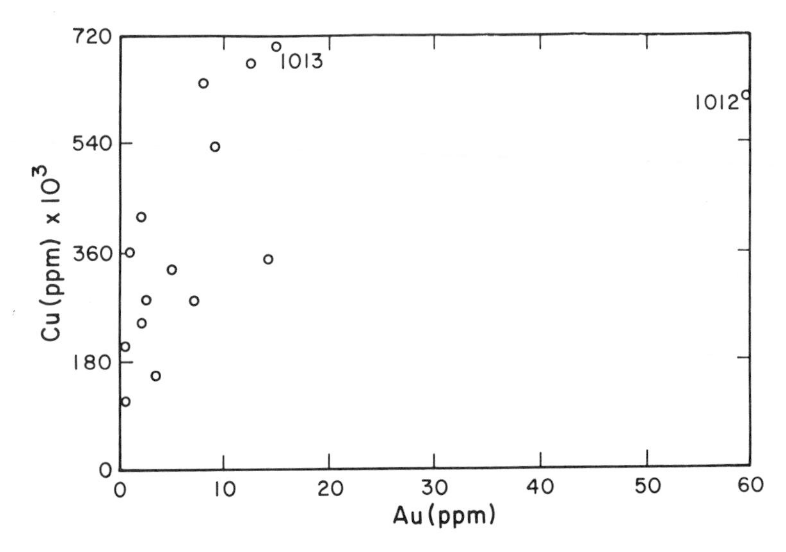

Figure 4. Correlation plot of gold relative to antimony concentration.

copper ingot, the high gold, arsenic, and antimony concentrations (despite the lack of tin), would strengthen the idea that the gold content is not related to the tin content and therefore cannot be used to infer a change in tin source.

The high level of arsenic seen in these samples and in the Laurion samples (*23*), in forms in which it may have been incorporated into copper metal, indicates copper, rather than tin, as the alloying metal source most responsible for the unusual data observed. A much smaller amount of gold could have been a characteristic of the tin, but it would have been masked by the higher concentration apparently residing with the arsenic–copper component.

Conclusion

The overall data may not be used in such an absolute sense as in analysis of a bronze casting, but they do provide some interesting information. Substantial concentrations of gold and arsenic were observed that appear to correlate with one another and with antimony concentration. Further work is needed to find authentic sources of the copper, tin, fluxes, and other materials used at these sites. Apparently, measurement of gold concentrations cannot be used to answer questions about a change in the sources of tin during this period.

Examination of these samples to determine additional trace element concentrations continues. Further examination of the nature of the ceramic materials is proceeding.

Acknowledgment

We are grateful to Avraham Biran from the Hebrew Union College in Jerusalem who, as director of excavations at Tel Dan, granted permission to remove and analyze these materials. We also thank Moshe Dothan of Haifa University, Director of Excavations at Tel Akko for permission to remove and analyze materials. This work was supported, in part, by a Reactor Sharing grant provided by the U.S. Department of Energy to UCI.

References

1. Jacobs, J. W.; Korotev, R. L.; Blanchard, D. P.; Haskin, L. A. *J. Radioanal. Chem.* **1977,** *40,* 93–114.
2. Maddin, R.; Wheeler, T. S.; Muhly, J. D. *Expedition* **1977,** *19*(2), 35.
3. Muhly, J. D. *Am. Sci.* **1973,** *66,* 404.
4. Goffer, Z. *Archaeological Chemistry*; Wiley: New York, 1980; p 200.
5. Wertime, T. A. In *The Search For Ancient Tin*; Franklin, A. D.; Olin, J. S.; Wertime, T. A., Eds.; Smithsonian Institution: Washington, DC, 1977; p 1.
6. Heidebrecht, N., Ph.D. Thesis, University of California at Los Angeles, to be submitted.
7. Heidebrecht, N. Presented at the American Oriental Society Meeting, University of Michigan; Ann Arbor, MI, 1985.
8. Tylecote, R. F. *A History of Metallurgy*; The Metals Society: London, 1976.
9. Tylecote, R. F.; Lupu, A.; Rothenberg, B. *J. Inst. Metals* **1967,** *95,* 235.
10. Mazar, B. *Isr. Exploration J.* **1950-1951,** *1,* 123; *7,* 265; *8,* 180.
11. Tylecote, R. F. *The Prehistory of Metallurgy in the British Isles*; The Institute of Metals; London, 1986.
12. Tylecote, R. F. In *Archeological Ceramics*; Olin, J. S.; Franklin, A. D., Eds.; Smithsonian Institution: Washington, DC, 1982; 231–242.
13. Craddock, P. A., personal communication.
14. Charles, J. A. In *The Search For Ancient Tin*; Franklin, A. D.; Olin, J. S.; Wertime, T. S., Eds.; Smithsonian Institution: Washington, DC, 1977; p 25.
15. Charles, J. A. *Antiquity* **1975,** *49,* 19.
16. Schuiling, R. D. *Econ. Geol.* **1967,** *62,* 540.
17. Heltzer, M. *Goods, Prices, and The Organization of Trade in Ugarit*; Reichert: Weisbaden, 1978.
18. Crofoot, T. A.;, Miller, G. E. *Spreadsheet Programs for Neutron Activation Analysis*; to be published.
19. Tylecote, R. F. In *Aspects of Early Metallurgy*; Oddy, W. A., Ed.; British Museum Occasional Paper No. 17, 1980.
20. McKerrell, H. In *The Search For Ancient Tin*; Franklin, A. D.; Olin, J. S.; Wertime, T. A., Eds.; Smithsonian Institution: Washington, DC, 1977; p 7.
21. Tylecote, R. F. In *The Search For Ancient Tin*; Franklin, A. D.; Olin, J. S.; Wertime, T. A., Eds.; Smithsonian Institution: Washington, DC, 1977; p. 49
22. Rapp, G. In *The Search for Ancient Tin*; Franklin, A. D.; Olin, J. S.; Wertime, T. A., Eds.; Smithsonian Institution: Washington, DC, 1977; p 59.
23. Gale, N. H.; Stos-Gale, Z. A. *Science,* **1982,** *216,* 11.
24. Yener, K. A. Presented at the 196th Meeting of the American Oriental Society, New Haven, CT, March 1986.
25. Bass, G. *Trans. Am. Phil. Soc.* - New Series **1967** *57 part 8,* 44.

RECEIVED for review June 11, 1987. ACCEPTED revised manuscript March 29, 1988.

11

Chemical Composition of Copper-Based Coins of the Roman Republic, 217–31 B.C.

Giles F. Carter and Hossein Razi

Department of Chemistry, Eastern Michigan University, Ypsilanti, MI 48197

Twenty-two coins from the years 217 to 31 B.C. of the Roman Republic were analyzed by X-ray fluorescence for Fe, Co, Ni, Cu, Zn, As, Ag, Sn, Sb, and Pb. No evidence was found for widespread remelting of coins. The early coins are remarkable for their relatively high Co contents. Several coins have exceptionally high Pb, As, or Sb contents. Generally, the compositions of these Roman Republican coins are very different from those of Roman Imperial coins. Although few coins were analyzed, their compositions correlate reasonably well with time. Further analyses are required to determine whether composition varies with denomination and whether coins may be dated to within a few years by their chemical compositions. Microstructures of two Roman Republican coins containing lead are presented.

CHEMICAL ANALYSES OF SEVERAL HUNDRED COPPER-BASED COINS of the Roman Empire have contributed to the solutions of numismatic problems such as debasement and quality control (*see* refs. 1–6). For example, debasement of the brass coinage of the Roman Empire has been studied by Caley and Riederer (*5*, *6*), who found a progressive decrease in zinc contents over the course of more than 100 years beginning with the reign of Nero. Carter has been able to arrange four issues of Augustan quadrantes in relative chronological order on the basis of chemical compositions (*2*). Detailed knowledge of the chemical compositions may be used to date an unknown issue (*4*). In addition, information may be obtained concerning ancient technology on the basis of the knowledge of chemical compositions and metallurgical microstructure of ancient coins.

0065–2393/89/0220–0213$06.00/0

Unfortunately, only scattered, incomplete, and sometimes unreliable analyses have been reported for copper-based coins of the Roman Republic (*7–10*). Many Republican copper-based coins were large and rather ugly. They obviously were not made of high purity copper, bronze, or brass. When Augustus, the first Roman emperor, began striking copper-based coins from relatively pure copper or from the new alloy, brass (ca. 23 B.C.), his coins must have made a deep psychological impression on the populace. The Romans probably considered that such a vast improvement in the copper-based coinage indicated a better form of government.

Experimental Procedure

Twenty-two copper-based coins of the Roman Republic were analyzed for Fe, Co, Ni, Cu, Zn, As, Ag, Sn, Sb, and Pb by using X-ray fluorescence according to the procedures described by Carter and Booth (*11*). Generally, X-ray fluorescence determines elements only in a thin surface layer, about 5–10 μm deep, so it was necessary to clean coins for analysis in such a way that the surface layer was as representative of the entire coin as possible. First, the coins were cleaned by electrolytic reduction in a hot solution of sodium carbonate. Next, the coins were abraded in an air stream containing finely divided aluminum oxide powder to remove about 10 to 15 μm of metal. Carter and Booth described the cleaning procedure in detail as well as the X-ray fluorescence parameters (*11*).

Fluorescing X-rays from most elements in a coin are almost completely absorbed by 5–10 μm of copper. If the composition of the surface layer is significantly different from that of the interior, inaccurate results will be obtained by X-ray fluorescence analysis. Carter and Kimiatek (*12*) reported that concentration gradients apparently are uncommon in copper and brass coins. However, leaded bronzes often have concentration gradients because of segregation of lead-rich or tin-rich phases during solidification. The lead-rich phase is liquid to a much lower temperature than copper. Upon cooling of a liquid copper–lead alloy, the lead can move away from the surface of the metal as it freezes. However, if the coin flan solidifies quickly, the gradient may be minimal.

A second problem with surface inhomogeneities in metals is the depletion of a more chemically active phase in contact with a less chemically active phase. When two metallic phases are in electrical contact, the more chemically active phase of the alloy corrodes much more rapidly than normal, even when it is exposed to even a mildly corrosive environment. If a coin has two phases exposed at the surface, such as copper and lead, then the more active metal, namely lead, will corrode preferentially. However, corrosion stops locally as soon as a given lead particle is consumed. The copper matrix then prevents further corrosion of lead because the copper phase must first corrode before additional lead is exposed at the surface. Hence, removal of a thin layer of metal at the surface of a coin usually exposes metal that is representative of the composition of the entire coin (*12*). The primary exception is leaded bronze, of which there are several coins in this study. The analyses of leaded bronzes by X-ray fluorescence are likely to be somewhat inaccurate. However, we feel that useful information is still obtained by the X-ray fluorescence analysis of such coins. The lead and tin concentrations, although not accurate, do provide a fair indication of the concentrations of these elements. Furthermore, knowledge of the concentrations of minor and trace elements in leaded bronzes is useful.

Because surface roughness reduces the number of fluorescing X-rays that leave a surface, the concentrations were normalized to 100%. Elements such as oxygen and silicon were not determined, but we assume that these concentrations were very low (probably less than a total of 0.5%). Normalization to 100% resulted in a slightly high value for copper.

National Bureau of Standards (NBS) alloys with certified compositions were used as standards along with Centre Technique des Industries de la Fonderie (CTIF) standards. These standards cover a fairly wide range of tin and lead concentrations. When no standards were available that had compositions close to those of the coins, corrections for the effect of lead and tin contents were calculated on the basis of the presence of these elements in various standards.

Physical Measurements

Table I lists the weights, densities, maximum and minimum diameters, maximum thicknesses, and die orientations of the coins. The coins are identified by Crawford number (*10*). Because the coins have been nondestructively analyzed, their physical measurements will identify the coins at any time unless the coins are deliberately altered.

Because Republican coins were made for such a long time, the denominations of coins changed with respect to weight and metal content. Most denominations were made very sporadically. Although asses were made for a long time, in some instances there were gaps in production of many years. The weights of asses decreased as a function of time by at least a factor of 2. The chemical compositions also changed markedly with time. Poor quality control must have been exercised because the coins varied in weight and chemical composition even within a few years. During the 200 years of coin production in the Roman Republic, the variations in the weight and composition of asses were so great that in some instances, one can scarcely be confident that the denomination has been correctly assigned. The late brass asses are probably worth twice the value of an as because of the premium placed on brass. Crawford has stated that these coins have a value of at least a dupondius, a coin of the Roman Empire with a value of two asses (*10*).

Table I. Physical Measurements of Copper-Based Coins of the Roman Republic

Coin[a]	Date (B.C.)	Weight (g)	Max. Thick. (cm)	Max. Diam. (cm)	Min. Diam. (cm)	Die Orientation	Density (g/cm^3)	Coin Identification
X–45	217–215	28.3140	0.520	3.03	2.91	3:00	8.81	38/5
M–46	217–215	6.2824	0.520	2.19	1.98	11:00	8.49	38/7
T–48	211–210	7.2719	0.318	2.22	2.10	6:00	8.36	87/3
S–49	ca. 211	10.3008	0.288	2.64	2.46	10:00	9.10	56/3
A–273	189–180	37.6548	0.565	3.42	3.12	3:00	9.27	143/1
A–51	169–158	19.7607	0.406	2.97	2.58	3:00	8.86	187/2
S–52	149	12.9326	0.377	2.57	2.33	7:00	9.08	210/3
A–53	148	13.0030[b]	0.304	3.09	2.84	–	8.64	218/2a
Q–80	134	3.3510	0.226	1.99	1.82	5:00	8.73	244/3
U–89	108–107	3.6734	0.265	1.64	1.59	12:00	8.82	308/4b
S–91	90	3.3760	0.200	2.14	1.97	1:30	8.71	340/5a
A–93	90	11.1772	0.341	2.91	2.58	1:00	7.83	342/7c
A–92	88	11.7774	0.319	2.89	2.61	10:30	8.46	346/3
Q–54	86	2.8260	0.254	1.56	1.47	2:30	8.77	350B/3a
A–130	45	12.0402	0.309	2.79	2.69	12:30	8.53	4776/1a
A–75	32–31	10.1068	0.401	2.42	2.23	12:00	7.80	1277
A–601	?	19.2998	0.415	2.94	2.81	8:00	9.06	535/1
A–275	?	16.9688	0.384	3.50	2.66	?	8.90	?
A–271	?	20.3070	0.357	3.20	3.09	4:00?	9.28	?
S–1601	?	5.5235	0.260	2.08	2.02	4:00?	8.40	?
T–1600	?	10.7326	0.396	2.49	2.40	6:00	7.79	?
S–274	?	22.4730	0.439	3.00	2.84	?	8.86	?

[a]The coin denominations are given by the letters: X, Sextans; M, Semuncia; U, Uncia; T, Triens; S, Semis; A, As; and Q, Quadrans.
[b]Reverse deeply abraded.

The densities of many of the lead-containing coins are higher than the density of copper (8.93 g/cm^3). However, bronze (Cu–Sn alloy) and brass (Cu–Zn alloy) both have densities below that of pure copper, and indeed, the measured densities of our coins reflect this fact. However, some of the cast coins have densities somewhat lower than expected from their chemical compositions. This low density is probably caused by interior porosity, which is rare in Roman Imperial coins (the hot striking of these coins would tend to close internal pores). The density of coin T-1600 is 7.79 g/cm^3. This very low density is caused by extensive interior corrosion of the coin. In fact, the coin seems to have noticeably expanded because of corrosion along grain boundaries into the interior of the coin.

The weight, diameter, and thickness of an unknown coin may be used to help date the coin. Furthermore, the chemical composition can independently be used to approximately date coins of the Republic.

Figure 1 shows two cast asses (Janus on the obverses and a prow of a ship on the reverse of A-51 and three prows on the reverse of A-931). These coins are known to have been cast in a long row because junctions between coins are visible along the tops and bottoms of the coin obverses (*see* Figure 1). A-93 has many small bubbles on the surface; this bubbling is typical of many cast coins. The two brass coins, A-130 (struck by Julius Caesar) and A-75, are shown in Figure 2.

Chemical Compositions

The chemical compositions of 22 Republican coins are given in Table II. Although many more analyses are needed before a full understanding of Republican coinage is obtained, some general observations are possible.

Figure 1. Cast coins A-51 (left) and A-93 (right). The length of the smallest division on the scale is 1.0 mm.

Figure 2. Brass coins A-75 (left) and A-130 (right). The length of the smallest division on the scale is 1.0 mm.

Table II. Chemical Compositions of Copper-Based Coins of the Roman Republic

Coin[a]	*Date (B.C.)*	*Fe*	*Co*	*Ni*	*Cu*	*Zn*	*As*	*Ac*	*Sn*	*Sb*	*Pb*
X–45	217–215	0.056	0.117	0.041	95.3	0.02	0.17	0.032	3.75	0.023	0.51
H–46	217–215	0.037	0.064	0.071	91.0	0.02	0.37	0.033	8.2	0.065	0.13
T–48	211–210	0.116	0.181	0.053	87.5	0.04	0.28	0.041	9.0	0.106	2.7
S–49	After 211	0.212	0.175	0.048	82.9	0.02	0.27	0.037	5.4	0.060	10.9
A–273	189–180	0.054	0.154	0.056	86.1	0.01	0.18	0.028	6.1	0.037	7.3
A–51	169–158	0.126	0.122	0.040	83.6	0.01	0.48	0.052	3.9	0.124	11.5
S–52	149	0.047	0.032	0.055	81.5	0.01	2.3	0.18	3.0	2.4	10.5
A–53	148	0.025	0.018	0.060	72.1	N	2.5	0.10	1.9	2.5	20.8
Q–90	134	0.066	N	0.036	99.1	0.03	0.01	0.33	0.13	0.15	0.12
U–89	108–107	0.091	0.013	0.072	93.3	0.03	0.07	0.081	6.1	0.075	0.14
S–91	90	0.183	0.021	0.67	85.0	0.01	0.32	0.21	7.3	1.4	4.9
A–93	90	0.021	N	0.106	96.8	0.01	N	0.105	2.7	0.18	0.01
A–92	88	0.055	N	0.074	95.9	0.01	N	0.28	1.8	0.53	1.4
Q–54	86	0.015	N	0.25	98.4	N	N	0.15	0.52	0.29	0.36
A–130	45	0.120	N	0.007	78.8	20.8	N	0.066	0.081	N	0.12
A–75	32–31	0.188	N	0.006	82.9	15.7	0.01	0.091	1.07	N	0.02
A–601	?	0.20	0.077	0.095	87.5	0.01	0.14	0.068	7.7	0.17	4.0
A–275	?	0.34	0.056	0.040	86.8	0.09	0.93	0.094	3.2	0.50	7.9
A–271	?	0.044	0.080	0.055	81.2	0.02	0.84	0.078	3.3	0.61	13.8
S–1601	?	0.019	0.027	0.026	88.3	0.01	0.16	0.034	11.5	0.018	0.02
T–1600	?	0.023	0.084	0.022	69.1	0.01	0.28	0.013	2.7	0.034	27.7
A–274	?	0.080	0.062	0.040	83.8	0.02	2.4	0.29	1.9	1.3	10.1

NOTE: All element concentrations are given in weight percent. N means none detected (less than 0.01% for all elements).

[a]The coin denominations are given by the letter: X, Sextans; M, Semuncia; U, Uncia; T, Triens; S, Semis; A, As; and Q, Quadrans.

Iron concentrations are given as a function of time in Figure 3. Only the compositions of coins of known or approximately known date are presented in Figures 3–10. The iron concentrations are relatively high in the two brass coins, A-75 and A-130. Brass coins of the Roman Empire usually have higher iron contents than copper or bronze coins, presumably either because of the ores used or the procedures used for producing the alloys. Many of the iron concentrations are comparatively low. Iron concentrations are highly variable and apparently do not correlate with other elements, date of manufacture, or denomination. Ancient bronzes often contain rather low concentrations of iron, possibly because lower temperatures were used to prepare coin flans (*4*).

The amounts of cobalt found in Roman Republican coins are extremely interesting because they are so different from those of Roman Imperial coins, which rarely contain detectable cobalt. (The lower detection limit of cobalt by X-ray fluorescence analysis is about 0.002%). Republican coins made before 135 B.C. all contain detectable cobalt, often in relatively high con-

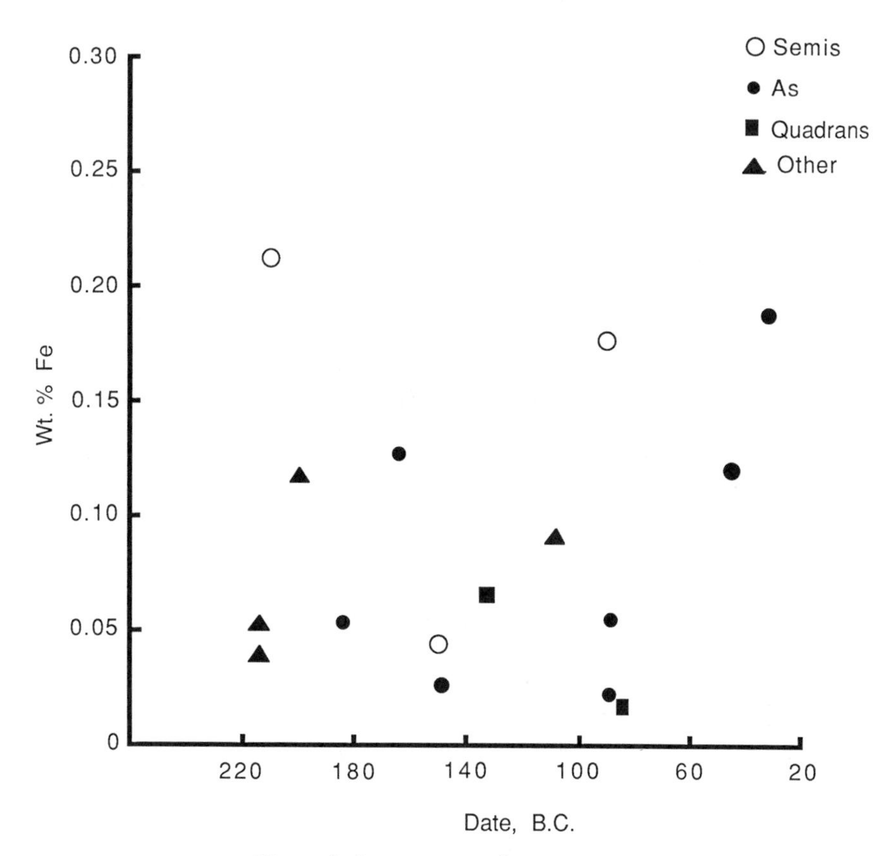

Figure 3. Iron content of various coins.

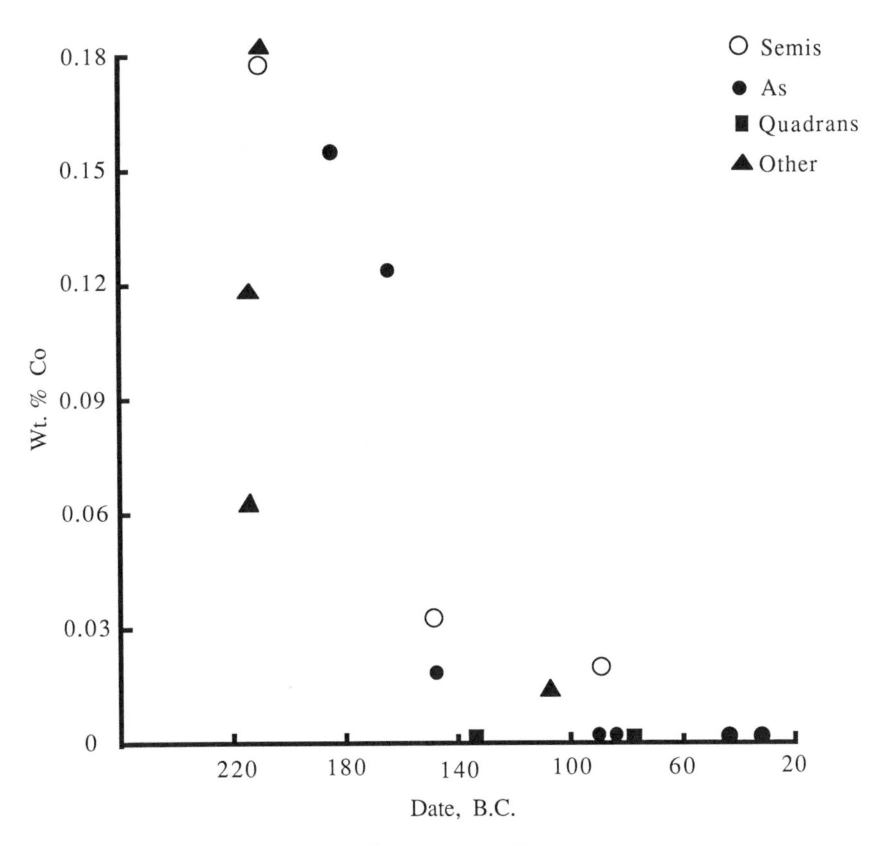

Figure 4. Cobalt content of various coins.

centrations (i.e. greater than 0.04%; *see* Figure 4). Coins of the first century B.C. rarely contain detectable cobalt. Cobalt in coins of the Roman Republic came from a particular ore source. The absence of cobalt from most later coins indicates that they were made from a new ore source, rather than from remelted coins or other metals.

Cobalt is a useful element for dating coins. For instance, coin A-601, which ostensibly was made in 38 B.C., contains cobalt. On close inspection, this coin was found to be an overstrike of an older coin. We believe that the coin was overstruck on an as that was probably at least 100 years old at the time.

Before 100 B.C., the nickel concentrations in coins usually varied within a fairly narrow range from 0.03 to 0.07% (*see* Figure 5). However, the two coins struck shortly after 100 B.C. are relatively high in Ni, a fact possibly indicating the that the Roman mint used a different ore source for a short period of time. Carter has reported two other periods of time (ca. 23–15 B.C. and A.D. 22–30), during which the nickel contents of Imperial coins are greater than 0.2% (*1*). With the two exceptions just mentioned, the

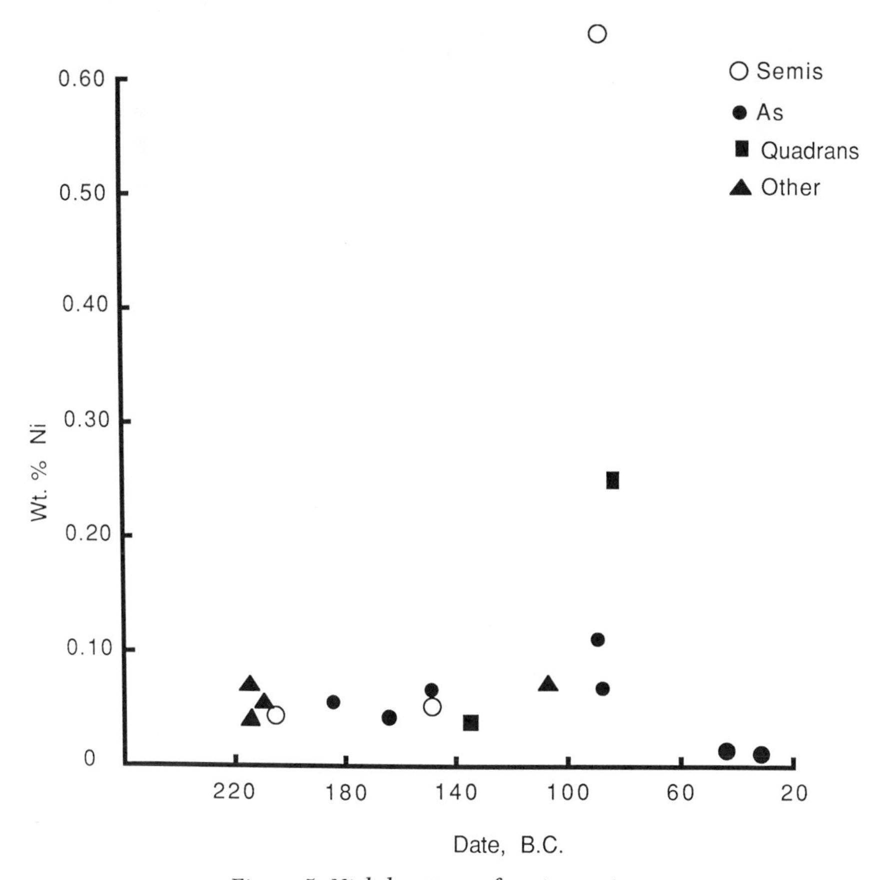

Figure 5. Nickel content of various coins.

nickel contents of the two late brass coins are quite low, as are the nickel contents of early Imperial coins (23 B.C. to A.D. 66).

The copper contents are not very meaningful because copper was the matrix metal. Any standards of production probably specified how many parts by weight of an alloying element (or ore) such as tin or lead would be added to copper.

Zinc was detected only at very low concentrations in most Republican coins. It was usually present as a trace impurity. However, two coins were deliberately made of brass; the earlier one was made by Julius Caesar. These are the earliest known Roman coins made of brass (*13*). Brass was a new and rather expensive alloy because it was necessary to make it by diffusion of zinc vapor into copper. Zinc was not prepared in metallic form by the Romans or other ancient peoples because of its volatility.

Undoubtedly, the similarity of the colors of brass and gold made brass unusually valuable to ancient people. Carter has calculated the value added by zinc to the Neronian coins. These coins showed the most complex use

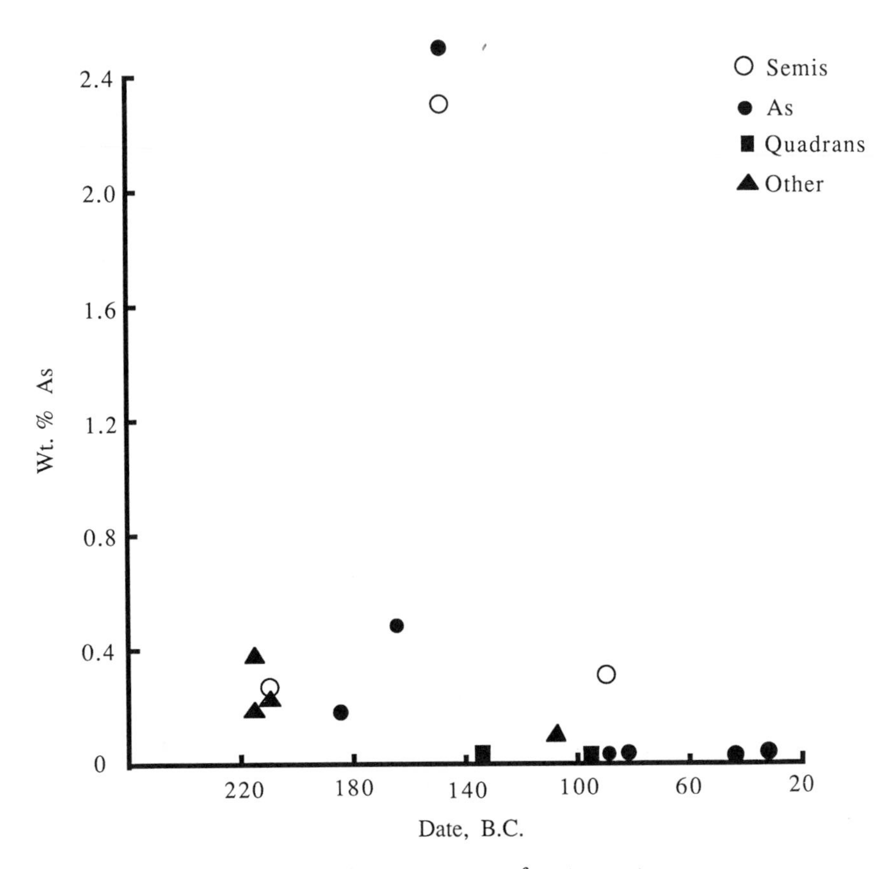

Figure 6. Arsenic content of various coins.

of various denominations of copper-based coins made by any Roman emperor (*14*).

Arsenic occurred in rather high concentrations (0.2% to about 0.4%; *see* Figure 6) in the coins struck before about 135 B.C. From about 160 to 135 B.C. (an approximate date that marks a rather important change in coin compositions in general), the arsenic contents of four coins are greater than 0.8% (assuming that coins A-275 and A-271, which have the prow of ships as found on other Roman Republican asses but which cannot be specifically identified, are dated within this period). Probably arsenic was deliberately added to these four coins (or perhaps an ore exceptionally rich in arsenic may have been used). Almost all of the coins made after 135 B.C. contain relatively low concentrations of arsenic.

From about 150 to 80 B.C., most of the coins contained much higher silver concentrations than normally found in copper-based Roman coins (*see* Figure 7). There is no indication that silver was deliberately added to any

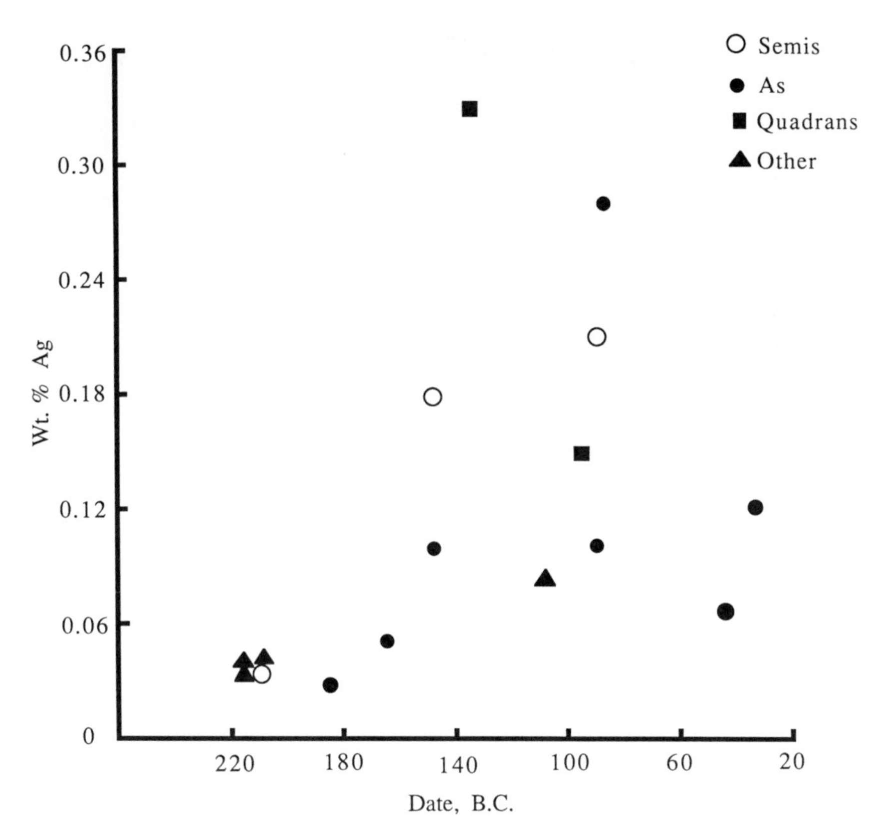

Figure 7. Silver content of various coins.

of the coins that were analyzed. All the silver levels are consistent with the silver contents of copper ores. If scrap metallic objects were occasionally melted to supply part of the metal for the coins, then once in a great while some silver-containing metal may inadvertently have been melted with copper, and a somewhat higher than normal concentration of silver resulted.

Tin was deliberately added to all the coins, with three probable exceptions (*see* Figure 8). Hence, most of the coins are true bronzes. Tin content generally ranged from about 3% to 9%. The tin determinations by X-ray fluorescence are of questionable accuracy because of inhomogeneities in the coins. One coin with a low tin content (to which tin was deliberately added) contains high concentrations of antimony and arsenic. The antimony and arsenic were either inadvertent or intended substitutes for tin. It is impossible to conclude from these data whether tin contents were set at different levels for various denominations of coins. However, both quadrantes contain no deliberately added tin.

Antimony contents are moderate to very high compared with the contents of Imperial coins. Four coins from the middle of the second century

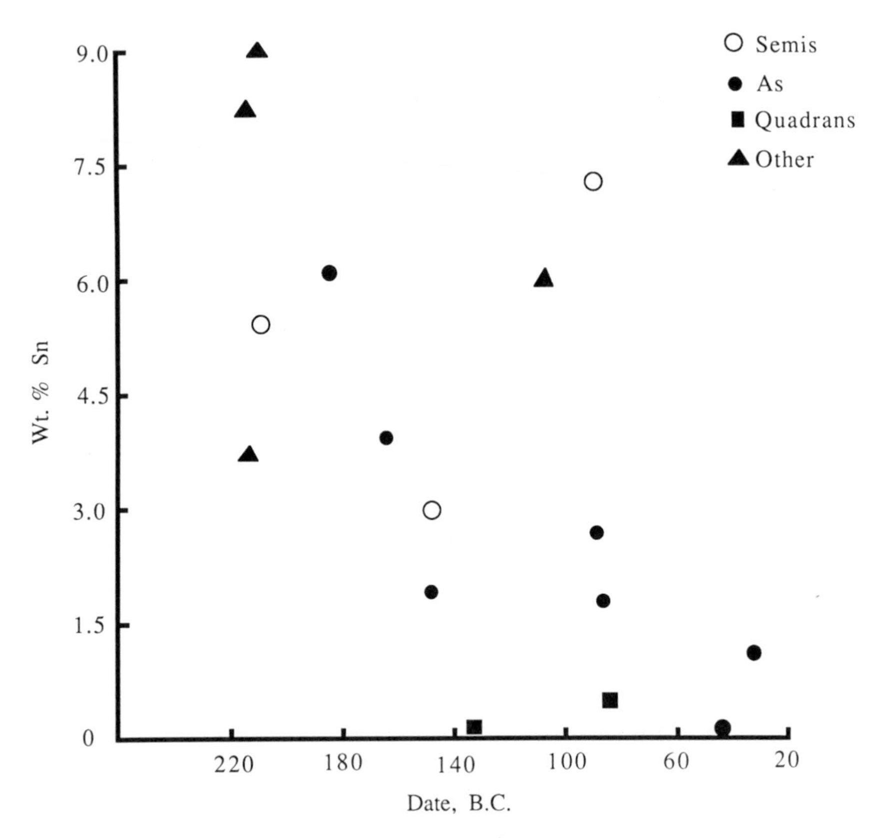

Figure 8. Tin content of various coins.

B.C. contain very high antimony concentrations, and two of these are semisses (only four semisses were analyzed). The coins from the earliest period to about 160 B.C. contain moderate amounts of antimony (*see* Figure 9). Most of the coins made from around 160 to 80 B.C. have relatively high antimony contents; these same coins also were relatively high in silver. The latest two coins have the very low antimony contents typical of coins minted in the early Roman Empire.

The coins made before 135 B.C. are almost all leaded bronzes (Figure 10). However, the two earliest coins, made in the late third century B.C., have low lead contents. The lead contents increased rapidly with passing time, until exceptionally high lead contents were found in the coins of three denominations: semisses, asses, and a triens, all of the middle of the second century B.C. The reported lead contents of some of these coins are probably lower than the actual contents because of the segregation and surface depletion of lead. Appreciable amounts of lead were not frequently added to coins after roughly 100 B.C. A marked change in the composition of coins occurred both before 135 B.C. and after about 100 B.C. (Only one

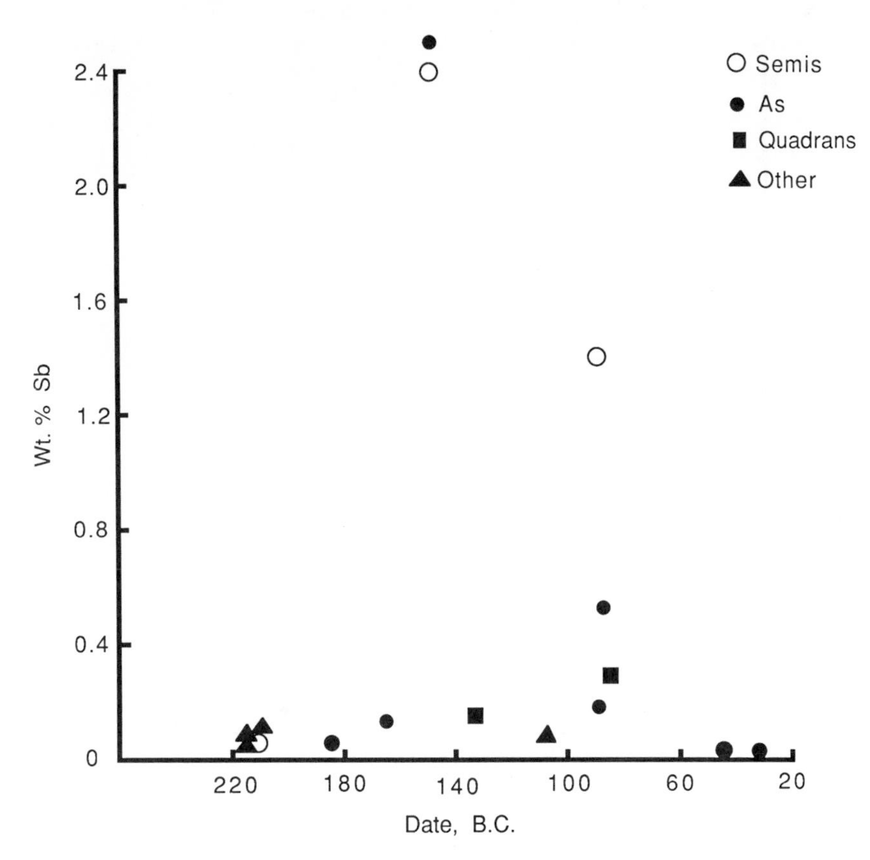

Figure 9. Antimony content of various coins.

coin from 135 to 100 B.C. was analyzed, and information from this period is unavailable).

Metallographic Examination of Two Coins

Coin A-275 was ground, polished with aluminum oxide, and then etched by continuous swabbing with a solution of 20 mL of 1:1 ammonia and 5 drops of hydrogen peroxide. The cross section was several millimeters from the edge of the coin. Twins are alternate light and dark etched crystals in the same grain. Twins are crystals of closely related orientation separated by a boundary surface which is planar in nondeformed twins. As shown in Figure 11, twins were present in almost every grain. The large gray particles are a lead-rich phase. Because the twins are not bent, the coin is not in the as-struck condition. The copper matrix of this coin is fairly soft: 74 to 102 Vickers hardness number (VHN). Although there were many small lead particles present, a few large lead particles were evenly distributed through-

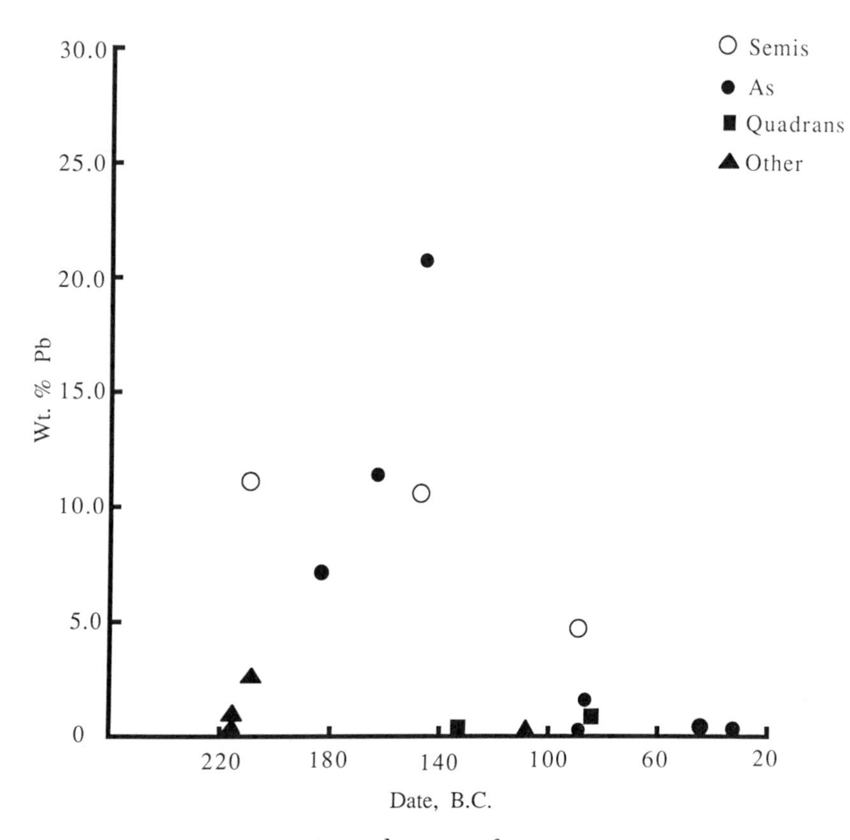

Figure 10. Lead content of various coins.

out the coin. The hardness of the lead-rich phase was less than 17 VHN (extremely soft). The microhardnesses measured at the edge of the coin were the same as interior microhardnesses. Either this coin was cast or it was annealed after striking.

Coin A-271 (*see* Figure 12) contained a few relatively large lead particles, as well as some small lead particles. The large lead particles tended to concentrate at the interior of the coin; no large particles were at the surface of the coin. The large lead particles contain small copper dendrites that precipitated as the liquid lead cooled. Figure 13 shows the light colored copper dendrites. The microhardness of the copper-rich phase ranged from 105 to 130 VHN. The lead was extremely soft at 17 VHN. The lead in this coin is undoubtedly present in a higher concentration than that reported by X-ray fluorescence. Coin A-271 apparently was cast and not subsequently struck, otherwise the large lead particles would have been appreciably flattened.

Figure 11. Microstructure of coin A-275. The length of the smallest division on the scale is 2.5 μm.

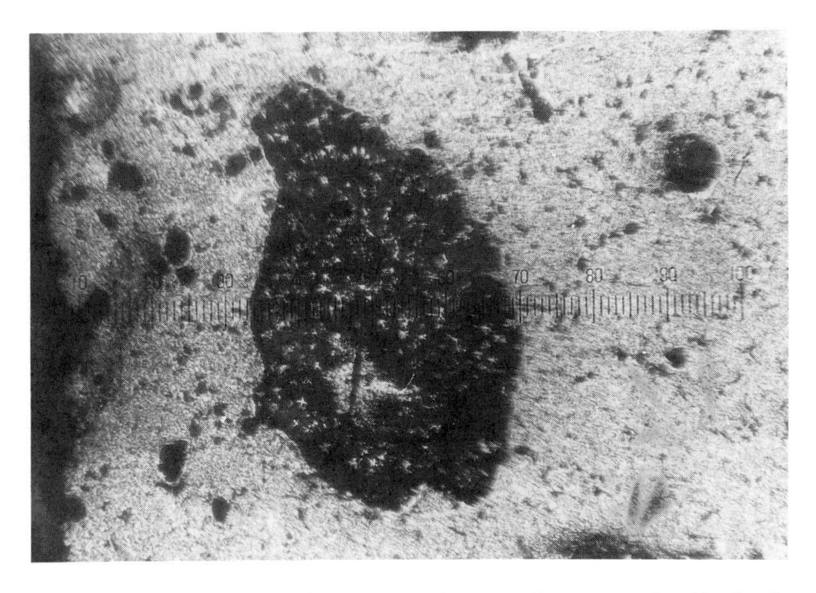

Figure 12. Microstructure of coin A-271 showing a large particle of lead. The length of the smallest division on the scale is 20 μm.

Figure 13. Copper dendrites inside a large particle of lead in coin A-271. The length of the smallest division on the scale is 2.5 μm.

Comments on Individual Coins

The brass as of Q. Oppius (A-75) shown in Figure 2, has not been dated with any certainty. It has been assigned a date of 32–31 A.D., but Crawford (*10*) is inclined to ascribe this issue to about 88 B.C. The chemical analyses support the later date. The very low antimony and the low nickel content both would place this coin in the late Republic, not in 88 B.C. It is possible that the coin could have been produced in a mint far from Rome, but the advanced technology necessary to produce brass would make this unlikely.

The as of Augustus and Julius Caesar (A-601) is an overstrike on a much older coin. The overstrike accounts for the presence of cobalt as well as other anomalies in the composition of this coin (e.g., relatively high nickel and antimony contents).

The dates of the coins, A-271, A-274, A-275, S-1601, and T-1600, are unknown. The following dates are estimated on the basis of their chemical compositions and weights: A-271 (ca. 160 B.C.), A-274 (ca. 150 B.C.), A-275 (ca. 160 B.C.), S-1601 (ca. 160–180 B.C.), T-1600 (ca. 180–200 B.C.).

Conclusions

1. **No evidence supports the widespread remelting of coins.** The later coins contain much lower concentrations of several elements than earlier coins. Certainly most of these elements, especially cobalt, would have appeared in relatively high con-

centrations in remelted coins. However, coins that contained low cobalt concentrations could have been made from a mixture of old coins that contained cobalt and new metal made from ore that contained no cobalt.

2. **Chemical compositions correlate reasonably well with time.** Coins may be dated approximately from their compositions. The compositions of coins made before about 135–100 B.C. are considerably different from all later coins. Coins made around 80 B.C. are different with respect to their contents of several elements when compared with compositions of coins from other times. The late Republican coins also have unique compositions. Five coins were approximately dated on the basis of their chemical compositions and weights.
3. **Chemical compositions of Republican copper-based coins are markedly different from those of Imperial coins.** In the time of Augustus, the coin compositions changed considerably from those of earlier Republican coins. The coins of Augustus that were minted in Rome were much purer; they were struck in two materials: in essentially pure copper and in brass (*15*).
4. **Early Republican coins are remarkable for their high cobalt contents and many for their exceptionally high lead, arsenic, or antimony contents.** The cobalt contents are uniquely high in early Republican coins compared with all Roman coins made from about 135 B.C. to A.D. 400. Some late Imperial Roman coins contain high amounts of lead, but not as high as in the early Republican semisses and asses.
5. **Additional analyses are required before one may reliably conclude that chemical composition correlates with denomination.** There are indications that composition correlates with denomination (e.g., the exceptionally high lead contents of mid-second-century B.C. semisses and asses and the low lead contents of quadrantes). However, larger numbers of coins must be chemically analyzed to obtain a thorough understanding of the Roman Republican copper-based coinage.

Denominations

The as was the largest denomination of copper-based coins of the Roman Republic. The values of the other coins were as follows:

- two semis equal one as
- three triens equal one as
- four quadrans equal one as
- six sextans equal one as
- twelve uncia equal one as

Various denominations were not continually struck; there were long periods during which no coins were struck for each of the denominations.

Acknowledgment

We thank Robert McCorkle for the opportunity to analyze his coins, S-1601 and T-1600.

References

1. Carter, G. F.; Buttrey, T. V. *Am. Numismatic Soc. Mus. Notes* **1977,** *22,* 49–65.
2. Carter, G. F. In *Archaeological Chemistry II*; Carter, G. F., Ed.; Advances in Chemistry Series No. 171; American Chemical Society: Washington, D.C., 1978; pp 347–377.
3. Carter, G. F. In *Archaeological Chemistry III*; Lambert, J. B., Ed.; Advances in Chemistry Series No. 205; American Chemical Society: Washington, D.C., 1984; pp 311–329.
4. Carter, G. F.; King, C. E. In *Metallurgy in Numismatics I*; Royal Numismatic Society: London, 1980; pp 157–167.
5. Caley, E. R. *Orichalcum and Related Ancient Alloys*; Numismatic Notes and Monographs No. 151; American Numismatic Society: New York, 1964.
6. Riederer, J. *Jahrbuch Numismatik Geldgeschichte* **1974,** *24,* 73–98.
7. Phillips, J. A. *Quarterly J. Chem. Soc. of London* **1852,** 265.
8. Caley, E. R. In *The Composition of Ancient Greek Bronze Coins*; American Philosophical Society: Philadelphia, 1939.
9. Bahrfeldt, M. *Numismatik Z.* **1905,** 9–56.
10. Crawford, M. In *Roman Republican Coinage*; Cambridge Univ. Press: Cambridge, 1974, Vol. II, 573–577.
11. Carter, G. F.; Booth, M. M. In *Problems of Medieval Coinage in the Iberian Area*; Instituto Politecnico de Santarem: Santarem, Portugal, 1984, 49–69.
12. Carter, G. F.; Kimiatek, M. H. *Archaeo-Physika* **1979,** *10,* 82–96.
13. Burnett, A. M.; Craddock, P. T.; Preston, K. In *Proceedings of the Int. Congress of Numismatics*; Hackens, T.; Weiler, R., Eds.; Louvain-la-Neuve, Luxemburg, 1982, 263–268.
14. Carter, G. F. *Am. Numismatic Soc. Mus. Notes,* **1988,** *33* in press.
15. Carter, G. F. In *Science and Archaeology*; MIT Press: Boston, 1971, 114–130.

RECEIVED for review December 29, 1987. ACCEPTED revised manuscript July 19, 1988.

ART OBJECTS

12

Two Medieval Enameled Objects Studied by X-ray Fluorescence

Philip K. Hopke[1], Wendell S. Williams[2], Henry Maguire[3], and David Farris[3]

Program on Ancient Technologies and Archaeological Materials, University of Illinois at Urbana-Champaign, Urbana, IL 61801

The composition of colored enamels used on late 12th-century copper articles associated with Limoges, France, is of interest to ceramic scientists and art historians. We analyzed a Limoges enamel cross from the World Heritage Museum of the University of Illinois by using X-ray fluorescence (XRF). The cross comprises a cast copper corpus and an enameled cross-shaped copper plaque that serves as a ground to which the corpus is attached. Enamel colors are light blue, dark blue, green, red, white, and yellow. Nondestructive elemental analysis of the enamel was carried out with a 1-Ci ^{241}Am source in the secondary mode with a Mo target. Metal masks of Sn isolated specific colored regions. The resulting XRF spectra indicated the presence of copper in all the colored enamels after the Cu background was subtracted from the substrate. The dark blue and green areas also showed strong peaks for lead, and the light blue, red, and white areas showed weak lead peaks. If we assume that the Pb is a constituent of the glass, all the colors can be attributed to various mineral compounds of copper. A similar study of reliquary from the Indianapolis Museum of Art showed quite different enamel compositions.

[1]Also with Institute for Environmental Studies, University of Illinois at Urbana-Champaign.
[2]Current address: Department of Materials Science and Engineering, Case Western Reserve University, University Circle, Cleveland, OH 44106.
[3]Also with Art History Division, School of Art and Design, University of Illinois at Urbana-Champaign.

0065-2393/89/0220-0233$06.00/0

THE GLORIOUS MEDIEVAL STAINED-GLASS WINDOWS of European cathedrals have been the subject of several chemical analyses to support conservation projects, to identify workshops and their practices, and to attempt the association of colors with specific metal ions in various oxidation states. However, little information of this sort has been generated from the beautifully decorated, enameled, and gilded copper ecclesiastical objects used in medieval cathedrals that are recognized under the generic name of Limoges enamels.

A notable exception is an important technical study of medieval enamels from the Boston Museum of Fine Arts contributed recently by England (*1*). The versatility of copper as a coloring agent was noted as follows: "Copper is responsible for a range of colors—blue, turquoise, green and red...." This statement supports our earlier finding (*2*) that copper is the principal coloring agent in the enamels of a Limoges cross at the University of Illinois. This chapter gives the pertinent details of that study.

We carried out an X-ray fluorescence (XRF) study of two museum objects, a cross and a reliquary, said to be medieval Limoges enamels. From the perspective of archaeological chemistry, the primary objective of the study was to identify the metal ions in the glass matrix responsible for the characteristic and intense blues, greens, reds, and yellows of the enamels. Another objective was to determine whether the enamel composition of the reliquary was sufficiently different from that of Limoges enamels to allow a discrimination to be made. The findings will supplement scholarly information on the period and the workshop of a given object and aid in the detection of modern forgeries.

Examples of Limoges Enamels

The objects examined in this study were fabricated from flat sheets of cast and hammered copper partially scooped out to receive the glass frit (champlevé process). Firing in air oxidizes the copper surface and fuses and bonds the glass to produce the characteristic enameled surfaces. The objects were gilded over the unenameled copper, as well. These objects are rich in color and in symbolism.

Where were these objects made? Documentary evidence from the Middle Ages has established that the city of Limoges, in southwest France, was an active center of European enamel workers as early as the 12th century (*3*). However, it is difficult to determine from the written evidence the precise form taken by the industry that produced Limoges enamels, *opus lemovicensis* in medieval sources. We do not know, for example, to what extent Limoges was a center of manufacturing or a center for the training of artisans who worked elsewhere. Nor do we know how much trade there was in the colored glasses used by enamelers. Even the chemical composition of these enamels has not been well documented until recently, and the study of enamel composition is the central concern of the present investigation.

Objects Studied

Cross. The principal object studied in this investigation is shown in Figure 1, a previously unpublished photograph of a champlevé-enameled copper cross with a relief figure of Christ. The cross is in the collection of the World Heritage Museum of the University of Illinois at Urbana-Champaign.

This object belongs to an important group of crosses made in Western Europe during the high Middle Ages (*4*). These crosses were first classified by Thoby (*5*) in 1953, using a methodology based on style, shape, technique of manufacture, use of artistic medium, and choice of motif.

The cross measures 25.5 by 18.5 cm. The principal colors of the enamel are dark blue, green, light blue, yellow, and red. The enamel ground is divided into three sectors: an inner cross of green, an outer cross of dark blue, and an outside border of blue and white. Within the inner cross is the figure of Christ. Above Christ's head is the *dextera Domini* in reserve. It extends downward to the monogram of Christ (IHS XPS), also in reserve. Behind the head of Christ is a nimbus of red, light blue, and white. Below Christ's feet is an abstracted white skull, and around his arms and head, there is a scrolling vine in reserve.

Within the outer cross are three roundels formed by concentric circles of yellow, green, dark blue, and red and filling three of the angles of the cross. The remaining field of the outer cross is punctuated by smaller circles composed of white, blue, yellow, and green enamel. The appliqué figure of Christ is made of cast and chiseled copper; it retains traces of gilding around the neck, forehead, hair, torso, and loin cloth. Further gilding remains randomly around the border of the cross. The figure is bolted to the enamel plaque by four copper nails at the hands and feet.

The iconographic program of the group of enamel crosses to which our example belongs demonstrates the results of serial production. By the late 12th century, Limoges enamels assumed a more industrial character in style and iconography. The earlier crosses depicted the crucified Christ on a stylized green tree, as for example in the cross at the Metropolitan Museum of Art, New York, executed between 1185 and 1190 (*4*). At the base of the earlier crosses, there is an image of either Adam's skull or a half figure of Adam emerging from a sarcophagus. The tree refers to the Tree of Life, and the skull refers to Adam's death. These images symbolize the rebirth of Adam through Christ. In some crosses, the green tree becomes a vine motif, as evident in an example at the Louvre. In other examples, the enameled leaves of the vine are replaced by circles or rosettes. On the cross in the World Heritage Museum, the reserve strip of copper in the green inner

Figure 1. Limoges enamel cross, late 12th century A.D., World Heritage Museum, University of Illinois, Urbana.

cross represents the vine on the Tree of Life, and the skull of Adam appears in a simplified form beneath the feet of Christ.

Reliquary. The other object studied is a hinged copper box, gilded and enameled on the outside with two cast, gilded figures riveted to one side (Figure 2). A pierced, flat crown surmounts the ridge of the peaked lid. The object is on loan from the Indianapolis Museum of Art. The reliquary is 27 cm long, 6.5 cm wide, and 13 cm high. The pierced decoration on top adds another 4.5 cm to the height.

The enameling was done in red, light blue, dark blue, green, and white. Considerable gilding remains on the unenameled copper. The enamel layers are appreciably thinner than those on the cross just described.

Chemistry of Colored Medieval Glass

England (*1*) reviewed the literature of the chemistry of medieval glass and enamel, including excerpts from the writings of Theophilus, the medieval scholar. Copper is mentioned as an important colorant.

Brill (*6*) summarized the information on the chemical basis of medieval glass colors as follows: red/orange opaque enamels, cuprite (Cu_2O) with lead additions beneficial for opacity; white opaque enamels, either tin oxide opa-

Figure 2. Medieval enameled reliquary, Indianapolis Museum of Art, Indianapolis, IN.

cifier or calcium antimonide at 5–8%; light blue, 1% copper or a small amount of cobalt (0.05%); yellow, minute flakes of standard yellow opacifiers ($Pb_2Sb_2O_7$ or $PbSnO_3$ at 5–8%); green, opaque yellow plus copper; and dark purple, Mn.

In an early XRF study of the chemical composition of cathedral glass, much of it medieval, Brill (*7*) obtained data for the elements Mn, Fe, Cu, Zn, Rb, Sr, Zr, Ag, Sn, Ba, and Pb. The measured values were converted into weight percentages of the corresponding oxides that would have been the form in which the element was introduced into the glass composition. The concentrations of the oxides ranged from 0.01 to 1% among the several colors sampled: green, ruby, amber, pale blue, dark blue, and yellow amber. PbO was present at 0.1% for all colors, but presumably, it was associated with the basic glass composition. However, this value went up to 2–3% for ruby glass from Ulm made circa A.D. 1400 Green glass from Canterbury, circa A.D. 1200–1225, contained 2.1% MnO, 1.1% Fe_2O_3, and 0.05% CuO, whereas ruby glass from the same source contained 0.9% MnO, 0.6% Fe_2O_3, and 0.4% CuO.

A study of ancient transparent and opaque glasses by Brill and Moll (*8*), using electron probe microanalysis, led to the association of metal oxides and color given in Table I. Although not marked as a colorant or opacifier, PbO gave a strong signal for opaque red (29.0) and opaque yellow (21.5%). These values are much greater than the general proportion of lead oxide in glass (0.2%).

An extensive study of ancient glass (first millennium, from Britain) was carried out by Sanderson, Hunter, and Warren (*9*) using XRF. Because these specimens were only lightly tinted, the interests of the authors were focused on the chemistry of the glass composition and the applicability of XRF to its determination rather than to the identification of the coloring agents.

In a study of medieval glass, Olin and Sayre (*10*) noted that different levels of alkalis and even a different specific alkali—potassium rather than sodium, from beechwood ash—distinguish medieval glasses from glasses of

Table I. Metal Oxides Associated with Colors of Medieval Glasses

Color	*Metal Oxides*
White opaque	Sb_2O_5, 8.9%
Dark blue transparent	CuO, 0.3%; CoO, 0.13%
Dark blue opaque	Sb_2O_5, 8.1%; CoO, 0.05%
Light blue opaque	Sb_2O_5, 5.8%; CuO, 3.7%
Red opaque	Cu_2O, 1.7%
Yellow opaque	Sb_2O_5, 2.2%

NOTE: The composition of the hypothetical parent glass is as follows: SiO_2, 67%; CaO, 5.0%; K_2O, 0.4%; MgO, 0.5%; Al_2O_3, 2.0%; Fe_2O_3, 0.4%; Sb_2O_5, 1.0%; and PbO, 0.2%.

the Roman period. However, after comparing the average compositions of colorless, amber, and blue glasses from the Rouen Chateau, Olin and Sayre (*10*) concluded that "For most elements the standard deviation ranges of concentrations encountered in all three colors overlap, and hence the average concentrations in each of the colors were not significantly different from each other." The authors found that cobalt, iron, thorium, and chromium concentrations were higher for the blue glasses than for the other two types of glasses and suggested that a Co-containing material had been deliberately added for coloration. The elements copper and lead were not included in their study.

According to Bamford (*11*), copper oxide normally imparts to soda lime–silicate glass a blue coloration that is associated with divalent copper ions. He also notes that "Lead silicate glasses containing copper are green, their shade deepening with increasing lead oxide content.... The most common blue color is achieved by the addition of cobalt oxide to the clear base glass, requiring approximately 330 ppm of Co_3O_4. An alternative shade is possible with copper oxide (approximately 1% by weight) melting under oxidizing conditions to ensure complete conversion to the cupric form." Explanations for the observed optical absorption associated with various metal ions in glass are given by Bamford (*11*) in terms of ligand field theory.

Elemental Analysis

Method. In the present study, XRF was used to identify major elements in the enamel and the substrate of the cross and reliquary described in a previous section. We used an XRF system with a 1-Ci ^{241}Am source in a New England Nuclear holder. The source was used in the secondary mode with, in this case, a molybdenum target. The output of the Si–Li detector was analyzed by a Nuclear Data 6620 computer-based multichannel analyzer. The downfacing detector–source–exciter system was positioned above the objects to permit nondestructive analysis while the objects remained horizontal.

A preliminary analysis of the multicolored area above the right arm of the cross revealed primarily copper and lead in the enamel. The substrate of the cross (Figure 3) and the copper corpus (Figure 4) showed no detectable zinc or tin.

A more detailed examination of the composition of the variously colored enameled areas was then carried out by masking all but the regions under study with double layers of 0.007-in. (0.18-mm) tin sheet of high purity. The tin by itself showed no peaks other than that of tin in the XRF spectrum, even when placed on a copper substrate. Hence the tin masks cannot have contributed to the observed peaks, nor can any of the signals detected have come from areas under the tin masks. Similar masking was performed for regions on the reliquary.

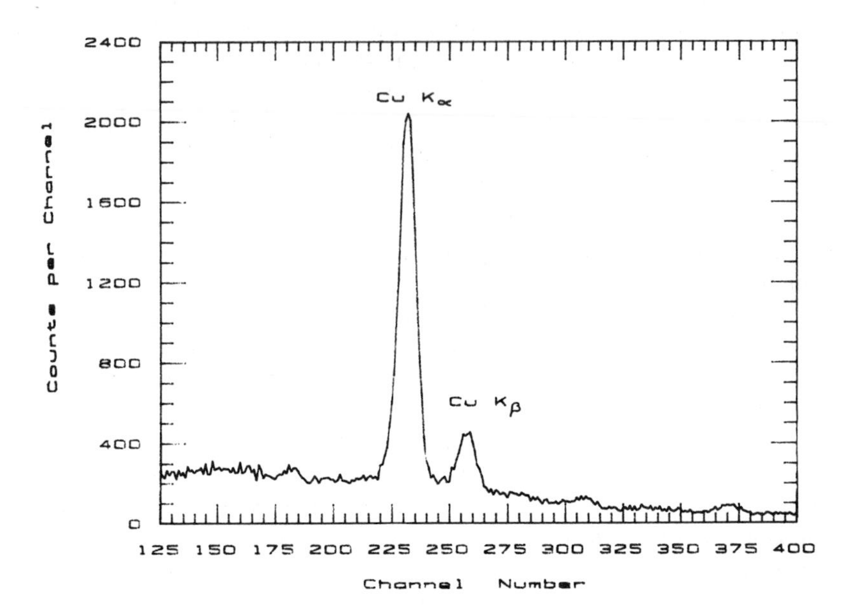

Figure 3. XRF spectrum from the cross. The exposed area of copper showed only copper peaks.

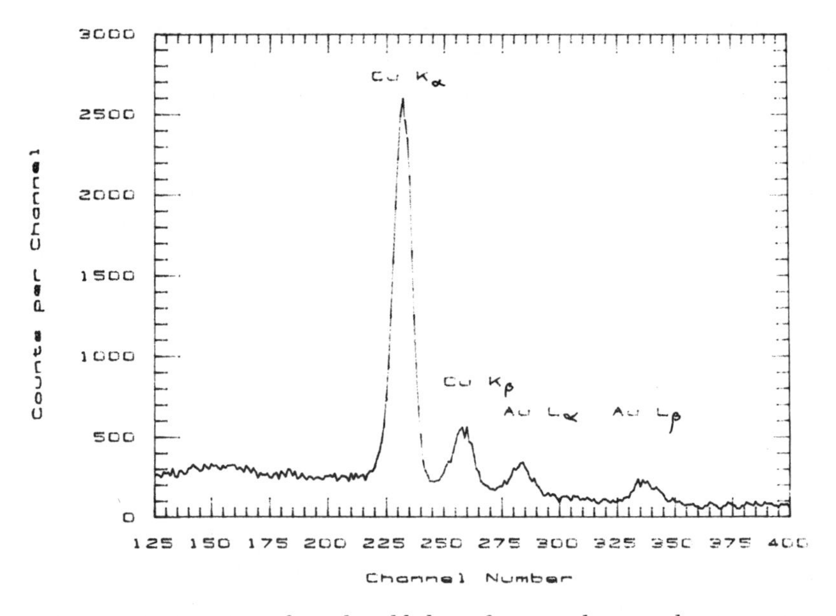

Figure 4. XRF spectrum from the gilded cast figure on the cross showing major copper and gold peaks and small lead peaks.

Table II summarizes the results of the XRF study of specific regions of the cross. The X-ray emission spectra for these several regions are shown graphically in Figures 5–10. Two spectra of regions on the reliquary are shown in Figures 11 and 12.

Results. All of the enameled regions of the cross investigated—red, yellow, green, dark blue, light blue, and white—gave strong copper peaks. The dark blue and green areas also showed strong peaks for lead. Weak

Table II. X-ray Fluorescence Analysis of Champlevé Limoges Enamel Cross in the World Heritage Museum, University of Illinois at Urbana–Champaign.

Sample	*Cu*	*Pb*	*Cr/Mn*	*Au*
		Enamel		
Red	X	x	x	
Yellow	X	x		
Green	X	X		
Dark blue	X	X		
Light blue	X	x		
White	X	x		
		Metal		
Figure	X	x		X[a]
Cross	X	x		X[a]

[a]Surface gliding still available.
NOTE: Abbreviations used: X, major amounts; x, minor amounts.

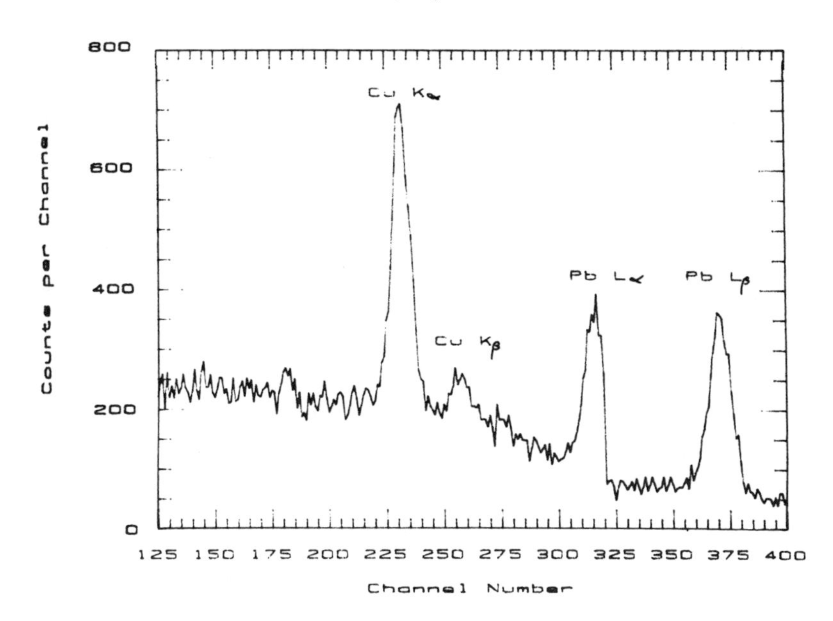

Figure 5. XRF spectrum from the cross. The green enamel near the upper arm showed major copper and lead peaks.

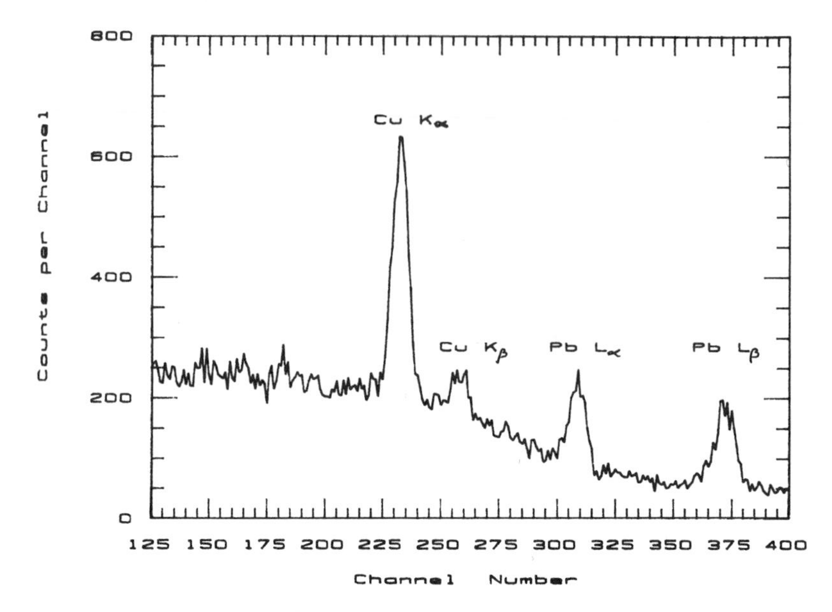

Figure 6. XRF spectrum from the cross. The dark blue enamel above the left arm showed major copper and lead peaks.

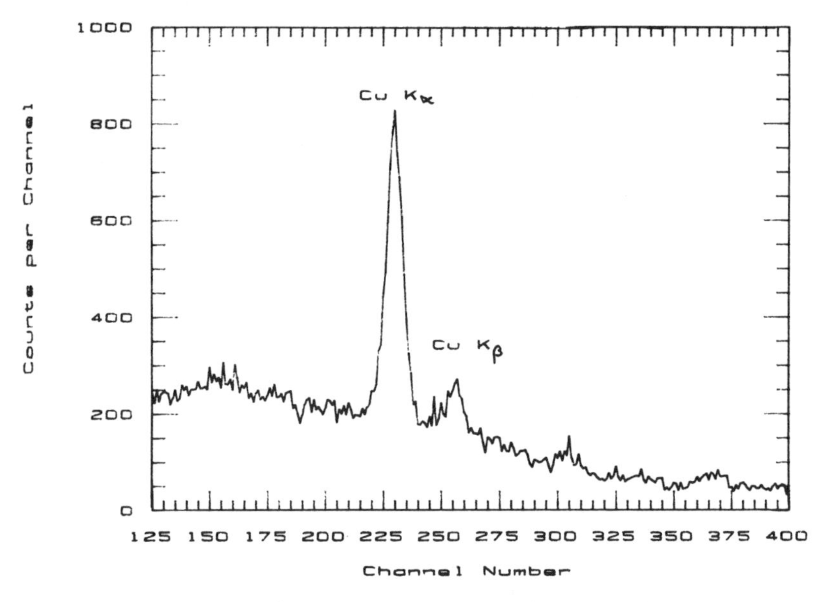

Figure 7. XRF spectrum from the cross. The light blue enamel below the left arm showed only copper peaks.

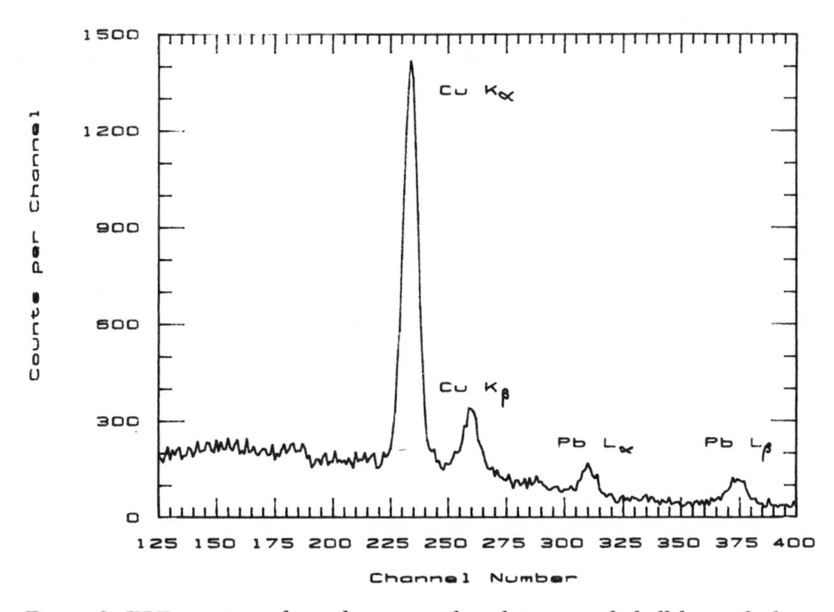

Figure 8. XRF spectrum from the cross. The white enamel skull beneath the feet showed major copper peaks and minor lead peaks.

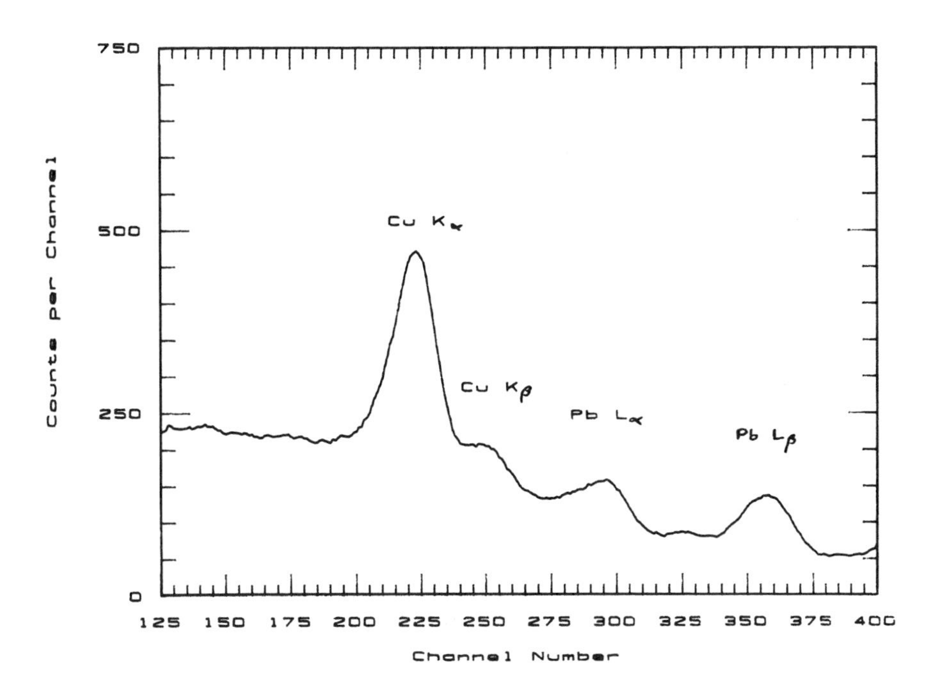

Figure 9. XRF spectrum from the cross. The yellow enamel circle below the arm showed major copper peaks and minor lead peaks.

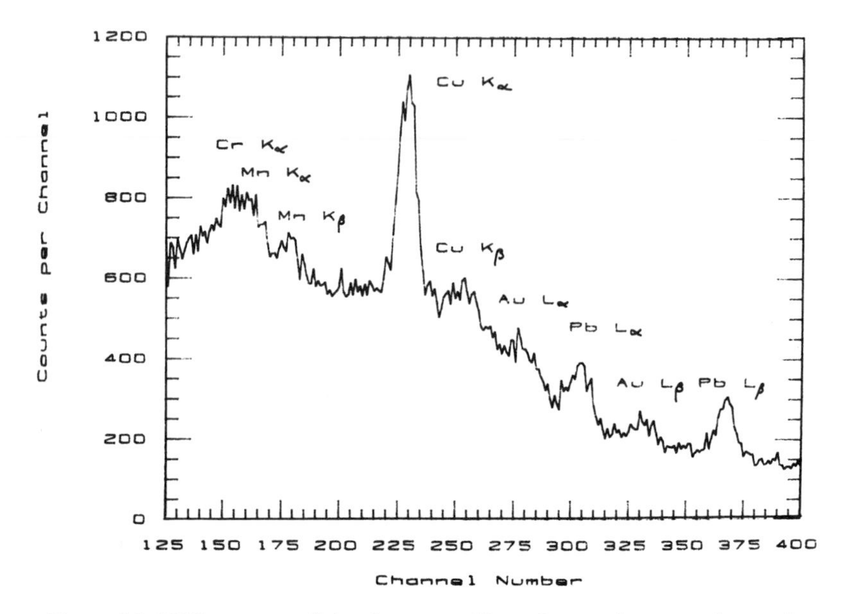

Figure 10. XRF spectrum from the cross. The red enamel area in the nimbus behind the head of Christ showed major copper peaks, minor lead peaks, and, possibly, minor peaks for chromium or manganese.

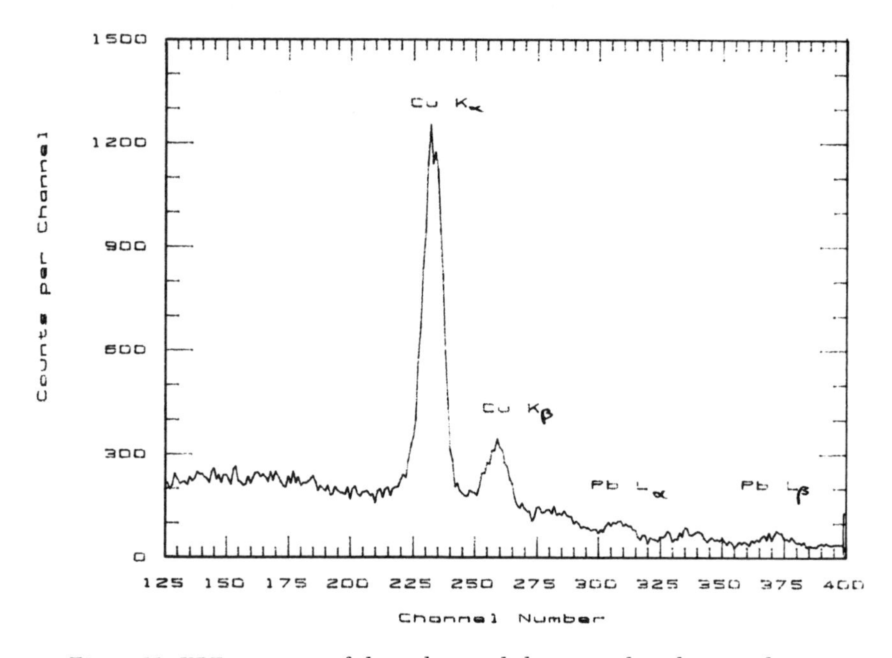

Figure 11. XRF spectrum of the red enameled spot on the reliquary showing only the major copper peaks.

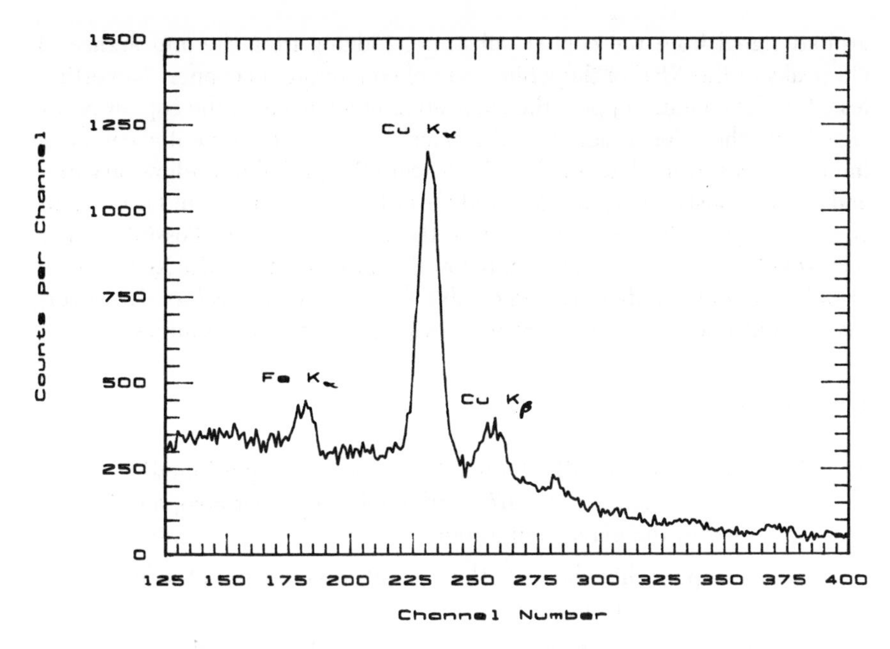

Figure 12. XRF spectrum of a light blue enameled region on the reliquary showing the major copper peaks and the K_α peak of iron.

peaks for lead were obtained from the light blue, red, and white areas. In addition, the red region showed broad peaks in the region of energy appropriate for manganese and chromium (Figure 10). No cobalt was detected in any of the spectra.

XRF spectra were also taken from the copper corpus attached to the cross and from an exposed region of the copper substrate from which a small piece of enamel had fallen away (Figures 3 and 6). Only copper and a small amount of lead were detected in the corpus, whereas the exposed substrate contained a high amount of copper with a very small amount of lead (about 1%).

The enamel on the reliquary, which is thinner than that on the cross, gave strong copper signals in all cases. The red spot showed only copper, and the light blue enamel contained iron; the spectrum of the green enamel showed small peaks for nickel and gold. No lead was detected in any of the enameled areas on the reliquary.

Discussion

The present study suggests that the coloring agents in the enameled copper cross examined, judged to be medieval from stylistic considerations, may be primarily compounds of copper. However, there was also a strong background signal from the copper substrate, judging from the appearance of Cu peaks in the XRF of the white enamel containing no copper. Nevertheless, two arguments support the association of a portion of the copper X-ray signal with the coloring agents in the enamel: (1) No other potential colorants (metal species) were detected by XRF (except Pb, probably a yellow opacifier and a glass constitutent, and Mn and Cr in the red enamel); and (2) various copper compounds can account for all the colors observed. Possibly, other metal oxides in low concentration, below the limit of detection of XRF (~1%), or of low atomic number are responsible for the observed colors. However, the present finding is consistent with evidence from other sources.

Conclusions

1. On the basis of XRF studies, the colors of a typical medieval Limoges enamel cross are attributable to copper compounds added to the glass composition.
2. The copper sheet beneath the enamel cross contains little lead; zinc and tin are not present.
3. On the basis of XRF studies, the enamels of a medieval reliquary are recognizably different, a fact suggesting that the object was made in a different workshop tradition, although its the stylistic features are typical of enameled medieval objects.

Acknowledgments

We thank Sarah Wisseman, Cynthia Wolfe, and Suzanne Hoban for their assistance in this project and the Graduate College of the University of Illinois for financial support. We also thank Barbara Bohen, Director of the World Heritage Museum of the University of Illinois, and the Indianapolis Museum of Art for their cooperation.

Literature Cited

1. England, P. In *Catalogue of Medieval Objects in the Museum of Fine Arts;* Swarzenski, H.; Netzer, N., Eds.; *Boston Museum of Fine Arts: Boston, 1986; pp xix–xxiv.*
2. Williams, W. S.; Hopke, P.; Maguire, H., *Composition of Medieval Limoges Enamel,* 1984 Symposium on Archaeometry; Smithsonian Institution: Washington, DC, 1984; Abstract, p 157.
3. Gauthier, M. M. *Bulletin du la Societe Nationale des Antiquaries de France,* 1976 (Proceedings of the 13th October, 1976); pp. 176–191.
4. Gauthier, M. M. *Emaux de moyen age occidental*; Office du livre: Paris, 1972; 443 pages.

5. Thoby, P. *Les Croix Limousins de la Fin du XIIe Siecle au Debut du XIVe Siecle;* A. et. J. Picard: Paris, 1953; 175 pp, 75 plates.
6. Brill, R. H. 1987, personal communication.
7. Brill, R. H. Eighth International Congress on Glass; Benham: Colchester, 1969 (Proceedings of 8th Congress in London, July 1968); pp. 47-68.
8. Brill, R. H.; Moll, S. *Advances in Glass Technology, Part 2*; Plenum: New York, 1963 (Sixth International Congress on Glass, Washington, DC. 1962); p 293.
9. Sanderson, D. C. W.; Hunter, J. R.; Warren, S. E. *J. Archaeological Sci.* **1984**, *11*, 53–69.
10. Olin, J. S.; Sayre, E. V. In *Archaeological Chemistry*; Beck, C. W., Ed., Advances in Chemistry Series No. 138 American Chemical Society: Washington, DC, 1974; pp 100–123.
11. Bamford, C. R. *Colour Generation and Control in Glass*; Elsevier: New York, 1977; pp 50–60,150.

RECEIVED FOR REVIEW September 15, 1987. ACCEPTED REVISED MANUSCRIPT February 8, 1988.

13

Ancient Gold Solders: What Was Chrysocolla?

G. Demortier

LARN, Facultés Universitaires Notre-Dame de la Paix 22, Rue Muzet, B-5000 Namur, Belgium

Arguments are presented for reconsidering the meaning of the name "chrysocolla," used in ancient textbooks on metallurgy, and for identifying chrysocolla as cadmium sulfide (greenockite), a yellow natural mineral. Several results from thousands of nondestructive analyses by proton-induced X-ray emission (PIXE) of solders on gold artifacts cannot be interpreted if chrysocolla is either green copper carbonate (malachite), as generally assumed until now, or blue hydrated copper silicate, the mineral now called chrysocolla. Several paragraphs of the 33rd book of Natural History *of the Elder Pliny (1st century A.D.) are critically analyzed in light of these analytical results.*

EARLY ARTISANS USED SEVERAL RECIPES to prepare gold solders. These recipes included natural chrysocolla (from the Greek *krysos* and *kolla*, or gold solder) as well as other compositions prepared from alloys containing metals such as copper, silver, and gold. The use of chrysocolla was described in considerable detail in the 1st century A.D. by Pliny the Elder in his *Natural History*. As can be expected from its etymological origin, the name chrysocolla meant in antiquity any material used to join parts of gold jewelry. At the beginning of his 33rd volume, Pliny wrote, "Gold is dug out and, with it, also chrysocolla, which continues to bear a name derived from the term gold in order that its value may appear greater." This special attention to chrysocolla indicates its importance in ancient goldsmithery.

Brazing alloys that contain cadmium have been used extensively since mid-19th century. A small amount of cadmium in a gold matrix can lower the melting point below that obtained by adding equivalent concentrations

0065–2393/89/0220–0249$06.00/0

of copper or silver to the same gold matrix. It is commonly believed that cadmium or cadmium compounds were not used in antiquity (cadmium ores are indeed rare in nature), and that objects containing cadmium, either in the solder or as an impurity in the gold alloy, were recently manufactured or restored. Nevertheless, cadmium was detected in a solder on a monetary medallion excavated several years ago in Houmeau, France by an official staff of French archaeologists (*1*). Several experts in ancient jewelry question the authenticity of the gold jewelry that we analyzed, because the artifacts were not found during official excavations and several of the artifacts have been declared by us as fakes or recent restorations on the basis of our analyses and our criteria of authentification.

Although it is usually present at lower concentrations, cadmium is sometimes present in concentrations up to several percent in regions where soldering was necessary to join elements in pieces of apparently ancient jewelry. We have shown (*2–7*) that the amount of cadmium found in such jewelry (generally from Iran, Syria, or southern Italy) is related to the amounts of copper and silver present, but in completely different proportions from those observed in modern soldering alloys. Analytical results (*2, 4, 6, 7*) collected during analyses of numerous artifacts, from museums as well as from private owners, have led us to propose several criteria upon which to distinguish modern from ancient joining procedures that used cadmium-based materials and to indicate a new interpretation of old metallurgical descriptions. We suggest that ancient Iranian goldsmiths smelted greenockite (a natural yellow cadmium ore) simultaneously with copper ores and gold to obtain an alloy suitable as a solder at low temperatures.

Iranian Goldsmithery from the 4th Century B.C.

The high level of workmanship of ancient goldsmiths can scarcely be matched even today. Among the several hundred items of Iranian and Syrian jewelry that we have studied (*2–7*), the wonderful Achemenide pendant of the 4th century B.C., which belongs to the Department of Iranian Antiquities of the Museé du Louvre, clearly illustrates the skill of ancient Iranian goldsmiths. The pendant is shown in Figure 1 and is accepted as genuine (*8*). The total width of the disk decorated in repoussé is 5.5 cm. The surrounding ornaments were made with 28 identical motifs. Each of them includes a hollow cylinder and two hemispherical caps. The detail of the top right part (Figure 2) shows clearly visible solders. These areas were nondestructively analyzed for composition, and the results are summarized in Table I. All measurements were performed with proton-induced X-ray emission (PIXE) in a microprobe assembly (*9*).

Our PIXE microprobe facility, the Laboratoire d'Analyse par Réactions Nucliaires (LARN), allows the irradiation of small regions of a sample in a vacuum. The vacuum chamber in which the artifact is placed was designed

Figure 1. Achemenide pendant of the 4th century B.C.

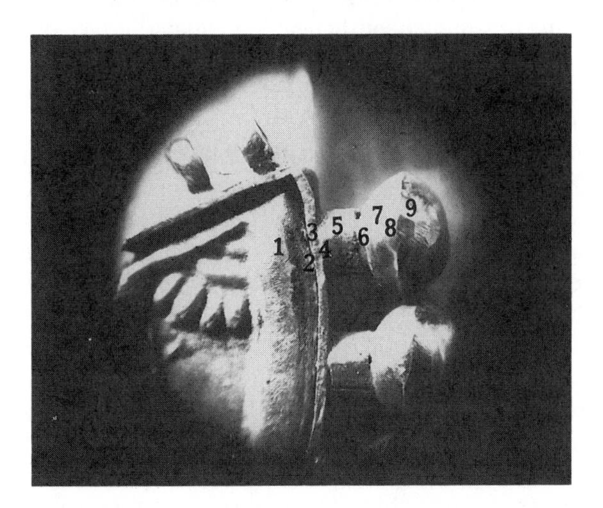

Figure 2. Detail of the Achemenide pendant showing regions of analysis.

to accept pieces of jewelry whose dimensions may reach 30 cm. The item is fixed on an *X–Y* frame that can be moved in the proton beam by stepping motors. A small computer controls mechanical displacement (with a reproducibility better than 2.5 μm after a translation of several centimeters) and collection of data from the photon and electron detectors and carries out

Table I. Composition of Different Regions of the Achemenide Pendant

Region	*Cu* %	*Ag* %	*Au* %	*Description*
1	5.6 ± 0.3	11.6 ± 0.6	82.8 ± 1.2	repoussé area
2	↑ 9.8 ± 0.6	↑ 15.6 ± 0.8	↓ 74.6 ± 1.5	brazing
3	5.2 ± 0.3	12.6 ± 0.7	82.2 ± 1.0	vertical sheet
4	↑ 7.5 ± 0.5	↑ 13.9 ± 0.8	↓ 78.6 ± 1.1	brazing
5	5.7 ± 0.4	11.7 ± 0.7	82.6 ± 0.9	cylinder
6	↓ 3.1 ± 0.3	↓ 6.8 ± 0.3	↑ 90.1 ± 1.5	local melting
7	7.0 ± 0.4	16.6 ± 0.9	76.4 ± 1.3	internal hemisphere
8	↑ 12.7 ± 0.7	↓ 14.6 ± 0.8	↓ 72.7 ± 1.3	copper diffusion
9	7.2 ± 0.4	15.6 ± 0.9	77.2 ± 1.4	external hemisphere

NOTE: The arrows indicate increased (or decreased) concentrations at solders

numerous calculations. At present, up to 128 simultaneous images can be collected, each involving up to 1600 analyzed points. The typical accumulation time is 1–5 s per point.

Four joins were made from regions 2 to 8 of Figure 2, over a distance extending less than 5 mm. The elemental compositions at solder sites 2, 4, 6, and 8 show that three different joining procedures were performed. At site 6, the gold content is significantly greater than in the neighboring regions and indicates that the join was made by welding without the addition of external material. The end of the cylinder and the bottom of the first hemispherical cap were simultaneously heated to an early stage of fusion. In this process, metals like copper and silver were selectively eliminated, mainly by oxidation, so that gold content was enhanced. The temperature was around 1000 °C.

At site 8, an increase in copper concentration was observed simultaneously with a decrease in gold and silver concentrations. The two caps, previously well-fitted, were joined by the process known as solid-state diffusion bonding with copper salts in a reducing atmosphere. The process took place at about 890 °C (*10*). The relatively low temperature permitted joining at site 8 without desoldering the join at site 6, only 1.5 mm distant. This joining method was widely used by the Etruscans in their famous granulation work.

An increase in copper and silver and a decrease in gold were simultaneously observed at sites 2 and 4, a result indicating that at both sites a brazing alloy had been used. The lower concentration of gold indicated that the temperature of fusion of the alloy used at site 2 was lower than that used at site 4 (possibly around 820 and 860 °C, respectively). These temperatures are close to 890 °C, the soldering temperature at site 8; however, the latter join did not desolder. The process of diffusion bonding produces a join that cannot be desoldered. Reheating after joining results in further copper diffusion and gives a local decrease in copper content with a consequent increase in local melting temperature. The temperature difference of sites 2 and 4 seems to indicate that the 28 attached elements were soldered to the gold sheet before the sheet was joined to the decorated disk, because the alloy used at site 4 had a higher melting point than that at site 2.

The high degree of workmanship displayed in a piece as complex as the Achemenide pendant shows that the goldsmith of 2500 years ago recognized and used different temperatures for various joining processes.

Later Iranian Goldsmithery

In any human activity, alternate means are sought to achieve a given end. Early goldsmiths must have tested and used a number of different joining materials and methods. Among Iranian and Syrian items from the 1st to the 9th century A.D., we have, at times, found significant amounts of cadmium in solders. For most people concerned with archaeological jewelry, the presence of cadmium suggests forgery or modern repair. However, in several investigations conducted since 1983, we have shown that ancient solders containing cadmium may sometimes be differentiated from modern solders (*4, 6, 9, 11*).

If several closely spaced joins are present on a piece of jewelry, solders with different melting points may be necessary, depending on the order of construction. Modern commercial solders are characterized by compositions in which silver and copper concentrations are retained in a constant proportion (direct correlation) during the soldering procedure (Figure 3). On the other hand, for ancient solders, a strong correlation exists between copper and cadmium concentrations (Figure 4). Our principal criterion by which we distinguish modern from ancient solders containing cadmium applies only to jewelry artifacts that show several regions of solders. The artifact is recognized as ancient if relative concentrations of copper, silver, and cadmium in all solders are correlated like those in Figure 4. The artifact is considered to be a forgery or a restored piece if these relative concentrations are correlated like those in Figure 3. No criterion for authenticity is available if the number of solders is not sufficient to draw a line across the points.

Other criteria may be used to distinguish ancient from modern cadmium-containing solders. These criteria include

a. The frequent presence of iron in ancient solders. Modern brazing alloys (for fine gold jewelry) sold by precious metal suppliers or made by jewelry craftsmen in their own shops contain only gold, silver, copper, and sometimes zinc and cadmium. These metals are alloyed from pure metal ingots, whereas ancient solders were made with rough materials, generally by direct smelting of ores like chalcopyrite (an iron–copper ore with a color close to that of gold). Iron is therefore expected to be present in ancient artifacts but not in modern jewelry.

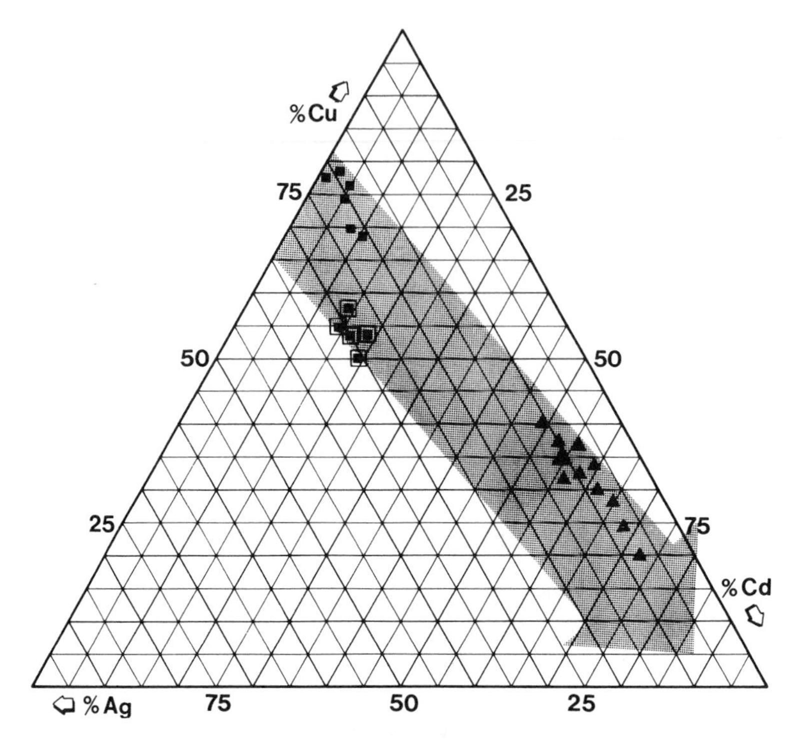

Figure 3. Ternary composition diagram for Cu–Ag–Cd in modern soldering alloys. Analytical results indicate that when the cadmium amount varies, the ratio of copper on silver remains constant.

b. A luminescence can be induced by the PIXE proton beam on ancient solders. This effect is probably due to the presence of microinclusions of slag retained by the metal produced by a crude smelting technique. This optical luminescence is induced only in insulators, not in metals, and does not appear in modern goldsmithery in which only metals are involved.

c. Differences in relative concentrations of zinc and cadmium

d. The presence of sulfur

Application of these criteria is illustrated in the following study of a hollow gold pearl (Figure 5) found in Syria along with other more prestigious objects (necklaces, pendants, phials) studied earlier (*3, 4, 6*).

Most of these artifacts were found during excavations in Hauran for the draining of Lake Orantes in the 1950s. Some of this jewelry was acquired by museums, such as the Staatlishe Museen in Berlin. Several pieces of the treasure, first considered as authentic by curators and archaeologists, were then classified as fakes or restored items after cadmium had been detected

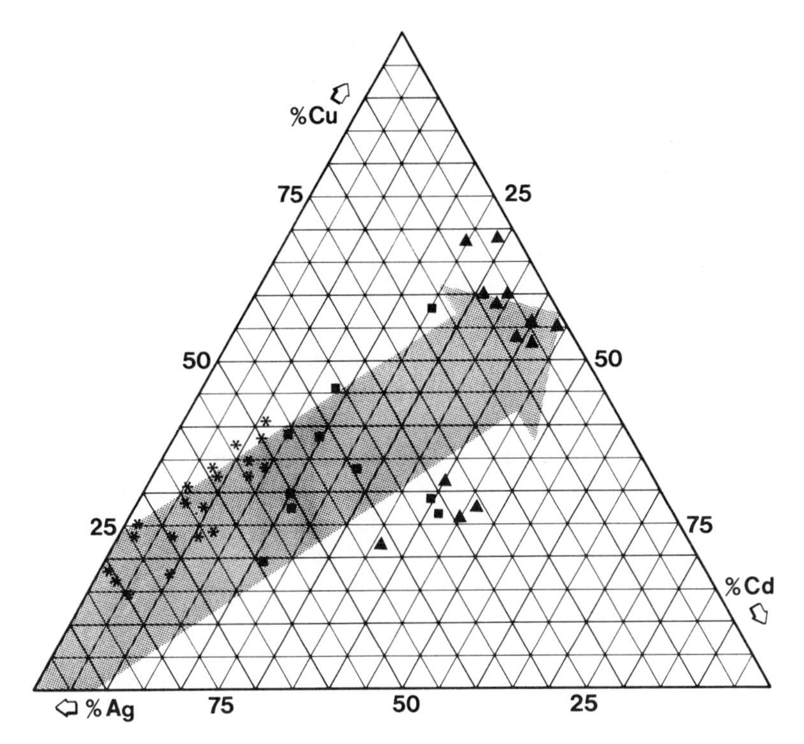

Figure 4. Ternary composition diagram for Cu–Ag–Cd in regions of solders on an Iranian necklace of the 1st century A.D. (See ref. 2 for details). Here, the copper concentration increases with the cadmium concentration.

Figure 5. Hollow pearl of the 6th century A.D. excavated with other more prestigious objects of Byzantine jewelry. Total width is 8.5 mm.

by our PIXE method and analyzed according to our criteria to distinguish modern from ancient pieces (*6, 7*). This identification of modern workmanship led to suspicion of other genuine artifacts of the treasure.

An electron micrograph (Figure 6) of a region near two soldered granules clearly shows that a brazing procedure had been used to join both granules to a thin gold base that was then soldered to the main hollow sphere. Scans across large areas were performed with the LARN proton probe (50 μm wide); narrower scans in characteristic regions were also obtained (*11*). Maps of copper, gold, silver, and cadmium (Figure 7) clearly indicate the presence of cadmium only in solders between the granules and between the main hollow sphere and each granule. The local decrease of silver and gold at each site where cadmium was present, and a slight increase of copper at that site, point out a direct correlation between copper and cadmium contents. The corresponding ternary diagram indicates that the results are located in the left bottom corner, far away from the regions characteristic of modern jewelry (*see* Figure 3).

When a solder such as the one just mentioned is irradiated in a cadmium-rich region with 1.9-MeV deuterons, the proton spectrum indicates traces of sulfur (around 30 ppm) (*12*). The presence of sulfur suggests that cadmium sulfide could have been used as a component of the brazing material. This possibility has been examined in our laboratory and has been confirmed (*13, 14*). Additional solder characterization was accomplished by using the se-

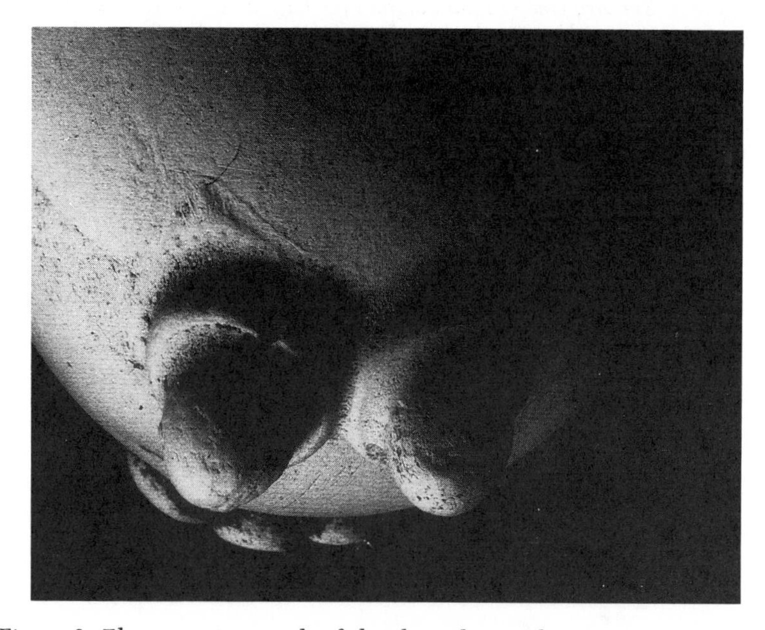

Figure 6. Electron micrograph of detail on the pearl in Figure 5. Brazing procedures are clearly indicated.

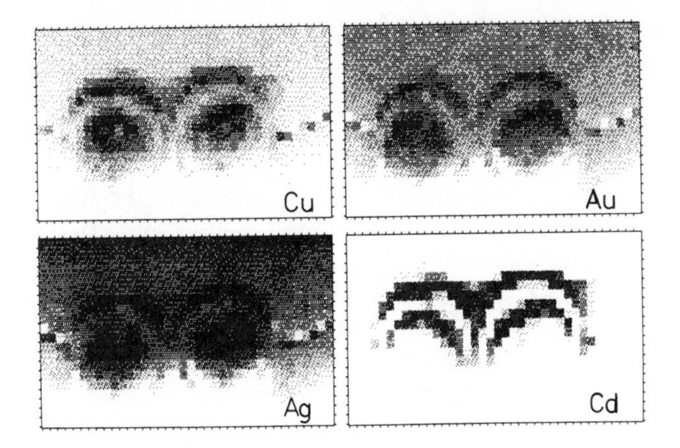

Figure 7. Maps of Cu, Ag, Au, and Cd obtained by PIXE microprobe analysis. Correlation of copper and cadmium is obvious.

lective excitation of Zn K X-rays without excitation of gold L X-rays. It was found that zinc is also present, at trace levels (about 300 ppm) (*15*).

Sulfur and zinc determinations may be added to the main criteria of characterization of ancient solders: copper–cadmium correlation and presence of iron, to give a total of four differences between solders on the ancient hollow pearl and solders made with commercial brazing alloys containing cadmium. Sulfur at a 30-ppm level in a region where cadmium is present at 3% indicates that, if CdS had been used as alloying material with gold, at least 99% of the sulfur would have been eliminated during the alloying process. This proportion was also observed in our experimental archaeological tests (*15*, *19*), which are summarized in the next section.

Fine modern cadmium-based brazing alloys are zinc-free or contain zinc and cadmium in a ratio of about 1:5. When Cu–Ag–Zn–Cd–Au alloys are melted to make solders, cadmium is selectively eliminated because of its high volatility. The final relative concentration of zinc to cadmium in modern joins is then still zero or higher than 1:5. In the hollow pearl (and also in other jewelry) this ratio lies between 1:25 and 1:300. The range of these ratios is either too low to be attributed to the use of modern alloys containing zinc or too high to be interpreted as impurities in metals used for alloying. These amounts of zinc in ancient jewelry items are understandable because greenockite (a natural cadmium ore) appears as a yellow coating on zinc blendes. The zinc content in ancient soldering processes is attributed to the imperfect procedure of separation of greenockite from the zinc blende.

Rudimentary Preparations of Low-Melting Brazing Alloys

Early attempts to prepare low-melting alloys with cadmium sulfide (its natural form, greenockite, is orange–yellow) encouraged further investigation to check whether this process might have been used in antiquity. Preliminary results (*13*) indicate that the production of brazing alloys by a procedure of "soft metallurgy," which can give a soldering whose composition can be more easily controlled than the one frequently used by modern jewelers (*14*).

A few milligrams of gold was melted to form a small sphere. Powdered cadmium sulfide was then poured onto the melted gold. The powder dissolved rapidly and gave a small sphere of alloy that was then analyzed microscopically. Each sphere of alloy appeared approximately homogeneous. The centers of the spheres contained less cadmium than the surfaces, but the difference in concentration between the center and the surface was less than 20%. Depending on several conditions (i.e., temperature, crucible shape, addition of copper and iron ores), the solubility of cadmium from cadmium sulfide reached saturation when the alloyed cadmium content was between 1% and 10%. In all cases, the addition of copper or copper ores to molten gold enhanced the solubility of cadmium. This observation leads to a direct correlation between copper and cadmium contents. When powdered cadmium sulfide is mixed with chips of gold and the mixture is heated to melting, the dissolution takes longer than when greenockite is added to liquid gold. Furthermore, the alloy obtained by this last procedure is less homogeneous than the alloy made with cadmium sulfide poured onto the molten gold. The sphere of gold–cadmium alloy sometimes appeared to consist of two different parts: a gold-rich alloy formed at the bottom of the crucible and a black coating that appeared at the top (Figure 8). Cadmium and copper correlations were observed in the gold-rich material as well as in the black earthy coating (*see* ref. 13).

Cadmium sulfide was used in this work as part of an archaeological argument: the mixture by ancients of materials of nearly the same color. Direct alloying of metallic cadmium with melted gold, a modern procedure, results in a violent interaction when the metallic cadmium is placed in the melt because the boiling point of cadmium (765 °C) is far below the melting point of gold (1063 °C). The melting point of cadmium sulfide is so high (above 1750 °C) that its incorporation in the molten gold is not violent. No cadmium vapor is produced when the amount of greenockite added to gold is below the saturation concentration. Above this limit, wisps with the color of cadmium sulfide appear floating above the crucible.

Analysis of the Elder Pliny's Natural History

None of the original manuscripts of Pliny is now available. The modern translation of Pliny's book often refers to the Bambergensis manuscript (10th century) and less often to the Parisinus Latinus (12th century). The two versions are very similar, but several terms are different. All comments in the English (*16*) and French (*17*) translations refer to the Bambergensis

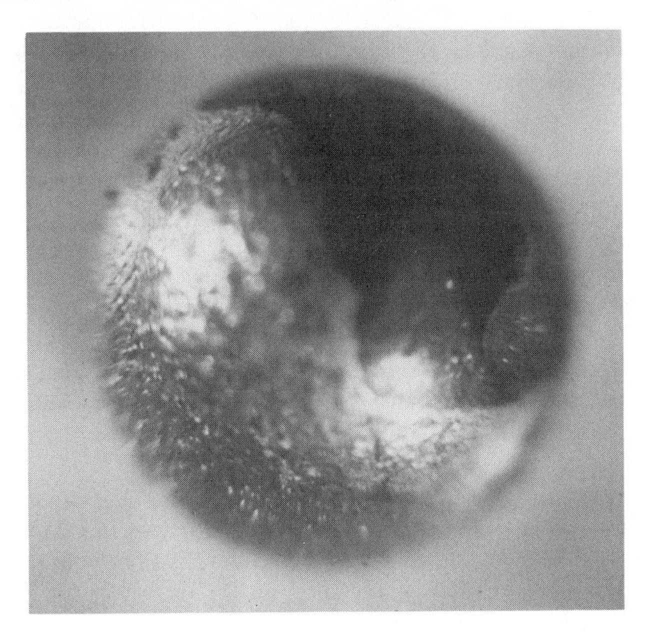

Figure 8. Gold–cadmium–copper alloy obtained by "dissolution" of CdS in molten Au–Cu alloy. The black earthy deposit is easily separated from the golden region.

manuscript, for which the English chemist K. C. Bailey gave technical comments on all scientific subjects (*18*).

The 26th paragraph of the 33rd book in both versions (Bambergensis and Parisinus Latinus) describes the origin of chrysocolla:

> Gold solder is a liquid found in the shafts we spoke of, flowing down along a vein of gold, with a slime that is solidified by the cold of winter even to the hardness of pumice stone. A more highly spoken of variety of the same metal has been ascertained to be formed in copper mines, and the next best in silver mines. A less valuable sort also with an element of gold is also found in lead mines. In all these mines however an artificial variety is produced that is much inferior to the natural kind referred to: the method is to introduce a gentle flow of water into the vein all winter and go on till the beginning of June and then to dry it off in June and July, clearly showing that gold solder is nothing else than the putrefaction of a vein of metal

The next sentence in the Bambergensis is, in the Latin version, "*Nativa duritia maxime distat; **uvam** vocant,*" which is translated as "Natural gold solder, known as '**grapes**' differs very greatly from the artificial in hardness"

(*16*). On the other hand, in the Parisinus Latinus one finds: "*Nativa duritia maxime distat,* ***luteam*** *vocant,*" which is now translated as "Natural gold solder is the best as far as its hardness is concerned; it is called **yellow chrysocolla.**"

The next sentence in both versions is "***Et tamen*** *illa quoque herba, quam* ***lutum*** *appellant,* ***tingitur,***" which is translated as "**Nevertheless** it is still **dyed** with a plant called **yellow-weed**"

If, in the translation of the Bambergensis, ***uvam*** refers to the **shape** of grapes, one cannot understand why Pliny's text continues with ***Et tamen*** (**Nevertheless**). This sentence refers to the color of the material, and "nevertheless" does not make sense, if the reference to grapes concerns shape. This interpretation has led translators and chemical archaeologists (*18*) to identify chrysocolla with green malachite, which sometimes occurs as green "grapes."

In the Parisinus Latinus version, ***uvam*** becomes ***luteam***, and ***luteam*** means **yellow**, yellow like gold (*19*). We think that in the Bambergensis version, "uvam" can be connected not with the shape but with the color of grapes; specifically the color of Italian grapes (muscat), which are golden yellow in full maturity. Chrysocolla is then yellow (like gold).

Color seems to be one of the reasons for the use of chrysocolla not only in gold jewelry, but also as a dye and in medicine. This use of a yellow plant to improve the yellow color of an alloy is difficult to understand with our present knowledge of materials science. Even ancient goldsmiths had observed that vegetables submitted to temperature treatments would undergo complete change in their shape and color. The usefulness of this yellow weed in the soldering procedure may probably be better understood in the context of the following paragraph of Pliny's description of urine being added to a recipe of goldsmithery. Reasons for adding urine may have been

a. to liberate carbon as reductor

b. to liberate nitrogen, which increases the solubility of cadmium from CdS in gold as observed in a previous work (*13*)

c. to emit a yellow substance in nonmetallurgical applications (27th and 28th paragraphs of Pliny's book).

According to the foregoing interpretation, chrysocolla of ancient times is neither malachite nor a blue copper silicate (the mineral now called chrysocolla) but is a yellow substance, possibly the yellow mineral cadmium sulfide, which appears as a coating on other minerals, chiefly zinc sulfide. This description fits Pliny's text, which describes gold solder (chrysocolla) as a liquid that "flows from several mines to give a solid deposit."

The 29th paragraph of Pliny's book deals with the use of chrysocolla by goldsmiths. In the Bambergensis version:

"*Chrysocollam et aurifices sibi vindicant adglutinando auro, et inde omnes appellatas similiter* ***virentes*** *dicunt*", or "The goldsmiths also use a special gold solder of their own for soldering gold, and according to them it is from this that all the other substances with a **similar green color** take the name." The translation of "virentes" as green is the most evident, but "virentes" may be also understood not as a color, but as the property of brightness that can be attributed to all metals that reflect light.

In the Parisinus Latinus, this sentence is "*Chrysocollam et aurifices sibi vindicant adglutinando auro: et inde omnes appellatam similiter* ***utentes*** *dicunt*" or "The goldsmiths also use gold solder for soldering gold, and **all the users** [goldsmiths] call with the same name any substance suitable for this purpose." This sentence gives no new information, but rather simply recalls that the meaning of chrysocolla refers to its use as gold solder.

On this use of chrysocolla by goldsmiths, both versions are identical at the end of the paragraph:

> ***Temperatur*** *autem ea* ***cypria aerugine****, et pueri impubis urina, addito nitro. Teritur cyprio aere in cypriis mortariis:* ***santernam*** vocant nostri. ***Ita ferruminatur aurum, quod argentosum vocant. Signumque est, si addita santerna nitescit. E diverso aerosum contrahit se, hebetaturque, et difficulter ferruminatur.*** *Ad id glutinum fit, auro, et septima parte argenti ad supradicta additis, unaque contritis*"

which is translated as

> They [goldsmiths] make the **mixture with copper verdigris** and with urine of a boy who has not yet reached puberty and some soda [sodium carbonate]. It is ground with a copper pestle in a copper mortar. They call this preparation **santerna. In this way they can solder argenteous gold. A sign of its having been so treated is if the application of santerna gives a brilliant colour. On the other hand coppery gold shrinks in size and becomes dull and is difficult to solder.** For this purpose a solder is made by adding gold with one seventh of silver into the above material and they grind them together.

Surprisingly, we have found no comment on this sentence. If chrysocolla is malachite (copper carbonate), why can cuprous gold not be soldered with a brazing alloy containing copper, when copper can be dissolved in gold in any proportion? If chrysocolla is malachite, why this dual use of the copper minerals verdigris and malachite? If chrysocolla is greenockite (cadmium sulfide); however, this situation is explained by our results on soldering alloys prepared with copper salts, cadmium sulfide, and gold showing a black earthy deposit in addition to the golden alloy (*13*, *14*).

Except for our investigations, no systematic study of cadmium in jewelry has been undertaken. Several relevant experiments by other investigators using the electron microprobe demonstrated an increase of copper and sometimes silver at solders (*20*). Silver was determined with L X-ray lines. We have demonstrated, however, that cadmium cannot be determined with an electron microprobe (*21*) in the presence of a high concentration of silver; thus, an actual simultaneous increase in copper and cadmium in solders may be interpreted as a simple increase in copper in experiments where cadmium cannot be determined. Results of electron microprobe studies neither confirm nor rule out the supposition that cadmium means a modern origin.

Conclusions

Copper minerals were used by early goldsmiths to perform intricate soldering operations on jewelry. Other procedures, perhaps involving cadmium sulfide, may have been used later. Cadmium sulfide seems to have been used only in Iran and Syria, from the 1st to the 9th century A.D. We analyzed about 100 pieces of jewelry from those regions, and of that period. About 30 of them contain cadmium. The criteria to distinguish restorations and fakes from ancient processes can only be applied on about 20 of those items. The other items contain cadmium in too few solders to identify their authenticity by the appropriate ternary Cu–Ag–Cd diagram. About 150 other items of Roman, Greek, Merovingian, Celtic, Byzantine, and American jewelry from varied periods also were investigated with the same analytical method, but cadmium, even at trace level, was never detected in any of them. Until now, only a small part of the results on these pieces were published (*22*, *23*).

Further study is indicated, especially under experimental conditions that allow the detection of cadmium in the presence of high concentrations of silver. The controversy surrounding the possibility of the use of cadmium compounds by goldsmiths of antiquity could more easily be elucidated if other analytical research teams were involved in analyses of jewelry of unquestionable authenticity from the same period and origin as those reported here. A more extensive study of ancient textbooks by both linguists and scientists would probably shed additional light on the subject.

References

1. Flouret, G.; Nicolini, G.; Metzger, C. *Gallia* **1981,** *39*, 85–101.
2. Demortier, G.; Hackens, T. *Nucl. Instrum. Methods* **1982,** *197*, 223–236.

3. Demortier, G. *Archéologia* **1983,** *176*, 41–51.
4. Demortier, G. *Gold Bull.* **1984,** *17*, 27–38.
5. Demortier, G. *Nucl. Instrum. Methods Phys. Res., Sect. B* **1986,** *14*, 152–155.
6. Demortier, G. In *Proceedings of the International Precious Metals Institute, Brussels*; Vermeylen, G.; Verbeeck, R, Eds.; IPMI Press: Allentown, PA, 1987; pp 55–73.
7. Demortier, G. *Archaeometry*, **1987,** 29(2), 275–288.
8. Ghirshman, R. *Perse: Proto-Iraniens, - Mèdes, Achéménides*; Gallimard: Paris, 1963; p 264.
9. Demortier, G. In *Applications of Nuclear Microprobes*; Watt, F.; Grime G. W., Eds.; Adam Hilger: Bristol (UK), 1987; pp 334–376.
10. Littledale, H. A. P. U.K. Patent 415 181, 1936.
11. Piette, M.; Demortier, G.; Bodart, F. In *Microbeam Analysis*; Romig, Jr., A. D.; Chambers W. F., Eds; San Francisco Press: San Francisco, 1986; pp 333–336.
12. Demortier, G.; Gilson, A. *Nucl. Instrum. Methods Phys. Res., Sect. B* **1987,** *18*, 286–290.
13. Demortier, G.; Van Oystaeyen, B.; Boullar, A. *Nucl. Instrum. Methods Phys. Res., Sect. B* **1984,** *3*, 399–403.
14. Demortier, G.; Decroupet, D.; Mathot, S. In *Proceedings of the International Precious Metals Institute, Brussels*; Vermeylen, G.; Verbeeck, R, Eds.; IPMI Press: Allentown, PA, 1987; pp 335–352.
15. Demortier, G. In *Proceedings of the XIth ICXOM Conference, London (Canada); University of Western Ontario Press: London, Ontario, 1987; pp 147–150.*
16. Racklam, H. *Pliny's Natural History, Loeb ed., Book 33*; Heinemann: London, 1968; pp 4–72.
17. Zehnacker, H. *Pline l'Ancien, Histoire Naturelle, Livre 33*; Les Belles Lettres: Paris, 1983; pp 47–85.
18. Bailey, K. C. *The Elder Pliny's Chapter on Chemical Subjects*; Arnold: London, 1929; pp 59–109; 175–209.
19. *Oxford Latin Dictionary*; Glare, P. G. N., Ed.; Clarendon: Oxford, 1976.
20. Parrini, P.; Formigli, E.; Mello, E. *Am. J. Archeol.* **1982,** *86*, 118–121.
21. Demortier, G.; Houbion, Y. *Oxford J. Archaeol.* **1987,** *6(1)*, 109–114.
22. Demortier, G. In *Gold Jewelry—Craft, Style and Meaning from Mycenae to Constantinopolis*; Hackens, T.; Winkes, R., Eds.; UCL Press: Louvain-la-Neuve, 1983; pp 215–227.
23. De Cuyper, F.; Demortier, G.; Dumoulin, J.; Pycke, J. *La croix Byzantine du Trésor de la Cathédrale de Tournai*; UCL Press: Louvain-la-Neuve, 1987; pp 1–88.

RECEIVED for review June 11, 1987. ACCEPTED revised manuscript September 6, 1988.

14

Applications of Infrared Microspectroscopy to Art Historical Questions about Medieval Manuscripts

Mary Virginia Orna[1], Patricia L. Lang[2], J. E. Katon[3], Thomas F. Mathews[4], and Robert S. Nelson[5]

[1]**Department of Chemistry, College of New Rochelle, New Rochelle, NY 10801**
[2]**Department of Chemistry, Ball State University, Muncie, IN 47306**
[3]**Molecular Microspectroscopy Laboratory, Miami University, Oxford, OH 45056**
[4]**Institute of Fine Arts, New York University, New York, NY 10021**
[5]**Department of Art, University of Chicago, Chicago, IL 60637**

Fourier transform infrared microspectroscopy applied to Byzantine manuscripts in the Special Collections Department of the University of Chicago Library revealed the use of numerous additives to the paint mixture including kaolin, hide glue, egg yolk, and other proteinaceous materials. Some evidence suggests that cochineal was used as a red pigment.

Pigment Analysis

Aims of Pigment Analysis. For the art historian, the chemical analysis of pigments serves two main purposes: for authentication, that is, to confirm or deny the alleged attribution or dating of a painting on the basis of comparison with the known painting practices of the artist or period; and as a key to understanding the work of a given artist or period. The latter use of pigment analysis is less glamorous (nothing makes better news than to declare an accepted masterpiece a fake), but in the long run, it is the more important enterprise. Although art historians rely primarily on style to describe the interrelationships of artists and schools, this task can also be approached from the point of view of the artist's materials.

Much progress has been made in analyzing the paintings of the Renaissance and later periods in terms of materials. It has been demonstrated,

0065-2393/89/0220-0265$07.00/0

for example, that Tintoretto's reputation as an innovative colorist has a very real foundation in the way he used pigments (*1*). Not only did Tintoretto use the widest range of pigments available in Venice at that time—four different blues, for example—but he also used pigments in mixtures and layers quite unparalleled in the work of his contemporaries. Thus, through the chemical analysis of pigment samples, it is possible to infer how the paints were prepared and, consequently, gain an insight into the artistic technique itself. Our knowledge of medieval pigments, however, is still very slight.

Rationale for the Approach. Like the medieval naturalist who preferred to speculate on the number of teeth in a horse's jaw rather than to inspect the beast, the modern art historian has taken a decidedly theoretical approach to the study of artists' materials in the Middle Ages. Relying on artists' recipe books, several of which survive from those times, art historians have generally left to speculation the question of which particular pigments were used in a given work of art.

An approach more empirical than the use of such recipe books was devised by Roosen-Runge and Werner (*2*, *3*) in their examination of the 8th-century Lindisfarne gospels. Their method involved a visual comparison of the various pigment surfaces contained in the manuscript with pigments manufactured according to the medieval painters' manuals. The prepared pigments were painted on small pieces of parchment, and the microscopic structures of the pigments in the manuscript were then compared with the known samples by using polarized light, supplemented with visual observation in ultraviolet light.

Another approach is whole-manuscript neutron activation analysis (NAA) in a nuclear reactor, followed by gamma-ray analysis and autoradiography (*4*). Gamma-ray analysis permits identification of the elements contained in the manuscript, and autoradiographic analysis indicates the specific locations of these elements. A similar approach applied to art objects of historical interest is energy-dispersive X-ray fluorescence (XRF) analysis, which permits nondestructive semiquantitative elemental analysis (*5*).

Although these methods are nondestructive in the sense that a manuscript remains intact after analysis, they have some serious shortcomings. The method of Roosen-Runge and Werner involves the use of reasonably intelligent guesses made on the basis of several assumptions. The other methods involve removing the manuscript from its location and subjecting it to multiple handlings and neutron or X-ray bombardment. But the most serious drawback is the fact that NAA and XRF provide information only regarding the elements contained in the pigments but no information regarding chemical formulas of compounds.

Given the availability of samples from a manuscript, there is no substitute for the standard methods of analysis of microscopic particles. Granted that the methods just described are nondestructive, the removal of pigment samples from a manuscript can be equally nondestructive. The use of pigment samples from manuscripts for analysis is preferable for various reasons.

1. Sampling does not require the removal of the manuscript from its permanent location.
2. A well-planned sampling is a one-time operation that need never be repeated.
3. Sampling involves much less handling of the manuscript than the other methods.
4. Only the samples, and not the entire manuscript, are subjected to high-energy irradiation.
5. The samples can be taken in such a way that the lacunae left by excision of the pigment particles are not discernible by the naked eye.

However, the most compelling reason for taking samples is the wealth of information that can be obtained from their analysis, including unambiguous chemical identification of pigment components. Of course, it is also necessary to identify cooperative curators.

Analysis of Medieval Manuscripts. An ongoing project (*6–8*) dealing with medieval manuscripts has had as its objective the application of small-particle-analysis techniques to the study of pigments in medieval Armenian (Greater Armenian), Cilician (Lesser Armenian), and Byzantine manuscripts. The Middle Eastern materials are an ideal starting point because many of the manuscripts are dated and can be located by colophons and inscriptions. This work has begun to shed more light on several art historical problems, including the tracing of lines of influence or interconnection between medieval centers of manuscript production and the clarification of periods of known usage of several important artists' pigments.

At present, several dozen manuscripts from museums and centers such as the Walters Art Gallery; the Freer Gallery of Art; the Pierpont Morgan Library; the Special Collections of the University of California at Los Angeles; the Spencer Collection of the New York City Public Library; the Armenian Patriarchate of St. James, Jerusalem; and the Monastery of San Lazzaro, Venice, have been sampled and analyzed. Some of the data and results have been published in representative journals.

This chapter is a progress report on the analysis of particles taken from Byzantine illuminated manuscripts from the University of Chicago Library Special Collections. Because Byzantine manuscripts have, heretofore, never been analyzed chemically (*9*), this report is the result of initial inquiry into a large and complex subject.

Another objective of this chapter is to introduce the use of Fourier transform infrared (FT–IR) microspectroscopy for the analysis of pigments and to provide some background about the technique.

FT–IR Microspectroscopy

Background. IR spectroscopy has long been used as an analytical tool for the identification of unknown materials because it provides selective, molecular information. However, IR spectroscopy has not been widely used to analyze historic artwork in the past primarily because the minimum sample size needed to obtain transmission spectra was in the order of microgram quantities (*10*). The development of FT–IR spectrometers, with their inherent advantages over grating instruments (*11*), has increased the usefulness of this technique for the analysis of works of art. The enhanced spectral throughput of FT–IR instruments and their ability to average many scans have allowed the analysis of much smaller samples, in the order of nanogram quantities.

In 1983, Shearer and co-workers (*10*) used FT–IR in conjunction with a beam-condensing accessory to analyze the paint chips removed from *Virgin and Child*, an Italian painting in the Clark Institute collection. Their spectral data supported the art conservator's opinion that the painting was either an imitation of an earlier work or a heavily reworked fragment. The recent coupling of FT–IR spectrometers with microscope accessories has further allowed the analyst to obtain spectra from picogram quantities of samples and, moreover, to obtain the spectra quickly and easily (*12*, *13*).

Microscope Accessory for FT–IR Analysis. The basic experimental design of a microscope accessory is shown in Figure 1. A beam of IR radiation is focused by a mirror to a spot approximately 0.5 mm in diameter. The focus is fixed on top of a 1-mm KCl window. The transmitted IR radiation is collected and magnified 15 times by a cassegrainian lens. This lens is made of reflective materials, and it is used because traditional microscope objectives, which are made of glass, absorb IR radiation. An iris located in the focal plane of the magnified image (called an adjustable field stop) allows the analyst to define the sample area under observation. The transmitted radiation is finally focused by another mirror onto a mercury–cadmium telluride (MCT) detector. The high sensitivity of the MCT detector increases the overall performance of the system.

A flip mirror (not shown) allows the sample to be viewed through a 10× ocular and brought into focus. The optical elements of this accessory have been aligned so that the visible focus is coincident with the focus of IR radiation.

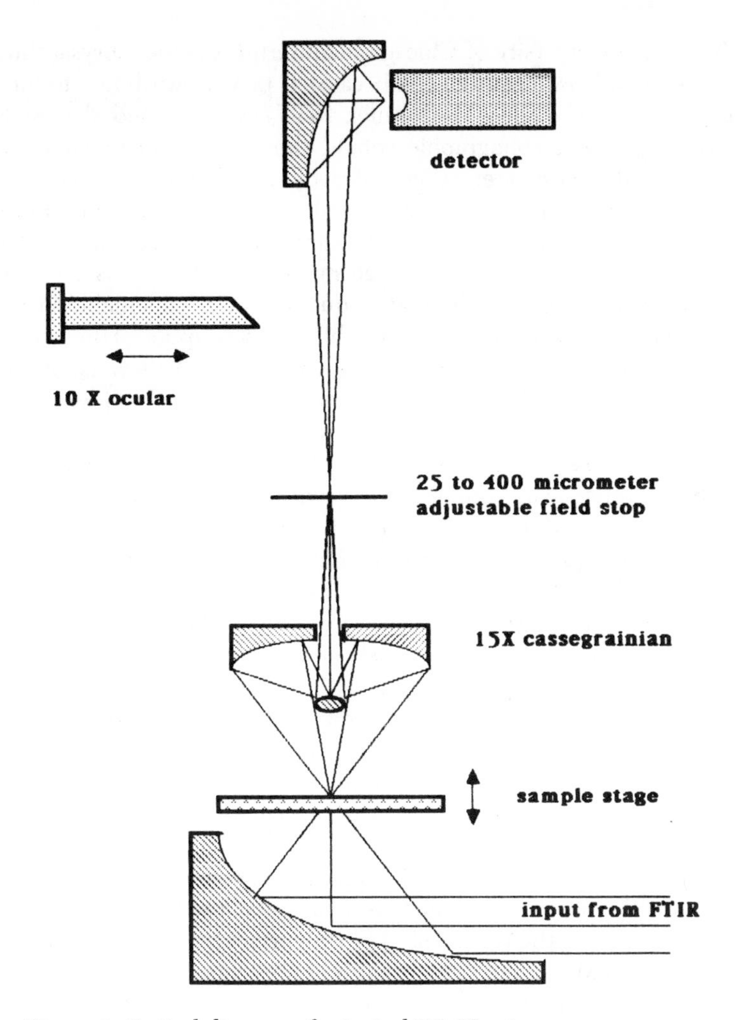

Figure 1. Optical diagram of a typical FT–IR microscope accessory.

Materials and Methods

Materials. At the University of Chicago, 10 decorated manuscripts were sampled by processes previously described (*6–8*). These manuscripts represent a broad chronological span ranging from the 10th century until the post-Byzantine era (16th century or later). The earliest book is the Elfleda Bond Goodspeed *Gospels*, decorated solely with headpieces and initials and linked with the Chrysostomos manuscripts in Paris and Rome (*14*). The second manuscript sampled was the celebrated Rockefeller–McCormick *New Testament*, a book redated to the second half of the 12th century by Carr (*15*) and associated with manuscripts from Cyprus or Palestine.

The third University of Chicago manuscript was the Chrysanthus gospels, a potentially interesting book because it may be attributed to the same period as the Rockefeller–McCormick *New Testament* and shares with it various stylistic and iconographic details, thus making a comparison of the pigments used a most interesting endeavor. In addition, the manuscript is valuable because miniatures were added to it during the post-Byzantine period, possibly because the original illuminations had begun to flake.

The fourth University of Chicago manuscript, known as the "Archaic Mark", is a book of decidedly modest quality but of undeniable interest to philologists. As Allison noted in an unpublished description of the manuscript (*16*), some readings find parallels in the early *Codex Vaticanus*, and others are unique. One scholar was so dazzled by the textual evidence that he thought that the University of Chicago Mark might contain the text of the Gospel of Mark "in a more primitive form than any other known manuscript" (*17*). On the other hand, others have suggested that its rare text may have been taken from a 19th-century printed version of the Greek gospels, and some have even questioned the authenticity of the manuscript.

The problem of the Archaic Mark is quite complex. Its miniatures are based on the cycle in a late 12th-century gospel book in the National Library in Athens, codex 93 (*18*). There exists also a set of fragmentary gospels in the Hermitage Museum in Leningrad that must have been made by the same scribe and illuminator, because the similarities in script, initials, and painting style can hardly be fortuitous. The Leningrad manuscript is illustrated in reference 19, and its contents and unusual text are described in reference 20. Both the University of Chicago and the Leningrad manuscripts are fragments.

The Leningrad codex has only eight folios from the end of the Gospel of Mark and the beginning of the Gospel of Luke; the University of Chicago manuscript contains the Gospel of Mark only. Both have the last few chapters of the Gospel of Mark, and consequently, these passages could be profitably studied to determine if they are identical. Because of the shared texts, the two manuscripts, made by the same scribe and illuminator, cannot be fragments from a single gospel book. Furthermore, the page sizes do not agree. The Leningrad volume measures 170 by 126 mm, whereas the Archaic Mark measures 119 by 84 mm. These numbers suggest that the smaller University of Chicago manuscript is to the larger Leningrad manuscript as an octavo is to a quarto. Thus the two manuscripts may have been made out of the same batch of parchment sheets, and the folios were formed by folding and refolding.

The six remaining University of Chicago manuscripts sampled were the Haskell gospels (attributed to Theodore Hagiopetrites) and the Georgius gospels, both of the late 13th century; the lectionary of Constantine the reader and the lectionary of St. Menas the wonder worker, both of which are late-12th-century works in the Cypriot style; a 12th-century Greek gospel book; and the early-12th-century Nicolaus gospels, dated A.D. 1133

The pertinent information about the manuscripts reported in this chapter are summarized in Table I.

Sampling Procedure. The manuscripts just described were sampled by abstraction of pigment samples with a fine surgical scalpel and several sampling needles of various thicknesses under a 40× binocular microscope. In some cases, samples were taken from offsets of pigment that had transferred onto the opposite page; thus a relatively large amount of sample could be removed without disturbing the corresponding miniature. In several other cases, pigment flakes that had fallen into the gutter of the manuscript were taken. In those cases when samples were excised directly from the paintings, the lacunae were hardly discernible even at 40× magnification.

Each sample was coded according to hue, folio number, and x–y coordinates in millimeters, with the bottom outside corner of the manuscript page as the origin. For example, a code of R.45v(35,65) identifies a red sample taken from folio 45 verso, 35 mm from the left edge and 65 mm from the bottom of the page. The locations of the samples were also marked on photographs made of each page of the manuscripts (Figure 2).

The average size of the offset samples was 50–100 μm; the sizes of other samples ranged from 4 to 40 μm. An attempt was made to sample the complete range of hues in each manuscript. Hues were differentiated by both visual assessment and by comparison with color chips from *The Munsell Book of Colors* (*21*).

Sample Analysis. The samples were initially analyzed by polarized light microscopy and X-ray diffraction by the methods outlined by McCrone and Delly (*22, 23*). The microscope was an Olympus POS-1 equipped for photomicrography in transmitted light. The opaque samples were photographed at 180× magnification in reflected light on a Reichert Me F 2 metallograph. The X-ray diffraction patterns were obtained by mounting

Table I. University of Chicago Special Collections Manuscripts Analyzed by FT–IR

Manuscript Number	*Name*	*Date*
972	Archaic Mark	?
1054	Elfleda Bond Goodspeed *Gospels*	10th century
965	Rockefeller–McCormick *New Testament*	late 12th century
131	Chrysanthus gospels	late 12th century
232	Greek gospels	12th century
46	Haskell gospels	late 13th century
129	Nicolaus gospels	1133
727	Georgius gospels	late 13th century
879	Lectionary of Constantine the reader	late 12th century
948	Lectionary of St. Menas the wonder worker	late 12th century

Figure 2. Moses Receiving the Law, *Rockefeller–McCormick* New Testament, *University of Chicago MS 965, folio 6 verso. The open circle indicates the place where the miniature was sampled. BR6v(127,108) indicates the code for brown sample, page 6v, 127 mm to the right and 108 mm from the bottom left page corner.*

the sample particles on a glass filament in a 114.59-mm-diameter Debye–Scherrer powder camera and irradiating with Cu K_α X-rays for 24 h at 30 kV and 15 mA. Although these methods enabled the specific identification of the mineral pigments in the manuscripts, the organic pigments could be classified only generally.

Samples that were organic in nature were subjected to the microchemical tests described by Hofenk-de Graaff (*24*) and subsequently analyzed by FT–IR microspectroscopy.

The microspectroscopic analysis proceeded as follows. Sample particles were transferred from the precleaned glass microscope slides with a fine pointed metal probe under a 30× binocular microscope to a KCl window measuring 6 mm in diameter and 1 mm thick. The KCl window was supported by a center-holed metal microscope slide that fits readily onto the stage of the FT–IR microscope accessory. Once on the stage, the sample was visually brought into focus, and the circular field stop was closed down around the sample image. The ratio of the sample spectrum to that of a background spectrum previously obtained at a similar aperture setting was obtained.

However, sampling in the manner just described sometimes resulted in low-quality spectra with many absorption bands that could not be distinguished from the background noise. This effect is largely caused by the inherent tendency of many pigment samples to scatter much of the transmitted IR signal away from the optical path leading to the detector. In such cases, spectral quality was improved by mixing a few particles of the pigment sample with a small crystal of KBr and flattening the mixture with a probe tip into a microscopic KBr plate measuring approximately 30–50 μm across. The quantities of the pigment samples were insufficient to allow the use of a conventional KBr disk or even a microdisk accessory.

An Analect AQS-20M system consisting of an fx-6260 FT–IR spectrometer and an fXa-515 microscope module was used to obtain 4-cm^{-1}-resolution spectra. Replicate spectra of each pigment sample were obtained whenever sample quantity allowed. The number of signal-averaged scans was varied depending on the apparent scattering ability of each pigment. In general, the better the scattering ability, the more scans were needed to reduce the noise in the spectrum. Peaks were located by using a cursor and then recorded.

Reference spectra were obtained from standard organic pigments from the Forbes collection of the Conservation Center, Institute of Fine Arts, New York University, and from a few organic and inorganic pigments from various sources. Additional reference spectra were obtained from compounds reportedly used as medieval paint additives, such as dried egg yolk, egg white, calcium carbonate, and kaolin. The spectroscopic identification of compounds in pigment samples from the manuscripts that we report herein was based on a comparison of sample spectra with these reference spectra.

Results and Discussion

The preliminary results of the X-ray, microscopic, and microspectroscopic analyses are summarized in Tables II and III.

Interpretation of IR spectra was complicated by the fact that the pigment samples often contained several components in addition to the pigment. These additives often served as binding media—thickening agents or extenders, for example. Because the amount of pigment is presumably small relative to the amounts of these nonpigment components, the spectral bands of the pigment are often difficult to distinguish from interfering spectral bands of nonpigment components. Chromatographic separation prior to spectroscopic analysis was not feasible because of the insolubility and limited quantity of the samples. Spectral subtraction was used in many cases as a means to "separate" the components.

For example, Figure 3 shows a representative spectrum obtained from a yellow pigment sample from manuscript (MS) 46 (Haskell gospels). Several bands can be attributed to egg yolk, a typical medium used in medieval times (*25*), as well as calcium carbonate and kaolin. Subtraction of the spectra of egg yolk and calcium carbonate from the spectrum of a replicate yellow pigment yielded the spectrum shown in Figure 4. (No kaolin was observed in the replicate run.) The absorption bands at 3382, 3298, and 1600 cm^{-1} suggest that a primary amine is present. Putrescine (1,4-diaminobutane), an amine found in decaying proteinaceous matter, has similar spectral features. Because parchment (being of animal skin origin) would be subject to putrefaction if not properly preserved, the amine in Figure 4 could be a product of bacteriological degradation. Obviously, any such sample degradation would complicate the interpretation of spectral data even further.

Although the pigment was not successfully identified in this case, the information that egg tempera is present as the binding medium is a valuable characterization by itself. Egg yolk was also identified in spectra obtained from a different-colored pigment sample from the same manuscript.

A different binding medium was identified in MS 965 (Rockefeller–McCormick *New Testament*), however. Figure 5 shows the spectrum obtained from a sample of the manuscript's sizing material. Although the spectrum is clearly not that of egg yolk, it has the features of a predominantly protein component. The protein-containing binders used in that period are presumed to include casein, egg white, and hide glue (*25*). The spectrum closely matches that of hide glue (*26*).

The spectrum obtained from a magenta sample removed from the binding of the same manuscript indicates a protein substance and additional unidentified component(s). The additional component(s) hinder identification of the protein. Although the protein present could be hide glue, it could also originate from pigments extracted from insect sources.

Pigment components have also been identified from their IR spectra. Figure 6 shows the spectrum obtained from a purple pigment removed from

MS 965. One can readily assign bands to ultramarine blue and white lead. In addition, an occasional, isolated blue chip was found in the sample. The spectra of these blue chips matched reference spectra of ultramarine blue obtained from the Forbes Collection. Therefore, a red pigment must be present with the blue to obtain a purple color.

The spectrum in Figure 7 is the result of the subtraction of the spectra ultramarine blue and white lead. Similar spectral features are present in Figure 8, a reference spectrum of crude cochineal (from Peruvian *Coccus cacti*). This reference spectrum was taken from the darker red, almost black, part of the dried insect. An important consideration is that if the region from 1400 to 1800 cm^{-1} is expanded, a pair of doublets become evident at 1465 and 1472 cm^{-1} and at 1708 and 1734 cm^{-1}. These same bands appear in the reference spectrum at 1463 and 1472 cm^{-1} and at 1708 and 1734 cm^{-1}. The fact that the frequencies of these doublets match suggests that the red pigment may very well be cochineal.

The frequencies of all absorption bands observed could not be matched, and the spectral differences may be caused by the presence of an additional component(s), the chemical differences between New World (reference) cochineal and the Old World variety (the red pigment obtained from *Coccus ilicis* and known as "kermes" [25]), or differences between current extraction procedures and those of medieval times. Spectra obtained from the brighter red insect parts and those obtained from the dried extracted cochineal are notably different from each other and from that in Figure 8. This difference is due to the inhomogeneity of the reference sample, a fact indicating that extra care in sampling is required. The darker red part of the insect that gave rise to the spectrum in Figure 8 is definitely a minor constituent of the whole insect. For the sake of comparison, the spectrum of the ground whole insect is shown in Figure 9.

Flesh-colored and red pigments removed from MS 232 (Greek gospel book) have spectral features similar to those in Figures 7 and 8. This similarity suggests that cochineal was also used as the red pigment in this manuscript.

Another example of pigment identification by IR microspectroscopy is shown in Figure 10. The bottom spectrum was obtained from a blue pigment from MS 972 (Archaic Mark); the top spectrum is a reference spectrum of Prussian blue. The band corresponding to the C≡N of ferric ferrocyanide is common to both spectra. Replicate spectra of blue pigments removed from different locations in MS 972 indicate that the average frequency of this band is 2083 ± 6 cm^{-1}. The ubiquitousness of an iron blue in this manuscript raises doubts about the authenticity of this manuscript.

The iron blues are the first of the artificial pigments with a known history and an established date of first preparation. The color was made by the Berlin color maker Diesbach in or around 1704. Moreover, according to Gettens and Stout (25), the material is so complex in composition and method of manufacture that there is practically no possibility that it was invented independently in other times or places. This fact, in addition to the evidence

Table II. Pigment Components of Manuscripts

Color	*Pigment*	*Manuscript Number*									
		972	*1054*	*965*	*131*	*232*	*46*	*129*	*727*	*879*	*948*
Black	Charcoal	+									
	Carbon black					+	+		+		
Blue	Azurite						+				
	Ultramarine	+	+	+	+	+	+	+			
	Verdigris								+		
	Prussian blue	o									
Blue green	Organic	+									
Brown	Organic			+	+						
	Vermilion					+					
Flesh	White lead, vermilion, and silicates	+									
	White lead and organic yellow					+					
	White lead, cochineal, and silicates					o					
Green	Organic green		+	+							
	Ultramarine, terre verte, and organic yellow	+									

	Ultramarine and organic yellow		+				+				
	Ultramarine and gamboge				+						
	Ultramarine and orpiment										
Magenta	Organic magenta			+	+						+
	White lead (?)				o						
Orange	Organic orange			+							
Pink	Organic red										
Purple	Organic purple	+		+				+			
	Ultramarine and organic red		+								
	Ultramarine, cochineal, and white lead			o							
Red	Organic red		+	+	+	+					
	Vermilion	+			+	+	+	+		+	
	Cochineal					o					
White	White lead	+		+		+					
Yellow	Gamboge				+						
	Organic yellow	+		+	+	+	+				
	Orpiment				+						

NOTE: Colors marked + were detected by microscopic and X-ray analysis, and colors marked o were detected by FT–IR microspectroscopic analysis.

Table III. Nonpigment Components of Manuscripts

Color or Component	*Pigment*	*Manuscript Number*									
		972	*1054*	*965*	*131*	*232*	*46*	*129*	*727*	*879*	*948*
Black	Formic acid salts						o				
	Egg yolk						o				
Blue	Formic acid salts	o									
	Calcium Sulfate	o									
Brown	Formic acid salts					o					
	Kaolin			o	o	o					
	Calcium carbonate			o							
Magenta	Protein			o							
	Formic acid salts										o
Red	Formic acid salts		o	o							
	Calcium carbonate					o					
Yellow	Egg yolk						o				
	Kaolin				o		o				
	Calcium carbonate						o				
	Formic acid salts			o							
Ink	Protein						o				
Size	Hide glue			o							

NOTE: The components were analyzed by FT–IR microspectroscopic analysis.

indicating that both the Archaic Mark and the Leningrad gospel fragment were copies of the Athens codex 93, suggests that these manuscripts originated some time much later than their purported 12th-century fabrication. Furthermore, neither of these manuscripts has a genealogy that can be traced prior to about 1930, a fact suggesting that their origin may very well be during the flurry of Athenian forgeries that came to the market in the 1920s.

Bands due to calcium sulfate and sodium formate are also evident in Figure 10. The sodium formate bands are at 773, 1362, 1591, 2716, 2830, and 2954 cm^{-1} (27). Sodium formate bands are evident in various spectra from almost all of the manuscripts that were analyzed by FT–IR microspectroscopy. In addition, the presence of sodium formate could not be associated with any one particular color of pigment. Because formic acid is a compound used in the more-modern tanning processes, we initially speculated that the formate in the samples originated from the medieval parchment-making process as well, very likely serving as a preservative. However, we recently identified sodium formate as a residue on precleaned glass microscope slides, presumably left from the manufacturer's cleaning process.

Conclusions and Further Work

The analyses cited previously (6–8) indicate that the palette of the Armenian artist was largely of mineral origin and contained ultramarine, orpiment, vermilion, white lead, and whiting as the staple pigments. The present work on Byzantine manuscripts, although far from conclusive because of the limited number of manuscripts examined, suggests that Byzantine illuminators relied more heavily on organic pigments and that ultramarine and vermilion

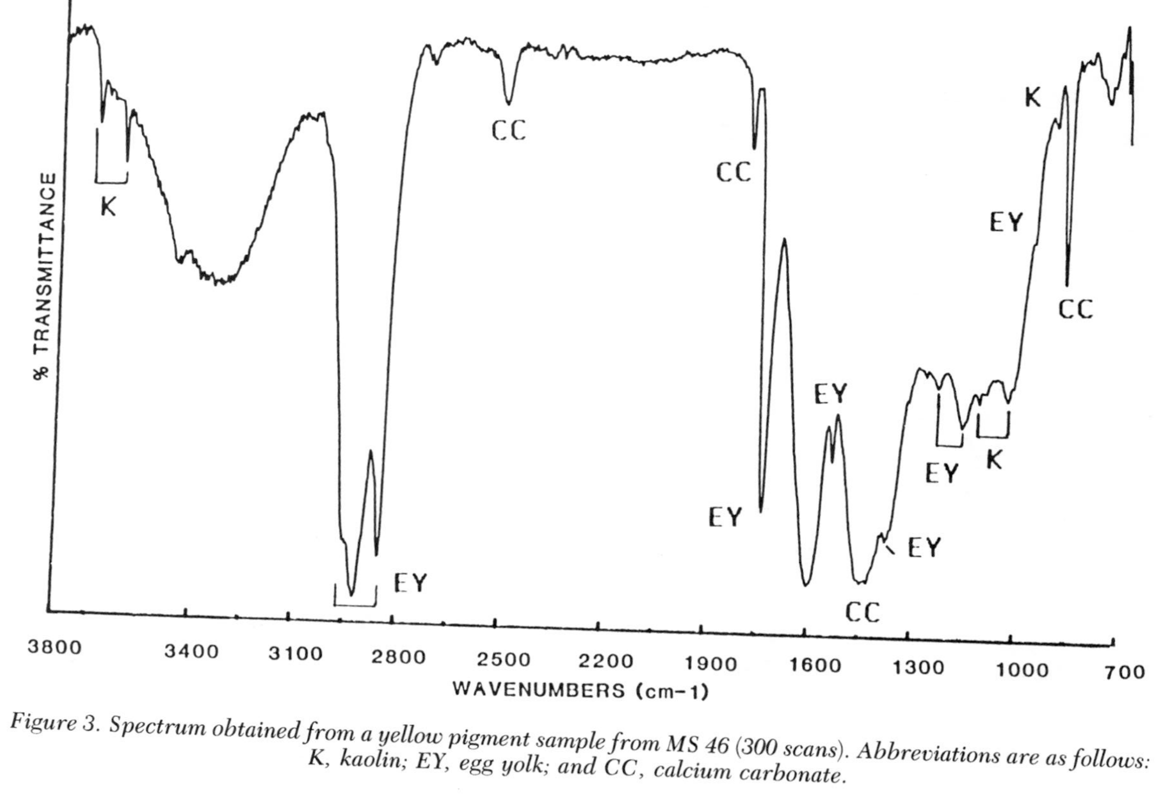

Figure 3. Spectrum obtained from a yellow pigment sample from MS 46 (300 scans). Abbreviations are as follows: K, kaolin; EY, egg yolk; and CC, calcium carbonate.

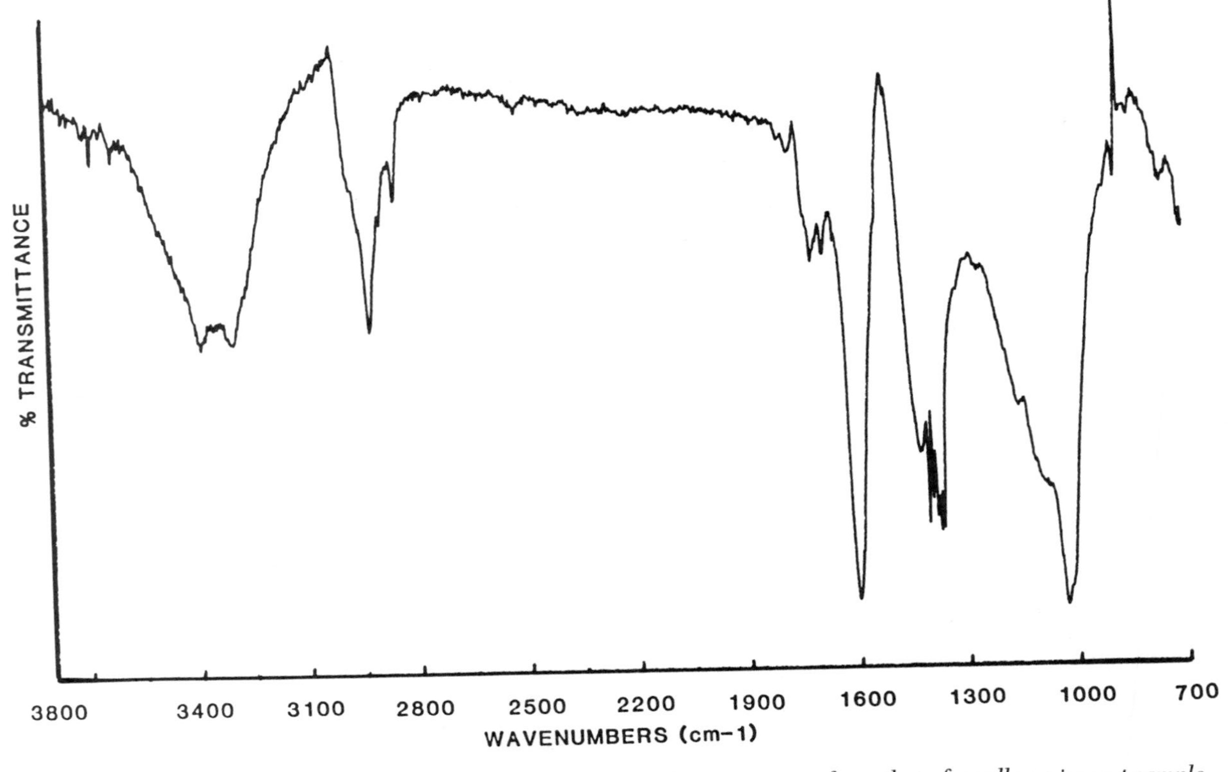

Figure 4. Results of subtraction of egg yolk and calcium carbonate spectra from that of a yellow pigment sample from MS 46.

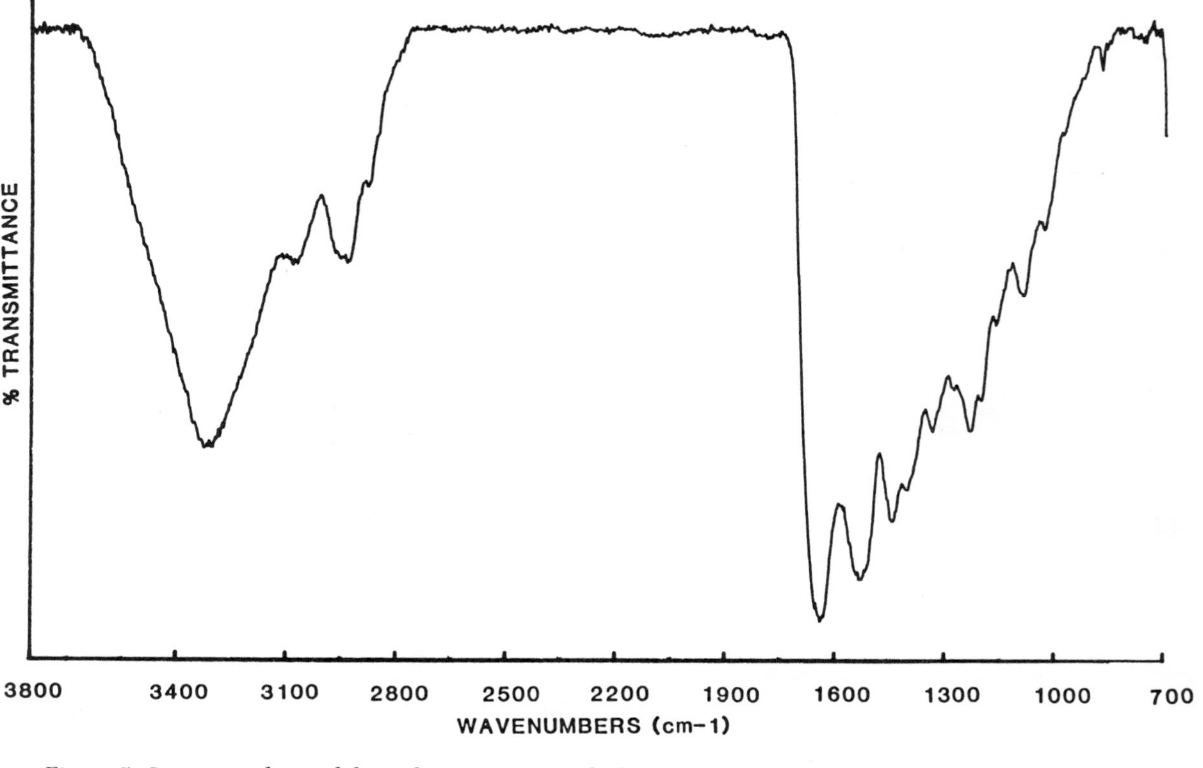

Figure 5. Spectrum obtained from the sizing material of MS 965 that was identified as hide glue (300 scans).

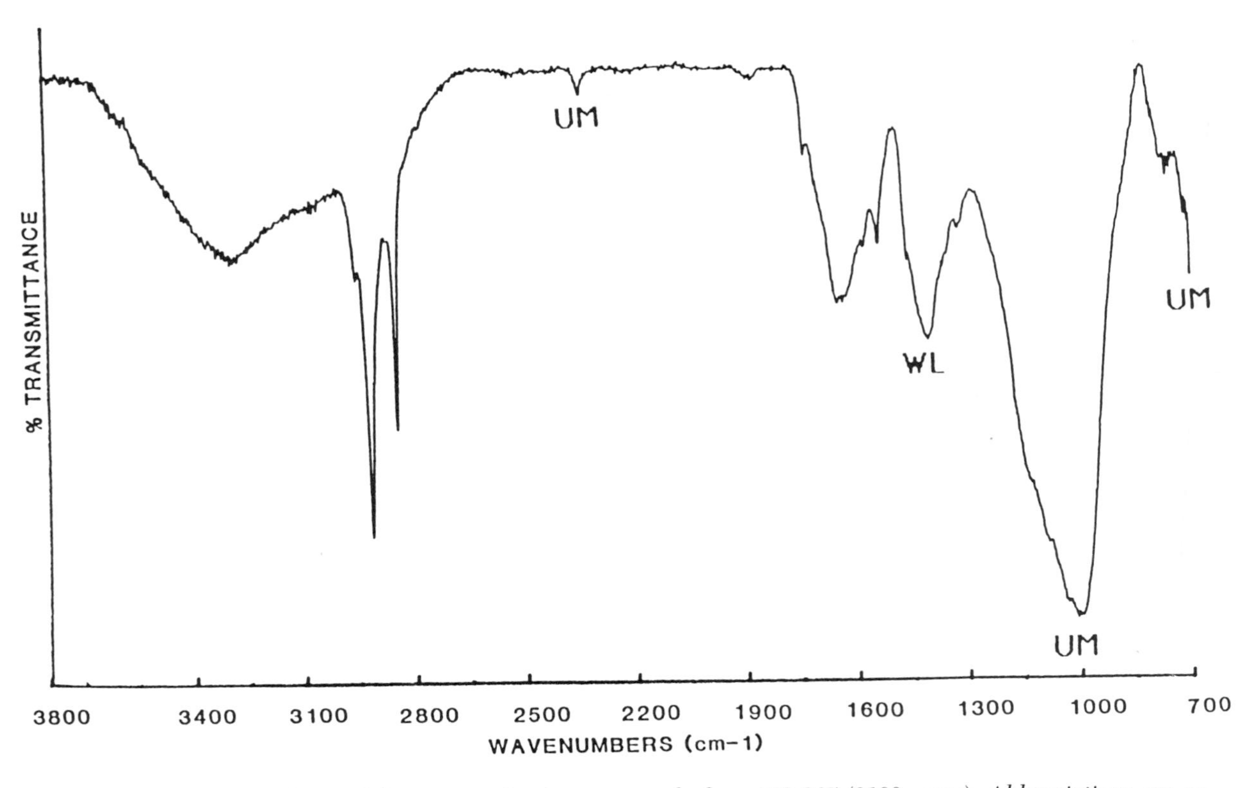

Figure 6. Spectrum obtained from a purple pigment sample from MS 965 (2100 scans). Abbreviations are as follows: UM, ultramarine; and WL, white lead.

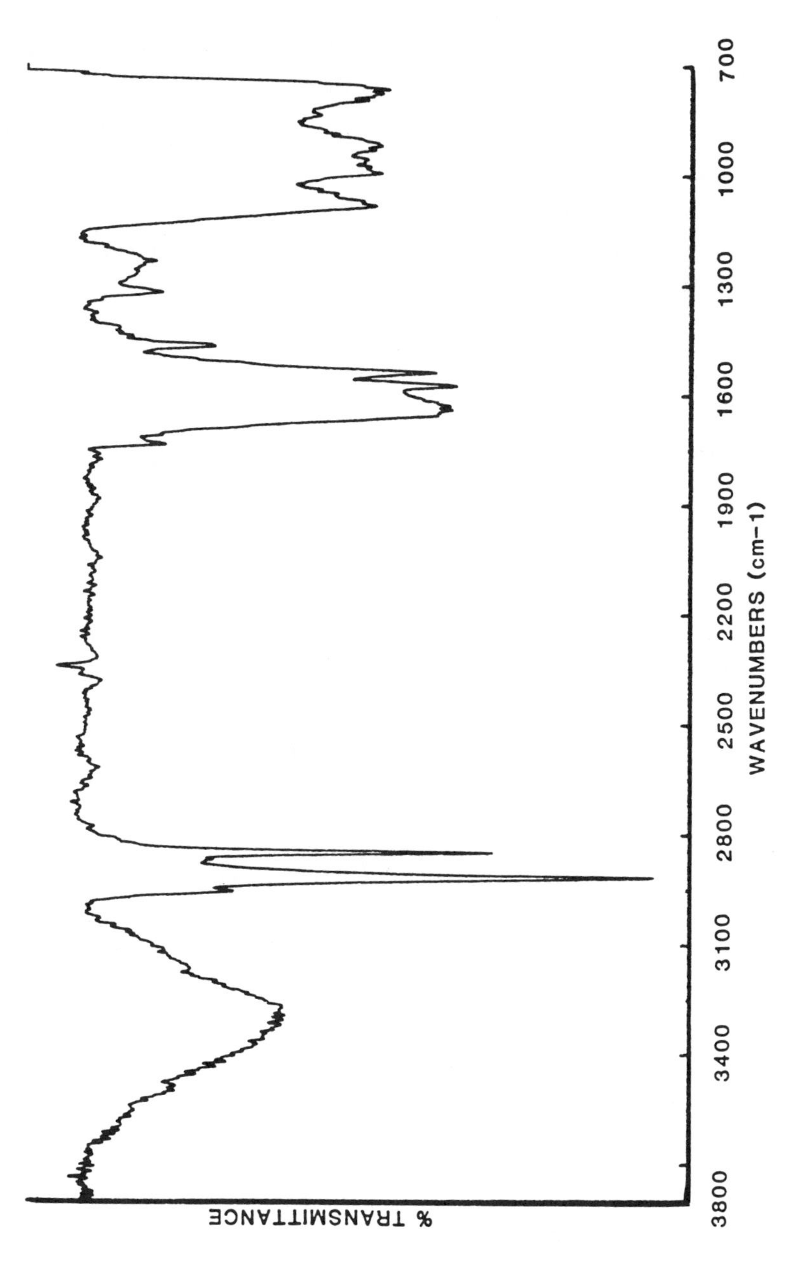

Figure 7. Result of subtraction of ultramarine blue and white lead spectra from the spectrum in Figure 6.

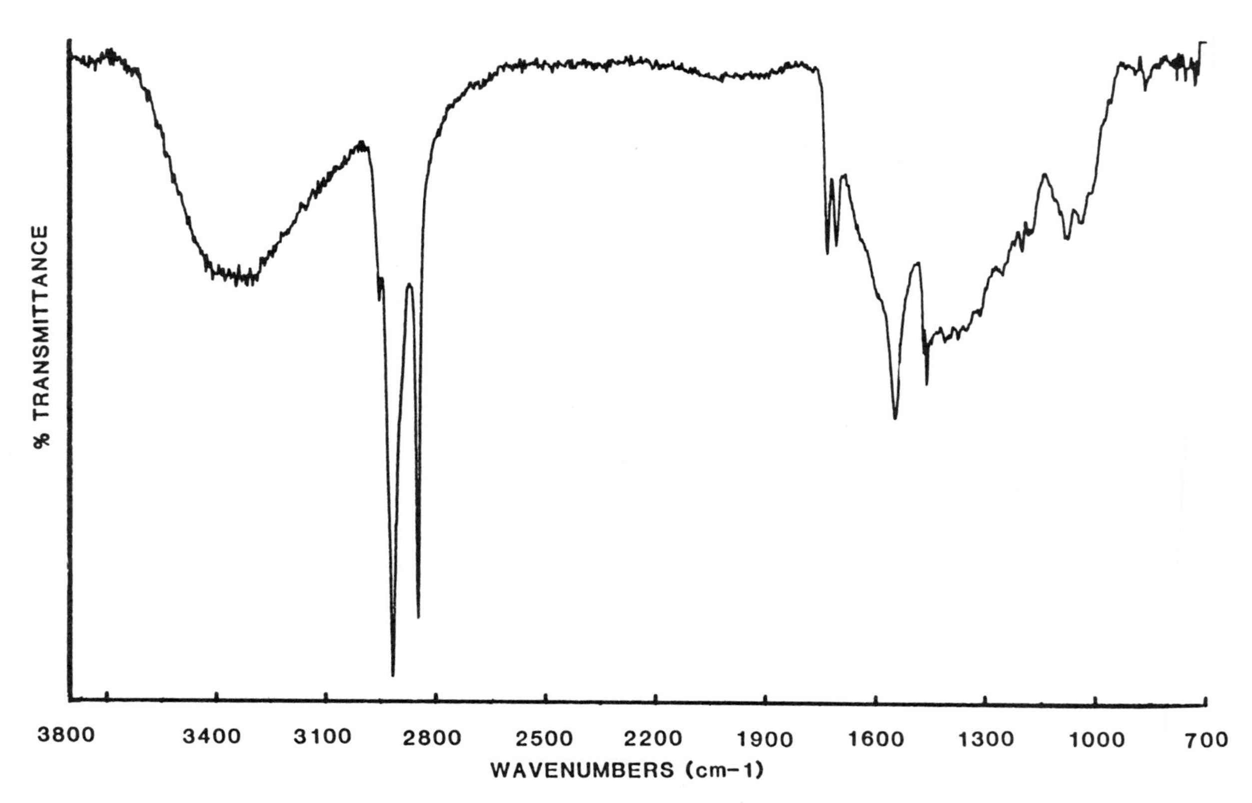

Figure 8. Reference spectrum obtained from dark red, almost black, part of cochineal insect (438 scans).

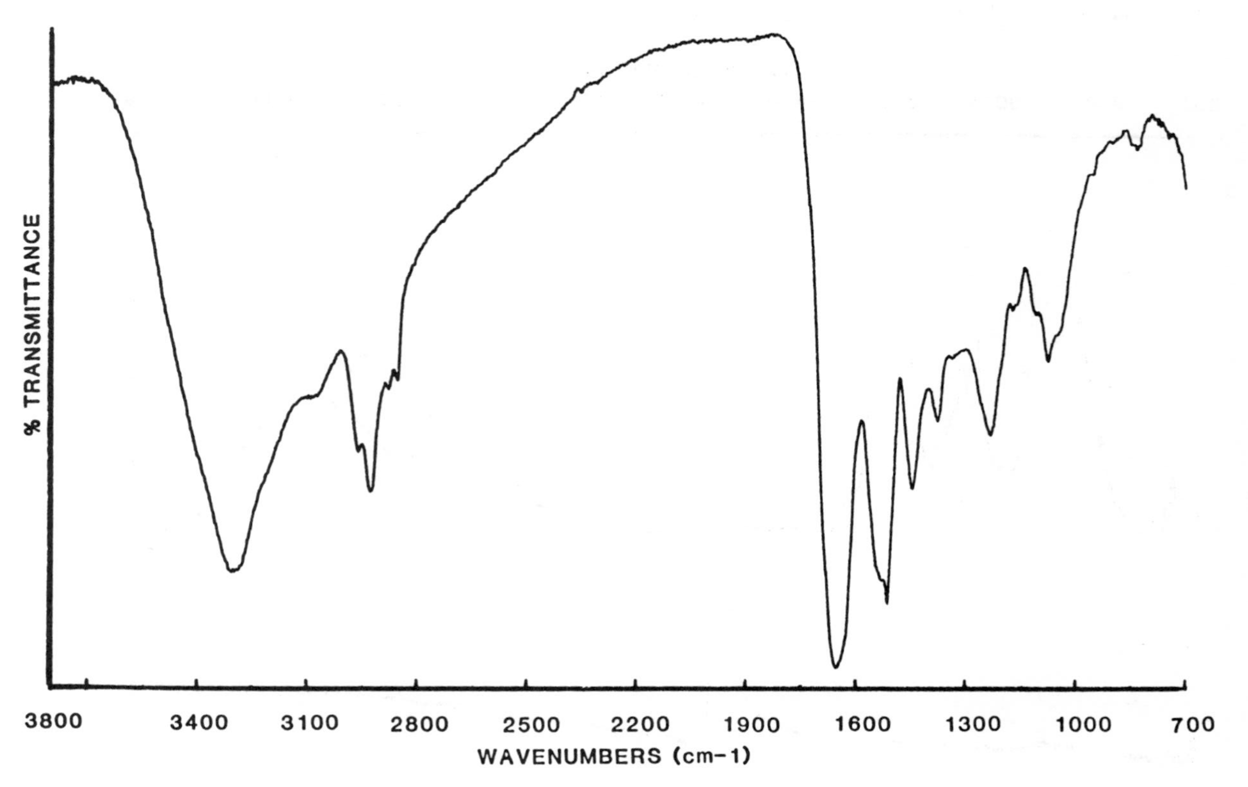

Figure 9. Reference spectrum of whole, ground Coccus cacti *(251 scans).*

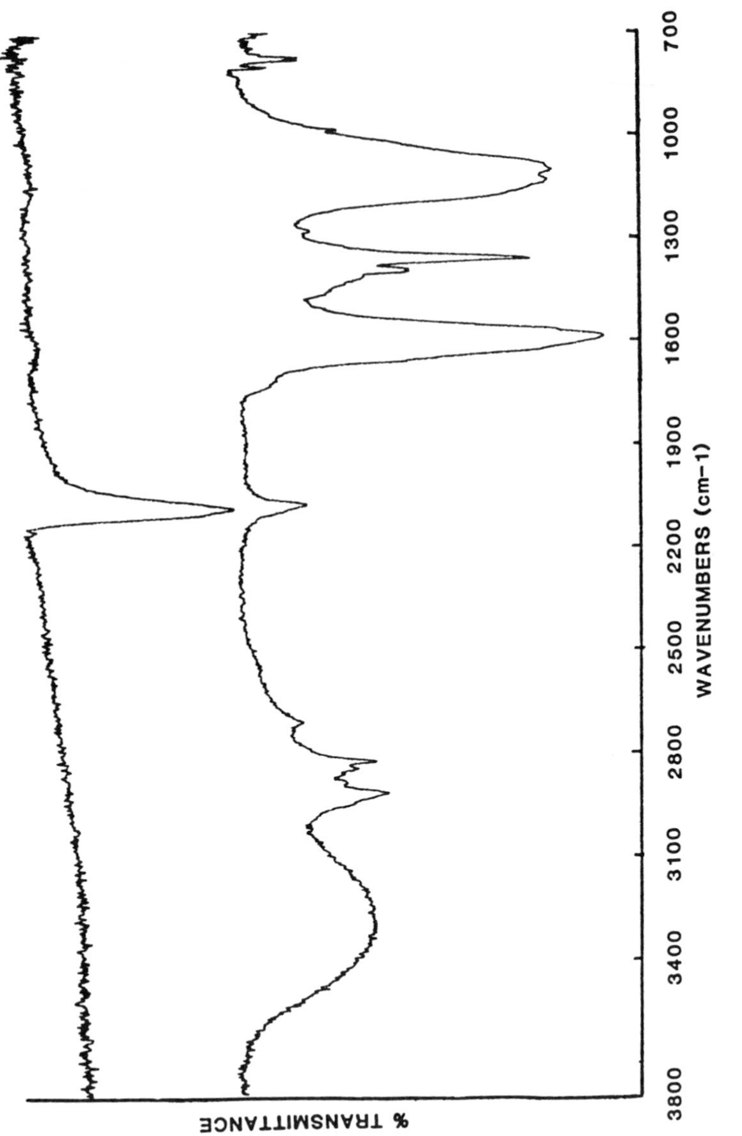

Figure 10. Top: Reference spectrum of Prussian blue (100 scans). Bottom: Spectrum of a blue pigment sample from MS 972, B34v (47,70) (247 scans).

were the only two mineral pigments used consistently. Additional Byzantine manuscripts will have to be examined to verify this pattern of usage. In addition, other artists of the Mediterranean, particularly those who practiced in the Islamic and Crusader kingdoms, developed their own traditions and techniques. An examination of manuscripts from these later sources may shed light on patterns of pigment usage throughout the medieval Middle East.

Acknowledgments

This work was performed at the Molecular Microspectroscopy Laboratory, Miami University, Oxford, OH.

This work was supported in part by grants from the Hagop Kevorkian Fund, National Museums Act, the Faculty Fund of the College of New Rochelle, and the New York University Research Challenge Fund.

We also thank Norbert S. Baer and Lawrence J. Majewski of the Conservation Center, Institute of Fine Arts, New York University, for supplying many of the reference materials from the Forbes Collection and Andre Sommer, Assistant Director of the Molecular Microspectroscopy Laboratory, Miami University, for his technical assistance and expertise.

References

1. Plesters, J.; Lazzarini, L. In *Conservation of Paintings and the Graphic Arts*, Lisbon Congress, 1972; International Institute of Conservation of Historic and Artistic Works: London, 1972; pp 153–160.
2. Roosen-Runge, H.; Werner, A. E. A. In *Evangeliorum Quattuor Codex Lindisfarnensis*; Kendrick, T. D., Ed.; Urs Gras: Lausanne, Switzerland, 1960; Vol. 2, pp 263–272.
3. Roosen-Runge, H. *Farbgebung und Technik Frühmittelalterlicher Buchmalerei*; Deutscher Kunstverlag: Munich, Federal Republic of Germany, 1967; Vol. 1, pp 11–17.
4. Muether, H. R.; Balazs, N. L.; Cotter, M. J. *Abstracts of Papers*, 68th Annual Meeting of the College Art Association of America, New Orleans, LA; College of Art Association: New Orleans, 1980; p 133.
5. Carriveau, G. In *Archaeological Chemistry—III*; Lambert, J. B., Ed.; Advances in Chemistry 205; American Chemical Society: Washington, DC, 1984; pp 395–402.
6. Orna, M. V.; Mathews, T. F. *Stud. Conserv.* **1981,** *26*, 57–72.
7. Cabelli, D.; Mathews, T. F. *J. Walters Art Gallery* **1982,** *40*, 37–40.
8. Cabelli, D.; Orna, M. V.; Mathews, T. F. In *Archaeological Chemistry—III*; Lambert, J. B., Ed.; Advances in Chemistry 205; American Chemical Society: Washington, DC, 1984; pp 243–254.
9. Demus, O. *Die Palette* **1967,** *26*, 3.
10. Shearer, J. C.; Peters, D. C.; Newton, G. H. T.; *Anal. Chem.* **1983,** *55*, 874A–880A.
11. Griffiths, P. R.; deHaseth, J. A. *Fourier Transform Infrared Spectrometry*; Wiley; New York, 1986; Vol. 83.

12. Katon, J. E.; Pacey, G. E.; O'Keefe, J. F. *Anal. Chem.* **1986,** *58,* 465A–481A.
13. Lang, P. L.; Katon, J. E.; O'Keefe, J. F.; Schiering, D. W. *Microchem. J.* **1986,** *35,* 319–331.
14. Madigan, S. *Abstract of Papers,* 8th Annual Byzantine Studies Conference; Byzantine Studies Conference: Chicago, 1982; p 49.
15. Carr, A. W. *Dumbarton Oaks Papers,* **1982,** *36,* 44.
16. Department of Special Collections, Joseph Regenstein Library, University of Chicago; unpublished notes.
17. Ibid., p. 9.
18. Marava-Chatzinicolaou, A.; Toufexi-Paschou, C. *Catalogue of the Illuminated Byzantine Manuscripts of the National Library of Greece; Publications Bureau of the Academy of Athens: Athens, 1978; Vol. I, pp 224–243.*
19. *Iskusstvo vizantii v sobranijah SSSR*; Sovetskij Xudoznik: Moscow, 1977; Vol. 2, p 58.
20. Treu, K. *Die Grieschischen Handschriften des Neuen Testaments in der USSR*; Akademie–Verlag: Berlin, 1966; pp 229–230.
21. *The Munsell Book of Color,* cabinet ed.; Macbeth Division of Kollmorgen Corp.: Baltimore, 1958.
22. McCrone, W. C.; Delly, J. G. *The Particle Atlas,* 2nd ed.; Ann Arbor Science: Ann Arbor, 1974 (Vols. I–IV) and 1978 (Vols. V and VI).
23. McCrone, W. C.; McCrone, L. B.; Delly, J. G. *Polarized Light Microscopy*; Ann Arbor Science: Ann Arbor, 1978.
24. Hofenk-De Graaff, J. H. *Natural Dyestuffs for Textile Materials: Origin, Chemical Constitution, Identification*; Central Research Laboratory for Objects of Art and Science: Amsterdam, The Netherlands, 1967; pp 13–15.
25. Gettens, R. J.; Stout, G. L. *Painting Materials: A Short Encyclopaedia*; Dover Publications: New York, 1966; pp 20–21.
26. Ibid., p 123.
27. Hummel, D. O.; Scholl, F. *Atlas of Polymer and Plastic Analysis*; Carl Harser Verlag: Munich, Federal Republic of Germany, 1984; Vol. 2, Part a/II.
28. *The Coblentz Society Desk Book of Infrared Spectra,* 2nd ed.; Craver, C. D., Ed.; Coblentz Society: Kirkwood, MO, 1977.

RECEIVED for review June 11, 1987. ACCEPTED revised manuscript March 8, 1988.

15

Archaeological Sites as Physicochemical Systems

Macroarchaeometry of the Tomb of Nefertari, Valley of the Queens, Egypt

G. Burns, K. M. Wilson-Yang, and J. E. Smeaton

Department of Chemistry, University of Toronto, Toronto, Ontario, M5S 1A1, Canada

Major archaeological monuments are considered macroarchaeometrically as physicochemical systems coupled to their environment and surroundings. These edifices can be studied with the tools of analytical, environmental, and physical chemistry, as well as those of related disciplines. One of the aims of macroarchaeological chemistry is to identify the principal physicochemical processes in archaeological systems. This approach is illustrated by studies in the tomb of Nefertari, Valley of the Queens, Egypt, which now exists in a desiccated, fragile state. As a first step, a computerized hygrothermograph coupled to a minicomputer was developed and placed in the closed tomb. This instrument can make measurements for 1 year. Four processes that affect the humidity of the tomb were identified. In a parallel set of experiments, samples collected from the tomb were analyzed. The results were compared with those of an analysis of plaster from the tomb of Horemheb, Valley of the Kings. A tentative plan for the reconstruction, restoration, and conservation of Nefertari's tomb is proposed.

THE TECHNIQUES OF CONSERVATION and conservation science and the techniques of anthropology, archaeology, and art history are often used to solve very specific problems such as the removal of efflorescent materials from the surface of a wall painting or the determination of the provenance

0065-2393/89/0220-0289$06.50/0

of pottery. Although this approach is often very effective, specific problems are only components of a larger picture. Each monument, site, object, or any other archaeological system exists in dynamic physicochemical interaction with its environment, as defined by a complex and changing set of conditions. These conditions may be chemical, climatic, geological, geographical, or physical. In some cases, changes in these conditions may be dramatic, such as those occurring immediately after excavation of an archaeological monument when it undergoes a severe ecological shock or as a result of some natural or human-made catastrophe.

The identification of these conditions is one of the primary tasks of archaeometry, a discipline in which the methods of the natural sciences are used to interpret archaeological or conservation science data. In a multidisciplinary field such as archaeometry, archaeological chemistry occupies a central role because its various branches include analytical, computational, and physical chemistry, as well as chemical physics (Figure 1).

Macroarchaeometry

Macroarchaeometry, the study of archaeological–physicochemical systems with their surroundings (*1*), provides a more comprehensive approach to investigations of ancient materials. Macroarchaeometric investigations, with their central core of macroarchaeological chemistry, are particularly useful if the physical size of an archaeological system is large; a physically large system makes it possible to identify time- and space-dependent physicochemical gradients within the system.

However, no sharp dividing line exists between archaeometry and macroarchaeometry, or between archaeological chemistry and macroarchaeological chemistry, because the effect of the environment may be more

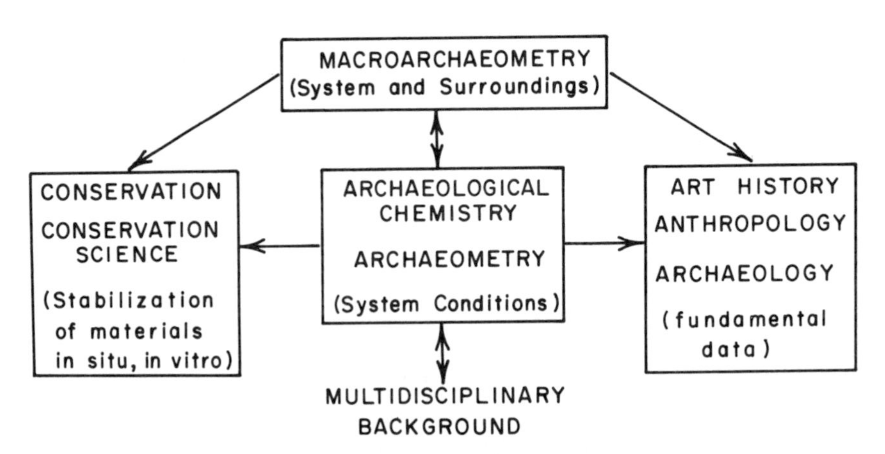

Figure 1. Interrelationship between archaeological chemistry and other fields of study.

important in one system than in another. In some cases, the interaction of an archaeological system with its environment may be negligible, and standard archaeometric techniques can be readily applied. In other cases, such as the study of large archaeological monuments that cover square kilometers of area and that interact in a complex manner with the environment, a comprehensive macroarchaeometric approach is needed.

In any case, macroarchaeometry provides understanding of the fundamental mechanisms in historical monuments; these mechanisms manifest themselves as specific problems. Problems such as accelerated degradation or long-term chemical reactions leading to compositional change are explained and can therefore be dealt with more successfully.

The macroarchaeometric approach is evident in various chapters of this volume; the importance of the macroarchaeometric component varies from case to case. The processes of degradation in bones and teeth (*2*) and in fibers (*3*) are discussed with respect to burial environments. The processes of radiogenic and thermally derived free-radical production in solids (*4, 5*) and amino acid racemization (*6, 7*) also depend on the system and the environment in which they are studied.

The macroarchaeometric approach has wide applicability. Climatic studies in the caves of Lascaux, France (*8*), and Alta Mira, Spain (*9*), are but two examples. This approach is now particularly useful in studies of the excavated and unexcavated ancient sites of the Nile River Valley because of the recent, accelerated deterioration of many of its monuments. It has already been applied to several ancient Egyptian archaeological sites (*10–21*).

Recent Environmental Changes in the Nile Valley

The climate of the Nile Valley was stable and predictable prior to the construction of the Aswan high dam. Although the effects of annual inundations were removed with the construction of the dam, the ensuing climatic changes ushered in a new set of phenomena that threaten the Nile Basin and its people, from Alexandria to Aswan. Our investigations (to be published) indicate that, because of overall increased irrigation, this threat now extends as far south as Khartoum.

These adverse phenomena include increases in soil salinization (*11, 12*), relative humidity, and rainfall. Such changes make ancient Egyptian sites more vulnerable to the ever-present danger of seismic activity, which is expected to increase because of the interaction of annual water level changes in the Aswan reservoir and the fault zones of the Nile Valley (*22*). Consequently, the establishment of the macroarchaeometric physicochemical conditions of Egyptian archaeological sites located at a distance of more than 1500 km along the Nile River is becoming increasingly important. In this chapter, we illustrate these points with a report on our studies, using macroarchaeological chemistry methods, of the physicochemical processes in the tomb of Nefertari and their consequences for its conservation.

Physical Condition of Nefertari's Tomb

The tomb of Nefertari (number 66, Valley of the Queens, Thebes, Upper Egypt) is a designated Egyptian national treasure. It was constructed approximately 3250 years ago for the favorite queen of Ramesses II (*23*). Its wall paintings are noted internationally for their striking design and are an excellent example of Egyptian art and technology at the height of their development during the 19th dynasty. In a large portion of the tomb, the murals are still relatively intact, unperturbed over the millennia. They have retained their colors and appear more attractive than the murals in many other ancient tombs in Egypt; even in the Valley of the Kings, many murals have suffered severe damage because of the changing environment and human intrusion.

After its discovery and early repair in 1904 (*24*), the tomb became a well-known tourist attraction, and reproductions of its murals were prominently displayed in early guidebooks (*25, 26*). It acquired such eminence that today Egyptian tourist posters are often reproductions of its various murals.

The tomb was cut from very poor quality limestone that is veined with sodium chloride. Some of these veins are several meters long and about 2 cm thick. The walls of the tomb were plastered with a 3–5-cm-thick mortar, carved in low relief, and coated with a thin white layer (*27*). From recent carbonate analyses, we found that this thin white layer consists of calcium carbonate whitewash. This whitewash was painted during the final stages of the tomb's construction.

Almost all decoration remains in the upper chambers of the tomb, whereas there are areas of massive loss in the lower chambers. The paint film, where it exists, appears to be in good condition, but in damaged areas, the layer is either broken or destroyed (Figure 2). The supporting plaster has lost its cohesion, and sodium chloride crystals are visible throughout the tomb. Some crystals are as large as 1 cm (Figure 3). Some parts of the tomb are covered by smaller crystals that sparkle under illumination. The gypsum-based plaster was dehydrated through the loss of chemically bound water (*27–29*). Because the tomb was discovered in an already fragile state, it is clear that this dehydration took place over an extended period prior to 1904. The tomb of Nefertari is therefore sensitive to perturbations in temperature and especially sensitive to changes in humidity, which affect the movement of sodium chloride and may induce changes in the dehydrated plaster.

On the basis of written descriptions (*24, 29, 30*), photographs (*23, 24, 30*), and physical evidence, the evolution of the tomb's condition as a function of time can be traced. Preliminary progress in this direction has been made (*28*). There is no evidence of smoke deposits on the walls, a fact that implies that the tomb was never inhabited. The complete absence of any Coptic or Arabic graffiti suggests that the tomb was not often visited at the time when many Egyptian tombs underwent particularly pronounced deterioration be-

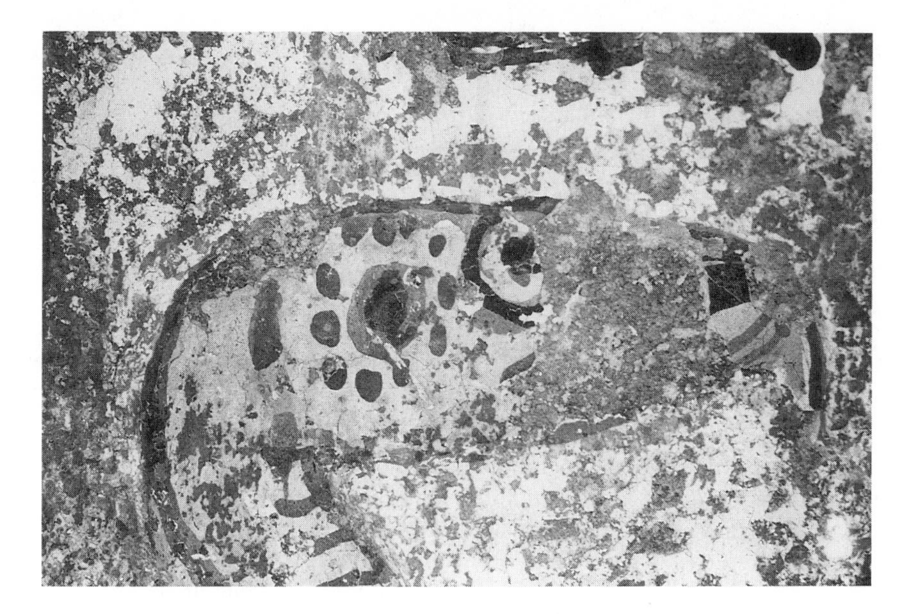

Figure 2. Paint layer damage in the tomb of Nefertari. The image recorded on a 35-mm negative is one-third that of the original.

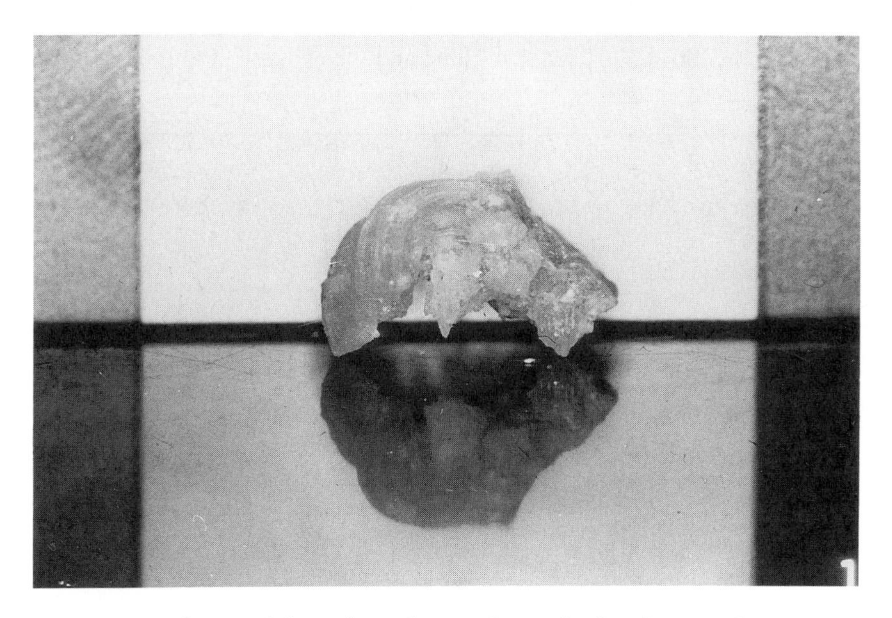

Figure 3. Salt crystal from the ceiling in the tomb of Nefertari. The image recorded on a 35-mm negative is the actual size.

cause of such human activities. The tomb has been robbed in antiquity and flooded probably more than once in its history, but it then remained largely isolated, physically and climatically (*28*), from the external environment until its discovery in 1904 (*24*).

Progressive Deterioration of the Tomb

The condition of the murals at the time of the tomb's discovery in 1904 (*24*) appeared deceptively similar to their condition in 1980, when our group began its studies (*28*). However, the tomb exists in a desiccated state, is very fragile, and is therefore particularly susceptible to ecological changes; loss and fragmentation of the paintings in the Nefertari tomb are evident (Figure 2). The tomb now requires detailed macroarchaeometric studies because its future is very much influenced not only by the physicochemical processes within, but also by minute, subtle interactions with the environment, including nearby irrigated fields and the Nile River situated about 5 km away (Figure 4).

Progressive deterioration occurred in the tomb after its discovery, and one of us (G. B.) found evidence in the tomb that partial restoration was attempted in 1935 to retard this deterioration. In about 1940, the Egyptian Antiquities Organization (EAO) assessed the condition of the tomb because of startling changes within it. Salt crystals were evident, no mention of which was made in 1904 (*24*). Plaster swelling and the collapse of mural fragments

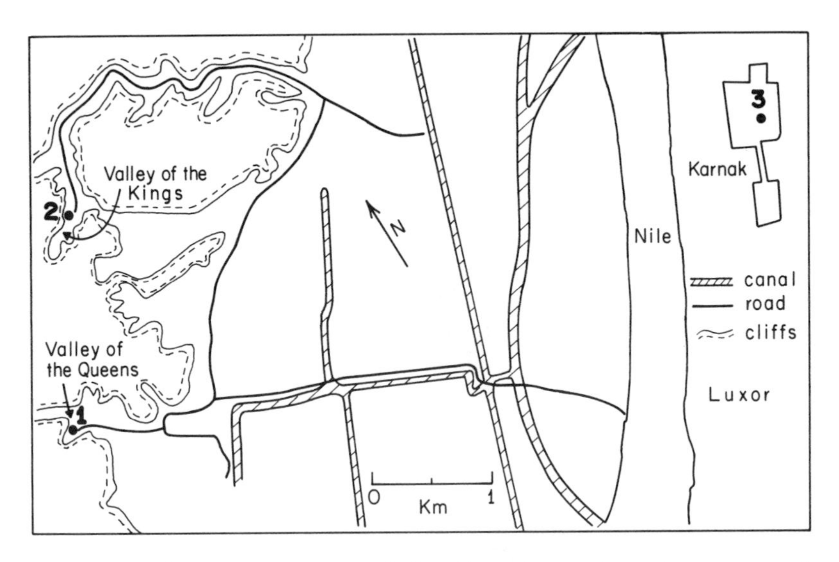

Figure 4. Schematic map of the Thebes–Karnak area. Point 1, the tomb of Nefertari; point 2, the tomb of Horemheb; and point 3, the position of the external hygrothermograph.

were also seen (*30*). A series of photographs and carefully annotated drawings was published in 1942 (*30*) that highlighted the areas of postdiscovery loss and potential loss on the basis of the original photographs of 1904 (*24*). This evidence indicated that deterioration had accelerated during the period of frequent human entry after 1904. Soon after this assessment, the tomb of Nefertari was closed, locked, and sealed as a result of the justifiable concern of the EAO. Thereafter, the deterioration of the tomb slowed, as evident from a comparison of observations made 1971 (*23*) and photographs taken in 1981 (*28, 31*) with photographs taken in 1904 (*24*) and 1942 (*30*).

There are historic reports of floods in the area (*32*; J. Rutherford, personal communication) and of an earthquake in 18 B.C. (*33*). The movement of the expansive Esna shale bed formation that underlies the whole area is also a source of disruption (*32*). The Esna shale bed formation, which contains montmorillonite clay, swells upon hydration by about 12.5% (*32*). Therefore, if an adequate amount of water is available, for example from increased irrigation, humidification, or flooding, the movement of this shale formation accelerates. Collectively, these dramatic natural events are the most probable causes of the loss of plaster and painted murals in the lower chamber of the tomb prior to its discovery, and they increasingly (*32*) threaten tombs in the Thebes area.

To assess the physicochemical processes in Nefertari's tomb, to identify the mechanisms of its deterioration, and to propose a feasible scheme for its conservation and restoration, the Inspector General of the EAO, Hishmat Messiha, urged our group to begin investigations. A permit to obtain representative samples from the tomb and its environment was issued to us by the EAO in 1980, and subsequently, our work was initiated (*28*).

Long-Term Studies of Temperature and Humidity

The physical condition of the tomb indicates that the primary physicochemical processes of the tomb are intimately connected to its internal climatic conditions (*28*). In an initial study in December 1977, a set of humidity measurements taken in the closed tomb for 1 week with an unattended clockwork hygrothermograph showed a remarkably stable internal temperature of 27 ± 1 °C; the internal relative humidity (RH) was 31 ± 2%. Small increases in humidity that were correlated with entry were also reported (*28, 31*).

One of the immediate implications of this latter finding is that any measurement of humidity taken while the individual taking the measurement is physically present in the tomb does not reflect the true climatic conditions in the undisturbed tomb. Therefore, hygrothermographic data taken concurrently with conservation work do not provide information about the conditions that should be achieved and subsequently maintained in the conserved tomb. Consequently, before any conservation work is begun, it would first be necessary to gather physicochemical data with minimum human interference.

As a first step toward data collection with little human interference, an automatic hygrothermograph interfaced with a minicomputer was developed in our laboratory to study the internal climate of the tomb. This instrument can take temperature and humidity measurements every 33 min for 1 year and store the data in a computer memory. Its sensitivity to humidity fluctuation exceeds that of conventional chart hygrothermographs by a factor of 5. Therefore, the instrument has an important advantage over a clockwork hygrothermograph because it provides information on long-term climatic fluctuations in the tomb.

Thermal Stability of the Tomb. The self-contained hygrothermograph was left in the tomb of Nefertari in February 1981 for 1 year (*34*). The temperature and humidity data for the 1-year period were correlated (Figures 5 and 6) with external data taken near the Nile River at Karnak (Figure 4). The Nile River data were obtained in 1981 by M. C. Traunecker, a research chemist with the Franco–Egyptian Centre for Studies of the Temples at Karnak.

Figure 5 shows average monthly high and low temperatures near the Nile River at Karnak and the monthly average temperatures inside the tomb. The results dramatically confirm the temperature stability of the tomb re-

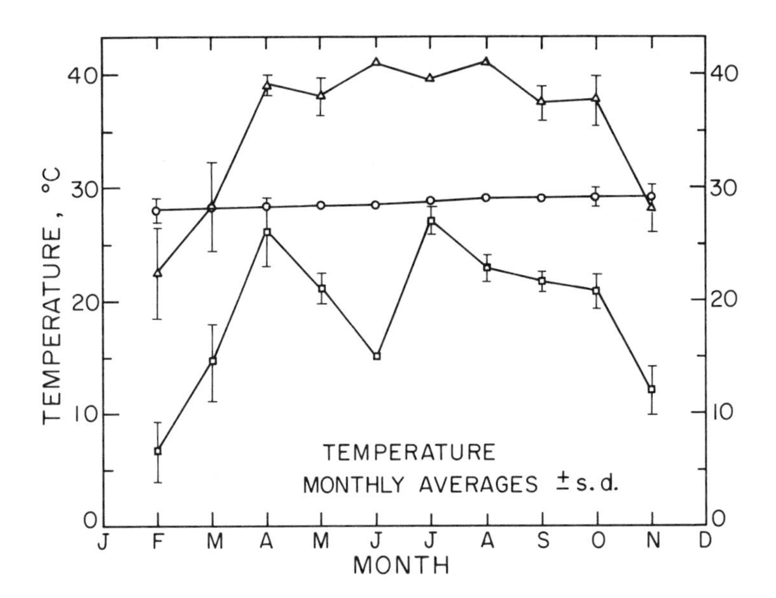

Figure 5. Average monthly high (Δ) and low (□) temperatures in the Nile Valley and average monthly temperature (○) in the tomb. Where error bars are not shown, the standard deviation is less than or equal to ± 0.5 °C.

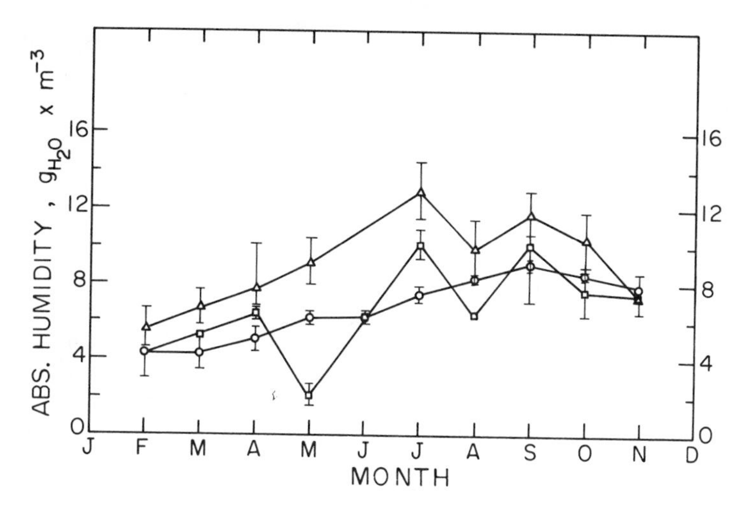

Figure 6. Average monthly high (Δ) and low (□) values of absolute humidity in the Nile Valley and average monthly values of absolute humidity (○) in the tomb.

ported earlier (*28*) and now show that the stable temperature is maintained over a longer period. The yearly average temperature inside the unperturbed tomb is 28.5 ± 0.5 °C, and the yearly average humidity is 7 ± 2 g of H_2O/m^3 or 25 ± 6% RH at 28 °C. (Values are means ± standard deviations, unless otherwise stated.) (*34*, *35*).

The inhospitable internal climate of this tomb has been thereby quantified. The ventilation in the tomb is poor; it is difficult to work in this closed underground monument for more than 1 or 2 h because of the lack of fresh air and the accumulation of carbon dioxide produced by respiration. These difficult conditions have proscribed prolonged human visitation of the tomb in the past and at present and explains how such a fragile monument survived into this century.

Cyclical Humidity Changes in the Tomb. The isothermal nature of the tomb simplifies its physicochemical description because one key variable is now fixed. The humidity data did not show the same degree of stability as the data for temperature did. Figure 6 shows that the external humidity peaks in July. The internal humidity follows this trend but peaks 1.5 months later at a value of 9.1 g of H_2O/m^3 or 32% RH at 28 °C. These data suggest an annual cycle, and these results indicate that the tomb of Nefertari is not totally isolated, even if it is closed, but that it is weakly coupled to the external environment. Consequently, the tomb is progressively humidified, as is the entire Nile Valley.

Figure 7 is a plot of individual temperature and humidity measurements for March 1981. The average temperature for this month was 28.2 ± 0.3 °C,

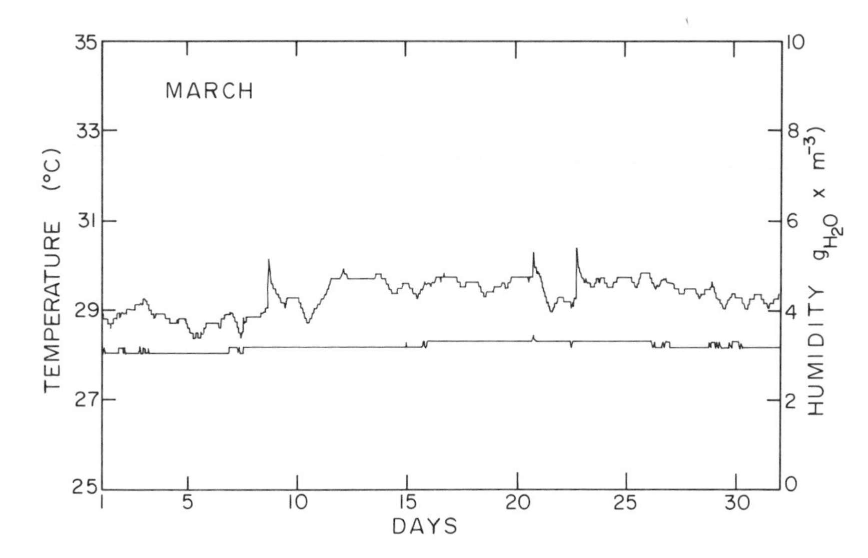

Figure 7. Temperature and humidity inside the tomb of Nefertari for March 1981.

and the average humidity was 16 ± 1% RH or 4.3 ± 0.4 g of H_2O/m^3. The temperature curve shows the remarkable time invariance already noted in the yearly data, however, the humidity curve is complex. Two distinct features are evident in the humidity data. The first is the appearance of cyclical humidity variations. These variations have a period of about 1 day and lag behind the external cycles by about 12 h. These cycles are evident throughout the full set of accumulated humidity data, and they indicate that the closed and locked tomb of Nefertari is not totally isolated from the outside, even on a daily basis.

The second feature is the appearance in the data of sharp peaks followed by slow decays. We have definitively correlated most of these peaks with human entry, and circumstantial evidence strongly suggests that all are, in fact, caused by human entry. The entry of even a single person produces a pronounced and abrupt change in humidity. Seventeen such peaks were recorded on the hygrothermograph over a period of 8.5 months. There were barely detectable positive temperature peaks of approximately 0.14 °C, lasting for 1 h on average, that were associated with these entries. None of the temperature peaks was of sufficient absolute magnitude or length to alter the basic isothermal condition of the tomb.

A preliminary value for the rate of water vapor removal in the closed tomb was calculated from these peaks by fitting the first five or six points of the slow decays to a pseudo-first-order rate law. The rate constant was approximately 5×10^{-3} min^{-1}. From these data, it is possible to estimate quantitatively the effects of human entry on the atmospheric water vapor content of the tomb. We have estimated the rate of water vapor production per person, from the initial increases of the entry curves, to be 0.013 g of H_2O/m^3 per person per min at 28 °C. If 10 people entered the tomb and stayed for 30 min, the increase in water vapor would be 4.0 g of H_2O/m^3. This value represents an increase of 14% RH at 28 °C. It would take 12 h for the absolute water vapor content to fall to within 0.1 g of H_2O/m^3 of its initial value if the water were simply removed at the measured rate.

The exact mechanisms for the removal of artificially introduced water vapor involve several physicochemical processes and have not yet been firmly defined. We are currently investigating the water adsorption characteristics of the tomb as a whole, by using rates and rate constants for the data on water removal that we gathered from 1981 to 1982. The existing ventilation rates of the tomb with its door closed are also being calculated.

Chemical Analyses

The effect of adsorbed water vapor on the tomb of Nefertari depends, in part, on the materials used in its construction. Thirteen small, well-documented samples of ancient plasters taken from throughout the tomb were analyzed in detail to determine the composition of the plaster material. Some results are summarized in Table I.

Analysis of Major Elements. Major elements were determined quantitatively by neutron activation analysis at the University of Toronto SLOWPOKE reactor. This technique was chosen because the available sample size was small. Differential scanning calorimetry with a Perkin–Elmer DSC-1B was used to establish the plaster composition further and to investigate its thermal behavior.

The analysis did not reveal the dehydration peaks that are expected if gypsum ($CaSO_4 \cdot 2H_2O$) were present. Analysis with a Dionex QIC Analyzer ion chromatograph revealed the presence of a significant amount of sulfate; therefore the plaster once did, indeed, contain gypsum, which subsequently became dehydrated. This finding confirms suggestions made in previous studies (*27–29*). Ion chromatography of the plaster also revealed a high chloride content, which is likely due to the presence of the sodium chloride mentioned previously.

Comparison with Horemheb's Tomb. The results were compared with those obtained from the analysis of the plaster of the royal tomb of Horemheb (number 57, Valley of the Kings, Thebes, Upper Egypt) (Figure 4). The tomb of Horemheb, approximately 80 years older than that of Nefertari, has been recently restored and is now relatively well preserved (*36*). It was excavated from limestone of excellent quality, and its original plaster is strong and cohesive. A comparison of the chloride and sulfate contents of

Table I. Analysis of Plasters from the Tombs of Nefertari and Horemheb

Sample	SO_4^{2-}	Cl^-	CO_3^{2-}	*Ca*	*Mg*	*Cl*	*Al*	*Na*
Nefertari								
36-II-N	8.4 ± 0.8	0.16 ± 0.08	32 ± 5	26.7 ± 0.8	3.3 ± 0.3	0.18 ± 0.01	0.44 ± 0.02	0.29 ± 0.03
	$n=9$	$n=9$	$n=3$	$n=5$	$n=5$	$n=5$	$n=5$	$n=5$
36-IV-N	30 ± 1	0.44 ± 0.03	NA[a]	23 ± 2	0.7 ± 0.2	0.42 ± 0.06	1.6 ± 0.1	0.58 ± 0.02
	$n=9$	$n=9$		$n=5$	$n=5$	$n=4$	$n=5$	$n=5$
38-III-N3	0.4 ± 0.2	0.121 ± 0.005	NA	30 ± 2	4.2 ± 0.1	0.141 ± 0.008	0.130 ± 0.006	0.20 ± 0.02
	$n=3$	$n=9$		$n=5$	$n=5$	$n=4$	$n=5$	$n=5$
38-IV-N	21 ± 1	0.17 ± 0.06	28 ± 2	25 ± 2	2.5 ± 0.2	0.08 ± 0.02	0.75 ± 0.07	0.27 ± 0.03
	$n=9$	$n=9$	$n=3$	$n=5$	$n=5$	$n=5$	$n=5$	$n=5$
Horemheb								
42-I-H	42 ± 2	0.029 ± 0.004	6.3 ± 0.7	20.7 ± 0.8	0.6 ± 0.1	<0.03	1.37 ± 0.04	0.10 ± 0.02
	$n=3$	$n=3$	$n=3$	$n=5$	$n=5$	$n=5$	$n=5$	$n=5$

NOTE: All values are given as weight percent (±95% confidence limit). Sulfate and chloride were analyzed by ion chromatography; carbonate was analyzed by classical titration; and calcium, magnesium, chlorine, aluminum, and sodium were analyzed by neutron activation analysis.

[a]NA, not analyzed.

the original Horemheb and Nefertari plasters is given in Table II. These values show that the plaster from Horemheb's tomb was of higher quality, with more than twice the sulfate content and one-tenth the chloride content of the plaster of the tomb of Nefertari. Furthermore, analysis by differential scanning calorimetry showed that the plaster of Horemheb's tomb exists primarily in the fully hydrated gypsum form. This result was confirmed by X-ray diffraction analysis (XRD) with a Philips X-ray diffractometer, Cu K_{α} radiation, and a Ni filter. XRD analysis showed that samples from the tomb of Nefertari consist largely of orthorhombic $CaSO_4$ and $CaCO_3$ (calcite), whereas samples from the tomb of Horemheb consist largely of $CaSO_4 \cdot 2H_2O$ and $CaCO_3$ (calcite), with only a small amount (<10%) of orthorhombic $CaSO_4$.

The $CaSO_4$–H_2O System

The preparation of gypsum-based plasters was well known to the ancient Egyptians (37). Gypsum-based plaster is formed by partially dehydrating (calcining) the mineral gypsum ($CaSO_4 \cdot 2H_2O$) to the hemihydrate form ($CaSO_4 \cdot ½H_2O$; plaster of Paris). Water is added to this material, and the resulting mixture is used as a wall preparation. Upon evaporation of excess water, a gypsum-based plaster remains. This plaster is a strong material with good binding properties and, in the pure form, has the same chemical formula and structure as the mineral gypsum. The purity of the final product is defined by the purity of the raw gypsum and the presence of any admixed components, such as $CaCO_3$.

In this chapter the terms "plaster" and "plaster material" refer to the existing wall preparation regardless of its analyzed composition. We assume that the original plaster material in the tomb of Nefertari was predominantly gypsum based, not only because this material would have the desired binding properties, but also because the formation of calcium sulfate dihydrate is kinetically favored under the conditions of formation just described (38).

In the previous section, we confirmed that the Nefertari plaster underwent dehydration, probably over the course of millennia, and it is not likely that this dehydration was the result of human interference. This information is important for two reasons. First, it means that the plaster is mechanically

Table II. Sulfate and Chloride Levels in Plasters from the Tombs of Nefertari and Horemheb

Ion	*Nefertari*[a]	*Horemheb*[b]
SO_4^{2-}	16 ± 3%	42 ± 2%
Cl^-	0.3 ± 0.1%	0.029 ± 0.004%

[a]Average is based on the analysis of 13 samples in triplicate; error is reported as 95% C.I.
[b]Average is based on the analysis of one sample done in triplicate; error is reported as 95% C.I.

unstable, and consequently, its exposure to changes in environmental conditions could be disastrous, and second, it provides evidence that the total dehydration of gypsum, a reaction that is too slow to follow under laboratory conditions of ambient temperature and pressure, actually took place.

The $CaSO_4$–H_2O system has been studied at various temperatures and pressures and in the presence or absence of a liquid phase (*38–41*). Its fundamental characteristics are reasonably well known. Anhydrite deposits have been observed in nature; however, their formation is not well understood (*42*). The formation of orthorhombic anhydrite is kinetically hindered, and we are aware of only one instance of orthorhombic anhydrite, which involved the incorporation of suspected catalysts, being produced at ambient temperatures and pressures (*43*). Therefore, a detailed study of the chemistry of the plaster material in the tomb of Nefertari is interesting from the viewpoint not only of chemists and conservators, but of other scientists as well.

The dehydration process through which gypsum proceeds can be described (*44*) as follows:

$CaSO_4 \cdot 2H_2O \rightarrow$	$CaSO_4 \cdot \frac{1}{2}H_2O \rightarrow$	$CaSO_4$ $\rightarrow$	$CaSO_4$
gypsum	hemihydrate	anhydrite	anhydrite
(monoclinic)	(hexagonal)	(hexagonal)	(orthorhombic)

The first and last forms are the most stable forms (*38*); the two hexagonal forms are metastable. The transformation from monoclinic gypsum to orthorhombic anhydrite involves a decrease in specific volume of approximately 22%, which contributes greatly to the physical weakening of the painting substrate and the fragility of the murals in Nefertari's tomb.

Effect of Climatic Conditions on Plaster Composition. The temperature and humidity of a tomb determine whether the gypsum form or the orthorhombic anhydrite form is most thermodynamically stable. At 28 °C and 25% RH, orthorhombic anhydrite is the stable form in the tomb of Nefertari (*38*). Early measurements of the temperature and humidity in six royal tombs in the Valley of the Kings indicate that the climatic conditions in all tombs inspected also favor the formation of orthorhombic anhydrite (*45*). Although Horemheb's tomb was not one of those studied, it was built at the same depth, in the same rock, and with a plan similar to that of the tomb of Seti I for which measurements of temperature and humidity exist (20 °C and 40% RH at the door and 25 °C and 34% RH at the far end) (*45*). Therefore, on the basis of this evidence alone, we expect that the climatic conditions within Horemheb's tomb also favor the formation of orthorhombic anhydrite from the original plaster. On the basis of our analyses, this reaction did not occur as completely as it did in the tomb of Nefertari (*see* previous section).

Effect of Impurities on $CaSO_4$ Transformation. The transition from gypsum to orthorhombic anhydrite is slow but occurs even at ambient temperatures (*44*). The relatively large concentration of finely divided sodium chloride present in the plaster in the tomb of Nefertari may have facilitated the dehydration process. The presence of a hygroscopic material, such as sodium chloride, can help promote dehydration reactions. Also, impurities within the lattice of a crystalline structure can weaken the lattice (*46, 47*) and thereby accelerate thermodynamically favored reactions. These points suggest a strong correlation between the extent to which the dehydration reaction proceeds and sodium chloride concentration, but they do not exclude the possibility that dehydration can take place in the absence of salt.

Although the orthorhombic $CaSO_4$ in Horemheb's tomb (*see* previous section) may be present as an impurity, it could also indicate that the plaster in Horemheb's tomb is now in its initial stages of deterioration. On the other hand, possibly the climate in the tomb of Horemheb, prior to its discovery, was cooler and more humid before it was discovered. The external climate of the Valley of the Queens may have been hotter and drier than that of the Valley of the Kings. Indeed, the interaction of all these climatic events may have contributed to the preferential dehydration of the tomb of Nefertari. More on-site hydroclimatological data from both the Valley of the Queens and the Valley of the Kings are now needed. The EAO has informed us that some of these measurements are already in progress.

The plaster in the tomb of Nefertari may remain in a thermodynamically stable, desiccated state as long as the conditions of temperature and humidity in which it now exists do not change. We are now studying the effects of changing humidity on the ancient plaster materials through a series of simulation experiments that involve the use of synthetic plasters and controlled humidities. Initial results have shown that sodium chloride plays a key role in the water uptake process in the tomb (*see* Long-Term Studies of Temperature and Humidity).

Reconstruction, Restoration, and Conservation Planning

Since 1904, maintaining the climatic status quo in the tomb of Nefertari has proved difficult, as has been described in previous sections. In addition to the humidification processes occurring in the tomb, well-meaning plaster repairs are a major source of additional water. Threatened murals have been fortified periodically with new plasters, the most recent repairs having been made in 1983. By March 1984, the humidity had increased considerably to an estimated 50 ± 10% RH compared with 31 ± 1% RH in December 1977 (*28*) and 16 ± 1% RH in March 1981.

In the closed tomb, increased humidity alone cannot lead to rehydration of the orthorhombic anhydrite, which is a thermodynamically unfavored reaction. Nevertheless, absorption and adsorption of water by the original plaster especially in the presence of sodium chloride will occur as a result of the increased humidity or the direct application of wet plasters to the walls during repairs. This water intake may lead to localized rehydration of the orthorhombic anhydrite with subsequent swelling in the areas near the application of wet plaster. The extent of this swelling may be significant. It is a complex function of the concentration of water brought into the tomb as wet plaster and the nonuniformity of the sodium chloride concentration. Therefore, there may be insidious gradients in plaster swelling that may be difficult to detect.

The migration of moisture will be accompanied by the migration of water-soluble salts, which may lead to hydration–rehydration cycles over much wider areas. All these processes definitely weaken this already fragile tomb and will lead to its further degradation. Therefore, traditional methods must clearly be reevaluated in the development of a conservation scheme for this site.

Reconstruction and Restoration. Enough information is available to indicate that the careful, far-sighted, scientific conservation of the tomb of Nefertari will be at least as slow and painstaking as the conservation of other great monuments in many other parts of the world. This situation is especially so because of the extreme fragility of the tomb. Very careful work must be done at this site because of the anticipated large influx of visitors. The construction of an exact replica of the tomb of Nefertari near the original site may be an excellent alternative to fast and untested conservation efforts and should seriously be considered. It is both practical and possible to build an exact and subterranean replica near the tomb without altering the character of the Valley of the Queens as a historic site. The careful construction of a replica 100 m east of the entrance to the valley will reduce the impact of vibrations created by any necessary preparatory excavation.

A free-standing replica of the tomb is another alternative. If such a replica is designated as the first stop on tours of the Valley of the Queens, tourist bus traffic into the valley proper could be forestalled; the hazardous vibrations and exhaust produced by these vehicles would be removed. A replica of Nefertari's tomb will allow more visitors to see a faithful reproduction of this unique monument.

The construction of a replica will not require further examination in the tomb. Surveys (*23, 48*) and several sets of photographs (*23, 24, 30*) of the tomb are available, the most recent having appeared in press in 1971 (*23*). The 1971 set of photographs is the only published complete compilation in color. Because the photographing of the tomb in 1971 was concerned with documenting the geometry of the tomb, the color reproduction in these photographs is not sufficiently accurate to serve as a guide to color replication. In 1982, the EAO, in collaboration with the United Nations Educational,

Scientific, and Cultural Organization (UNESCO), rephotographed the tomb, using the most modern techniques. This series of photographs will soon become available.

The techniques used in ancient Egyptian paintings are also known (*27, 36*), and there are talented Egyptian specialists who can recreate them. The importance of this monument justifies such an approach; respected precedents exist, notably at Lascaux (*8*). By taking this path, conservation work on the original tomb can deal with the stabilization of the monument and not with documentation or the accommodation of interested visitors.

Conservation Planning. The unique physicochemical, climatic, and geological conditions of the tomb of Nefertari should provide the basis of any proposed conservation scheme. These studies must be conducted meticulously and will necessarily be time consuming. Predictive studies on every aspect of this site must precede any conservation work. A series of initial recommendations (*31, 34, 35*) was submitted to the EAO in this regard, and some of these recommendations are discussed in the following sections.

MAINTENANCE OF TEMPERATURE AND HUMIDITY. As a first step towards stabilization of the whole tomb, the identification, improvement, and maintenance of its climatically most stable condition should be undertaken. The excessive harmful humidity must be eliminated, and the tomb should be dehumidified to at least its pre-1983 level (25% RH). The temperature and humidity in the closed tomb must be measured continuously.

The construction of an air-conditioning system is the most essential step of the conservation scheme, especially during the implementation of conservation treatments. A system that is rugged, reliable, and unobtrusive can be custom designed. The system must have sufficient capacity to keep the temperature and humidity stable within the tomb regardless of all external conditions and also be able to respond quickly to changes caused by entry (*49, 50*). The air-conditioning system must have safeguard mechanisms against extreme fluctuations caused by humans or natural events. Indeed, such a mechanism may also serve as a means of controlling access to the tomb, which will be based on the ability of the system to accommodate changes in temperature and humidity. A back-up system is also necessary in the event of failure of the primary system.

For most of the year, the external humidity is higher than that in the unventilated tomb (Figure 6). Therefore, the air supply to the tomb must be dehumidified to an acceptable average value, and the temperature must be automatically readjusted to match the optimal internal temperature. A cooler–dehumidifier coupled with a heater is a possible, but not necessarily the best, configuration, which would require 29,000 kJ/h for dehumidification and 22,000 kJ/h for heating. These values are based on two air changes per hour, with air entering at 35 °C and 35% RH and leaving at 28 °C and

25% RH. The volume of the tomb is approximately 450 m^3; two changes of treated air per hour will remove the water vapor produced by 60 people in 1 h and maintain the CO_2 produced by 43 people in 1 h at 0.1% v/v CO_2. These numbers are presented as useful guidelines; they are not meant to imply the recommendation of unrestricted access.

CONTROL OF CO_2 AND DUST. Ventilation of the tomb must be modulated to ensure a constant and appropriate CO_2 partial pressure. Changes in CO_2 concentrations in the air may be crucial to the survival of the paint layer because the underlying $CaCO_3$ whitewash layers are affected by the presence of CO_2. At the Beni Hasan tombs, Middle Egypt, CO_2 is suspected to contribute to deterioration (*17–21*). We are performing a study of this mechanism of paint layer and pigment deterioration in Nefertari's tomb.

Particle size analysis of airborne dust is also necessary in the consideration of appropriate air-cleaning devices for use in a complete air-conditioning system at this site. All components of this air-conditioning system must be installed so that the architectural and artistic character of the tomb is not disrupted or endangered in any way.

The construction of a double-door climatic seal, that is, an entrance foyer with a door between it and the outside and a second door between the tomb and the foyer, should be considered (*35*). This seal will prevent the direct exposure of the tomb to the environment. When the climatic conditions are stable, essential work within the tomb, preceded by careful testing, can be considered.

The first step will be dust removal. The influx of dust and other desert debris is a natural consequence of increased access to the tomb. Furthermore, the tomb floor is a reservoir of dust. We have examined the role of dust in the deterioration of mural surfaces (*17–21*). After the removal of dust, a dust-free, preformed cement, or better still, granite floor should be installed. Parts of the tombs in which paintings are already irretrievably lost must be covered with polished granite slabs to prevent dust generation from these obvious dust sources. Structural consolidation, cleaning of paint layers, and protection can then follow.

STRUCTURAL CONSOLIDATION. Structural consolidation of the plaster layer is necessary for protection against mechanical stresses, including seismic activity and vibrations generated by buses that bring visitors to the Valley of the Queens. The use of injectable polymers is a viable solution. The most stable polymers must be used, and it may be necessary to custom design polymers for this purpose. The integrity of the plaster layer, before, during, and after treatment must be determined and monitored by ultrasonic, piezoelectric, or photoacoustic methods. This procedure and the use of polymers for paint layer protection will require extensive controlled experimentation, especially on the physicochemical nature of the various pigments used in the tomb. Current knowledge of ancient pigment composition and chemistry must be extended to include the interaction of the pigments with the total environment of the tomb of Nefertari. Areas of discoloration are known in this tomb, some of which were caused by previous conservation experiments (*27*).

The polymers used must protect murals from moisture, must be totally inert, and must not alter the appearance of the paint layer or cause structural changes within it. Proper ventilation requirements for polymer use during conservation must also be considered. When new and better techniques become available, it must be possible to remove these polymers and the polymers used in the structural consolidation of the plasters with minimal or no damage to the tomb, in the tradition of reversible conservation treatments.

A revision of the lighting practices in this tomb, for example the use of readily available, sharp-cut-off UV filters, must be undertaken to exclude possible photodegradation of pigments. At present, a system of fluorescent lights illuminates the walls from the base of each mural. This system is adequate now because of its infrequent use. However, the situation will be different if increased visitation is allowed. We are studying all these aspects of the conservation scheme with existing, well-documented samples taken from this tomb and other nearby tombs.

The wholesale removal of salts should not be considered at this time. The salts are essentially a structural component in their own right. The isolation of the tomb from the influx of water and humidity will arrest the movement of these salts.

DRAINAGE. Like many of its neighbors, the tomb of Nefertari is threatened by increasingly frequent torrential downpours. A canopy on top of the site with appropriate drainage should be constructed to divert this source of water (*34*). Although historically, degradation of tomb environments may have been influenced by groundwater flow, this is not the case with the Valley of the Queens or with most of the tombs of the Valley of the Kings. These sites were specifically chosen for tomb construction because of their extremely stable and dry environments.

Although irrigation in the Nile Valley has increased more recently, the Valley of the Queens is still situated approximately 2 km from irrigated fields. Therefore, direct and upward water flow in the limestone rock is not likely to affect the tomb of Nefertari because the water table must be considerably below the lower levels of the tomb. However, the flow of water through the surrounding rock should be evaluated by appropriate geological studies, as well as by computer simulation. A trench system around the tomb may be considered as a means of isolating the walls from any incoming water. Consolidation of the plasters must precede this step, and the construction of these trenches must be done without causing vibrations in the tomb.

REVIEW OF CONSERVATION EFFORTS. Conservation efforts are often conducted by individuals and groups who do not release details of their work even when protection of forthcoming patents is not involved. Such confidentiality is unnecessary and is, indeed, suspect (*51*). This unwarranted secrecy must not accompany the conservation efforts for such a national treasure as the tomb of Nefertari. Therefore, publication of detailed conservation work in respected peer-reviewed journals is mandatory. Not only will this well-accepted scientific practice result in more responsible conservation work, but it will ensure international deliberation, which can lead only to a more rigorous study and evaluation of optimum solutions to conservation problems.

Conclusions

The ongoing macroarchaeometric study of the tomb of Nefertari allows us to examine the unique chemistry occurring in the archaeological materials used in its construction. We hope that this work will also contribute to the art history of the tomb and to a more detailed understanding of the technology used by its builder 3250 years ago. This work provides an example of a general approach to some modern conservation problems. As a corollary, it demonstrates the ability of large archaeological structures to act as sensitive probes of environmental change.

Acknowledgments

This chapter is based on the progress report submitted to the Egyptian Antiquities Organization, Cairo, January 1987. We acknowledge the Egyptian Antiquities Organization for providing permits to work in the tomb of Nefertari.

We are grateful to Hishmat Messiha who first suggested the study of this tomb to us. Our interest in the project was considerably enhanced by discussions with the late Zaki Iskander. We are grateful to Ali el Khouli, Director of Excavations, Egypt, whose absolute and contagious dedication to the cause of the preservation of Egyptian antiquities ensured our continuous interest in this project. The Franco-Egyptian Centre for the Studies of the Temples of Karnak, as part of their collaborative agreement with the University of Toronto, provided us with the climatic data from the Nile Valley. R. G. V. Hancock of the University of Toronto SLOWPOKE reactor provided technical advice and assistance in the neutron activation analyses.

K. M. Wilson-Yang acknowledges the receipt of a grant-in-aid from Sigma Xi. This research was sponsored by the University of Toronto, the Natural Sciences and Engineering Research Council of Canada, and the Canadian Commission of UNESCO/CIDA Assistance Programme.

References

1. Billard, T. C.; Burns, G.; Wilson-Yang, K. M. Presented at the XXIth International Conference on Archaeometry, Brookhaven National Laboratories, Stoney Brook, NY, May 1981.
2. El-Kammar, A.; Hancock, R. G. V.; Allen, R. O. In Chapter 18 of this book.
3. Jakes, K. A.; Angel, A. In Chapter 26 of this book.
4. Sales, K. D.; Oduwole, A. D.; Convert, J.; Robins, G. V. *Archaeometry* **1987,** *29,* 103–109.
5. Sales, K. D.; Robins, G. V.; Oduwole, D. In Chapter 19 of this book.
6. Bada, J. L. *ISR Interdisc. Sci. Rev.* **1982,** *7,* 30–46.
7. Bada, J. L.; Gillespie, R.; Gowlett, J. A. J.; Hedges, R. E. M. *Nature (London)* **1984,** *312,* 442–444.
8. Vouvé, J.; Brunet, J.; Vidal, P.; Marsal, J. *Stud. Conserv.* **1983,** *28,* 107–116.
9. Quindos, L. S.; Bonet, A.; Diaz-Caneja, N.; Fernandez, I.; Gutierrez, J. R. *Atmos. Environ.* **1987,** *21,* 551–560.
10. Billard, T. C.; Burns, G. "On Deterioration of Egyptian Antiquities, Causes and Remedies"; progress report to the Egyptian Antiquities Organization; University of Toronto: Toronto, Canada, 1979.
11. Billard, T. C.; Burns, G. *Nature (London)* **1980,** *285,* 654–655.
12. Billard, T. C.; Burns, G.; Wilson-Yang, K. M. "Salinization of the Nile Valley: The Karnak Area"; report to the Egyptian Antiquities Organization; University of Toronto: Toronto, Canada, 1981.
13. Billard, T. C.; Burns, G.; Wilson-Yang, K. M. *J. Am. Res. Cent. Egypt* **1982,** *19,* 115–117.
14. Burns, G.; Wilson-Yang, K. M. "Degradation in Egyptian Black Granite"; report to the Egyptian Antiquities Organization; University of Toronto: Toronto, Canada, 1982.
15. McFarlane, J.; Wilson-Yang, K. M.; Joseph, S.; Frey, C.; Burns, G. *Can. J. Chem.* **1983,** *61,* 718–723.
16. Wilson-Yang, K. M.; McFarlane, J.; Burns, G. *J. Soc. Stud. Egypt. Antiq. Assoc.* **1985,** *15,* 52–54.
17. Burns, G.; Wilson-Yang, K. M. "The Great Tombs at Beni Hasan and Their Conservation"; preliminary report to the Egyptian Antiquities Organization; University of Toronto: Toronto, Canada, 1981.
18. Wilson-Yang, K. M.; Burns, G. *J. Am. Res. Cent. Egypt* **1982,** *19,* 119–121.
19. Burns, G.; Wilson-Yang, K. M. "Chemistry of the Deteriorating Murals in the Beni Hasan Tombs"; progress report to the Egyptian Antiquities Organization; University of Toronto: Toronto, Canada, 1984.
20. Burns, G.; Wilson-Yang, K. M. "The X-ray Photoelectron Spectroscopy of the Ancient Murals in the Tombs at Beni Hasan, Egypt"; progress report to the Egyptian Antiquities Organization; University of Toronto: Toronto, Canada, 1987.
21. Wilson-Yang, K. M.; Burns, G. *Can. J. Chem.* **1987,** *65,* 1058–1064.
22. Smith, S. E. *J. Field Archaeol.* **1986,** *13,* 503–510.
23. Goedicke, H.; Thausing, G. *Nofretari*; Akademische Druck u. Verlagsanstalt: Graz, Austria, 1971.
24. Schiaparelli E. *Relazione sui Lavori della Missione Archaeologica Italiana in Egitto (1903–1922)*; 1922; Vol. 1, pp 51–104.
25. Baikie, J. *Egyptian Antiquities in the Nile Valley*; Methuen: London, 1932; pp 517–521.
26. Capart, J. *Thèbes*; Vromant et Cie: Brussels, 1925; p 153.

27. Iskander, Z. In *Recent Advances in Science and Technology of Materials*; Bishay, A., Ed.; Plenum: New York, 1974; Vol. 3; p 1.
28. Wilson-Yang, K. M.; Billard, T. C.; Burns, G. *J. Soc. Stud. Egypt. Antiq.* **1982,** *12,* 9–11.
29. Plenderlith, H. J.; Mora, P.; Torraca, G.; de Guichen, G. "Conservation Problems in Egypt"; UNESCO consultant report, Contract 33.591, 1970.
30. Stoppelere, A. *Ann. Serv. Antiq. Egypt* **1942,** *40,* 941–950.
31. Burns, G.; Wilson-Yang, K. M. "The Tomb of Nefertari, Valley of the Queens and its Conservation Problems"; preliminary report to the Egyptian Antiquities Organization; University of Toronto: Toronto, Canada, 1981.
32. Curtis, G.; Rutherford, *J. Soil Mech. Found. Div. Am. Soc. Civ. Eng.* **1981,** *3,* 71–74.
33. Strabo; cited in Heizer, R. F.; Stross, F.; Hester, T. R.; Albee, A.; Perlman, I.; Asaro, F.; Bowman, H. *Science* **1973,** *182,* 1219–1224.
34. Burns, G.; Wilson-Yang, K. M. "The Internal Climate of the Tomb of Nefertari"; progress report to the Egyptian Antiquities Organization: University of Toronto: Toronto, Canada, 1982.
35. Burns, G.; Wilson-Yang, K. M. "The Continuous Measurement of Temperature and Humidity in the Tomb of Nefertari"; progress report to the Egyptian Antiquities Organization; University of Toronto: Toronto, Canada, 1984.
36. Hornung, E. *Das Grab des Haremhab im Tal der Königen*; Francke Verlag: Berne, 1971.
37. Lucas, A.; Harris, J. R. *Ancient Egyptian Materials and Industries,* 4th ed.; Harris and Arnold: London, 1962; p 78.
38. Blount, C. W.; Dickson, F. W. *Am. Mineral.* **1973,** *58,* 323–331.
39. Posnjak, E. *Am. J. Sci.* **1938,** *35A,* 247–272.
40. Zen, E.-A. *J. Petrol.* **1965,** *6,* 124–164.
41. Hardie, L. A. *Am. Mineral.* **1967,** *52,* 171–200.
42. Cody, R. D.; Hull, A. B. *Geology* **1980,** *8,* 505–509.
43. Kinsman, D. J. J. N. *Ohio Geol. Soc 2nd Proc. Symp. on Salt* **1966,** *1,* 302–326.
44. Ridge, M. J.; Beretka, J. *Rev. Pure, Appl. Chem.* **1969,** *19,* 17–44.
45. Lucas, A. *Ann. Serv. Antiq. Egypt* **1924,** *24,* 12–14.
46. Kittel, C. *Introduction to Solid State Physics,* 5th ed.; Wiley: Toronto, 1976.
47. Hannay, N. B. *Solid State Chemistry*; Prentice–Hall: Toronto, 1967; Chapter 8.
48. Weeks, K. *Newslett. Am. Res. Cent. Egypt* **1981,** *Winter.*
49. Croome-Gale, D. J.; Roberts, B. M. *Airconditioning and Ventilation in Buildings*; Pergamon: Toronto, 1975.
50. McQuiston, F. C. *Heating, Ventilation and Air-conditioning,* 2nd ed.; Wiley: Toronto, 1982.
51. Dutton, D. *Nature (London)* **1987,** 327, 10.

RECEIVED for review June 11, 1987. ACCEPTED revised manuscript February 10, 1988.

ORGANIC MATERIALS

16

Carbon Dating the Shroud of Turin

A Test of Recent Improvements in the Technique

Garman Harbottle[1] and Walden Heino[2]

[1]Chemistry Department, Brookhaven National Laboratory, Upton, NY 11973
[2]Chemistry Department, Luther College, Decorah, IA 52101

Recent improvements in reducing the sample size needed for carbon dating have opened up new possibilities for the archaeologist and museum curator. In particular, it has become feasible to carbon date the Shroud of Turin, and the results were announced recently.

Radiocarbon Measurements

Since its introduction approximately 40 years ago, radiocarbon dating has assumed extraordinary importance in archaeology because of the widespread occurrence of carbon in archaeological contexts. ^{14}C measurement techniques have been refined to the point where dates can now be determined for milligram-scale samples and with materials up to 80,000 years old.

Detailed expositions of radiocarbon techniques and applications are found elsewhere (*1–8*), but for completeness, a brief introduction is given in this chapter. The purpose of this introduction is to summarize the current status of the analytical methods and to show how advances in radiocarbon dating have made it possible to carry out a program for measuring the age of the Shroud of Turin.

The Shroud of Turin, perhaps the best-known and most deeply revered of all religious relics, is a linen textile 4.5 m long by 1.1 m wide and is preserved in a special chapel within the Cathedral of Turin in Italy. It carries the image of the back and front of a naked man, whose body is marked by reddish splashes resembling bloodstains, that suggest by their location that the man portrayed was the victim of crucifixion. The idea that the man represented might be Christ is reinforced by marks on the back that strongly suggest that the body was scourged by a flagellum of Roman design.

0065–2393/89/0220–0313$06.00/0

The face is that of a bearded man and is rather unimpressive when seen on the Shroud cloth itself. It has one astonishing, inexplicable property, however, that was first noted in 1898 when the Italian photographer Secondo Pia took the very first photographs of the Shroud. The image on the linen cloth is apparently a negative. When it is reversed, by making a photographic negative, what emerges is "a hauntingly majestic countenance, with eyes closed in death" (9). If it is a painting, it does not closely resemble the work of any known artist.

The Shroud has been known since the middle 1300s, and as early as 1389 the Bishop of Troyes, in France, in a letter to Pope Clement VII, denounced it as a forgery, by saying that in a local church at Lirey there was "a certain cloth cunningly painted, upon which by clever sleight of hand was depicted the twofold image of one man, that is to say the back and front, they falsely declaring and pretending that this was the actual Shroud in which our Savior Christ was enfolded in the tomb." Despite this denunciation, the Shroud, in the hands of the House of Savoy, who became the rulers of Italy, was more and more fervently venerated through the centuries, and today stands at the very pinnacle of the whole group of holy relics (crowns of thorns, pieces of the true cross, etc.) associated with the life and death of the Saviour.

A number of scientific studies of the Shroud (examination of the textile; microscopy; photography; investigations of adherent pollen, dust and "blood" fragments, etc.) were made by members of the Shroud of Turin Research Project (STURP) in 1978 (*10*). None of the results obtained by STURP have, however, had much bearing on the question of authenticity. As Ian Wilson (*The Mysterious Shroud*, Doubleday, p. 129) says, ". . . without doubt, the Shroud test, all too long awaited, is that of independently dating the cloth's linen by the method known as carbon-14". Although ^{14}C techniques have been available for 40 years, the devices used to carry out the measurement of ^{14}C, called proportional counters, demanded grams of carbon, and called for a proportionate quantity of the Shroud to be destroyed. (Actually, a piece roughly the size of a pocket handkerchief would have to be burned for each measurement, and no scientific study would have been considered adequate with only a single measurement.) The dating of the Shroud of Turin would seem, then, to be a good example of what could be done with improved and miniaturized radiocarbon techniques.

Calibration of Radiocarbon Measurement To Yield the Calendar Age. It was previously thought that if the half-life of ^{14}C was accurately known, then a simple exponential decay equation could be applied to learn the age of archaeological organic materials. The measurement would be subject only to errors in measurement of the ratio of the radiocarbon in the sample, A, to the equivalent quantity in modern carbon, A_0. However, it is now known that A_0 has been far from constant in the past few thousand years. Also, the half-life of carbon-14 is not known to great accuracy. Finally, isotopic fractionation, which can vary the ^{14}C enrichment among samples of carbon with identical ages, must be taken into account.

All these problems have led to a procedure among radiocarbon scientists that is quite different from that envisioned by Libby, who invented ^{14}C dating, and described by a simple decay equation. The procedure is:

1. The ratio A/A_0 is measured either by β^- counting or with the tandem accelerator mass spectrometer (TAMS) in which direct atom counting is used.
2. The observed ratio is corrected for isotopic fractionation by measuring the $^{13}C/^{12}C$ ratio, and assuming that its change (from standard value) is one-half the $^{14}C/^{12}C$ change. In the TAMS, the $^{13}C/^{12}C$ ratio is measured simultaneously with $^{14}C/^{12}C$.
3. A conventional "Libby" half-life of 5568 years (known to be incorrect) is used to obtain a Libby or "conventional" radiocarbon age in years, *t*.
4. This conventional age *t* is converted to a calendar date by calibration curves that were obtained by very precise measurement of the A/A_0 values of tree-ring wood of exactly known calendar age (*11–13*).

The calibration curve is rather precisely measured back to several millennia B.C. The error in conventional radiocarbon age can, in principle, be as small as ±20–30 years. The calibration curve, however, does not fall monotonically with time. At times, two or three dates could give the same conventional radiocarbon age. For this reason, radiocarbon dating is particularly unworkable for the period from the mid 1600s (A.D.) to the present.

Fractionation and Contamination. The ratio $^{14}C/^{12}C$ in certain materials may be affected by isotopic fractionation. For example, the uptake of carbon dioxide and its incorporation into plant tissue may be accompanied by substantial fractionation that depends on the plant species. With marine organisms, fractionation may also be important, especially when inorganic carbonate and bicarbonate are involved. Corrections for fractionation must be made for precise radiocarbon dating.

Errors in the $^{14}C/^{12}C$ ratio can also be produced by the injection of environmental materials into the archaeological sample. Intrusion of modern carbon, in the form of rootlets, molds, and bacteria will increase the ratio A/A_0 and yield ages that are too recent, whereas percolation of groundwater containing ancient "dead" carbonate may give ages that are too old. In bone, it is advisable to date the separated collagen fraction. To a large extent, these problems, when recognized, can be overcome by sample pretreatment. Fractionation and contamination are discussed by Parkes (*3*) and by Taylor (*14*).

Measurement of Carbon-14. Until about 1985, radiocarbon concentrations (*A*) were routinely determined by measuring the radioactivity of ^{14}C. For very old or very small samples, long counting times were necessary in order to obtain *A* with reasonable precision. Special counting systems have been constructed to accommodate these very long counting times (*20,21*). The small-sized "minicounters" used for dating 10-mg samples still require long counts, but errors caused by background radiation are reduced by the careful choice of construction materials, pulse height analysis, and active cosmic ray shielding.

New mass spectrometric techniques have also been developed that permit the radiocarbon concentration ($^{14}C/^{12}C$ ratio) to be measured directly, in a relatively short time, and for very small samples. Sensitivity is better than for radioactive decay counting. With this approach, age determinations can be carried out with 1-mg samples or with materials that are as much as 80,000 years old, although isotopic preenrichment will be necessary in such cases. Beyond this age, operational interferences such as contamination by carbon in reagents, or by the instrument itself, may be limiting. Instruments of somewhat different configurations have been used, for high-energy mass spectrometry (HEMS), accelerator mass spectrometry (AMS) (*15,16*), or tandem accelerator mass spectrometry (TAMS) (*17*). In each case, there have been significant improvements over older mass spectrometers to differentiate ^{14}C from the much more abundant ^{14}N. Accelerator dating is reviewed in reports by Taylor (*18*) and Taylor et al. (*19*).

Sampling. Concentrations of carbon in archaeological materials range from very large in charcoal to less than 1% in metals, foundry slags, and pottery. The size of the sample needed for analysis, thus, depends on the nature of the material as well as its age. The analytical procedure used to isolate the carbon may result in significant losses during extraction and chemical conversion. Samples should always be taken in sufficient quantity for replicate determinations and comparison with control specimens.

Archaeologists should be sure the samples they take are representative and homogeneous with respect to the questions being asked. Contextual state is important in this regard, particularly when disturbance and intrusion are evident. Environmental factors such as soil composition, water content, and pH may also be relevant to the dating procedure. Sampling requirements and procedures for decay counting have been reviewed by Ralph (*5*) and for HEMS by Hedges (*15*).

Expression and Interpretation of Results. Archaeological interpretation of a radiocarbon age may depend critically on the error associated with that age. Errors are commonly expressed as a variance range attached to the central number (e.g., 2250 ± 80 years). The ±80 years in this example may correspond to the random error for a single analytical step. Both decay and direct-atom counting are statistical in nature, and lead to errors that vary as the square root of the number of counts. The error may also be expressed as the overall random experimental error (the sum of individual errors.) Overall random error can be determined only by analyzing replicate samples.

If a given error is random, a statistical significance can be attached to it. If the number ±80 represents one standard deviation (1σ), the measured age shows a 68% probability of lying in the range 2170 (2250 – 80) to 2330 (2250 + 80) years. Within-range probability is increased to 95% for a 2σ range, and to 99% for a 3σ range. In the literature, radiocarbon ages may carry any of these errors even though standard conventions have been suggested.

A third type of error is the systematic error that cannot be defined in statistical terms. Systematic error is treated by analyzing reference samples to establish corrections. The calibration procedure can produce calendar dates that have large error ranges even when the conventional ^{14}C age is known with very high precision (*13*).

Dating the Shroud of Turin

The ability to analyze small samples has allowed archaeologists to begin measuring ages in specimens that are of great interest, but where the amount of carbon available is very small. This is the case in studies of valuable museum objects. Also, in many archaeological investigations only tiny samples of carbon are available. (*23*). A case in point is the dating of single seeds that relate to the earliest agriculture.

In 1973 the techniques available for ^{14}C dating would have required a handkerchief-sized sample. The analytical procedures available now will allow the Shroud of Turin to be dated by using postage stamp-sized pieces of the sacred relic. The ability to determine dates by using such small samples has brought the project from practical impossibility to the brink of completion.

Early in 1979, Gove of the University of Rochester, one of the physicists involved in the development of the accelerator method, and Harbottle submitted to the Archbishop of Turin a joint proposal to carbon date the Shroud by both miniature techniques, by using small samples that had already been cut off the Shroud for a textile study in 1973. Although this proposal was not implemented, it did succeed in arousing the interest of the ecclesiastical authorities. Since that time, the most extraordinary sequence of events has transpired, comprising politics, both scientific and ecclesiastical; scientific meetings; papers; discussion and experimentation; polemics; press leaks; telegrams; and internecine maneuvering on a scale usually associated with research leading to competition for a Nobel Prize. The most significant event during this time was the first round robin intercomparison of minicarbon techniques, proposed by Otlet at Bradford in 1982, implemented by the British Museum, and reported at the Trondheim radiocarbon conference in 1985 (*24*). Six laboratories took part in the round robin.

In 1979, Umberto II, the ex-King of Italy, was the head of the House of Savoy, and, as such was the owner of the Shroud. Umberto favored the dating experiment; when he died in 1983, the Shroud was bequeathed to the Church. By this act, the Pope became a participant in all subsequent proposals for examination or dating. In late September 1986, a meeting was called in Turin by the Pontifical Academy of Sciences and the Archbishop of Turin and presided over by the President of the Pontifical Academy, Professor Carlos Chagas. Present were representatives of five "accelerator" (University of Rochester, University of Arizona, Oxford University, Saclay [France], and the Technische Hochschule [Zurich]) and two "counter" (Brookhaven and Harwell) laboratories (*24*), the British Museum, STURP, a distinguished textile specialist from Switzerland, the Science advisors of the Pope and the Archbishop of Turin, other scientists, archaeologists, conservators, and several Catholic clergymen. They were all cloistered for 3 days, and despite scientific differences that occasionally became intense, reached a consensus on a statistically defensible protocol for carbon dating the Shroud of Turin. This protocol, which was submitted to His Eminence Anastasio Cardinal Ballestrero of Turin and His Holiness Pope John Paul II, was based on the following general principles:

1. Samples were to be taken by an expert in textile conservation.
2. Representatives of the Pontiff, the Archbishop of Turin, and the British Museum would be present.
3. Six samples (called S) would be taken from the Shroud.
4. Samples would be provided from two dummy populations, A and B.
5. Sets of three samples each would be distributed to each of seven laboratories for ^{14}C measurement. They were to be coded by number.
6. No laboratory was to know what it was analyzing. The laboratory could have samples SAB, SAA, AAB, BBB, SSB, etc. The laboratory would report the $^{14}C/^{12}C$ and isotopic ratios that it measured for all three samples according to a predetermined format.

7. Reports were to go to the British Museum.
8. No announcements were to be made until all results were collected and analyzed by the British Museum.
9. Any publication would carry the names of all the participants, not merely those who analyzed S samples. In fact, there was to be no announcement of which laboratory handled which samples.

The provision of six samples to seven laboratories might seem to be excessive, but in several preliminary round-robin intercomparison measurements, outliers (results differing from the mean by many standard deviations) had turned up, and could only be discarded with certainty because there were enough other results with which to compare (*24*). If only two or three laboratories had produced those results, it would have been difficult or impossible to identify the outlying data. The whole basis of the protocol was the expressed desire of the religious authorities to produce the best possible scientific result and the effort of a number of ^{14}C scientists to meet that desire.

Despite this expressed wish and the protocol that had been designed to yield the desired results, on October 10, 1987, a letter from the Archbishop of Turin to all participants in the conference announced a very different plan (*25*). There would be only three laboratories (Oxford, Tucson and Zurich), and all would use TAMS. No provision was made for blind measurements, although control samples were to be distributed. Each laboratory knew with certainty which of its samples was the Shroud sample (*26*).

As of October 21, 1988, the ^{14}C measurements have been made, and the results sent in to the British Museum Research Laboratory. It has been announced that the date obtained falls within the 14th century. The agreement of dates from the three laboratories is an impressive demonstration of the new ^{14}C dating technology.

Acknowledgments

This research was carried out at Brookhaven National Laboratory under contract DE-AC02-76CH00016 with the U.S. Department of Energy.

References

1. Burleigh, R. *J. Archaeo.Sci.* **1974,** *1*(*1*), 69.
2. Ralph, E. K.; Michael, H. N. *Archaeometry* **1967,** *10,* 3.
3. Parkes, P. A. *Current Scientific Techniques in Archaeology*; Martin's: New York, 1986.
4. Fleming, S. J. *Dating in Archaeology: a Guide to Scientific Techniques*; Dent: London, 1978.

5. *Dating Techniques for the Archaeologist*; Michael, H. N.; Ralph, E. K., Eds.; MIT: Cambridge, MA, 1971.
6. Goffer, Z. *Archaeological Chemistry*; Wiley: New York, 1980.
7. *Archaeological Chemistry—III*; Lambert, J. B., Ed.; Advances in Chemistry Series 205; American Chemical Society: Washington, DC, 1984.
8. *Archaeological Chemistry II*; Carter, G. F., Ed.; Advances in Chemistry Series 171; American Chemical Society: Washington, DC, 1978.
9. Wilson, I. *The Mysterious Shroud*; Doubleday: Garden City, NY, 1986.
10. Heller, J. H. *Report on the Shroud of Turin*; Houghton Mifflin: Boston, MA, 1983.
11. *Archaeology, Dendrochronology, and the Radiocarbon Calibration Curve*; Ottaway, B. S., Ed.; University of Edinburgh: Edinburgh, 1983.
12. Pearson, G. W. *Antiquity*, **1987** *61*, 98.
13. Stuiver, M.; Pearson, G. W. *Radiocarbon*, **1986** *28(2B)*, 805.
14. Taylor, R. E.; cited in reference 8, p 33.
15. Hedges, R. E. M. *Archaeometry* **1981** *23*, 3.
16. Hedges, R. E. M.; Goulett, J. A. J. *Sci. Am.* **1986,** *1*, 100.
17. Bennet, C. L.; Beukens, R. P.; Clover, M. R.; Gove, H. E.; Liebert, R. B.; Litherland, A. E.; Purser, K. H.; Sandheim, W. E. *Science*, **1977,** *198*, 508.
18. Taylor, R. E. *Anal. Chem.* **1987** *59*, 317A.
19. Taylor, R. E.; Donahue, D. J.; Zabel, T. H.; Damon, P. E.; Jull, A. J. T; cited in reference 7, p 333.
20. Otlet, R. L.; Huxtable, G; Evans, G. V.; Humphreys, G. D.; Short, T. D.; Couchie, S. J. *Radiocarbon* **1983,** *25*, 565.
21. Harbottle, G.; Sayre, E. V.; Stoenner, R. W.; *Science* **1979,** *206*, 683.
22. Ralph, E. K.; cited in reference 5, p 1.
23. Hedges, R. E. M.; Housely, R. A.; Law, I. A.; Perry, C.; Gowlett, J. A. J. *Archaeometry* **1987,** *29*, 289. See also previous date lists in this series.
24. Burleigh, R.; Leese, M.; Tite, M. *Radiocarbon* **1986,** *28(2A)*, 571.
25. Letter of Anastasio Cardinal Ballestrero, Archbishop of Turin, to all participants in the conference, dated October 10, 1987.
26. Hall, E. T. private communication.

RECEIVED for review September 15, 1987. ACCEPTED revised manuscript August 18, 1988.

17

Radiocalcium Dating: Potential Applications in Archaeology and Paleoanthropology

R. E. Taylor[1], Peter J. Slota, Jr.[1], Walter Henning[2], Walter Kutschera[3], and Michael Paul[4]

[1]Radiocarbon Laboratory, Department of Anthropology, Institute of Geophysics and Planetary Physics, University of California, Riverside, Riverside, CA 92521
[2]Gesellschaft für Schwerionenforschung, D–6100, Darmstadt 11, Federal Republic of Germany
[3]Physics Division, Argonne National Laboratory, Argonne, IL 60439
[4]Racah Institute of Physics, Hebrew University of Jerusalem, Jerusalem, Israel

The usefulness and feasibility of using a long-lived ($t_{1/2}$ = *about 100,000 years) isotope of calcium* (^{41}Ca) *as a means of dating bone and other calcium-containing samples up to ca. 1 million years old are being evaluated. The* $^{41}Ca/^{40}Ca$ *ratio in natural terrestrial materials has been measured for the first time by using accelerator mass spectrometry. The* ^{41}Ca *method could potentially provide an independent temporal scale for the period centering on the Middle Pleistocene that is comparable to that provided by K/Ar values for the early portion of the Pleistocene and* ^{14}C *values for the terminal Pleistocene and Holocene.*

A VARIETY OF GEOCHRONOLOGICAL METHODS have been used over the last 3 decades to provide the temporal frameworks that have permitted archaeologists and paleoanthropologists to reconstruct temporal relationships among fossil hominid forms and to investigate rates of change in the evolution of hominid behavior (*1*). The reconstruction of basic chronological relationships among the late Miocene, Pliocene, and Early Pleistocene Hominidae has been accomplished by the analysis of K/Ar age estimates and inferences from the oxygen isotope and paleomagnetic record combined with litho-

0065–2393/89/0220–0321$06.00/0

stratigraphic and biostratigraphic studies. Chronological frameworks are reasonably complete for the last 30,000 to 50,000 years because objects from this time span can be dated by the ^{14}C method.

Although many problems remain, there are reasonably specific chronological frameworks for hominid evolution at both ends of the Quaternary period. By contrast, the dating frameworks for the period centered on the Middle Pleistocene are generally imprecise and ambiguous. This ambiguity was highlighted more than a decade ago when Glynn Isaac emphasized the critical need for chronological resolution for the time period beyond the conventional ^{14}C range and before the period for which the K/Ar method can be routinely used. There is very little chronometric data for this period, which lasted from about 1 million to 60,000 B.P., and much of what exists is of questionable validity (*2*).

When accelerator mass spectrometry (AMS) was introduced for direct ^{14}C counting, it was initially anticipated that the dating range of the ^{14}C method would expand rapidly and extend back as far as 100,000 years (*3*, *4*). Although it is possible that this limit may eventually be reached, current experimental conditions reduce this maximum to between 40,000 and 60,000 years. For conventional decay counting, limitations are imposed by the sample sizes generally available from archaeological contexts and the problems of removing contamination from sample preparations. These limitations reduce the maximum ages that can be obtained practically to between 40,000 and 50,000 years. Under special circumstances, with samples that are larger than are usually available from the typical prehistoric archaeological sites, the maximum range can be extended to about 60,000 years. With isotopic enrichment, again using relatively large samples, ages up to 75,000 years have been reported for a few samples (*5–7*).

Over the next decade, if the stringent requirements for the exclusion of modern carbon contamination in sample preparations can be met, developments in AMS may permit routine extension of the ^{14}C time frame beyond the current 40,000- to 60,000-year range for the typical archaeological sample (*8*). However, the potential of AMS to extend the ^{14}C time frame into the 70,000- to 100,000-year range for samples from archaeological contexts may be possible only with the development of practical methods of ^{14}C isotopic enrichment that can be routinely used for samples that contain less than 1 g of carbon. The possibility of using a laser-based approach for ^{14}C enrichment before AMS analysis is being studied (*9*, *10*).

Significant deficiencies are present in the physical dating methods used to infer temporal relationships for the paleoanthropological and archaeological record for the whole of the Middle Pleistocene (about 730,000 to 125,000 B.P.) (*11*), and for the interval of time up to the initiation of the ^{14}C time scale. There are no recognized major reversals of the earth's geomagnetic field after 730,000 years ago. Many important sites are not in areas of volcanism, and even for those that are, there are often serious analytical problems in working with samples from the more recent end of the K/Ar time scale. Uranium series dating can be used for sites with calcareous sediments, but contamination problems are sometimes acute, and the geochemical history of the typical terrestrial sample is often unclear (*12*).

Attempts to apply thermoluminescence (TL) and electron spin resonance (ESR) data to date soil horizons, and to use ESR to date bone are underway. As yet, there are no agreed-upon criteria on which to evaluate the overall reliability of age inferences based on ESR data (*13*). It has been suggested (*14*) that in some circumstances the obsidian hydration method can be used to infer chronological age over in excess of 100,000 years (*14*). However, obsidian is not a widespread natural resource and some of the hydration rate structures appear to yield problematic results. Amino acid racemization (AAR) values can be used under some conditions to infer accurate age values for bone, but seriously anomalous values can be obtained (*15, 16*). The conditions under which AAR values can be used to accurately infer age, particularly for bone samples, continue to be investigated.

Some Middle Pleistocene sites have been dated by assigning various occupation levels to one of the traditional major glacial–interglacial cycles. This approach to dating is still used even though oxygen isotope analysis of deep sea cores a decade ago (*17*) demonstrated that the climatic record is extremely complex and difficult to resolve for many intervals. Many of the sites in Eurasia are placed into the glacial–interglacial scheme on the basis of faunal correlations, loess stratigraphic cycles, and soil formation cycles.

Typically, sites containing a "warm" fauna are placed into an interglacial phase and sites with a "cold" fauna are placed into a glacial phase. The percentage of extent fauna and various taxa are used along with the stone tool technology to determine to which stadial phase or interstadial phase the site is to be assigned. The accuracy of the understanding of faunal successions, culture change, and environmental variability has not been tested by such methods (*18*). An isotopic dating technique that can directly assign age to bone samples dating to the Middle Pleistocene would be of major significance in the critical study of the processes involved in the biocultural evolution of humans.

Basis of Radiocalcium Dating Method

Although the idea of using ^{41}Ca for dating had been suggested earlier (*19*), Raisbeck and Yiou (*20*) provided the first detailed outline of a dating model for the radiocalcium method. Figure 1 outlines the method and is based on their discussion. Because it has a half-life of about 100,000 years, ^{41}Ca could potentially be used to infer the ages of calcium-containing samples (e.g., bone and $CaCO_3$ contained in soils) over about the last million years. Like ^{14}C, ^{41}Ca is produced by cosmic ray neutron secondaries; however, the bulk of ^{41}Ca is not made in the atmosphere like ^{14}C. ^{41}Ca is produced primarily

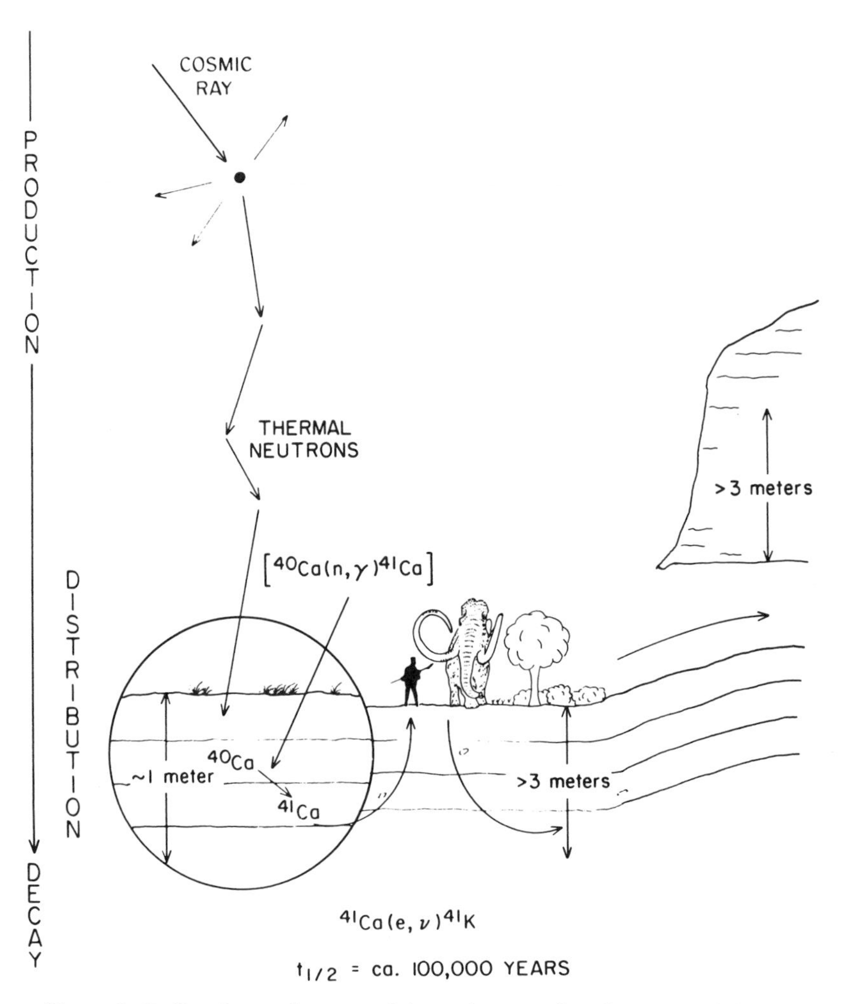

Figure 1. Radiocalcium dating model: production, distribution, and decay of ^{41}Ca. After Raisbeck and Yiou (20) *and Taylor* (21).

in the upper meter of the soil profile by neutron capture on ^{40}Ca. The cosmogenic ^{41}Ca is mixed with the other naturally occurring calcium isotopes into the surface soils through ground water action. Calcium is taken up into the plant tissue in the form of Ca^{2+} through ion absorption by the root system. Radiocalcium is then incorporated into bone through ingestion of plants (*21*).

A $^{41}Ca/^{40}Ca$ equilibrium ratio is maintained in living organisms by exchange and metabolic processes. In contrast to ^{14}C dating, where the death of an animal or plant and the isolation of a sample from one of the carbon reservoirs constitutes the $t = 0$ event (0 B.P.), the $t = 0$ event in the ^{41}Ca

method occurs when a sample is shielded from the effect of the cosmic-ray-produced neutron irradiation (e.g., by the placement of a sample deeper than about 3 m of soil or rock overburden). This shielding could be accomplished either by burial or placement in a cave–rock shelter environment. Age inferences are made on the basis of the measurement of the residual ^{41}Ca with respect to the stable isotopes of Ca. ^{41}Ca decays to ^{41}K by electron capture and the emission of a neutrino. To directly use changes in $^{41}Ca/Ca$ ratios to accurately infer an age for calcium-containing samples, certain fundamental assumptions must be made. These assumptions include

1. that the initial concentration of cosmogenic ^{41}Ca in samples has remained essentially constant over the projected ^{41}Ca time scale (or that appropriate corrections can be made for documented variations)
2. that the $^{41}Ca/Ca$ ratio in a sample has not been altered except by ^{41}Ca decay since the sample was shielded from the effects of neutron irradiation (i.e., no postdepositional exchange–contamination of the in situ ^{41}Ca has occurred or build up of new ^{41}Ca by neutrons from natural radioactivity [uranium and thorium])
3. that mixing of cosmogenic ^{41}Ca in the calcium source for a sample has occurred over a relatively short period of time in comparison to the half-life of ^{41}Ca
4. that the half-life of ^{41}Ca is accurately known
5. that natural levels of ^{41}Ca can be measured within reasonable levels of uncertainty.

From a technological point of view, the ^{41}Ca method currently stands at the same point in its development as the ^{14}C method did in about 1946–47. The favored mode of production for ^{14}C had been known for some time (thermal neutrons on ^{14}N) and, on the basis of this and several other considerations, Libby formulated a dating model. However, the half-life was still somewhat uncertain and the initial measurements of natural ^{14}C concentrations were just being made. Routine low-level counting was several years away. There were no experimental data to support several fundamental assumptions on which the practical use of ^{14}C would depend. Libby himself would later say (22) that he was initially concerned that his "notion" of ^{14}C dating was "beyond reasonable credence".

For all its comparisons to ^{14}C, the ^{41}Ca method has potentially serious deficits that could easily put quite rigid constraints on the types of depositional environments that could be expected to give straightforward results. For example, the fact that lithospheric rather than atmospheric production predominates raises the strong possibility that localized mixing and erosion may have caused significant variations in initial $^{41}Ca/Ca$ ratios in many environments. In addition, because samples that have not been buried deeply enough will continue to be subjected to ^{41}Ca formation from the cosmic ray-generated neutron secondaries, the burial histories of samples may affect ^{41}Ca concentrations. Also, it is reasonable to expect post-depositional exchange of calcium in some samples through ground water contact. These and other diagenetic factors suggest that samples from each site or locality might exhibit unique initial ^{41}Ca concentrations. An advantage of the projected ^{41}Ca method is that studies can be carried out on the samples themselves and do not necessarily depend on access to the actual burial environment, as is the case with TL, ESR, AAR, and obsidian hydration applications that require temperature and local radiation dose levels to be measured to rather close tolerances to achieve reasonable accuracy.

Major Issues in the Development of the ^{41}Ca Method. To develop a practical method of using ^{41}Ca concentrations in calcium-containing samples to infer age and to determine the degree of general applicability, a series of studies must be undertaken that, in broad outline, would parallel the initial set of experiments that established the general utility and accuracy of the ^{14}C method. All of the experiments that would be needed to demonstrate the usefulness of the proposed ^{41}Ca method would require a practical and effective means of routinely measuring natural ^{41}Ca values in terrestrial samples. A practical measurement technology and sample preparation methodology is being developed against the backdrop of the following considerations and experimental data.

Natural ^{41}Ca Equilibrium Concentrations in Terrestrial Samples. The equilibrium concentration (or "saturation" values) of ^{41}Ca in terrestrial (near-surface) samples was initially estimated by using the relationship expressed in equation 1 as given by Raisbeck and Yiou (*20*)

$$\frac{^{41}\mathrm{Ca}}{^{40}\mathrm{Ca}} = \frac{f\sigma t_{1/2}}{0.693}\left\{1 - \exp\left(\frac{-0.693t}{t_{1/2}}\right)\right\} \qquad (1)$$

where f is the thermal neutron flux, σ is the cross section of the reaction, $t_{1/2}$ is the half-life of ^{41}Ca, and t is the period of exposure. If it is assumed that t is much greater than $t_{1/2}$ (ca. 10^5 years), $\sigma = 4.4 \times 10^{-25}$ cm^2, and $f = 3 \times 10^{-3}$ neutrons cm^{-2} s^{-1} in the upper meter of soil (*23*), then $^{41}Ca/^{40}Ca$ is about 10^{-14} (*24*). This value indicates that for maximum usefulness in dating applications, it would be necessary to determine $^{41}Ca/^{40}Ca$ ratios in the range of 10^{-14} to 10^{-15}.

Half-life of ^{41}Ca. Published values for the half-life of ^{41}Ca range from 0.7×10^5 to 1.7×10^5 years with stated uncertainties ranging as high as 30% (*see* the references in references 25 and 26). Raisbeck and Yiou (*20*, *27*) used 1.3×10^5 years, whereas other researchers have used values of about 1.0×10^5 y. The most recent value of $t_{1/2} = 1.03 \pm 0.04 \times 10^5$ years was determined from the relative yields of $^{40}Ca(n, \gamma)^{41}Ca$ to $^{41}Ca(n, \gamma)^{45}Ca$ by Mabuchi et al. (*26*). Unfortunately, no direct measurement of the half-life via specific activity has been performed. This measurement would involve a mass spectrometric determination of the ^{41}Ca in a highly enriched sample and an activity measurement. The activity measurement poses the greater problem because the only detectable radiation in the decay of ^{41}Ca is soft X-rays and Auger electrons of around 3-keV energy. Efforts to achieve a direct measurement of the half-life of ^{41}Ca by measuring its specific activity are underway.

Measurement of Natural ^{41}Ca Concentrations in Terrestrial Samples

With the development of AMS methods of isotope detection, it has become feasible to examine the potential of ^{41}Ca dating of bone and other calcium-containing samples. Counting ^{41}Ca at natural levels (i.e., at low activities) is very difficult because, in addition to the long half-life and consequently low specific activity, decay is by electron capture that emits a low-energy X-ray and Auger-electron. However, by using AMS techniques, it is possible to measure a ^{41}Ca ion directly, even in the presence of a stable element of the same atomic mass (^{41}K, with 6.7% natural abundance). AMS measurements involve the acceleration of ions to sufficiently high energies that their atomic number can be determined. This acceleration usually involves a final energy of at least 5 MeV/atomic mass unit (amu), so that ions with adjacent values of Z have appreciably different energy losses in a thin detector, or that all the electrons can be removed from the ions so that they can be separated electromagnetically. For ^{41}Ca, the presence of the relatively abundant stable isobar means that separation at the final energy must be carried out very efficiently.

Since 1980, several groups have examined various approaches to the measurement of ^{41}Ca. Using the Alice accelerator facility at Orsay, Raisbeck and Yiou (27) accelerated Ca ions to about 1 MeV/amu in a linear accelerator, stripped off enough electrons to make Ca^{14+} ions, and then accelerated these ions to 7.5 MeV/amu in a cyclotron. These ions were further stripped to Ca^{20+}, analyzed in a magnetic spectrometer, and measured in a detector telescope that measures both their specific energy loss (ΔE) and total energy (E). As expected, the principal background was from ^{41}K. These experiments demonstrated that energy levels could be attained that correspond to $^{41}Ca/^{40}Ca = 10^{-12}$ in enriched samples. This method was not efficient enough to measure natural ^{41}Ca concentrations. In an alternative approach, Raisbeck et al. (28) determined that by starting with CaH_3^-, which is an easily formed negative molecular ion, it was possible to obtain enormously improved suppression of the ^{41}K background, because KH_3^- does not form readily, if it forms at all. The CaH_3^- ion could then be injected into a tandem electrostatic accelerator to achieve a high enough final energy that magnetic analysis–detection could separate ^{41}Ca from the remaining ^{41}K.

The approach just described has been evaluated by several groups. A detailed study was made by Fink et al. (*29*), who injected both $^{41}CaO^-$ and $^{41}CaH_3^-$ from artificially enriched samples ($^{41}Ca/^{40}Ca = 2 \times 10^{-12}$) into the Rehovot Pelletron tandem accelerator (*30*). Although a final ^{41}Ca energy of only 130 MeV was obtained, the ^{41}K interference was strongly suppressed by the $^{41}CaH_3^-$ injection. Fink et al. achieved sensitivities of $^{41}Ca/^{40}Ca = 1 \times 10^{-11}$ for CaO^- injection and 5×10^{-12} for CaH_3^-. This sensitivity was still not good enough for natural terrestrial samples because of the low negative ion yield and the presence of $^{42}CaH_2^-$ and $^{43}CaH^-$ in the injected ion beams that, in the absence of a velocity-sensitive element in the analysis system, resulted in high background count rates. More recently, $^{41}Ca/Ca$ ratios in the range of 10^{-12} have been measured in meteorites at the AMS tandem facilities of Rehovot(*31*) and Rochester (*32*).

The first direct measurement of the terrestrial ^{41}Ca concentration in natural samples was accomplished at the Argonne National Laboratory with the Argonne tandem linac accelerator system (ATLAS) following the preenrichment of the ^{41}Ca with a Calutron isotope separator at the Oak Ridge National Laboratory (*33*). A calibration sample with a known ^{41}Ca concentration (artificially enriched in ^{41}Ca by neutron irradiation) was included to check on the accuracy of the enrichment factors determined from the concentration of ^{42}Ca collected during the preenrichment process. On the basis of the data from the the calibration sample, it was determined that the preenrichment factors calculated on the basis of ^{42}Ca data were accurate to $\pm 15\%$.

Calcium extracted from a modern bovine bone and from surface and buried limestone was preenriched by about 2 orders of magnitude. The calcium as $^{41}CaH_3^-$ was accelerated in a negative-ion sputter source for injection into the tandem accelerator. Again, CaH_3^- ions were chosen to greatly reduce isobaric interference from ^{41}K. Very efficient suppression of neighboring stable Ca isotopes was achieved through the combined filtering action of the velocity-focussing linac and the magnetic beam-transport system. An Enge split-pole magnetic spectrometer filled with nitrogen was used for particle identification. This method very efficiently separated ^{41}Ca from the interfering stable ^{41}K isobar.

Table I lists the results of this experiment. A metallic calcium sample was prepared from limestone from an 11-m depth and was not preenriched. The sample was assumed to contain no cosmogenic ^{41}Ca at the instrumental detection limit of the AMS system used, because of its geological age (Mesozoic) and burial depth. No counts were observed for a time approximately three times longer than that for the bone sample. From the measurement

Table I. $^{41}Ca/Ca$ Concentrations in Natural Preenriched Terrestrial Samples

Sample	*Observed* $^{41}Ca/Ca$[a]	*Preenrichment Factor*[b]	*Inferred Original* $^{41}Ca/Ca$[c]
Limestone[d]	$<5.8 \times 10^{-14}$	1	–
(11 m)	$3.0 \pm 0.6 \times 10^{-12}$	151	$2.0 \pm 0.5 \times 10^{-14}$
Modern bone			
Limestone (surface)	$8.8 \pm 4.4 \times 10^{-13}$	116	$7.6 \pm 4.5 \times 10^{-15}$
Limestone (11 m)	$4.0 \pm 1.9 \times 10^{-13}$	117	$3.4 \pm 2.1 \times 10^{-15}$

[a]The quoted errors in this column reflect counting statistics only.
[b]Preenrichment factor estimated to be accurate to about ±15%.
[c]On the basis of previous experiments with other measurements on the same AMS system, the systematic errors are estimated to be about ±15%. The quoted errors in this column reflect both counting and systematic errors.
[d]Sample was not enriched.
SOURCE: Adapted from ref. 33.

of this sample, it was concluded that $^{41}Ca/Ca$ ratios as low as about 6×10^{-14} can be unambiguously measured with the AMS system. By using preenriched samples with more than 2 orders of magnitude enrichment, $^{41}Ca/Ca$ ratios as low as about 5×10^{-16} can be inferred. The preenriched bone and limestone samples show $^{41}Ca/Ca$ ratios well above the limit observed with the background sample. In our view, these results exclude instrumental effects and cross contamination in the ion source from the calibration sample as the origin of the observed ^{41}Ca. The inferred $^{41}Ca/Ca$ ratio actually measured in modern bone, $2.0 \pm 0.5 \times 10^{-14}$, is somewhat higher than the value estimated by Raisbeck and Yiou (*20*). It was expected that the surface limestone would exhibit about the same $^{41}Ca/Ca$ ratio as the modern bone sample and that the ratio for the 11-m limestone would be several orders of magnitude below this. However, both limestone samples (within error) showed ^{41}Ca concentrations that were two to three times lower than expected. Unfortunately, the large experimental errors on the limestone samples prevent any compelling interpretation for the results on the limestone. A number of possible explanations would account for these results.

Critical Examination of the ^{41}Ca Dating Model

The development of an effective method of measuring natural ^{41}Ca concentrations in terrestrial samples will permit a critical examination of the ^{41}Ca dating model. A series of interrelated issues needs to be addressed to determine the feasibility of the method. These issues would involve investigations relating to

1. variations in $^{41}Ca/Ca$ values between modern samples and samples of "infinite age" ($> 10^6$ years) as far as their cosmogenic ^{41}Ca content is concerned

2. variations in ^{41}Ca content in contemporary samples that have ideal preservation histories (i.e., no postdepositional exchange of calcium) from a variety of geographical areas
3. ^{41}Ca variability in samples that have been subjected to varying degrees of calcium exchange
4. differences in the ^{41}Ca content of contemporary samples recovered from a range of biogeochemical environments

Unfortunately, there appears to be little fractionation of Ca isotopes in the natural environment. This view was supported by $^{40}Ca/^{44}Ca$ measurements made on a wide range of contemporary marine and terrestrial samples. If this situation was also observed in fossil bone, it would mean that that there would be no clues from the distribution of the stable Ca isotopes to indicate if isotopic exchange had occurred in samples (*34*). By using techniques outlined in reference 34, Papanastassiou recently measured the $^{40}Ca/^{44}Ca$ ratios in two fossil bone samples supplied by Taylor. Within experimental error, the $^{40}Ca/^{44}Ca$ ratios of these two fossil bone samples are well within the narrow range observed in contemporary terrestrial samples We concur that Ca stable isotope analysis will not be useful in identifying postdepositional isotopic exchange in fossil bone samples.

If variations in contemporary ^{41}Ca concentrations fall within a relatively narrow range, then the age of samples may be inferred without a detailed knowledge of the geochemical status of the calcium source(s). The current expectation, however, is that there may be significant variations in contemporary ^{41}Ca content as well as variations in ^{41}Ca content as a result of diagenetic processes. Because of these variations, it may be necessary to treat each depositional situation (e.g., cave, rock shelter, sedimentary deposit) as a separate geochemical system. Because of the potential complexity of the factors that can influence the ^{41}Ca content of samples, we are currently operating under the assumption that the technique needs to be developed under conditions for which the accuracy of the ^{41}Ca age estimates can be rigorously gauged and tested. When the basic parameters of the method are defined, it will be necessary to examine a series of bone and other calcium-containing samples from long stratified sequences that extend from upper layers datable by the ^{14}C method (either by decay or direct counting) down into older portions of the deposit currently off the ^{14}C scale. An evaluation of results from several localities with contrasting ground water conditions, types of soils, and other geochemical environments would be required to permit critical judgments to be made of the general applicability of the method.

Current Studies

A systematic investigation of the ^{41}Ca dating model requires that a practical means be available to measure $^{41}Ca/Ca$ ratios at natural concentrations in terrestrial samples with a reasonable precision. We are involved in several types of studies to meet this goal. Our first study is concerned with the development of sample preparation methods that can be applied to appropriate terrestrial samples. The current focus of these studies is on improvements in procedures that can be used to extract calcium from bone samples and to prepare an appropriate chemical compound that can be directly used in the ion source of an AMS system. This research involves the preparation of metallic calcium. A second set of studies is concerned with further improvements in AMS operating conditions and efficiencies and is particularly directed at developing the capacity to obtain $^{41}Ca/Ca$ measurements in the range 10^{-14} to 10^{-15} without the need for preenrichment.

To reach the level of sensitivity required to examine natural ^{41}Ca concentrations in terrestrial materials, it has been necessary to use a preenrichment step in the analysis. Preenrichment is a lengthy and costly operation with inherent limitations on the achievable accuracy. These conditions just mentioned are hardly consistent with the large number of samples that will be needed to understand the systematics of ^{41}Ca distribution in the natural terrestrial environment.

A crucial problem in the development of the AMS method for direct $^{41}Ca/Ca$ measurements at natural terrestrial levels without preenrichment involves the improvement of the overall detection efficiency. Currently, the major limitation is the low yield of negative ions from the Cs-beam sputter source. However, encouraging results have been reported by R. Middleton and Sharma (*35, 36*) who observed CaH_3^- currents from freshly prepared CaH_2 materials that were 1 to 2 orders of magnitude higher than previously obtained (*30–33*). It will be interesting to see whether this material will be suitable for routine measurement of natural samples. The use of another negative ion, CaF^-, has been investigated at the Munich tandem accelerator facility (*37*). By boosting the energy of tandem-accelerated ^{41}Ca ions with a linear accelerator and by applying the full-stripping technique, natural $^{41}Ca/Ca$ ratios have recently been measured at a level of $^{41}Ca/Ca$ of 10^{-12} in meteoritic samples (*38*). ^{41}Ca concentrations in meteorites had been measured at this level before with modest beam currents from metallic ion source samples at two other laboratories (*31, 32*).

In a different approach, a positive-ion source was used in a first test of the heavy ion linear accelerator (UNILAC) at Gesellschaft für Schwerionenforschung (GSI) Darmstadt by Steinhof et al. (*39*). $^{41}Ca^{3+}$ ions were extracted from a penning-sputter source and accelerated to 14 MeV/amu. The $^{40}Ca^{3+}$ output current of the ion source was about 100 μA. The combination of a velocity-focussing accelerator and a magnetic ion-beam transport system completely eliminated background from the other calcium isotopes. Full stripping and detection of $^{41}Ca^{20+}$ ion with a magnetic spectrograph provided separation from isobaric ^{41}K and, at present, a sensitivity of $<2 \times 10^{-15}$.

At Argonne, a new positive ion injector for ATLAS based on an electron cyclotron resonance (ECR) plasma ion source is under construction (*40*). The

ECR source promises to provide highly charged positive Ca ions with high efficiency for injection into the superconducting linear accelerator. When this injector is ready, a ^{41}Ca detection like that described in the GSI experiment will be attempted with its goal to reach $^{41}Ca/Ca$ sensitivities for measurements at natural concentration levels.

Summary

The ability to provide an independent temporal scale for the Middle Pleistocene comparable to that established for the Early Pleistocene by K/Ar values and for the late Quaternary with ^{14}C data would be of major significance in providing critical data that would impinge on several important debates underway among paleoanthropologists. For example, did hominid evolution during the Middle Pleistocene occur as the result of slow, gradual, cumulative modifications in morphology and behavior (*41*), or were there long periods of relatively little variation, punctuated with rapid, major alterations (*42*). This controversy goes beyond the details of human evolution into theories of evolutionary process in general. Unfortunately, the present lack of precision in the chronological framework for the most critical intervals do not permit a definitive resolution of the arguments about hominid evolution, particularly, questions that deal with the evolution of anatomically modern *Homo sapiens* and its relationship to archaic *Homo sapiens*.

The taxon *Homo* is widely accepted as having evolved from *Homo erectus* precursors during the Middle Pleistocene. The oldest specimens generally classified as *Homo sapiens* are between 120,000 and 400,000 years old. However, the archaic forms of *Homo sapiens* all show skeletal morphologies that are outside the range of variations known in the living races of humans. Modern fossil skeletons did not become common until about 30,000 years ago. What was the nature of the process that gave rise to anatomically modern *Homo sapiens*? Did our own species originate in one specific region and from there spread rapidly, assimilating preexisting archaic populations elsewhere, or did the process occur in a number of separate regions essentially concurrently? Various intermediate mechanisms are also possible. The role of the Neanderthal variant in this process has long been debated (*43–45*).

Similar uncertainties surround the question of whether cultural developments involve gradual incremental progress or some abrupt transitions. An example of such a controversy is the relationship of the variants of the mousterian industrial pattern to the "Upper Paleolithic" industries of western Eurasia. Fundamental to these and other similar archaeological and paleoanthropological questions is the accuracy of the chronological framework for the relevant fossil and artifact record. The absence of reliable method to unambiguously assign specific ages to hominids known to be, or thought to be, more than 30,000 to 50,000 years old and younger than the current corpus of reliable K/Ar values that begin at about 750,000 to 1 million years B.P., inhibits the ability to reconstruct the details of the phylogenetic and cultural relationships among the Middle and Late Pleistocene Hominidae. The development of a reliable isotopic dating method that is applicable to bone and could document this critical period in hominid evolution would have a major impact on many important paleoanthropological and archaeological issues.

Acknowledgments

The research of the University of California, Riverside (UCR) Radiocarbon Laboratory is supported by a National Science Foundation grant BNS 8603478 (Anthropology Program) with additional funds supplied by UCR Chancellors Theodore L. Huller and Rosemary S. J. Schraer. Radiocalcium measurements are supported by the U.S. Department of Energy, Nuclear Physics Division, under contract W-31-109-ENG-38.

We are indebted to Roy Middleton of University of Pennsylvania for calling our attention to reference 19 and for T. A. Tombrello and D. A. Papanastassiou of the California Institute of Technology for their collaboration. We are also grateful for material provided by the late Glynn Issac of Harvard University as organized and revised by Martha J. Tappen and commented upon by Eric Trinkaus of the University of New Mexico. We also appreciate the comments of A. J. T. Jull of the University of Arizona and Eric Delson of the American Museum of Natural History, New York on an earlier draft. This is contribution 87/7 of the Institute of Geophysics and Planetary Physics, University of California, Riverside.

References

1. *Chronology of Hominoid Evolution*; Bishop, W. W.; Miller, J. A., Eds.; Scottish Academic: Edinburgh, 1972.
2. Isaac, G. L. In *Calibration of Hominoid Evolution*; Bishop, W. W.; Miller, J. A., Eds.; Scottish Academic: Edinburgh, 1972; p 382.
3. Muller, R. A. *Science* **1977,** *196,* 489–494.
4. Libby, W. F. *Environ. Int.* **1979,** *2,* 205–207.
5. Grootes, P. M.; Mook, W. G.; Vogel, J. C.; deVries, A. E.; Haring, A.; Kistmaker, J. Z. *Naturforsch.* **1975,** *A30,* 1–14.
6. Stuiver, M.; Heusser, C. H.; Yang, I. C. *Science* **1978,** *200,* 16–21.
7. Erlenheuser, H. In *Radiocarbon Dating*; Berger, R.; Suess, H. E., Eds.; University of California: Berkeley, 1979; pp 202–215.
8. Stuiver, M. *Science* **1978,** *202,* 881–883.
9. Hedges, R. E. M; Ho, P.; Moore, C. B. *Appl. Phys.* **1980,** *23,* 25–32.
10. Mannik, L.; Brown, S. K. *Appl. Phys.* **1985,** *B37,* 79–86.
11. Butzer, K.; Isaac, G. L. In *After the Australopithecines, Stratigraphy, Ecology, and Culture Change in the Middle Pleistocene*; Butzer, K. W.; Isaac, G. L., Eds.; Mouton: The Hague, 1975; pp 901–903.
12. Ku, T. L.; Liang, Z. C. *Nucl. Instrum. Methods Phys. Res.* **1985,** *223,* 563–571.

13. Ikeya, M. *ESR Dating and Dosimetry*; Ionics: Tokyo, 1985.
14. Michels, J. W.; Tsong, I. S. T.; Nelson, C. M. *Science* **1983,** *219,* 361–366.
15. Taylor, R. E. *Radiocarbon* **1983,** *25,* 647–654.
16. Taylor, R. E.; Payen, L. A.; Prior, C. A.; Slota, Jr., P. J.; Gillespie, R.; Gowlett, A. J.; Hedges, R. E. M.; Jull, A. J. T.; Zabel T. H.; Donahue, D. J.; Berger, R. *Am. Antiq.* **1985,** *50,* 136–140.
17. Shackleton, N. J. In *After the Australopithecines, Stratigraphy, Ecology and Culture Change in the Middle Pleistocene*; Butzer, K. W.; Isaac, G. L., Eds.; Mouton: The Hague, 1975; pp 1–24.
18. Isaac, G. L. In *After the Australopithecines Stratigraphy, Ecology and Culture Change in the Middle Pleistocene*; Butzer, K. W.; Isaacs, G. L., Eds.; Mouton: The Hague, 1975; pp 875–887.
19. Yamaguchi, Y. *Prog. Theor. Phys.* **1963,** *29,* 567.
20. Raisbeck, G. M.; Yiou, F. *Nature (London)* **1979,** *277,* 42–44.
21. Taylor, R. E. *Anal. Chem.* **1987,** *59,* 317A–331A.
22. Libby, W. F. *Radiocarbon* **1980,** *22,* 1017–1020.
23. O'Brien, K.; Sandmeier, H. A.; Hansen, G. E.; Campbell, J. E. *J. Geophys. Res.* **1978,** *83,* 114–120.
24. Tombrello, T. A., California Institute of Technology, personal communication, 1986.
25. Emery, J. F.; Reynolds, S. A.; Wyatt, E. I.; Gleason, G. I. *Nucl. Sci. Eng.* **1975,** *48,* 319–323.
26. Mabuchi, H.; Takahashi, H.; Nakamura, Y.; Notsu, K.; Hamaguchi, H. *J. Inorg. Nucl. Chem.* **1974,** *36,* 1687–1688.
27. Raisbeck, G. M.; Yiou, F. *Rev. Archeometrie* **1980,** *4,* 121–125.
28. Raisbeck, G. M.; Yiou, F. In *Proceedings of the Symposium on Accelerator Mass Spectrometry,* Argonne National Laboratory, Argonne, IL; Kutschera, W., Ed.; National Technical Information Service: Springfield, VA, 1981; pp 426–430.
29. Fink, D.; Paul, M.; Meirav, O. Presented at the 12th International Radiocarbon Conference, Trondheim, Norway, June 1985.
30. Fink, D.; Meirav, O.; Paul, M.; Ernst, H.; Henning, W.; Kutschera, K.; Kaim, R.; Kaufman, A.; Magaritz, M. *Nucl. Instrum. Methods Phys. Res.* **1984,** *233(B5),* 123–128.
31. Paul, M.; Fink, D.; Meirav, O.; Theis, S.; Englert, P. *Meteoritics* **1985,** *20,* 726.
32. Kubik, P. W.; Elmore, D.; Conrad, N. J.; Nishiizumi, K.; Arnold, J. R. *Nature (London)* **1986,** *319,* 568–570.
33. Henning, W.; Bell, W. A.; Billquist, P. J.; Glagola, B.; Kutschera, W.; Liu Z.; Lucas, H. F.; Paul, M.; Rehm, K. E.; Yntema, J. L. *Science* **1987,** *236,* 725.
34. Russell, W. A.; Papanastassiou, D. A.; Tombrello, T. A. *Geochim. Cosmochim. Acta* **1978,** *42,* 1075.
35. Middleton, R. In *Workshop on Techniques in Accelerator Mass Spectrometry,* Oxford, June 30–July 1, 1986; Hedges, R. E. M.; Hall, E. T., Eds.; Research Laboratory for Archaeology: Oxford, 1986; p 82.
36. Sharma, P.; Middleton, R. *Nucl. Instrum. Methods* **1987,** *B29,* 63–66.
37. Urban, A.; Korschinek, G.; Nolte, E. In *Workshop on Techniques in Accelerator Mass Spectrometry,* Oxford, June 30–July 1, 1986; Hedges, R. E. M; Hall, E. T., Eds.; Research Laboratory for Archaeology: Oxford, 1986; p 108.
38. Korschinek, G.; Morinaga, H.; Nolte, E.; Preisenberger, E.; Ratzinger, V.; Urban, A.; Dragovitsch, P.; Vogt, S. *Nucl. Instrum. Methods* **1987,** *B29,* 67–71.
39. Steinhof, A.; Henning, W.; Korschinek, G.; Muller, M.; Nolte, E.; Paul, M.; Roeckl, E.; Schill, D. *Nucl. Instrum. Methods* **1987,** *B29,* 59–62.
40. Pardo, R. C.; Bollinger, L. M.; Separd, K. W. *Nucl. Instrum. Methods* **1987,** *24/25B,* 746–751.

41. Cronin, J. E.; Boaz, N. T.; Stringer, C. B.; Rak, Y. *Nature (London)* **1981,** *292,* 113–122.
42. Gould, S. J.; Eldredge, N. *Paleobiology* **1977,** *3,* 115–151.
43. Howell, F. C. *Q. Rev. Biol.* **1957,** *32,* 330–347.
44. Trinkaus, E.; Howells, W. *Sci. Am.* **1979,** *241,* 118–133.
45. *The Mousterian Legacy*; Trinhaus, E., Ed.; British Archaeological Reports: Oxford, 1983.

RECEIVED for review June 6, 1987. ACCEPTED revised manuscript May 24, 1988.

18

Human Bones as Archaeological Samples

Changes Due to Contamination and Diagenesis

A. El-Kammar[1], R. G. V. Hancock[2], and Ralph O. Allen[3]

[1]**Faculty of Science, Cairo University, Giza, Egypt**
[2]**Department of Chemical Engineering & Applied Chemistry, and the SLOWPOKE Reactor Facility and the Department of Chemical Engineering and Applied Science, University of Toronto, Toronto, Ontario, Canada M5S 1A4**
[3]**Department of Chemistry, University of Virginia, Charlottesville, VA 22901**

While the inorganic matrix of human bones can survive, it can also be contaminated by the soil in which it was buried. Instrumental neutron activation analysis and X-ray fluorescence can be used to detect levels of contamination. Microscopic studies show that voids in the inorganic matrix can be filled with new mineral deposits that have resulted from diagenesis and contamination.

HUMAN BONES BECOME ARCHAEOLOGICAL SAMPLES because an inorganic mineral phase (poorly crystalline hydroxylapatite) develops in association with organic material such as collagen and lipids in the human body (*1*). After burial, the organic material gradually decomposes and leaves behind the inorganic mineral phase. There is a great variability in the rate at which the organic fraction in bone decreases. This variability is especially pronounced in buried bone. Although modern bones contain about 35% (by weight) organic material, buried bones obtained as archaeological samples can contain anywhere between 0% and 35% organic material.

Many studies have been directed at understanding the changes that occur in the organic portion of buried bone (*2*); however, the inorganic components of bone can also be affected by burial. One goal of this study was to investigate some archaeological bones to show how the inorganic components were affected by burial. The archaeological bones investigated

0065-2393/89/0220-0337$06.00/0

were all from Egypt and represented ages ranging from the Predynastic (ca. 5500 B.P.) to the Roman (ca. 2000 B.P.) periods. Some of the samples were from mummies that were protected from direct contact with soils. Other bones had been buried. Another goal of these studies was to investigate the value of two types of analysis (instrumental neutron activation analysis and X-ray fluorescence) for the rapid screening of bone samples to determine their state of contamination.

The idea that the inorganic composition of bone could yield information about ancient human diet has been pursued for over a decade (*3, 4*). These studies have shown that soil contamination can complicate and even invalidate the interpretation of trace element data obtained by analyzing archaeological bones (*5–7*).

If only the surface of a bone specimen were contaminated by the soil, then it might be possible to clean the surfaces before analysis (*5*). Problems of sample contamination can be overcome by determining which trace elements are least affected by the soils and the diagenesis of the bones themselves. For example, Sr and Zn have been found to be less sensitive to soil contamination than several other elements. Through the studies of ancient bones, the theory has evolved that the Sr concentration in bone is inversely proportional to the protein or meat intake in the human diet. Although Sr is generally found at low levels in meat (as compared to plants), the Zn levels in meat are high (*8*). Rheingold et al. (*9*), used this information to demonstrate that the concentration of Sr and Zn in the bones of modern animals depends upon diet. Research with animals fed controlled diets has been suggested to determine whether a series of trace elements could be used to provide a multidimensional pattern analysis that would discriminate between different diets (*10*).

Because of the well-recognized problems of diagenesis and contamination, the question of using trace elements to obtain information on diet remains uncertain. There have been new proposals on how to obtain additional information from archaeological bones. Taylor and coworkers (*11, 12*) have proposed using ^{41}Ca as a means of dating bones that have been buried (or placed in caves) deep enough to shield them from the cosmic ray bombardment. The introduction of "modern" material into ancient samples can seriously affect the accuracy of the isotopic data. To know the reliability of the proposed ^{41}Ca dating procedure, it is important to know the extent to which diagenesis can introduce modern Ca into the mineral matrix of ancient bones. For buried bones, modern Ca could be introduced by recrystallization of the hydroxylapatite. This recrystallization involves exchange with Ca in the environment or the filling of voids (left behind as the organic components weather away) by new Ca-containing minerals. These studies were also directed at observing the mechanism for this type of contamination or exchange with one of the major components.

Materials and Methods

Modern human bone samples were analyzed to test the instrumental neutron activation analysis (INAA) and to help establish more complete information on the natural levels of several trace elements in modern bones. The modern bones included seven samples of cortical bone and 16 samples of cancellous bone (eight of which were defatted). All samples were obtained from femoral heads. The data obtained for these 23 samples and a further description of the INAA technique have been reported elsewhere (*13*). For the purpose of comparison with the archaeological bone samples, this data obtained by using the same INAA technique is summarized in this chapter.

The archaeological bones included eight samples from mummies in the collection of the Royal Ontario Museum (Toronto, Canada). Three of these samples were about 3200 years old (from Nakht), one was about 3700 years old (from George), and four were from Roman period mummies and were about 4000 years old. Five samples of human bones that had been buried in the soil were collected at archaeological sites on the eastern side of the Nile Delta at Tel-Roba. One sample was analyzed from each of the following periods: the Predynastic period (ca. 5500 B.P.), the early Dynastic period (4000–3000 B.P.), the later Dynastic period (3000–2000 B.P.), and the Roman period (ca. 2000 B.P.). In addition, one tooth from an early Dynastic period (ca. 4000 B.P.) grave at Tel-Roba was also analyzed. These five samples were all cleaned to remove surface contamination by a drill fitted with a carborundum abrading bit. For one Roman period sample there was evidence of a fire above the grave and some soil was fused to the bone.

In an earlier paper (*13*), the INAA procedure was described and compared to the chemical analysis of bone by using inductively coupled plasma emission spectroscopy (ICPES). The ICPES technique required dissolution of the ashed sample before analysis, but otherwise the techniques proved to be quite comparable. In this study, some of the samples were analyzed by using X-ray fluorescence (XRF) to determine whether there was adequate sensitivity for the study of contamination in archaeological bone.

For the INAA procedure, dried samples (100–600 mg) were placed in polyethylene vials and irradiated for 1–3 min in the SLOWPOKE Reactor at the University of Toronto (flux of 10^{-11} n cm^{-2} s^{-1}). The P was measured with the ^{31}P (n, α) ^{28}Al reaction. To measure the actual concentration of P, the same samples were irradiated a few days later under identical conditions while wrapped in cadmium foil (*14*). This step allowed for a correction in samples for which there was significant Al contamination. Although the INAA procedure could provide a clear indication of soil contamination, the corrections were large whenever badly contaminated soils were encountered. If soil contamination was expected, the dried bone samples were analyzed by X-ray fluorescence. The analysis was performed with a wavelength dispersive X-ray spectrometer (XRD-700, Diano Corporation), and calculations were carried out with a fundamental parameters program (Corset).

For bones that had been buried in contact with the soil, a scanning electron microscope (SEM) (JSM-35) was used to examine the samples. A Kevex X-ray system (Si(Li) detector and associated analytical programs) was used to provide semiquantitative analysis of particles observed with the SEM. This analysis, along with the crystal morphology, were used to identify the minerals in the bone samples. X-ray diffraction with Cu–K_α radiation (Ni filter) provided information on the minerals in the bulk bone samples. Powder patterns were obtained with a Phillips diffractometer.

Results and Discussion

With the neutron activation procedure, the measurement of P is reliable only when there is very little Al (<20 ppm), or when an epithermal irradiation (Cd-covered) is used to correct for the direct activation of ^{28}Al from the Al. For badly contaminated samples, the X-ray fluorescence data are more suitable for the P determination. The sequential irradiation procedures (with and without cadmium) showed that many of the mummy samples were contaminated to some extent and contained more than 20 ppm of Al.

P and Ca were the major components of the inorganic mineral phase (presumably hydroxylapatite), although the concentrations in these Egyptian bones varied considerably (Table I). This was not surprising because it was anticipated that the bone samples could contain variable amounts of organic matter. The organic component dilutes the Ca and P. The dilution factor is usually determined by analyzing ashed bone samples. Because we were interested in knowing how well some of the samples were preserved, the archaeological samples were dried and not ashed before analysis.

The effect of ashing samples is shown in Table I and in Figure 1. Sample 13 (*) is the average of the concentrations of Ca and P in the 23 modern bone samples (values ranged from 9% to 24% Ca and 4% to 10.2% P). When the modern samples were ashed at 600 °C and reanalyzed, the values obtained by the INAA procedure clustered close to the average values of 39.9 ± 0.8% Ca and 18.8 ± 1.0% P, shown as 14 (□) in Figure 1. This result agrees with the theoretical values (40.3% Ca and 18.4% P) for these elements in hydroxylapatite. The line drawn through these points and the origin represents the dilution of the mineral phase by the organic portion of the bones. When archaeological samples are analyzed and plotted on this diagram, the position of the data point can be used to indirectly estimate the amount of organic material remaining in the sample.

Despite the range in the absolute concentrations of Ca and P, Figure 1 indicates that for the most part the Ca/P ratios are nearly those expected for the dilution of hydroxylapatite, which has a weight ratio of 2.16. The straight line in Figure 1 is that expected for the simple dilution of hydroxyl-

Table I. Analytical Results for Dried Archaeological Bone Samples from Egypt

Characteristic *Source* *Burial Content*	*Modern*[a] *Ashed* –	*~5500 B.P.* *Tel-Roba* *Soil*	*~4000 B.P.* *Tel-Roba* *Soil*	*~3700 B.P.* *George* *Mummy*	*~3200 B.P.* *Nakht* *Mummy*	*~3200 B.P.* *Nakht* *Mummy*	*~3200 B.P.* *Nakht* *Mummy*	*~2800 B.P.* *Tel-Roba* *Soil*
Ca (%)	39.9 ± 0.7	24.8	28.8	25.4	24.0	18.0	27.2	29.3
P (%)	18.8 ± 1.0	11.1	13.5	11.0	11.3	9.7	10.9	13.5
Ca/P Wt	2.12 ± .05	2.23	2.13	2.32	2.12	1.86	2.49	2.17
Mg (%)	0.41 ± .04	1.48	0.68	0.27	0.29	0.24	2.76	1.01
Na Total (%)	0.96 ± .28	0.68	0.96	0.98	0.53	0.90	0.05	2.2
Na Insoluble (%)	—	0.42	0.53	—	—	—	—	0.16
Al (%)	—	2.4	0.76	0.002	0.0067	0.0025	0.28	0.15
Cr (ppm)	—	24	11	—	—	—	—	8
SiO (%)	—	21.5	7.5	—	—	—	—	1.5
S (%)	—	0.71	1.36	—	—	—	—	2.18
Sr (ppm)	—	301	353	120	≤90	130	1200	398
Rb (ppm)	—	12	8	—	—	—	—	9
Ba (ppm)	—	168	31	44	≤18	≤44	100	72
Ti (ppm)	—	2900	480	60	≤50	≤65	≤150	300
V (ppm)	—	21	23	0.5	≤0.38	≤0.51	42	23
Loss in weight upon ignition	—	3.41	10.17	—	—	—	—	11.07
Sample No.[c]	14	1	2	3	6	5	4	7

Continued on next page.

Table I.—*Continued*

Characteristic *Source* *Burial Content*	*Modern*[b] *Dried* —	*~2000 B.P.* *Roman* *Mummy*	*~2000 B.P.* *Roman* *Mummy*	*~2000 B.P.* *Roman* *Mummy*	*~2000 B.P.* *Roman* *Mummy*	*~2000 B.P.* *Tel-Roba* *Soil*	*~4000 B.P.* *Tel-Roba* *Tooth-soil*
Ca (%)	18.7 ± 3.8	20.1	23.0	22.4	3.3	19.2	33.5
P (%)	8.45 ± 1.5	9.5	11.2	9.6	1.4	8.54	14.9
Ca/P Wt.	2.20 ± 0.11	2.12	2.05	2.33	2.36	2.25	2.24
Mg (%)	0.20 ± 0.04	0.35	0.27	0.26	≤0.16	2.02	0.73
Na Total (%)	0.58 ± 0.1	1.45	0.66	0.85	2.27	0.41	0.58
Na Insoluble (%)	—	—	—	—	—	0.36	0.47
Al (%)	≤0.002	0.016	<0.002	0.0025	0.23	3.6	0.46
Cr (ppm)	—	—	—	—	—	47.0	—
SiO (%)	—	—	—	—	—	32.1	3.8
S (%)	—	—	—	—	—	0.6	1.07
Sr (ppm)	≤110	220	140	210	≤50	282	—
Rb (ppm)	—	—	—	—	—	18	—
Ba (ppm)	≤40	≤43	≤24	≤28	≤25	250	—
Ti (ppm)	≤50	≤110	≤50	≤80	≤170	4500	180
V (ppm)	≤0.3	1	≤0.3	≤0.65	≤1.3	47	—
Loss in weight upon ignition	—	—	—	—	—	0.84	7.79
Sample No.[c]	13	8	10	11	9	12	15

NOTE: All mummy samples and the modern bone samples were analyzed by INAA; soil-buried samples were measured by XRF.
[a]Average of 23 modern human bone samples analyzed after ashing at 500 °C (for more detail *see* ref. 13).
[b]Average of 23 modern dried human bone samples described in more detail in ref. 16.
[c]Figure 1.

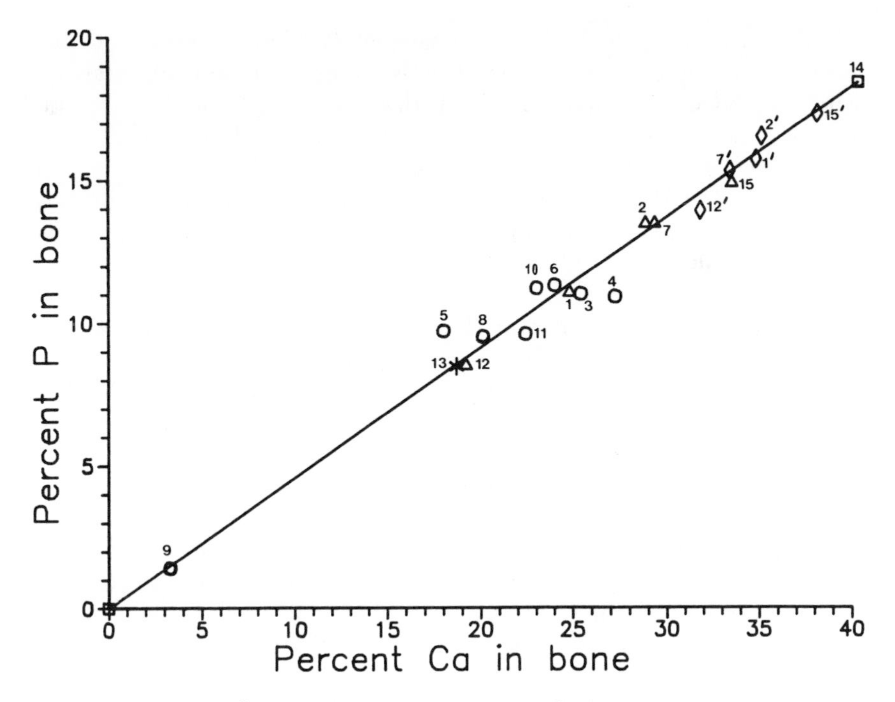

Figure 1. Ca and P concentrations in Egyptian bones. Key: ○, samples from mummies; △, buried bones; ◇ and ′ buried samples corrected for dilution by the organic component and soil contamination.

apatite ($Ca_{10}(PO_4)_6(OH)_2$). A plot of the analyses of the mummy samples (Figure 1, ○), shows that these bones still contain considerable organic material. All but one of the samples (No. 9 which was a vertebra sample from a Roman period mummy) show higher Ca and P concentrations than the average of the 23 modern bone samples. These concentrations suggest some loss of the organic fraction for these mummy bones. The reason for sample No. 9 having such low Ca and P concentrations is, in part, the result of mummification. The high levels of Na indicated the presence of salt, which was used in preparing the remains for burial.

For bones buried in the soil, the organic fraction is not the only diluent of hydroxylapatite. For example, the samples from Tel-Roba (△ in Figure 1) suggest that substantial amounts of organic material remain in soil-buried samples, although generally more organic material is found in in the mummy samples. However, when these Tel-Roba samples were ashed (to 500 °C), the weight loss (Table I) was not as large as suggested by the position of the data points in Figure 1. For example, the Ca and P concentrations in the Roman period sample (No. 12) suggest that samples from this period contained as much organic material as the modern samples (No. 13). However, only a 0.84% weight loss occurred on ignition.

The problem with these buried samples is dilution by the soil that remained despite careful cleaning of the surfaces. If these badly contaminated samples were analyzed by INAA, the corrections to the P concentrations would be very large because of the high Al levels. The X-ray fluorescence technique, thus, may screen bones that may be contaminated with soils. Si and Al concentrations were measured in the Tel-Roba samples, and the values were converted to the weights of SiO_2 and Al_2O_3. By assuming that all of the SiO_2 and Al_2O_3 were contaminants (clay and quartz), corrected concentrations of Ca and P in the bone were calculated. These corrected value [primed numbers in Figure 1], were calculated by adding the weight fractions of organic material (loss on ashing) and the fractions of SiO_2 and Al_2O_3. This approach does not change the Ca/P ratio and moves the data points nearer the theoretical hydroxylapatite and ashed bone (No. 14). Figure 1 suggests that correction using the SiO_2, Al_2O_3, and loss on ignition brings the badly contaminated sample No. 12 to about 79% of the Ca concentration expected for hydroxylapatite. Part of the problem may have been the presence of iron oxides, either in the soils or precipitated from solutions. For sample No. 12, the X-ray fluorescence results indicate that if all the iron present were Fe_2O_3, iron would represent about 3.9% of the weight, which would still not be enough to account for the low "corrected" values for Ca and P.

The corrected results that came closest to the theoretical values were those for sample No. 15. Sample No. 15 was a tooth from the same Early Dynastic skeleton from Tel-Roba as sample No. 12. Like the bone, this tooth was analyzed by the X-ray fluorescence. Because the tooth was protected by a hard, nonporous enamel, it should have been less sensitive to soil contamination. There was still 7.8% organic matter (based upon the weight loss upon ignition), but when the Ca and P concentrations were calculated on an organic-free basis, the values were still not as high as expected for hydroxylapatite, even though the Ca/P weight ratio was 2.24. The correction for the small amount of soil contamination (SiO_2 and Al_2O_3) gave a Ca value that was 94.5% of the expected value for a pure mineral sample.

The results in Figure 1 show how bones change upon burial. For protected samples, the organic fraction decreases by variable amounts. If the samples are buried, contamination also occurs. As the organic material is lost, voids left in the inorganic structure can be filled by the surrounding material. The question of how these voids are filled may be partially answered by understanding why the corrections for soil contamination and loss of the remaining organic material do not bring the Ca and P concentrations up to the levels expected for hydroxylapatite (No. 14 in Figure 1).

The tooth sample (No. 15) and the Late Dynastic period (ca. 4000 B.P.) bone from Tel-Roba (No. 2) showed fewer signs of contamination by the soils than any of the samples. Therefore, these two samples were used to calculate an apparent composition of the inorganic phase. Compared to the theoretical formula for hydroxylapatite,

$$Ca_{10}(PO_4)_6(OH)_2$$

the dynastic period tooth gave

$$Ca_{9.7}(Mg,\ Na,\ .\ .\ .)_{0.3}P_{5.6}(S\ .\ .\ .)_{0.4}O_{24}(OH)_2$$

This result suggests that there is some replacement of Ca and P in the apatite structure, even though the Ca/P ratio remains about the same. Some substitution of this type is natural in human bones because there are ion substitutions. Na and Mg frequently substitute for Ca. Wuthier (*15*) pointed out that biological apatites tend to be Ca-deficient and, as an example, he showed that for bovine cortical bone the weight ratio of Ca/Mg is about 61. Modern human bones have a Ca/Mg weight ratio range of between 91 and 102 (*13*). For the Dynastic period tooth, the Ca/Mg ratio was 96. This result suggested that the degree of substitution observed and the calculated formula probably reflect the natural system rather than any changes caused by burial.

The data for Dynastic period bone, however, are very different from the Postdynastic period bone (ca. 2800 B.P.) from Tel-Roba. Although the degree of soil contamination is small, the Ca/Mg ratio is much lower (29), and the calculated apatite formula is

$$Ca_{8.3}(Mg,\ Na\ .\ .\ .)_{1.7}P_{5.0}(S\ .\ .\ .)_{1.0}O_{24}(OH)_2$$

The substantial amount of substitution, especially for the Ca, suggests that the inorganic mineral phase has undergone substantial exchange and replacement when in contact with soil and with water percolating though it. Another explanation for these results is the precipitation of another mineral phase in the voids left by the loss of the organic material. That there may be other mineral phases filling the voids is suggested by the presence of more S (presumably as calcium or magnesium sulfate) in the Postdynastic period bone (2.2% S) than in the tooth (1.0% S).

All of the samples from Tel-Roba were examined with X-ray powder diffraction to determine the degree that they were contaminated by the soils in which they were buried. The diffraction pattern for the tooth had sharper peaks than any of the bone samples. This pattern suggested that there was a higher degree of crystallinity in the tooth. The sharpness of the X-ray diffraction patterns differ markedly for samples from the different parts of the Postdynastic period bone. The sharper X-ray diffraction patterns were obtained for the harder outer crust of the bone. This result suggested that there was a higher degree of crystallinity (perhaps due to diagenesis) near the outer surfaces of the bone. All of the samples showed reflections caused by hydroxylapatite, and in addition, the bone samples showed additional reflections (of varying intensity) at 2.898, 2.485, and 2.201 Å that were probably caused by another minor phosphate phase (perhaps whitlockite). Some diffraction reflections were caused by a number of soil contaminants such as quartz, feldspars, hematite, halite, and gypsum. These minerals were present in part because of the penetration of fine soil grains into the voids, but part were primarily the result of crystallization from solutions.

Examination of the soil-buried bones with the SEM and the semiquantitative analysis of grains by the X-rays emitted showed the filling of voids in the bone by a variety of minerals. Figure 2 shows a void in the Postdynastic bone from Tel-Roba along what was the wall of a blood vessel. Although this sample displays very little contamination by the soil minerals (only 1.45% SiO_2), the crystalline material along this void is some type of iron oxide that was presumably deposited from solution. Another part of this bone shows calcium sulfate (gypsum or anhydrite) coating the walls of these vessels.

Figure 3 shows a diagonal cut of the same bone (parallel to the former blood vessels) with a thin bright layer of calcium sulfate on the walls of the vessels (dark areas). At higher magnification (Figure 4), the same bone

Figure 2. SEM photomicrograph of bone from about 2800 B.P. found buried in the soil at Tel-Roba (No. 7 in Figure 1). The wall of the void left by a blood vessel is covered with a ferruginous material.

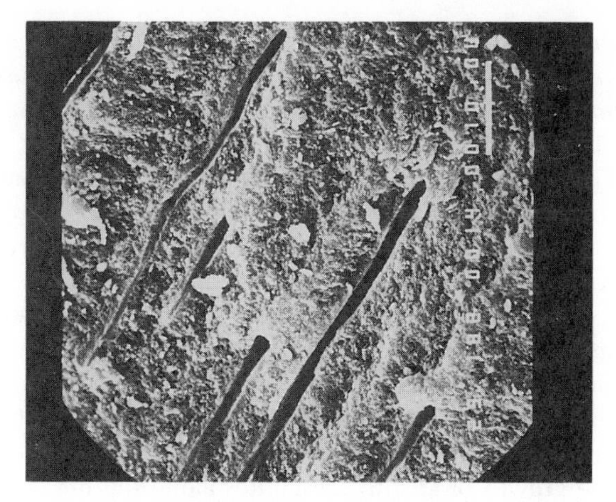

Figure 3. SEM photomicrograph of same bone as Figure 2 cut parallel to blood vessels (dark areas). The thin bright area along the edge of the voids left by the blood vessels are calcium sulfate (gypsum or anhydrite).

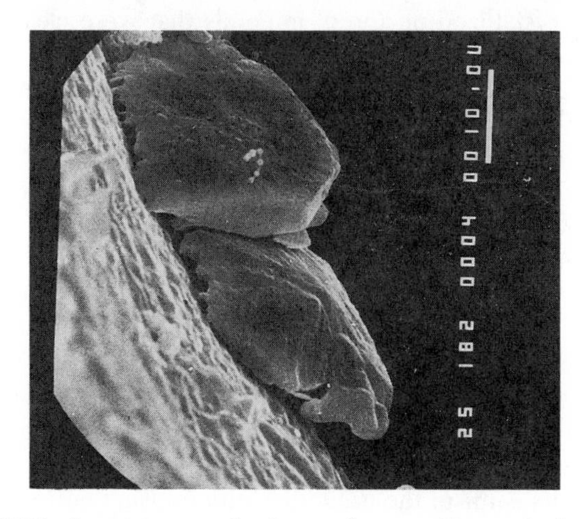

Figure 4. SEM photomicrograph of same bone as Figure 2 showing poorly crystallized $CaCO_3$ in a cavity.

showed what appears to be poorly crystallized calcium carbonate in one of the cavities. Although this sample appeared to be free of contamination by soil minerals (clay), apparently Ca had been added. The X-ray diffraction pattern showed gypsum, and the SEM examination showed at least two Ca-containing minerals that had presumably been deposited from solutions percolating through the bone.

Examination of the Predynastic (ca. 5500 B.P.) bone from Tel-Roba showed the effect of contamination from the soil and solutions. Near the surface of the bone (Figure 5), the apatite groundmass is covered by kaolinite and coarse-grained subhedral hematite. Further, within the same bone, the cavities left by the loss of organic material were filled by a variety of well-crystallized minerals that reached into the smallest voids. Figure 6 shows crystals whose composition and crystal habit suggest that they are carnallite ($KMgCl_3 \cdot 6H_2O$). Such distorted carnallite contamination that penetrates throughout these voids will be difficult or impossible to remove.

Because the prevailing conditions in Egypt favor the precipitation of several minerals containing Ca (e.g., gypsum and anhydrite) throughout the bone sample, most likely, Ca is present in the ground water solutions. It is not clear whether the original biological apatite undergoes any recrystallization upon burial. The indirect evidence that recrystallization with replacement of Ca by Mg occurs is enhanced by some of the crystalline apatite observed in the bones from the Early Dynastic period (ca. 4000 B.P.). In Figure 7A, the growth of apatite crystallites along the fibers of collagen is shown. In Figure 7B, the crystallites, shown in higher magnification, are usually poorly crystalline, but this sample showed several examples of well-crystallized hexagonal prisms of hydroxylapatite. Also, in some places, apatite appears to be crystallized on top of minerals that were clearly diagenetic. Figure 7C shows some type of apatite crystals that grew on the surfaces of well crystallized hematite and halite. Thus, even though the Ca is a major component of the bone, burial in the ground apparently allows some of the biological apatite to equilibrate with the percolating solutions.

The contamination of archaeological bone samples by the soil will affect many of the trace elements. For example, the bones buried in the soil (Table I) contain substantially more Ti, Si, Ba, and V than most of the mummy samples. That the Cr concentrations correlate well with the SiO_2 content suggests that Cr concentrations can be used as a measure of soil contamination. Hancock et al. found that, when using INAA, the Mn, Al, and V levels seemed to be the most sensitive indicators of soil contamination (*16*). Whether these same elements will indicate the effects of deposition of mineral phases from solution is not clear.

These results confirm the great difficulty in using buried bone samples in any type of trace element study. Even some of the mummy samples we studied were also contaminated. These results confirm that the analysis of the inorganic components of archaeological bone must be approached with great caution, especially when the bones have been in contact with soils or ground water solutions. Even isotopic studies may be affected by changes caused by recrystallization processes.

The INAA technique provides an analytically useful screening technique that is very sensitive to the Al contamination. For samples buried in the soil where the contamination is likely to be higher, the X-ray fluorescence and diffraction approach provides more data on the sources and types of contamination.

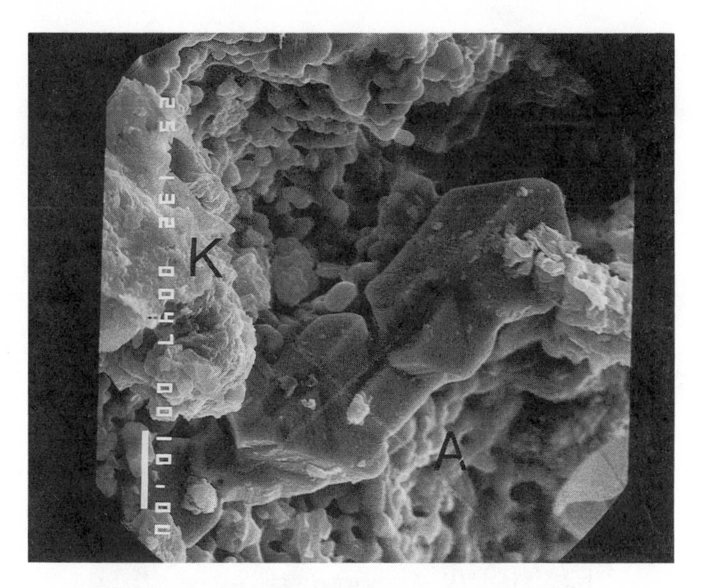

Figure 5. SEM photomicrograph of Predynastic (ca. 5500 B.P.) bone found buried in the soil at Tel-Roba. Apatite groundmass of bone (A) is covered with kaolinite clay (K) and coarse subhedral grains of hematite.

Figure 6. SEM photomicrograph of same bone sample as in Figure 5 showing extensive diagenetic filling of cavity by structurally distorted carnallite crystals that are probably derived from irrigation water.

A

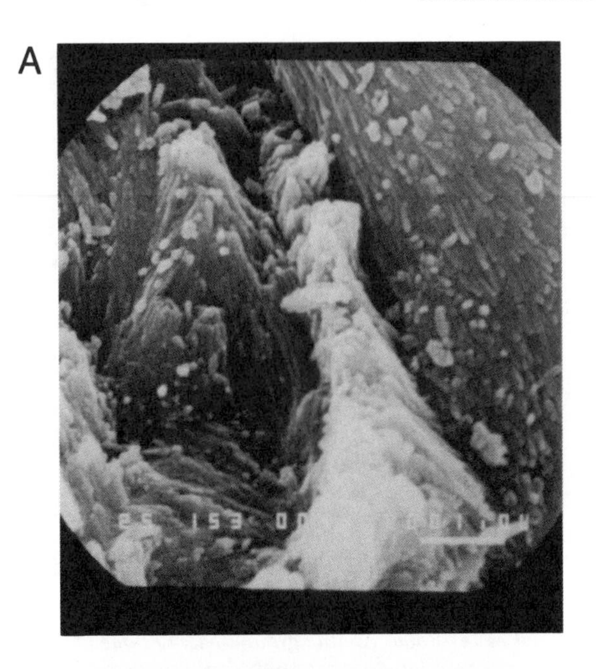

B

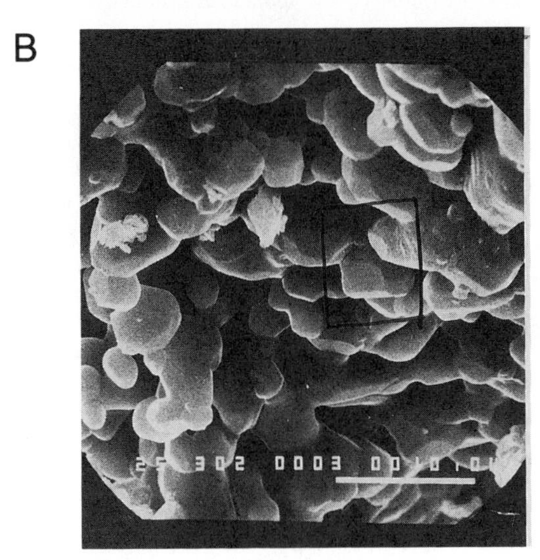

Figure 7. SEM photomicrographs of the bone from the Early Dynastic period (ca. 4000 B.P.) found buried at Tel-Roba. Micrograph A shows typical poorly crystalline hydroxylapatite in the bone along the collagen fibers while B shows the magnification of a hexagonal prism of hydroxylapatite suggesting diagenetic recrystallization.

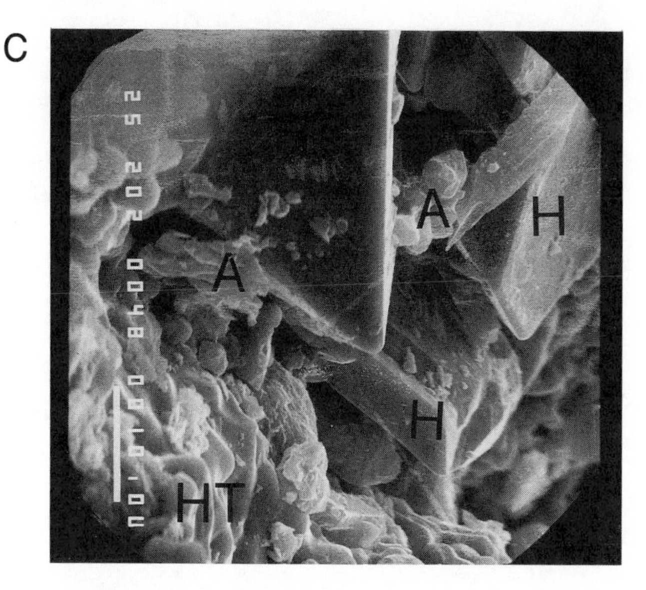

Figure 7.—Continued. Micrograph C shows more clearly that hydroxylapatite (A) has crystallized after a void in the bone was filled by halite (HT) and well-crystallized hematite (H).

References

1. Budy, A. M. *The Biology of Hard Tissues*; New York Academy of Sciences: New York, 1967.
2. Bada, J. L.; Shou, M. Y. In *Biogeochemistry of Amino Acids*; Hare, P. E.; Hoering, T. C.; King, Jr., K., Eds.; Wiley: New York, 1980; pp 235–255.
3. Brown, A. B. *Contrib. Geol.* **1974,** *13,* 47–48.
4. Lambert, J. B.; Simpson, S. V.; Szpunar, C. B.; Buikstra, J. E. *Acc. Chem. Res.* **1984,** *17,* 298–305.
5. Lambert, J. B.; Simpson, S. V.; Buikstra, J. E.; Charles, D. K. In *Archaeological Chemistry— III*; Lambert, J. B., Ed.; Advances in Chemistry 205; American Chemical Society: Washington, DC, 1984; pp 97–113.
6. Edward, J.; Fossey, J. M.; Yaffee, L. *J. Field Archaeol.* **1984,** *11,* 39.
7. Lambert, J. B.; Szpunar, C. B.; Buikstra, J. E. *Archaeometry* **1979,** *21,* 115.
8. Gilbert, Jr., R. I. In *Biocultural Adaptation in Prehistoric America*; Blakeley, R. L., Ed.; University of Georgia: Athens, GA, 1977; pp 85–100.
9. Rheingold, A. L.; Hues, S.; Cohen, M. N. *J. Chem. Educ.* **1983,** *60,* 233–234.
10. Lambert, J. B., personal communication.
11. Taylor, R. E. *Anal. Chem.* **1987,** *59,* 317A–331A.
12. Taylor, R. E.; Slota, P. J.; Henning, W.; Kutshcera, W.; Paul, M. In Chapter 17 of this book.
13. Gynpas, M. D.; Pritzker, K. P. H.; Hancock, R. G. V. *Biol. Trace Elem. Res.* **1987,** *13,* 333–344.
14. Gatschke, W.; Gawlik, D.; Kraft, D. *J. Clin. Chem. Clin. Biochem.* **1980,** *18,* 403.

15. Wuthier, R. E. In *Metal Ions in Biological Systems*; Sigel, H., Ed.; Marcel Dekker: New York, **1982**; Vol. 17, pp 441–460.
16. Hancock, R. G. V.; Grynpas, M. D.; Alpert, B. *J. Radioanal. Chem.*, in press.

RECEIVED for review December 29, 1987. ACCEPTED revised manuscript July 14, 1988.

19

Electron Spin Resonance Study of Bones from the Paleolithic Site at Zhoukoudian, China

K. D. Sales, G. V. Robins, and D. Oduwole

Chemistry Department, Queen Mary College, University of London, London E1 4NS England

A group of bones from the paleolithic site at Zhoukoudian, China, has been studied with electron spin resonance (ESR) spectroscopy. The spectra observed for these bones are complicated because of overlapping signals from several different species. Computer simulation was used to elucidate the signals present and give information about the thermal histories and ages of the bones. The major indication is that, with one exception, the samples have not been heated. The samples have an age range of at least 260,000 to 570,000 years, depending upon the layer from which they came. A complex organic radical signal, probably derived from protein residues in the bone, was used for the first time to indicate the antiquity of the bone and the retention of such residues.

THE SITE AT ZHOUKOUDIAN, CHINA, HAS BEEN THE SUBJECT of many archaeological investigations because of its association with *Homo erectus* and the early use of fire. Our interest in material from the site arose primarily through our development of electron spin resonance (ESR) spectroscopy as a tool for the study of thermal histories, with a view toward obtaining information about the use of fire in ancient heating and cooking. Earlier work (*1*) was concerned with cereal grains from the controversial site at Wadi Kubbaniya. At this site, the grains had not been heated, and were, therefore, unlikely to have existed for a long time. Other research (2) has dealt with

0065-2393/89/0220-0353$06.00/0

the stomach contents of the Lindow man (2), which yielded interesting information about the cooking of his last meal.

Previous ESR studies of bone have been primarily concerned with attempts to date the bone, but the complexity of the spectra obtained for the bones from Zhoukoudian (3) warranted further investigation. As this chapter is one of the first comprehensive accounts of such a study, it is appropriate to begin it with a brief survey of ESR spectroscopy and its application to bone. A more complete account can be found in reference 4.

ESR Spectroscopy

ESR spectroscopy was introduced in the mid 1940s, partly as a result of the expertise developed in microwave technology during World War II. The sample is placed between the poles of a magnet with a strong, homogeneous field in the range 0 to 1 T and irradiated with microwaves in the X-band region (i.e., with a frequency of about 9.7 GHz). In many forms of spectroscopy, the spectrum is obtained as a graph of power absorbed as a function of frequency. For ESR, the frequency is constant as the magnetic field is swept. Furthermore, because of the type of modulation used in the experiment, the first derivative of the absorbed power is detected. A typical single-line spectrum is shown in Figure 1, together with the parameters used to characterized it: the g-value (i.e. the position of the line, effectively corrected for the microwave frequency and given as a dimensionless number); the line width (measured in magnetic field units); and the size of the signal (in arbitrary units). Most of the spectra observed for archaeological samples

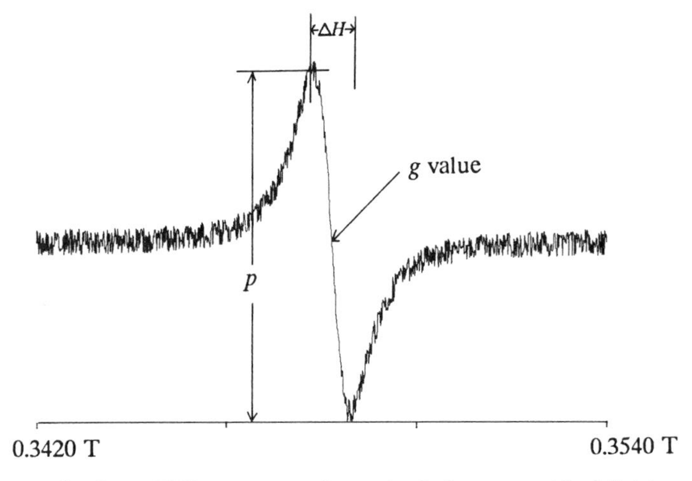

Figure 1. The basic ESR spectrum of unpaired electrons with definitions of the g value and the line width (ΔH); the signal intensity is proportional to $p(\Delta H)^2$. Nominal frequency: 9.765 GHz.

occur in a relatively narrow region centered on 0.3 T for a frequency of 9.7 GHz (usually called the $g = 2$ region). If there is more than one line in the spectrum due to a single species, then the distance between them is usually referred to as the hyperfine splitting and is measured directly in magnetic field units (Figure 2).

ESR spectra may be observed for a wide variety of species in the gaseous, liquid, or solid states. For archaeological applications, the solid state is common. A sample normally weighs 5–100 mg and is examined as a small piece or a coarse powder. If thermal stability or dating studies are not being undertaken, the sample can often remain undamaged.

Experimental Details

To obtain an ESR spectrum, the sample is placed in a synthetic fused silica (Spectrosil) tube with an internal diameter of 2 to 4 mm. The tube is then positioned between the poles of an electromagnet and irradiated with microwaves. For studies of thermal history, the sample, preferably along with modern equivalents, is heated in a standard muffle furnace to preset temperatures, and the spectrum recorded after each heating. For dating purposes, the sample must be subjected to γ-irradiation and the stability of the resulting signal checked. In this case a calibrated ^{60}Co source was used. We have found that the dating signal does not fade with time, but that various subsidiary features decrease over about 7 days.

The experiments reported in this chapter were performed with a Bruker 200D spectrometer with an 11-in. magnet and a TE_{011} cavity. The crushed samples, which weighed about 0.2 g, occupied about 3.5 cm of a 2-mm Spectrosil tube. Each sample was irradiated with 6 mW of microwave power, and 0.1 mT of 100-kHz modulation was applied.

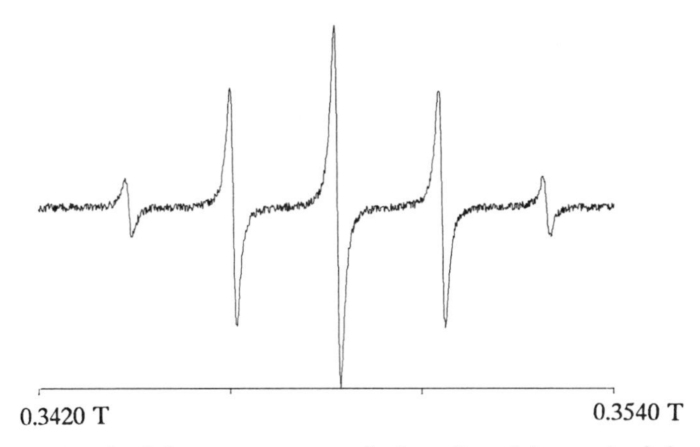

Figure 2. A multiple-line spectrum typical of coupling of the unpaired electron to four equivalent protons as in the alanine radical, CH_3CH_3COOH. Nominal frequency: 9.765 GHz.

ESR-Active Species

For a sample to give rise to a spectrum, the sample must be paramagnetic; that is, the individual molecules or species must contain electrons with unpaired spins. Electrons in molecules usually occur in pairs, and their spins mutually cancel. Organic species with unpaired electrons (radicals) tend to be unstable, but in some cases exist in isolation, and are always stabilized if contained within a suitable solid lattice. A good example is coal, which shows a very strong ESR signal.

In archaeological applications, the existence of a signal from an organic radical is caused by some previous treatment of the sample, such as heating; UV irradiation; α-, β-, or γ-rays; or by some combination of these treatments. Irradiation from radioactive sources can also give rise to a defect or dating signal (Figure 3) because of electrons being liberated in one part of a lattice and trapped, apparently indefinitely, in another part. This effect is responsible for thermoluminescence (TL), and there is a parallel between the TL and ESR methods of dating.

Another category of species containing unpaired electrons is transition metal ions, which are stable. The most common transition metal ions found in archaeological samples are iron (Figure 4) and manganese (Figure 5). These ions are often picked up from the surrounding soil. These elements (especially iron) can cause a coloration of the sample that can be confused with blackness due to heating.

The signals from organic, defect, and transition metal species can be distinguished because organic species have relatively narrow line widths (0.1–1 mT) and small hyperfine splittings (ca. 2 mT). Defect signals have anisotropic g values, and transition metals give rise to large line widths (10

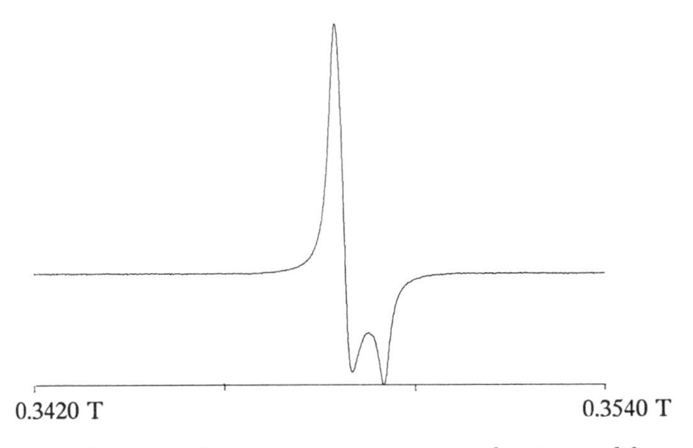

Figure 3. A defect signal showing an anisotropic g-*value. Nominal frequency: 9.765 GHz.*

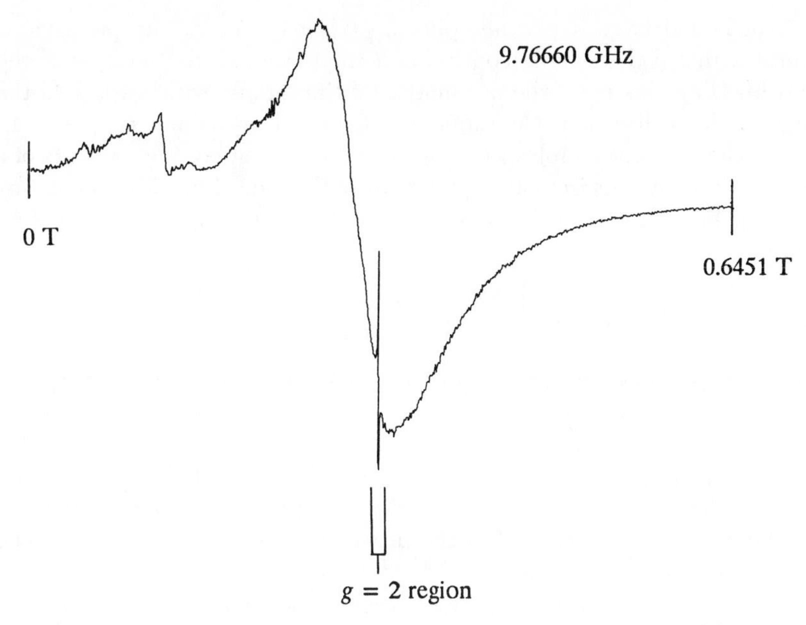

Figure 4. A signal due to Fe(III). Note the much wider magnetic field sweep and the very broad feature at g = 2.

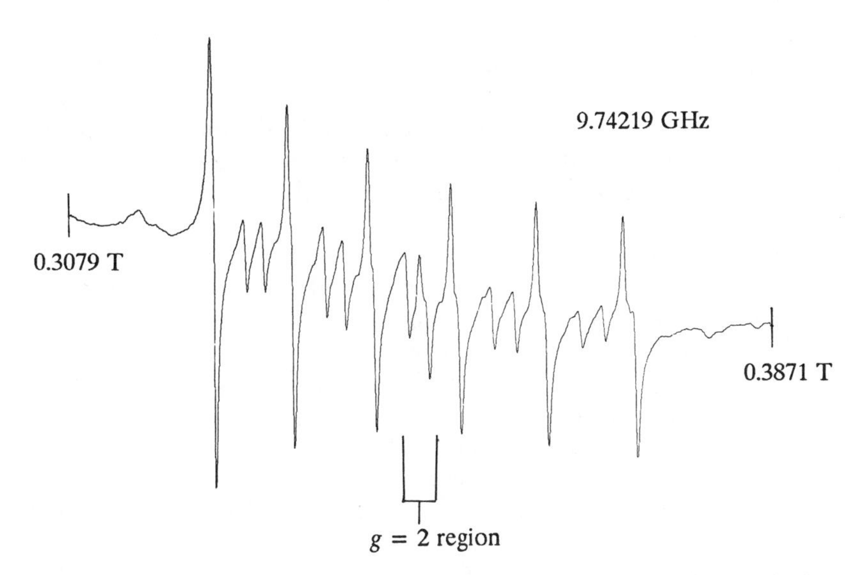

Figure 5. Mn(II) spectrum showing the six-line splitting together with the several smaller ("forbidden") lines. Note the field sweep and the size of the splitting between the main six lines.

to 50 mT) and large hyperfine splittings (10 mT). An anisotropic g value indicates that if the sample consists of a single crystal, the position of the ESR line depends upon the orientation of the sample with respect to the magnetic field direction. Consequently, for a polycrystalline sample (e.g., most archaeological samples), the signal is more complex, and consists of a superposition of lines from all orientations of the radical species. The dating signal in bone is a typical case of a spectrum caused by an anisotropic g value.

Bone

Bone is a complicated material that consists essentially of two interpenetrating lattices: one of proteinaceous material and one of complex calcium phosphates. As bone ages in the ground, it undergoes a great variety of chemical and physical changes that often result in its complete disintegration, but sometimes in its mineralization and therefore in a form of preservation. The types of signal described in the previous section may all be observed in bone, given the requisite burial conditions.

The phosphate lattice, apparently regardless of its precise composition, can trap electrons released by irradiation from naturally occurring radioactive elements in the environment of the bone. The dating of bone with ESR spectroscopy depends upon measuring the intensity of this defect signal in the sample as found, and then measuring again after each of several calibrated laboratory γ-irradiations. A plot of the signal intensity against laboratory γ-dose allows an extrapolation back to zero signal intensity, a point that is called the equivalent dose (ED) (5), that is, the dose that the sample appears to have received during its interment (*see* Figure 6). To convert the ED to an age, the annual dose rate of the sample must be determined. This conversation has many uncertainties; however, an annual dose rate of 2 mGy per year appears to give dates that agree reasonably with those obtained by other techniques (5).

ESR dating seems to be important because of its potential range, and thereby its coverage of the gaps left by other methods, and its ability to deal with bone. Furthermore, if only a rough idea of the age of a sample is required, the technique is nondestructive.

Bone may also show a stable signal due to carbon (*1*) derived from the pyrolysis of organic material in the bone. In principle, careful measurement of the parameters of this signal and calibration with modern equivalents allow the quantification of the thermal history of ancient samples that may provide insight into ancient cooking methods. In practice, overlap between the different signals makes accurate characterization difficult.

Fresh bone does not normally show spectra due to transition metal ions. It is, however, important to differentiate between the presence of an element and the observation of its ESR spectrum. As has been shown for chert

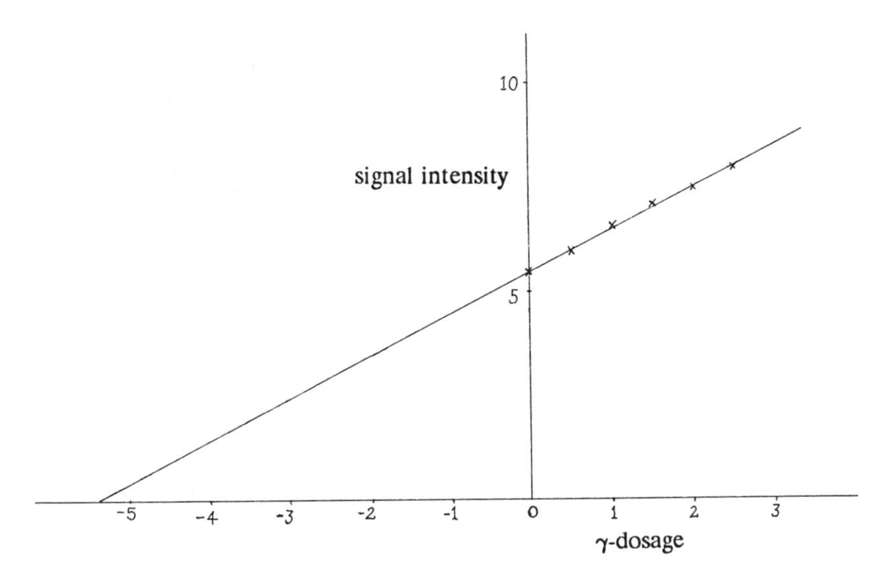

Figure 6. A plot of signal intensity against laboratory γ-dosage. The signal intensity is in arbitrary units; each unit of γ-dosage corresponds to 195 Gy, and this ED of 5.45 units is 1063 Gy (sample C).

samples (6), signals may or may not be observable because of the oxidation state or the lattice surroundings of the metal ion. Signals caused by Mn(II) tend to appear in old bone because of absorption of manganese from the surrounding soil. Whether a signal appears or not is related to the thermal history of the sample, as is the actual number of lines observed. Mn(II) spectra always consist of six lines. In many circumstances, however, smaller lines appear, and sometimes the six lines are partially split into doublets.

The Fe(III) spectra consist of two features: a line at much lower field than any other line (the $g = 4$ line) and a broad featureless line at $g = 2$. The latter line is sometimes so intense and broad that it can be difficult to measure accurately. The broad signal may even make it difficult to observe the other signals that appear at $g = 2$ (*see* Figure 4).

Results for the Zhoukoudian Bones

The samples obtained for this study and their ESR characteristics are given in Table I. The spectra are quite complex, and the presence or absence of some lines is difficult to ascertain by simple inspection because the lines from different species overlap. In an attempt to disentangle these different spectra, the experimental data were simulated by computer. A typical spectrum (Figure 7) shows a strong dating signal and a weak line to the low-field side of it, but the middle of the dating signal is slightly unusual, and it is not clear whether there is another signal. This distortion becomes more

Table I. ESR Signals in Zhoukoudian Bones

Sample	*Layer*	*Description*	*Alanine*[a]	*Mn*[a]	*Fe*[a]	*ED/Gy*	*Equivalence Age at 2 mGy/year* ($\times 10^6$ *years*)	*Equivalence Age at 1.2 mGy/year* ($\times 10^6$ *years*)
A	10	Cervus scapula	w	–	w	878	0.44	0.73
B	10	Equus sp. metacarpal	w	–	m	–	–	–
C	12	Equus scapula	w	–	m	1063	0.53	0.89
D	6	Equus metacarpal	?	w	w	517	0.26	0.43
E	10	Not given	w	–	w	1141	0.57	0.95
E′[b]	10	–	w	–	s	–	–	
F	10	Equus sp. femur	w	–	m	–	–	–
G	6	Bos sp. rib	w	–	–	–	–	
H	10	Equus sp. humerus	w	w	m	–	–	–
I	10	Equus sp. metacarpal	w	w	w	–	–	–
J	10	Rhinoceros sp. tibia	?	w	s	–	–	–
K	10	Equus sp. metacarpal	–	w	s	322	0.16	0.27
L	9	Cervus sp. unclear; appears burnt	–	–	m	439	0.22	0.37
L′	9	–	–	–	m	–	–	–

NOTES: All the samples were off white except for one E (dirty white), and the L samples, one of which was burnt brown and the other gray. The ages of the different layers are considered to be 350.000 years for layer 6, >400,000 years for layer 9, 520,000–610,000 years for layer 10, and >500,000 years for layer 12. These results were obtained by the U/Th method except for layer 10, for which the TL method was used. The ages have been calculated for two reasonable limits (3) for the annual dose rate. All samples showed a weak signal to the low-field side of the dating signal except for L and L′, as discussed in the text.

[a]Abbreviations: w, weak; m, medium; s, strong; ?, the spectra were too ill-defined in the regions of interest for a decision to be made as to the presence or not of an alanine signal.

[b]E′ is the scrapings from the surface of the sample.

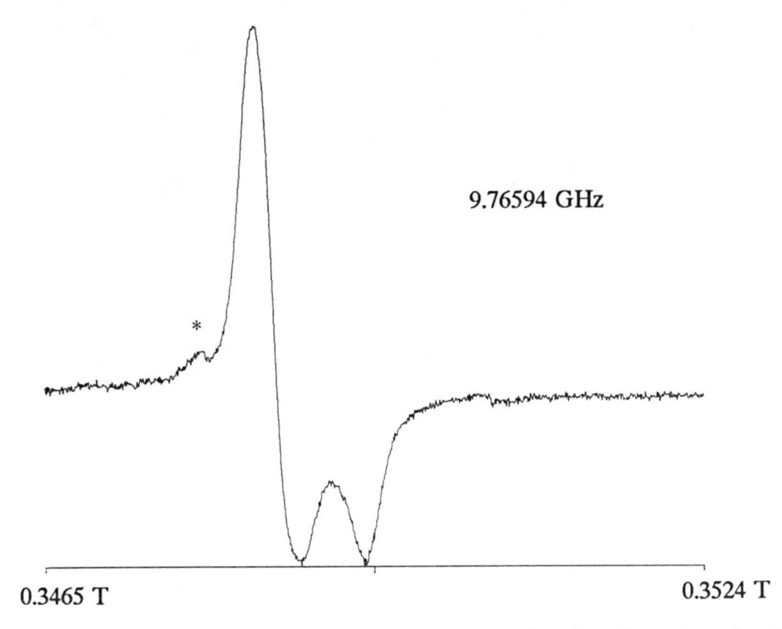

Figure 7. A typical spectrum obtained from Zhoukoudian bone (sample A) showing the large dating signal and the small signal to the low-field side of it at ca. g = 2.0055 (marked with an asterisk).

apparent after heating the bone (Figure 8), when an extra feature in the middle of the dating signal becomes evident. Compare this figure with the idealized dating signal in Figure 3.

The spectra were simulated by assuming a Lorentzian shape (with assumed g value, line width, and intensity) for each component, calculating the derivative, and adding the results together. The sum of the residuals from the calculated and experimental spectra was then minimized by using Powell's technique (7). The only complication was that the g-value for the dating signal was anisotropic so that this part of the spectrum had to be simulated by calculating the derivative for many different orientations of the species (i.e., for the g values obtained for the different directions in space) and averaging the result (8). Figure 9 shows the simulated spectrum for sample A. Figure 10 shows the simulated spectrum for sample A after it had been heated. The agreement between the experimental and simulated spectra was quite good. The results obtained for sample A, without heating and after two separate heatings, are presented in Table II.

Discussion

The dating signal was very strong in all the samples studied except for L′, which showed an unusually shaped signal in the same place as the dating

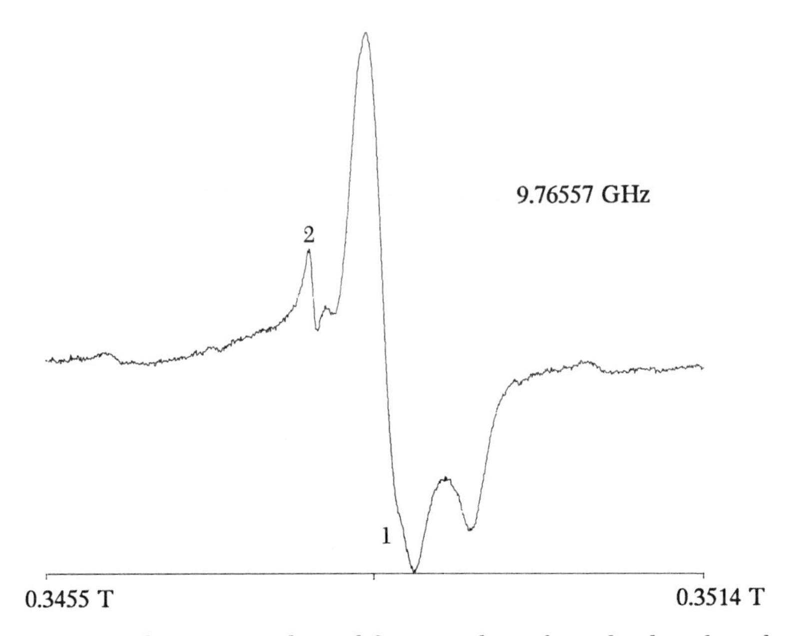

Figure 8. The spectrum obtained from sample A after it has been heated to 200 °C showing more clearly the distortion in the middle of the dating signal (marked 1). The enhancement of the g = 2.0055 feature (marked 2), and perhaps the appearance of another feature at ca. g ≃ 2.0044 (slightly to the right of the point marked 2).

signal. L′ was also unusual because it showed strong signals from simple organic radicals (RCH_2 and CH_3; *see* Figure 11). Signals of this type have been observed for chert (9), but not previously for bone. This sample looked rather different from the others, and we became doubtful whether it was bone. However, the infrared spectrum indicated that it had a similar phosphate content to the other samples. Sample E is interesting because, despite having been found on a hearth site, it appeared not to have been burnt. The color of the dark outermost layer was caused by heavy iron contamination. The interior gave the normal signals found in most of the other samples. Our results suggest that samples L and L′ are rather different from the other samples.

All the samples give a weak signal to the low-field side of the dating signal (at ca. $g = 2.0055$), except for L, which had a relatively strong line in this region (*see* Figure 12). The intensity of this signal increased when the sample was heated in the laboratory, but the g value did not change with temperature of heating. Such behavior was different from that of other organic materials (*1*), and probably indicates that this thermal marker is not radical carbon.

The signal appears to be different from the signal obtained for laboratory-heated modern and less ancient bone (*see* Table III). The signal for these

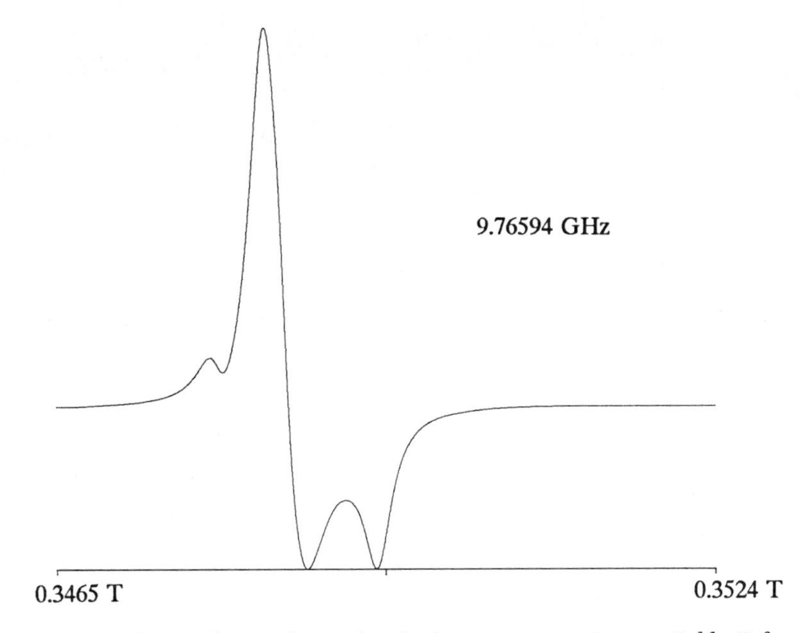

Figure 9. The simulation obtained with the parameters given in Table II for sample A. Compare with Figure 7.

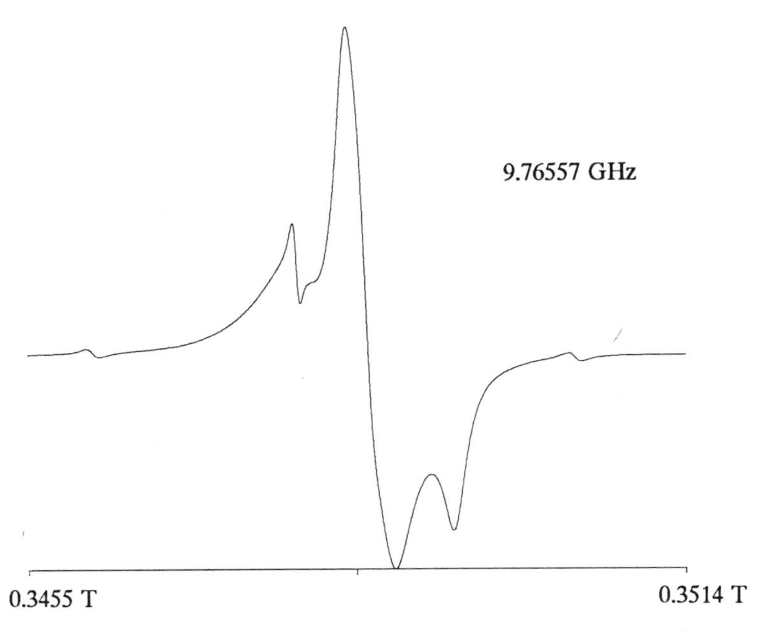

Figure 10. The simulation obtained with the parameters given in Table II for sample A2. Compare with Figure 8.

Table II. Results of Spectral Simulation for Sample A

Parameters	*A0*	*A2*	*A3*
Group 1			
g_x	2.0021	2.0021	2.0021
g_y	2.0032	2.0031	2.0031
g_z	1.9973	1.9975	1.9975
Line width	1.78	1.53	1.49
Intensity	1.44	1.74	1.90
Group 2			
g	2.0056	2.0057	2.0056
Line width	2.00	0.60	0.65
Intensity	3.36	6.16	6.83
Group 3			
g	2.0013	2.0011	2.0011
Line width	2.33	2.78	2.73
Intensity	4.62	12.40	13.65
Group 4			
g	–	2.0049	2.0046
Line width	–	5.32	5.54
Intensity	–	8.70	12.55

NOTES: The first group of parameters refers to the dating signal; the second to the signal to the low-field side of it; the third to a line in the middle of it, and the final set to a further line that appears to the low-field side of it after the sample has been heated. A0 is the unheated sample; A2 indicates heating to 200 °C; A3 indicates heating to 300 °C. A2 and A3 also had a small five-line spectrum added with the parameters: g = 2.0037, line width = 0.125 mT, intensity = 0.20, and a splitting of 2.182 mT.

samples has a g value of about 2.0046, which is sufficiently different from 2.0055 for the two lines to be classified as distinct, even though the g value hardly appears to have changed with temperature of heating. Further evidence that the lines are distinct is that a small hump appears to the low-field side of the 2.0046 signal (i.e., to higher g value) for the Abu Huyrera bone heated to about 250 °C, and for the modern sheep's bone when heated to about 500 °C. Structural changes occur at about 500 °C in bone (the manganese spectrum changes its character from second-order to first-order (*3*)), and in ivory (as indicated by its IR spectrum (*10*)). Finally, the results in Table II show that good simulations of the spectra for sample A, when heated to 200 °C and above, could be obtained only if both signals were included. The origins of these lines is not clear except that they appear to be related to the heat treatment of the sample. Nevertheless, the weakness of these signals in the Zhoukoudian bone and their enhancement on heating the bone to 200 °C suggests that any heating of these bones (except for sample L) in antiquity occurred at a far lower temperature than 200 °C.

The signal to the low-field side of the dating signal in sample L is much more intense for the outside than for the inside of the bone. (The dating experiments were performed on a mixture of inner and outer bone as the distinction was not then appreciated.) Laboratory heating of the inner bone

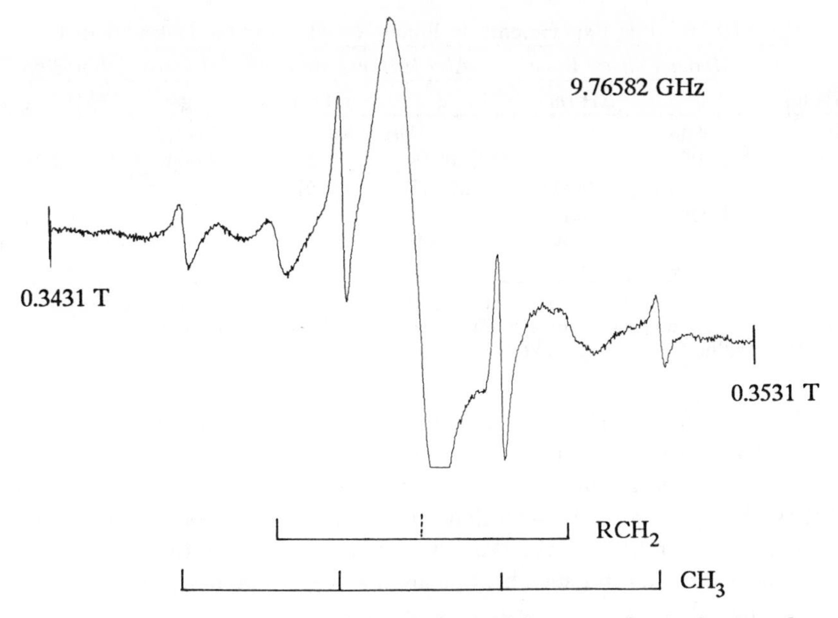

Figure 11. The rather unusual signal obtained from sample L′, showing also the lines due to RCH_2 and CH_3. The spectrum has been amplified to show more clearly the smaller features with the result that the dating signal has been clipped on the high-field side. This signal does not show the higher field component shown in Figure 3.

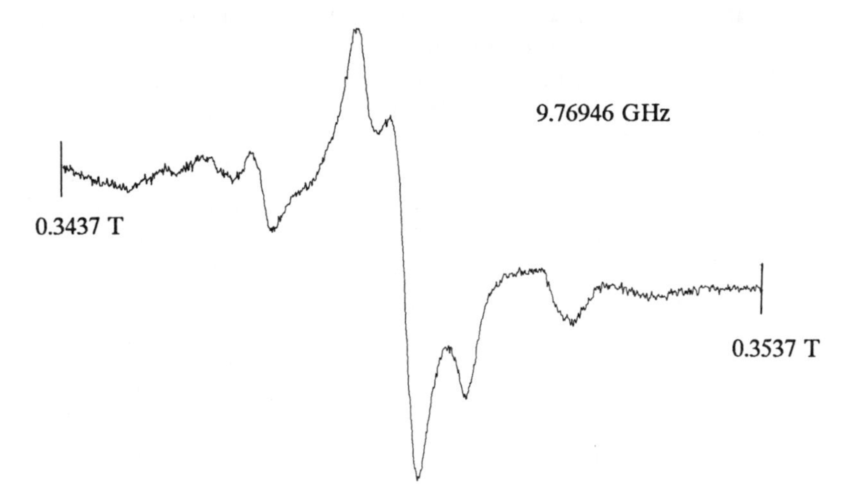

Figure 12. The spectrum due to sample L showing the relatively large line on the low-field side of the dating signal.

Table III. Heating Experiments on Bones Not Showing the Dating Signal

	Modern Sheep Bone		*Abu Hureyra Bone*		*Modern Rabbit Bone*	
T (°C)	g	ΔH (*mT*)	g	ΔH (*mT*)	g	ΔH (*mT*)
100	2.0048	0.78	Very weak		2.0047	0.78
200	2.0050	0.80	2.0048	0.59	2.0046	0.62
300	2.0045	0.61	2.0045	0.57	–	–
400	2.0045	0.48	2.0045	0.58	–	–
500	2.0045	0.50	2.0046	0.59	–	–
600	2.0044	0.48	–	–	–	–

NOTES: Abu Hureyra is the site in Syria (ca. 10,000 years B.P.) Each sample was held at the temperature indicated for about 4 h. The accuracy of the *g*-value measurements is about ±0.0002. Within the experimental error, all these lines are at about $g = 2.0046$.

to about 300 °C produced a signal similar to that in Figure 12. This temperature is consistent with heating on a hearth, but presumably only for a short time because the inner bone did not get very hot. Moreover, we suggest that the bone had already been defleshed when heated, because otherwise the exterior would not have attained such a high temperature so quickly. The practical reasons for such heating are not obvious, although it is possible that it might have had some ritual significance.

Table I shows that the ED for sample L is unusually low compared with those of the other samples, probably because it had been heated (3). (*See* the comment relating to the samples L and L′.) Sample K is anomalous because the very high concentration of iron in the sample makes accurate measurement of the dating signal difficult. The dates obtained for the other samples show reasonable agreement with other dating methods, given the assumption concerning the annual dose rate, and given that there is a tendency for the EDs to increase with layer number. The size of the dating signal indicates that the bone is extremely old. Furthermore, because the strength of the signal increases with γ-dosage, the defect traps are not saturated, and therefore the dating range of the ESR technique extends well beyond the ages of these samples.

Ikeya's (*11*) assignment of the quintet signal to the alanine radical is open to some doubt. Single-crystal studies of the alanine radical (*12–14*) show that the accidental equality of the methyl and methyne proton splittings is lost when the sample is cooled to liquid nitrogen temperatures. At these low temperatures, the rotation about the carbon–carbon bond is slowed and the lines in the spectrum are split into doublets. When the spectrum of a Zhoukoudian bone is recorded at lower temperature, the lines broaden somewhat but do not split. Furthermore, the hyperfine splitting (2.16 mT) is somewhat different from the average value obtained in such studies (2.35 mT). γ-Irradiation of heated powdered alanine gave the spectrum shown in Figure 13. The splitting (2.44 mT) was larger than that from the single crystal, and the extra splitting at room temperature of the five main lines was a new feature. Experiments with other amino acids gave spectra with a different

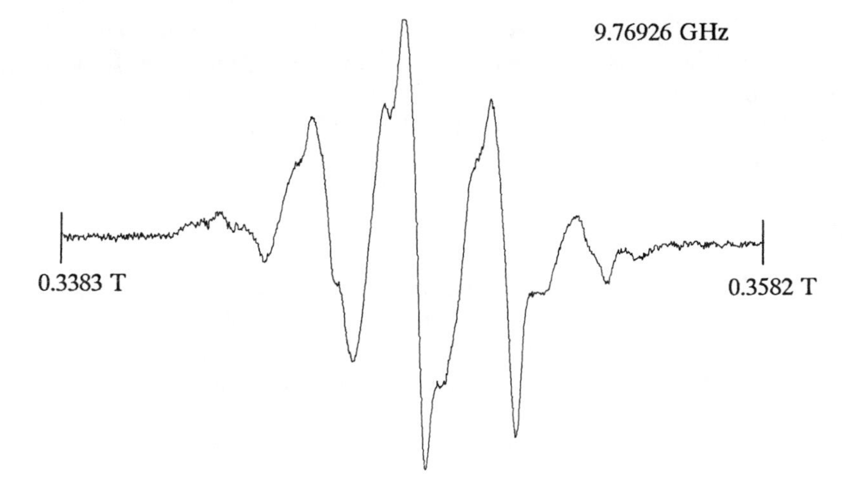

Figure 13. The spectrum obtained from powdered alanine after γ-irradiation.

number of lines. Attempts to modify the signal obtained from powdered alanine by mixing the alanine with silica and calcium silicate before γ-irradiation were not successful. Nevertheless, it is difficult to assign this signal to a species other than alanine, and the signal could be caused by alanine in some other form than a radical derived from the free amino-acid. This possibility is of such archaeological interest that we were prompted to examine it further.

The presence of the alanine signal in Zhoukoudian and in other ancient bone, and its absence in younger bone (i.e., <25,000 years old) suggest that there is a threshold for the signal's production. Earlier work on γ-irradiated heated bone had never produced the alanine signal. γ-Irradiation of a series of bones (modern, historical, and upper paleolithic) to various levels, up to the Zhoukoudian ED, produced only the dating signal. However, after gentle heating, the alanine signal appeared in all samples that had been γ-irradiated to a level corresponding to about 100,000 years. Further experiments showed that it is possible to generate the alanine signal in any bone, provided that the 100,000-year γ-dose is exceeded and that the sample is gently heated; the sequence of operations must be in this order.

The sensitivity of ESR spectroscopy is such that, in ancient bone, what we see may be a relict protein structure far beyond the limits of extractive techniques. The stability of the signal and the correspondence between ancient and modern bone suggests that the signal has not been caused by recent intrusions or a free amino acid. The alanine signal may be produced by an alanine moiety on the end of a protein chain that is denatured by the γ-irradiation. Presumably, heating the modern samples accelerates some chemical reaction that would otherwise take many years to complete; this hypothesis explains why the heating has to come after the γ-irradiation. The reason that alanine is the only amino acid observed is not clear. If the size of this signal could be quantitatively related to the age of the bone, then it might be used for dating purposes. For the moment, though, the observation of an alanine signal seems to indicate considerable antiquity.

Acknowledgments

We thank John Olsen of the Department of Anthropology, University of Arizona, and Professor Wu of the Institute for Vertebrate Paleontology, Beijing, for the Zhoukoudian bone samples. We thank Sandy Olsen of the Institute of Archaeology, London, and Lech Kaminski-Robins for the other bone samples, and Stuart Adams for the IR spectra. The ESR measurements were performed by the University of London Intercollegiate Research Service.

References

1. Hillman, G. C.; Robins, G. V.; Oduwole, D.; Sales, K. D.; McNeil, D. A. C. *J. Archaeol. Sci.* **1985,** *12,* 49.
2. Robins, G. V.; Sales, K. D.; Oduwole, D.; Holden, T.; Hillman, G. C. In *Lindow Man*; Stead, I. M.; Bourke, J. B.; Brothwell, D., Eds.; British Museum: London, 1986; p 140.
3. Sales, K. D.; Oduwole, A. D.; Robins, G. V.; Olsen, S. *Nucl. Tracks* **1985,** *10,* 845.
4. Robins, G. V.; Sales, K. D.; McNeil, D. A. C. *Chem. Br.* **1984,** *20,* 894.
5. Henning, G. J.; Grun, R. *Quat. Sci. Rev.* **1983,** *2,* 157.
6. Robins, G. V.; Seeley, N. J.; Symons, M. C. R.; McNeil, D. A. C. *Archaeometry* **1981,** *23,* 103.
7. Powell, M. J. D.; *Comput. J.* **1964,** *7,* 303.
8. Wasserman, E.; Snyder, L. C.; Yager, W. A. *J. Chem. Phys.* **1964,** *41,* 1763.
9. Griffiths, D. R.; Robins, G. V.; Seeley, N. J.; Chandra, H.; MacNeil, D. A. C.; Symons M. C. R. *Nature (London)* **1982,** *300,* 435.
10. Robins, G. V.; del Re, C.; Seeley, N. J.; Davis, A. G.; Hawari, J. A.-A. *J. Archaeol. Sci.* **1983,** *10,* 385.
11. Ikeya, M. *Naturwissenschaften* **1981,** *67,* 474.
12. Horsfield, A.; Morton, J. R.; Whiffen, D. H. *Mol. Phys.* **1961,** *4,* 425.
13. Miyagawa, I.; Gordy, W. *J. Chem. Phys.* **1960,** *32,* 255.
14. Morton, J. R.; Horsfield, A. *J. Chem. Phys.* **1961,** *35,* 1142.

RECEIVED for review June 11, 1987. ACCEPTED revised manuscript February 9, 1988.

20

Residues and Linings in Ancient Mediterranean Transport Amphoras

Curt W. Beck, Christopher J. Smart[1], and Dorreen J. Ossenkop

Amber Research Laboratory, Department of Chemistry, Vassar College, Poughkeepsie, NY 12601

Parallel analyses by IR, thin-layer chromatography (TLC), and gas chromatography–mass spectrometry (GC–MS) of organic remains adhering to shards of ancient amphoras excavated in the harbor of Carthage (Tunisia) identified these remains as pine pitches. Capillary GC of methylated acid fractions showed abietic acid, dehydroabietic acid, and 7-ketodehydroabietic acid as the principal components. Two-dimensional TLC of untreated ether extracts revealed abietic acid in 12 of 31 samples and dehydroabietic acid in 26 of 31 samples. IR spectra of solid, raw samples indicated the presence of isopropyl groups, characteristic of the abietane skeleton, in 80% of the samples. Rapid and convenient analysis by TLC and IR was, in most cases, sufficient to identify pine resin products even after extensive pyrolytic and oxidative degradation.

RESIDUES AND LININGS IN ARCHAELOGICAL CERAMICS, particularly in the unglazed transport amphoras that served as shipping containers of the Mediterranean Sea trade from the Bronze Age to historic times, have long been known. Until recently, the methods of analytical chemistry permitted only a cursory study and tentative identification of these complex organic mixtures. Linings are thin, continuous coatings applied to the interior surface of an unglazed vessel to render it impermeable; residues are the altered remains of the vessel's contents. In the archaeological record, both residues

[1]Present address: Department of Chemistry, Yale University, New Haven, CT 06520

0065–2393/89/0220–0369$06.00/0

and linings are loosely described as being resin, rosin, pitch, tar, asphalt, or bitumen without chemical analysis and with little regard to the precise meanings of these terms.

Definition of Terms

Resin applies to natural exudations of plants that have been used without intentional modification. In archaeological contexts, however, resins may undergo changes. The low-molecular-weight, volatile, monoterpenoid components of *oleoresins* are readily lost by evaporation, and the water-soluble carbohydrate components of *gum resins* will certainly dissolve if the object is exposed to water. Accidental exposure to fire leads to even more drastic pyrolytic transformations that may not be distinguishable from transformations caused by intentional heat treatment. Thus a sample that was a resin when originally used may have been converted to a pitch by a catastrophic fire.

Rosin is sometimes used to refer to certain resins, especially the natural exudates of fir and pine trees, and in such designations as "rosin-tree" for the South African shrub *Cineraria resinifera*. Strictly speaking, rosin is the residue after distillation of the volatile components of the whole resin, again, especially of fir and pine resins. The term is synonymous to *colophony*. In modern practice, colophony is obtained by vacuum distillation of the volatile constituents of resin in the absence of air. The product retains the typically yellow color of the original resin. In earlier times, the resin was heated in open vessels and the product was brown or black and partially pyrolyzed; it was, in fact, pitch. Because of this ambiguity, it is best to avoid the word *rosin* altogether.

Pitch is the residue after the distillation of volatile resin components in an open vessel. The material is typically black and is referred to as *pix* by Roman writers (*1*). When obtained from pine resin, it may be called *pine pitch*; when obtained from other plant resins, it may be called more generally *wood pitch*.

The distillate of resin distillation is called *tar*. *Pine tar* is from pine resin, and *wood tar* is from other plant resins. *Coal tar* is made from coal. The use of tar in antiquity is uncertain and unlikely (*2*).

Bitumen is a loosely used term (*3*). It is best defined to mean "native substances" (*4*) "composed principally of hydrocarbons and substantially free of oxygenated bodies" (*5*) that occur naturally as the heavy fraction of petroleum.

Asphalt suffers from the uncertainty of meaning of the Greek word *asphaltos* and of its correspondence to related words in ancient Near Eastern languages (*2*). In modern terminology the term refers to a natural or manufactured mixture of mineral fines and bitumen.

Previous Analyses

Of these materials, resin, pitch, and bitumen are mentioned in the classical literature and are likely to be found in Mediterranean archaeological contexts. Bitumen and asphalt from Near Eastern sites have been studied by Marschner and Wright (*6*). The identification of a wide variety of resins in archaeological and art historical contexts was pioneered by Mills and his coworkers (*7*) at the Research Laboratory of the National Gallery in London. The British group and Condamin and Formenti (*8*) in France used gas chromatography (GC) of methylated diterpene resin acids to identify pine pitch on the ram of a Carthaginian warship and in transport amphoras, respectively. Most recently, Shackley (*9*) reported the work done by Mills on resinous deposits in a 6th-century A.D. storage jar from Boqeq (Israel). GC showed dehydroabietic acid (structure **1**) and 7-ketodehydroabietic acid (structure **2**) but not abietic acid (structure **3**).

The identification of individual resin acids by GC is well established. The method requires a considerable investment of time and equipment and is unlikely to be used for the large number of amphoras and amphora shards that have come to light from numerous sites. A single shipwreck may well

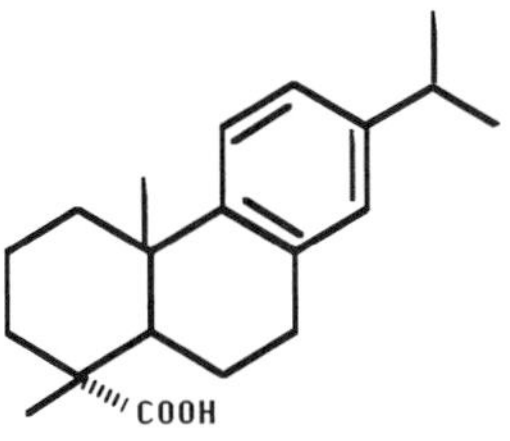

COOH

1

Dehydroabietic Acid

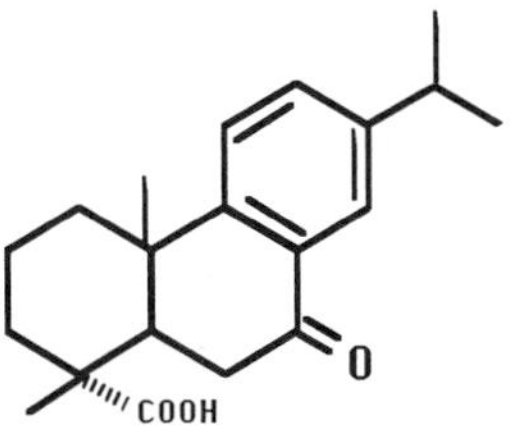

O

COOH

2

7-Ketodehydroabietic Acid

COOH

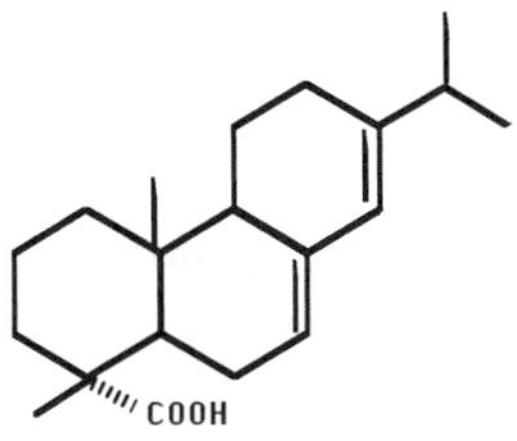

3

Abietic Acid

produce 500 to 1000 amphoras. Clearly, more rapid and simple methods of identifying resinous organic remains are needed.

A suite of shards of transport amphoras excavated by the Oriental Institute of the University of Chicago, under the direction of L. E. Stager, provided an opportunity to test infrared (IR) spectroscopy and thin-layer chromatography (TLC) as quick and easy methods of identifying the botanical sources of resins and to compare the results obtained with these methods with those obtained by GC and GC–mass spectrometry (GC–MS).

The Amphora Shards

Table I lists the 31 shards used in this study. Their fragmentary state made typological identification difficult and close dating impossible. According to Wolff (*10*), seven of the shards date to the Byzantine period (4th–7th centuries A.D.), one is Roman (1st–4th centuries A.D.), and the remaining 23 represent Punic, Greek, Graeco-Italic, Corinthian, and unidentified types, ranging in date from the 4th century B.C. to the end of the third Punic War in 146 B.C. As a group, the shards span more than a millennium.

The number of shards with organic remains was a small fraction of the total. Thousands of shards from the Punic period were collected, but fewer than 250 had any organic linings or residues. Among the shards of this period, only 8% of the Corinthian type, 5% of the Graeco-Italic type, and still smaller numbers for the other types had organic linings or residues (*10*).

Table I also shows the nature of the organic remains, as far as this could be ascertained. If the interior surface of a shard, and especially of a body shard, was covered with a thin, continuous coating of resinous material, the sample was classified as a lining. If there was a substantial, compact deposit at the bottom or "toe" of the vessel but no coating on adjacent interior surfaces, the sample was classified as a residue. Eight samples could not be classified because the evidence was ambiguous. Two samples from rims may have been seals used to lute the cover of the vessel opening.

Materials and Methods

IR Spectroscopy. Samples for IR spectroscopy were prepared by pressing 100-mg KBr pellets containing 1 mg of sample. Samples were run both on a Perkin–Elmer model 167 dispersive (grating) instrument and on a Perkin–Elmer model 1750 Fourier transform diffractometer–model 7300 laboratory computer system. Only the latter instrument afforded the resolution needed to identify the skeletal frequencies of isopropyl groups.

TLC Analysis. Two-dimensional TLC was carried out on Anasil-OF silica-on-glass plates. An ether solution of the crude sample was spotted on a plate, and the plate was developed twice with heptane to carry nonpolar components to the top edge of the plate. The acids remained at the origin and were separated by development of the plate perpendicular to the original direction with hep-

Table I. List of Samples and Results of TLC and IR Analyses

Sample No.	*Registration No.*	*Amphora Type*	*Amphora Part*	*Sample Type*	*TLC Result*		*IR Result*
					Abietic Acid	*Dehydroabietic Acid*	
1	12 627	Byzantine	Toe	Residue	−	+	Isopropyl
2	12 644	Late Punic?	Toe	Unidentified	+	+	No organic materials
3	12 628	Graeco-Italic	Toe	Residue	+	+	Isopropyl
4	12 625	Graeco-Italic	Body	Lining	+	+	No organic materials
5	12 641	Late Punic	Body	Unidentified	−	−	No organic materials
6	12 677-80	Byzantine	Toe and body	Lining	−	+	Isopropyl
7	12 635	Byzantine	Toe	Residue	+	+	Isopropyl
8	5 131	Corinthian	Toe	Lining	−	+	Isopropyl
9	12 640	Unidentified	Body	Lining	+	+	Isopropyl
10	12 637	Byzantine	Toe	Unidentified	−	+	Isopropyl
11	12 636	Byzantine	Toe	Residue	−	+	Isopropyl
12	12 639	Roman	Toe	Residue	−	+	Isopropyl
13	6 857	Graeco-Italic?	Toe	Residue	−	−	No organic materials
14	12 643	Corinthian	Upper body	Lining	+	+	Isopropyl
15	11 124	Flat-top triangular	Rim	Unidentified	+	+	Isopropyl
16	11 127	Graeco-Italic	Body	Lining	−	+	Isopropyl
17	12 645	Byzantine	Toe	Residue	−	+	Isopropyl
18	12 046	Graeco-Italic?	Toe	Unidentified	−	+	Isopropyl
19	11 957	Ridged	Body	Lining	+	+	Unidentifiable
20	12 089	Corinthian	Rim	Unidentified	−	+	Isopropyl
21	6 876	flat-top triangular	Rim	Sealant?	+	+	Isopropyl
22	12 626	Greek	Body	Lining	+	+	Isopropyl
23	12 629	Greek	Body	Lining	−	−	Unidentifiable
24	12 630	Graeco-Italic?	Body	Lining	+	+	Isopropyl
25	12 631	Greek	Body	Lining	−	+	Isopropyl
26	12 632	Greek	Body	Lining	−	−	Isopropyl
27	12 633	Greek	Body	Lining	−	−	Isopropyl
28	12 634	Greek	Body	Lining	+	+	Isopropyl
29	12 642	Late Punic	Toe	Unidentified	−	+	Isopropyl
30	12 638	Byzantine	Toe	Unidentified	−	+	Isopropyl
31	6 880	Corinthian	Rim	Sealant?	−	+	Isopropyl

tane–toluene–diethyl ether (1:1:1). The separated components were made visible by spraying the plates with 20% sulfuric acid and heating to 100 °C on a hotplate for 15 s. Partial charring of the organic acids on the plate provided an additional basis for identification; abietic acid spots turn reddish-brown, whereas dehydroabietic acid spots turn yellow-brown to greenish-brown.

GC Analysis. Samples for GC were prepared by extracting the crude sample with ether, extracting the ether solution (15 mL) three times with 5 mL of 5% sodium bicarbonate, acidifying the extract to pH 2 with concentrated hydrochloric acid, extracting the acids again with ether, drying the ether solution over anhydrous sodium sulfate, and methylating the acids with diazomethane. Both ether solutions were evaporated to dryness to determine the weights of organic materials and and acids in the sample.

Gas chromatograms were obtained by using a Hewlett–Packard model 5880A gas chromatograph equipped with flame-ionization detector and a data station. The chromatographic analyses were isothermal at 200 °C on a 12-m-length cross-linked methyl silicone capillary column. The quantitative data in Table II are derived from these analyses. Peak identities were confirmed on a Hewlett–Packard model 5995C/96A GC–MS system with model 5997A workstation with a 15-m-length RSL-150 polydimethylsiloxane capillary column at an initial temperature of 100 °C rising at 5 °C/min to 250 °C.

Reference Standards. Reference standards for TLC, GC, and MS were obtained as follows. Pure abietic acid (mp 172–173 °C) was used as received. Dehydroabietic acid was prepared by oxidation of abietic acid with selenium dioxide to hydroxyabietic acid and subsequent dehydration with glacial acetic acid (*11*). 7-Ketodehydroabietic acid was prepared by oxidation of dehydroabietic acid with potassium permanganate and isolation of the product by way of the Girard reagent T (*12*). Pyroabietic acid (a mixture of dehydro- and dihydroabietic acids) was prepared by heating abietic acid with 10% palladium on charcoal to 250 °C for 1 h (*13*).

Additional experimental details and analytical data are reported elsewhere by Smart (*14*).

Results and Discussion

IR Spectroscopy. IR spectroscopy is a simple and rapid method of identifying functional groups and skeletal types in milligram samples of solid organic materials. Dispersive spectra have been used to determine the

Table II. Analyses by Gas Chromatography–Mass Spectrometry

Sample Number	*Organic Materials*	*Acids*		*Abietic Acid*	*Dehydroabietic Acid*	*7-Ketodehydroabietic Acid*
		in Total Sample	*in Organic Fraction*			
1	84.41	28.05	33.23	1.0	7.8	10.4
7	76.78	13.46	17.53	20.9	51.8	2.9
10	73.14	3.16	4.33	none	10.5	22.0
12	71.52	5.78	8.07	none	12.4	22.9

NOTE: All values are percentages. Values for resin acids are given as percentages of the acid fraction.

provenance of fossil resins (*15*) and to identify organic remains in pottery from Quseir al-Qadim in Upper Egypt as pitch rather than bitumen (*16*). The high resolution of Fourier transform IR (FTIR) instruments permits a more detailed study of skeletal absorptions and, thus, the identification of structures characteristic of well-defined botanical sources.

About 80% of the diterpene acids of the genus *Pinus* have an abietane skeleton, with abietic acid (structure **3**) as the predominant component. Other diterpene acids have a pimarane skeleton (structure **4**). All compounds with the abietane skeleton are distinguished by the presence of an isopropyl group.

Extensive theoretical and empirical work (*17*) has shown that the isopropyl group is identifiable by (1) split symmetrical carbon–hydrogen deformations at 1382.5 ± 2.5 and 1367.5 ± 2.5 cm^{-1} and (2) by two skeletal vibrations. One of these vibrations occurs at a remarkably constant frequency of 1168.5 ± 1.5 cm^{-1}, whereas the frequency of the other vibration decreases as a function of the molecular weight (MW) of the rest of the molecule, from 1170 cm^{-1} for MW 15 (methyl group) to 1142 cm^{-1} for MW 99 (heptyl group). For an attached moiety of MW 259, as in abietic acid, the band whose frequency depends on the molecular weight of the rest of the molecule is found at still lower wavenumbers.

On aging, and more rapidly on heating, abietic acid is oxidized first by disproportionation (i.e., without need of oxygen) to dehydroabietic acid (structure **1**) and then (but only if exposed to oxygen) to 7-ketodehydroabietic acid (structure **2**). The skeletal frequencies of dehydroabietic acid lie at 1110, 1130, and 1175 cm^{-1} (*14*).

Of the 31 Carthaginian samples, 25 yielded IR spectra that indicate an isopropyl group (Table I), as illustrated by the spectrum of Carthage sample 6 of the lining of a 6th-century A.D. Byzantine amphora (Figure 1a). The carbon–hydrogen deformation bands lie at 1385 and 1367 cm^{-1} (Figure 1b), and the skeletal absorptions are at 1107, 1127, and 1175 cm^{-1} (Figure 1c), a very nearly perfect match for dehydroabietic acid. Six samples (2, 4, 5, 13, 19, and 23) yielded IR spectra that indicated either no organic constituents (samples 2, 4, 5, and 13) or a low concentration of

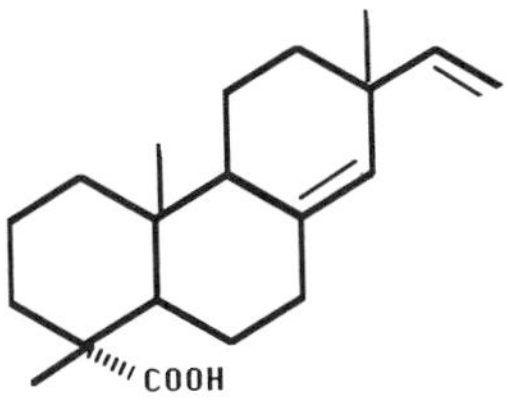

COOH

4

Pimaric Acid

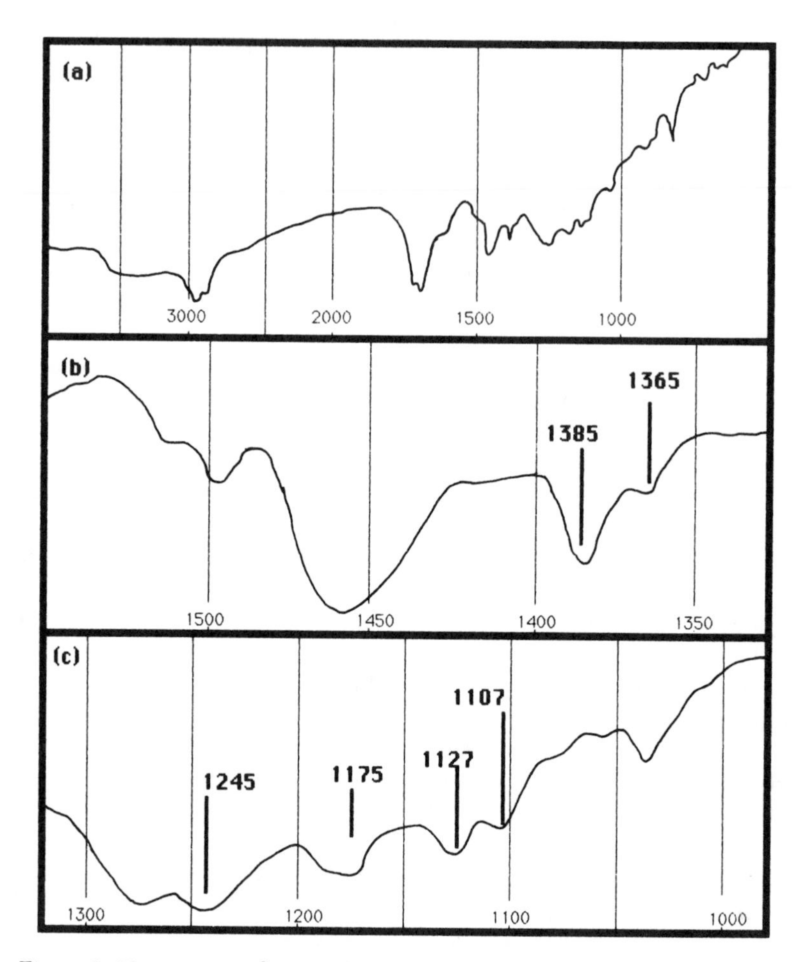

Figure 1. IR spectrum of an amphora lining (sample 6): complete spectrum (a) and partial spectra (b and c) showing isopropyl group absorptions.

unidentifiable organic compounds (samples 19 and 23). TLC later confirmed the absence of both abietic and dehydroabietic acids for three of these six samples (5, 13, and 23). For the remaining three samples (2, 4, and 19), both abietic and dehydroabietic acids were identified by TLC although these compounds were undetected by IR spectroscopy.

Eighty percent of all the Carthaginian samples and 90% of those that contain diterpene resins detectable by TLC were identified as pine resin derivatives by a single IR spectrum. Thus, IR spectroscopy is a useful rapid screening technique for large numbers of samples. However, its limitations are evident. Low concentrations of abietanes, if present in a strongly absorbing matrix, may be undetectable or unidentifiable by IR spectroscopy, and the limits of detectability depend on the absorbance of the matrix and cannot be categorically specified.

TLC Analysis. Small sample requirements, minimal sample preparation, high sensitivity, and low cost make TLC an attractive method for organic archaeometry. Its suitability for the detection of resin acids in complex mixtures was tested by subjecting the Carthaginian samples to a two-dimensional technique. Ether solutions of the organic material were spotted onto the plate and first freed from nonpolar components by elution in one direction with heptane. The residual carboxylic acids were then developed, with reference standards in adjacent tracks, in the second direction with heptane–toluene–ether (1:1:1). Under these conditions, 7-ketodehydroabietic acid remains at or very near the origin (maximum retardation factor [R_f] = 0.04), but abietic acid and dehydroabietic acid are readily identified.

The results of TLC analysis are directly compared with the IR results in Table I. Of the 31 Carthaginian samples, 26 contained dehydroabietic acid, but only 12 of these also contained unchanged abietic acid. This finding is not surprising, because abietic acid is converted to dehydroabietic acid as a function of time and temperature.

We cannot determine whether the abietic acid was lost during the manufacture of the pine pitch or during its depositional history. Loss during deposition is less likely, because almost all (six of seven) of the most recent samples (from the Byzantine period) were devoid of abietic acid. Of the five samples that showed neither of the resin acids by TLC (samples 5, 13, 23, 26, and 27), three (5, 13, and 23) yielded IR spectra that revealed no isopropyl group.

Three samples (4, 26, and 27) demonstrated the complementary rather than parallel functions of IR and TLC analyses in organic archaeometry. Organic materials were not revealed by IR in sample 4, but both abietic and dehydroabietic acids were detected by TLC in the same sample. Clearly, the resin acid concentrations were below the limit of detection of whole-resin IR spectroscopy but were sufficient for detection by TLC. Conversely, samples 26 and 27 showed neither acid by TLC but gave a clear indication of the isopropyl structure in the IR spectra. In this case, the IR technique was superior; although the resin acids have been lost evidently (most likely by decarboxylation), the remaining neutral compounds retained the isopropyl group, a fact that allowed the identification of the samples as pine pitches.

GC–MS Analysis. The acidic components of four Carthaginian samples (1, 7, 10, and 12) were subjected to a complete analysis by GC. Sample preparation is described in a previous section (*see* Materials and Methods). Results are given in Table II. The identities of the acids were established by (i) the retention times obtained from reference samples, (ii) the comigration of acids with reference standards during spiking experiments, and (iii) the mass spectra of the constituents (*18*).

The organic content of samples was consistently between 70 and 85% of the total sample weight, but the acid fraction varied almost by an order of magnitude, from 3 to 28% of the sample weight or from 4 to 33% of the organic fraction. This variation is probably due to differences in initial methods of preparation or, less likely, depositional history. The resin acids with an abietane skeleton in pine resins respond to heat by isomerization to abietic acid, disproportionation to dehydroabietic acid, and oxidation to 7-ketodehydroabietic. In addition, they are decarboxylated, a process that is catalyzed by active clay (*13*) and alkaline conditions (*19*). The variations in the acid composition of the four samples must reflect differences in preparation or history.

Sample 7 contained more than 20% unchanged abietic acid, more than 50% dehydroabietic acid, but very little 7-ketodehydroabietic acid. This composition is not expected of a pine pitch but of a pine resin that has been through slow disproportionation in an anaerobic environment. This residue is therefore that of a pine resin added to wine to make the retsina for which Greece was noted in antiquity as it is today. Sample 1 contained only 1% unchanged abietic acid, but more than half of the dehydroabietic acid was oxidized to 7-ketodehydroabietic acid, a finding that indicates the presence of air during the preparation of the pitch.

Samples 10 and 12 are virtually identical; neither contained any unchanged abietic acid, and both contained about twice as much 7-ketodehydroabietic acid as dehydroabietic acid. Both are extensively decarboxylated, as indicated by their low total acid content. Therefore, these samples are pitches that were prepared by heating a pine resin for a long time or at a high temperature with ready access to oxygen, as would be the case with stirring in a shallow vessel.

The GC data confirmed the TLC results. TLC detected abietic acid in sample 7 but not in samples 1, 10, and 12. This result indicates that a resin acid concentration of 1% is below the limit of detectability by TLC. TLC revealed dehydroabietic acid in all four samples, and this result was confirmed by GC. The GC data also corroborated the utility of IR spectroscopy. IR spectra showed the isopropyl structure in all four samples, including those for which the acid fraction was less than 10% (samples 10 and 12). Clearly, these isopropyl bands are not from the small amount of residual acid but are caused by skeletal absorptions of the isopropyl group in decarboxylated neutral decomposition products.

Conclusions

The analytical methods applied to the amphora shards from the harbor of Carthage each have distinct but overlapping utility. IR spectroscopy detects the carbon skeleton characteristic of pine resin or pitch with a reliability of 80–90% but without identifying individual constituents. It fails with small amounts of organic matter in a strongly absorbing matrix but has the advantage of detecting the isopropyl group in extensively decarboxylated samples. Because of the speed and ease of IR spectroscopy, it is an effective screening method when hundreds, or even thousands, of ceramic finds must be tested.

TLC is more sensitive than IR spectroscopy in detecting small amounts of resin acids because it is a separation technique, but it cannot identify pine resins that have been completely decarboxylated. However, completely decarboxylated resins are rare (two of 31 samples in the present study), and TLC can be used also for the rapid and inexpensive screening of large numbers of specimens with a reliability of >90%.

Both IR spectroscopy and TLC are therefore appropriate for determining whether an organic residue or lining is pinaceous. This information is far from trivial because it limits the materials that a transport amphora must have contained. Oils dissolve pine pitch and are discolored by it; a transport amphora lined with pine pitch cannot have been used to ship olive oil but may have contained wine, fruit, or preserved fish.

GC (with or without identification of constituents by MS) remains the most precise, quantitative method for the detailed analysis of resinous remains and the choice when a relatively small number of samples must be studied in depth. It provides information about the method of preparation, that is, about early resin and pitch technology. The analyses reported in this chapter and elsewhere (*20*) have been the impetus for current work in this laboratory on replication of ancient pitch manufacturing methods and identification of the types of pine resins that were used.

Acknowledgments

We thank Thomas F. Sanderson, Hercules Research Center, Wilmington, DE, for a sample of pure abietic acid and Roxanne J. Fine, Vassar College, for the preparation of the other standards.

This work was supported by an equipment grant towards the acquisition of an FT IR spectrophotometer from the National Science Foundation (Division of Anthropology; grant no. 84-01207). We are grateful to Vassar College for its continuing support of the Amber Research Laboratory (ARL); this report constitutes ARL Contribution No. 72.

References

1. Pliny *Historia Naturalis*, XVI:xvi–xxiii; Rackham, H., Ed.; Loeb Classical Library, Harvard University: Cambridge, 1968; pp 413–427.
2. Forbes, R. J. *Studies in Ancient Technology*, 3rd ed.; Brill: Leiden, 1964; Vol. I; pp 1–124.
3. Hanson, W. E. *Bituminous Materials: Asphalts, Tars and Pitches*; A. J. Hoiberg, Ed.; Wiley: New York, 1964; pp 1–24.
4. Tomkeieff, S. I. *Coals and Bitumens*; Pergamon: London, 1954; p 27.
5. Abraham, H. *Asphalts and Allied Substances*, 6th ed.; Van Nostrand: New York, 1960; Vol. I; p 54.

6. Marschner, R. F.; Wright, H. T. In *Archaeological Chemistry II*; Carter, G. F., Ed.; Advances in Chemistry 171; American Chemical Society: Washington, DC, 1978; pp 150–171.
7. Mills, J. S., White, R. *Stud. Conserv.* **1977,** *22,* 12–31.
8. Condamin, J.; Formenti, F. *Figlina* **1976,** *1,* 143–158.
9. Shackley, M. *J. Archaeol. Sci.* **1982,** *9,* 305–306.
10. Wolff, S. R. *Cahiers des Études Anciennes* **1986,** *19,* 136–153.
11. Fieser, L. F.; Campbell, W. P. *J. Am. Chem. Soc.* **1938,** *60,* 159–170.
12. Pratt, Y. T. *J. Am. Chem. Soc.* **1951,** *73,* 3803–3807.
13. Fleck, E. E.; Palkin, S. J. *J. Am. Chem. Soc.* **1937,** *59,* 1593–1595.
14. Smart, C. J., B.A. Thesis, Vassar College, Poughkeepsie, NY, 1983.
15. Beck, C. W. *Appl. Spectr. Rev.* **1986,** *22,* 57–110.
16. Beck, C. W.; Moray, L. In *Quseir al-Qadim 1978: Preliminary Report*; Whitcomb, D. S.; Johnson, J. H., Eds.; American Research Center in Egypt: Cairo, 1979; pp 253–256.
17. Bellamy, L. J. *The Infrared Spectra of Complex Molecules*; Wiley: New York, 1975; pp 21–29.
18. Zinkel, D. F.; Zank, L. C.; Weselowski, M. F. *Diterpene Resin Acids,* U.S. Department of Agriculture, Forest Service, Forest Products Laboratory: Madison, WI 1971.
19. Belov, V. N.; Kustova, S. D. *J. Gen. Chem. USSR* **1954,** *24,* 1083–1088.
20. Beck, C. W.; Ossenkop, D. J. In *M. Katzev, The Kyrenia Wreck,* in press.

RECEIVED for review June 11, 1987. ACCEPTED revised manuscript March 7, 1988.

21

Analysis of Mexican Amber by Carbon-13 NMR Spectroscopy

Joseph B. Lambert[1], James S. Frye[2], Thomas A. Lee, Jr.[3], Christopher J. Welch[4], and George O. Poinar, Jr.[5]

[1]Department of Chemistry, Northwestern University, Evanston, IL 60208
[2]Colorado State University, Regional NMR Center, Fort Collins, CO 80523
[3]New World Archaeological Foundation, Brigham Young University, Provo, UT 84602
[4]Diagnostic Division, Abbott Laboratories, North Chicago, IL 60064
[5]Department of Entomology and Parasitology, College of Natural Resources, University of California, Berkeley, CA 94720

Amber samples from two sites (Simojovel and Totolapa) in Chiapas, Mexico, were characterized by ^{13}C nuclear magnetic resonance spectra taken with magic angle spinning and cross polarization. Twelve samples from the two sites gave essentially identical spectra, despite a wide range of color from very light yellow to deep red. This result suggests that Chiapas amber comes from a single paleobotanical source. There are small but significant differences between the spectra of Mexican and Dominican ambers and even larger differences between both of those and the spectra of Baltic amber. Interrupted proton decoupling, which selects quaternary and methyl carbons, accentuated the differences between New World and Old World ambers.

AMBER IS A FOSSILIZED FORM of terpenoid resin found in many parts of the world and used by people of many different cultures. The scientific analysis of amber began as early as the 16th century (*1*), and the analysis of its chemical composition has been going on for over 100 years (*2*). Because the material is nearly insoluble, modern analysis is primarily by infrared spectroscopy (*3*). Additional structural information about amber is now avail-

0065-2393/89/0220-0381$06.00/0

able by using solid-state ^{13}C nuclear magnetic resonance (NMR) spectroscopy (*4*). With this method, we have characterized many amber samples from Europe (*5*) and the Dominican Republic (*6*). Such characterization provides information about trade routes, authenticity, chemical structure, and the paleobotanical origin of these samples.

Another prime source of amber is the Chiapas region of Mexico, where its use as early as 300 B.C. is documented. Amber is still used in Chiapas as jewelry and as a charm to prevent the "evil eye" from affecting children (*7*). The best known source of Mexican amber is the village of Simojovel (17° 8.9' N, 92° 42.8' W), but recently, an old source associated with the village of Totolapa (16° 31.3' N, 92° 41' W) but forgotten for centuries was rediscovered. The Totolapa source was reported in the late 1600s by the traveler and historian Vázquez (*8*) but was subsequently forgotten until 1982, when its exploitation by local campesinos became general knowledge (*9*).

Archaeological samples might have come from both locations. Consequently, methods of distinguishing the two sources are useful. To compare these two sources, we carried out the first ^{13}C NMR examination of Mexican amber. These studies can prove whether the two sources are paleobotanically identical; conversely, any differences that are found can be used to distinguish the two sources. In this chapter, we report a complete ^{13}C NMR study of amber from Simojovel and Totolapa. Furthermore, to increase our understanding of the polymeric structure of the resin, we carried out interrupted-proton-decoupling experiments as a measurement of the number of protons bonded to carbon atoms. We compared the results with those of similar experiments on amber from the Baltic Sea region, the traditional source of European archaeological amber.

Materials and Methods

Samples were obtained on site by T. A. Lee, Jr., C. J. Welch, and R. Lowe. Isolated samples of amber from Simojovel and the sample from the Dominican Republic were provided by G. O. Poinar, Jr. (University of California, Berkeley). The samples of Baltic amber were provided by C. W. Beck (Vassar College).

The typical sample size was 100 mg, but a sample as little as 35 mg was sufficient. Samples were crushed to a powder, although whole samples may be examined. Surface material was discarded if it appeared different from the bulk. Powdered specimens were examined directly by NMR; no processing was required. Spectral methods have been described previously (*4–6*).

Comparison of Mexican, Dominican, and Baltic Ambers

Mexican Ambers. Ten samples of amber from Simojovel and two samples from Totolapa were examined by ^{13}C NMR with magic angle spinning and cross polarization (MAS–CP). Figure 1 shows representative spectra. The spectra of the 10 Simojovel samples are extremely homogeneous, even though samples were chosen from a wide color range—very light yellow, light yellow, dark yellow (the traditional amber color), orange-red, and dark red. The two samples from Totolapa give spectra that were similar to each other and to those of samples from Simojovel. The spectrum of the Totolapa sample (Figure 1C) is almost indistinguishable from that of the second Simojovel sample.

The major difference between the spectra of the two Simojovel samples (Figures 1A and 1B) is in the alkene region (δ 100–160). In this spectral region are found all peaks for carbon atoms in C=C double bonds. The peaks at δ 110 and 150 are diagnostic for the $H_2C=C<$ (exomethylene) group, that is, an unsubstituted sp^2-hybridized carbon bonded to an sp^2-hybridized carbon with two carbon substituents. The exomethylene resonances were slightly stronger in the spectra of Totolapa samples. In the spectra of the Simojovel samples, these resonances range from very small to absent.

On the basis of these spectra, we conclude that Mexican amber from these two sources are essentially identical, or at least they are not distinguishable on the basis of the ^{13}C NMR spectra. These results suggest a common paleobotanical origin for the Chiapas ambers.

Baltic and Dominican Ambers. Figure 2 shows the ^{13}C NMR spectra of Baltic and Dominican (Figures 2A and 2B, respectively) ambers for comparison. Although the spectra show overall similarities, there are reliable distinctions. Baltic amber yields a very homogeneous set of NMR spectra (5), despite the enormous abundance and wide geographical sources of this material. The major differences between Baltic and Mexican ambers include the following features in the spectra of Baltic amber: (1) the stronger carboxyl resonances (δ 180–200), the dominance of the δ 173 ester peak, and the presence of an acid peak at δ 180; (2) the stronger exomethylene resonances at δ 110 and 150; and (3) the presence of a small alcohol peak at δ 67 and a strong methylene resonance near δ 28–30.

Mexican and Dominican Ambers. The Dominican ambers give spectra that are more similar to those of ambers from Mexico but with two significant differences. First, the exomethylene peaks in the spectra of Dominican amber vary in intensity but are always stronger than those of Mexican amber. Second, the spectra of Dominican amber always have an ether resonance near δ 75, which is very small in the spectra of Simojovel samples. The spectra of the Mexican and Dominican samples are very similar in the saturated region from δ 10 to 50.

On the basis of these comparisons, ^{13}C NMR spectroscopy can easily distinguish the two New World ambers from Baltic amber. Normally, Mexican and Dominican ambers are distinguishable on the basis of the exometh-

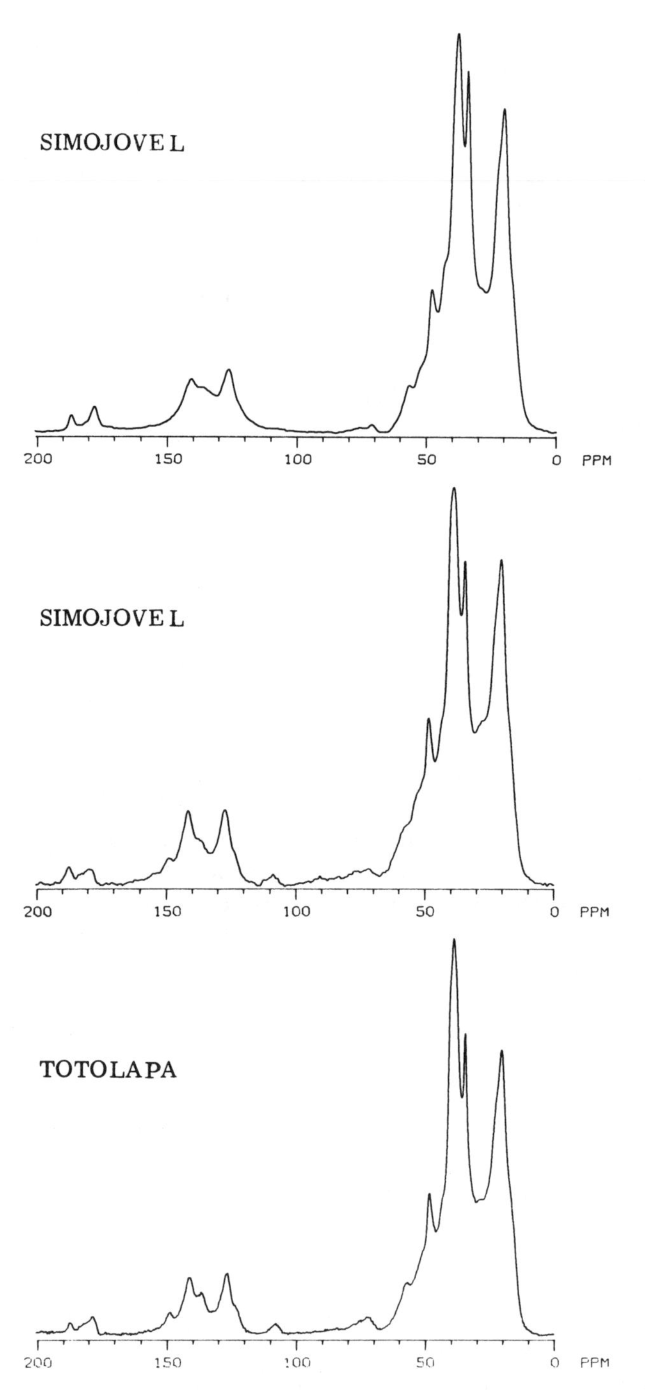

Figure 1. The fully proton-decoupled ^{13}C NMR spectra of amber samples from Simojovel and Totolapa, Chiapas, Mexico, taken on the solid samples with magic angle spinning and cross polarization. The top sample was orange-red and the middle sample yellow.

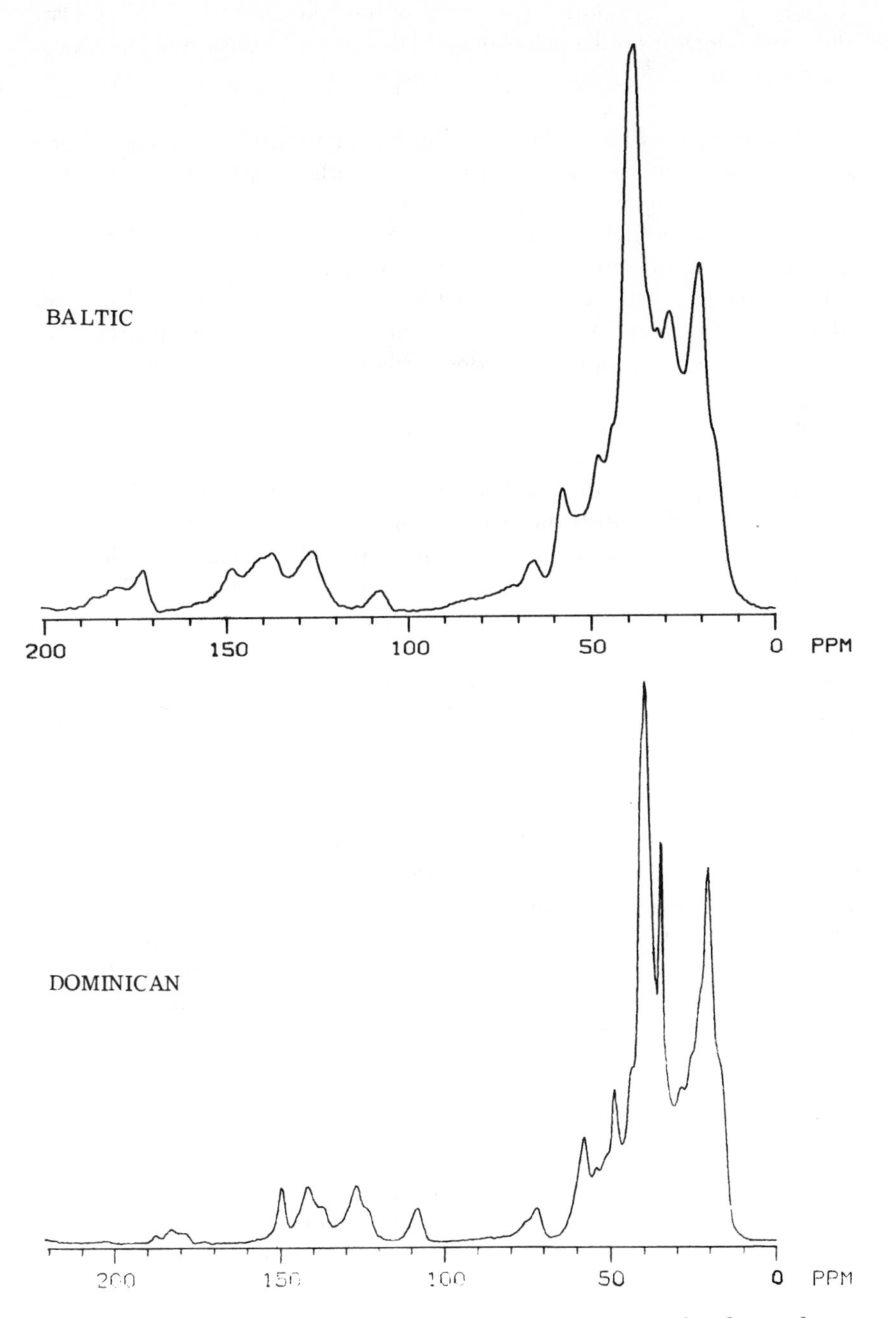

Figure 2. The fully proton-decoupled ^{13}C NMR spectra of Baltic amber (A, from Stettin) and Dominican amber (B, from Palo Alto).

ylene resonances, but outlying varieties of the two materials could overlap. The results suggest similar paleobotanical origins for Mexican and Dominican ambers.

Interrupted-Proton-Decoupling Experiments. Interrupted-proton-decoupling experiments were carried out with a 50-μs interruption. This technique selects quaternary carbons (those lacking an attached proton) and rapidly moving methyl carbons. Resonances from methylene and methine carbons are suppressed unless, for some reason, their motion within the solid occurs at a rate similar to that of methyl groups. In essence, the interrupted-proton-decoupling experiment, which has not been previously reported for amber samples, provides an alternative fingerprint for the samples.

Figure 3 gives the results of interrupted proton decoupling for the Simojovel sample with the ^{13}C NMR spectrum shown in Figure 1B and for the Baltic sample with the ^{13}C NMR spectrum shown in Figure 2A.

Changes in the spectrum of the Simojovel amber after interrupted proton decoupling are evident. The resonances of carbonyl carbons, which are quaternary, are clearly enhanced, and the resonances of CH or CH_2 groups bonded to electron-withdrawing atoms such as oxygen (between δ 50 and 70 in the original spectrum [Figure 1]) are completely suppressed. The resonances of quaternary carbons in the alkene region and of methyl carbons in the alkane region are enhanced. The sharp resonance at δ 50 is probably due to quaternary sp^3 carbons. The alkane and alkene regions still seem too highly populated. Possibly, rapidly moving CH_2 carbons are still represented.

The spectrum of the Baltic sample shows similar changes. The exomethylene $=CH_2$ peak at δ 110 is entirely suppressed, as are the peaks in the δ 50–70 region. Comparison of the two spectra in Figure 3 shows that interrupted proton decoupling provides more-differentiated spectra than does normal decoupling. The quaternary exomethylene resonance at δ 150 in the spectrum of the Baltic sample is enhanced, whereas the corresponding peak is absent in the spectrum of the Mexican sample. The remaining regions all show very clear differences.

Summary

Amber samples from Simojovel and Totolapa, in the state of Chiapas, Mexico, give ^{13}C NMR spectra that are essentially identical. Several samples of various colors from Simojovel also give nearly identical spectra. The results suggest that Mexican amber is a relatively homogeneous family of materials with a common paleobotanical source. Mexican and Baltic amber samples give distinguishable spectra. The spectra of Mexican and Dominican samples also are distinguishable, although there are considerable similarities. Inter-

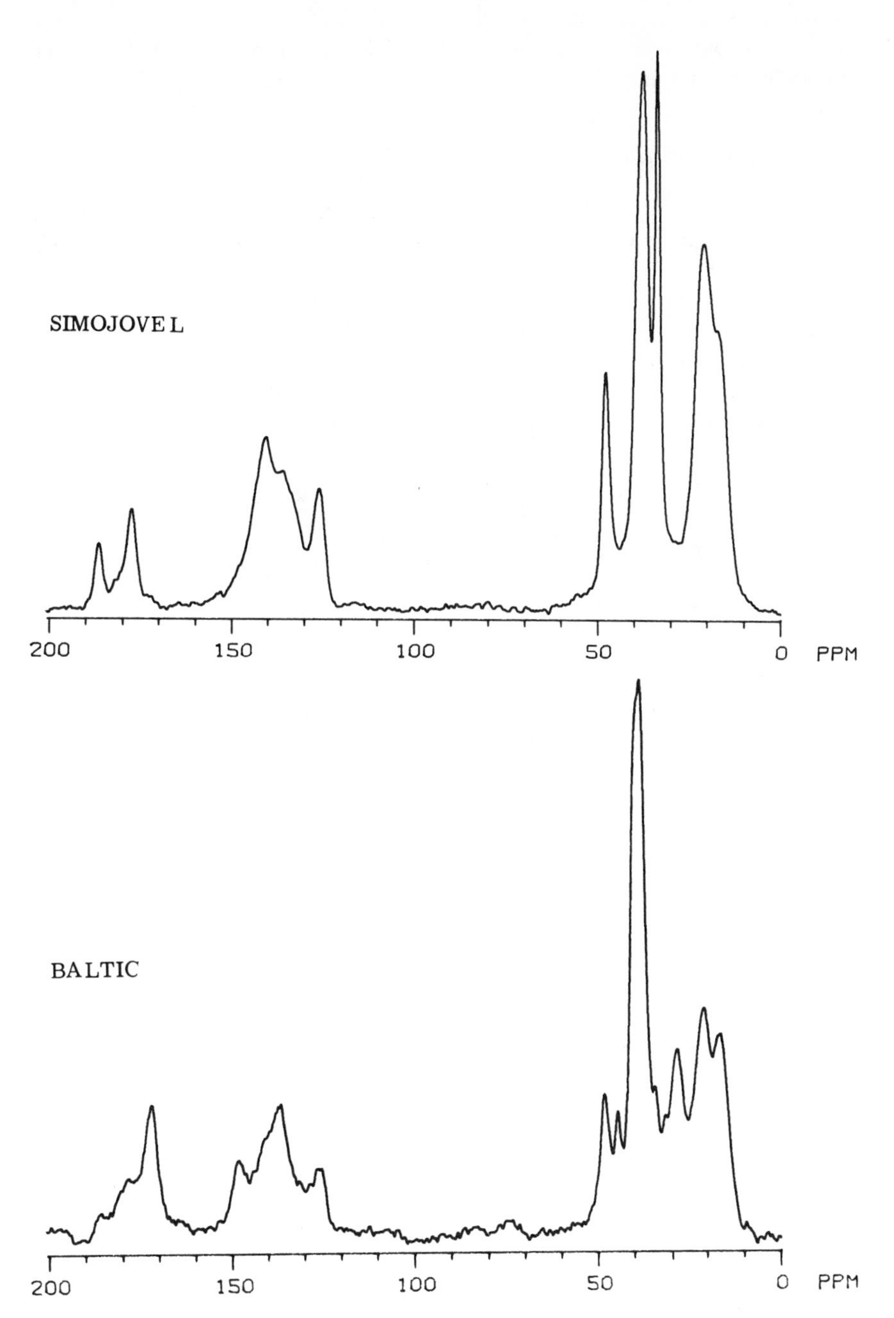

Figure 3. The ^{13}C NMR spectra of Mexican amber from Simojovel and Baltic amber taken with interrupted proton decoupling.

rupted proton decoupling, which preferentially selects quaternary and methyl carbons, accentuates the differences between the spectra of Mexican and Baltic amber samples.

Acknowledgments

This work was supported by the National Science Foundation via a grant to the Colorado State University Regional NMR Facility (grant no. CHE82-08821).

References

1. Agricola, G. *De Natura Fossilium*, English ed.; Bandy, M. A.; Bandy, J. A., Eds.; Special Paper No. 63; Geological Society of America: New York, 1955; original published in 1546.
2. Helm, O. *Schriften der naturforschenden Gessellschaft in Danzig N. F.* **1878,** *4*(3), 214–216.
3. Beck, C. W. *Appl. Spectrosc. Rev.* **1986,** 22, 57–110.
4. Lambert, J. B.; Frye, J. S. *Science* **1982,** *217*, 55–57.
5. Lambert, J. B.; Beck, C. W.; Frye, J. S. *Archaeometry* **1988,** *30*, 248–263.
6. Lambert, J. B.; Frye, J. S.; Poinar, G. O., Jr. *Archaeometry* **1985,** 27, 43–51.
7. Navarrete, C.; Lee, Jr., T. A., *Boletin Instituto Nacional de Antropologia e Historia*, **1969,** *35*, 13–19.
8. Vázquez de Espinosa, A. "Compendio y Descripción de las Indias Occidentales"; Smithsonian Institution Miscellaneous Collection 108; Washington, DC, 1948.
9. Bryant, D. O. *Am. Antiq.* **1983,** *48*, 354–357.

RECEIVED for review June 11, 1987. ACCEPTED revised manuscript February 8, 1988.

22

Detection of the Human Origin of Blood or Tissue

Use of a Monoclonal Antibody Specific to Human Albumin

John C. Herr[1], David C. Benjamin[2], and Michael P. Woodward[3]

[1]Department of Anatomy and Cell Biology, University of Virginia, Box 439, Medical Center, Charlottesville, VA 22908
[2]Department of Microbiology, University of Virginia, Box 439, Medical Center, Charlottesville, VA 22908
[3]Humagen, Inc., 1500 Avon Street, Charlottesville, VA 22901

This chapter describes the development and validation of monoclonal antibody probes specific to the human protein, albumin, which is a 66,000-dalton molecule present in serum and tissue. The monoclonal antibodies were employed in an enzyme-linked immunosorbent assay (ELISA) capable of detecting nanogram amounts of this protein. The assay was developed under contract from the Federal Bureau of Investigation (FBI) to detect the presence of human albumin in forensic specimens, and thus determine the species of origin of an unknown blood stain. The assay described in this chapter has not been validated on archaeological specimens. The method is described in this volume to acquaint the archaeological community with key concepts and research reagents that may be used in determining the species of origin of unknown, presumptive blood or tissue stains or fragments.

THE NEED FOR PRECISE METHODS of identifying the species of origin of a blood stain or tissue fragment found in an archaeological setting is well documented. In his analysis of stone tools of various types and materials, Loy (*1*) detected surface blood deposits on 86% of the samples. These artifacts

0065-2393/89/0220-0389$06.00/0

ranged in age from 1000 to 6000 years, and on many artifacts, blood proteins and red blood cells were detected. Two methods of determining the species of origin of blood deposits have been proposed: (1) crystallization of the hemoglobins extracted from the artifact and comparison of the hemoglobin crystal structure with that of hemoglobin crystals from contemporary species and (2) determination by immunological techniques (*1*).

The development of monoclonal antibody probes specific to human albumin is the subject of this chapter. These antibody probes and the test methods were developed and evaluated for use in forensic science situations. Although these methods work well on soluble extracts of dried blood stains that are several years old, they have not been applied to archaeological material. The successful development of monoclonal antibodies specific to the human albumin molecule suggests that an approach similar to that used for identifying human tissues and blood on forensic evidence could be applied to any species of interest to the archaeologist.

Protein Antigens

Proteins are the most abundant and diverse class of antigens to which the immune system can respond. This class includes toxins, allergens, products of infectious organisms, and transplantation antigens. Also included in this class are proteins that are of particular interest to archaeologists, such as blood and tissue proteins, which may be soluble or found on the cell surface, secreted or nonsecreted.

The use of immunological methods to detect specific proteins and to identify the species of origin of a sample depends on the ability of immunological reagents to specifically distinguish between similar, yet antigenically very complex, proteins from closely related species. The development of such methods has been difficult, and only recently, with the advent of monoclonal antibody and recombinant DNA technologies, have we been able to better understand the structural and genetic bases for immunogenicity and antigenicity and to produce reagents with the required specificity.

Antigen Structure

Immunogenicity is the ability to induce an immune response, whereas *antigenicity* is the ability to be recognized by the product of that immune response, for example, the antibody. Antibodies formed in response to a native protein often do not react with the denatured form of that protein (*2*, *3*), and yet antibodies can be raised against peptides of undefined conformation (*4–8*), for example, synthetic vaccines. Antigenic sites (also known as antigenic determinants or epitopes) are of two types: (1) assembled, that is, consisting of amino acids far apart in the primary sequence that are brought together on the surface of the antigen as it folds into its unique three-dimensional shape, or (2) segmental, that is, existing wholly within a continuous segment of the amino acid sequence (9) (Figure 1).

Although assembled antigenic sites are more numerous, both sites are topographic in that they are composed of structures on the protein surface. However, an antibody exerts high-affinity binding to segmental sites only when those sites are in a preferred conformation. Therefore, all determinants must be conformation specific.

There are two views of what constitutes an antigenic site. One view maintains that antigens possess a very limited number of sites that are immunogenic irrespective of the individual or species responding to that antigen (*10–12*). This concept suggests that the chemical or physical properties of certain parts of protein molecules make them intrinsically more immunogenic. The opposing view (9) states that most, if not all, of the accessible surface of a protein molecule is potentially immunogenic, that immunogenic sites are defined only with respect to a particular individual, and that the total antigenic structure of a protein is the sum of all sites recognized by a large variety of individuals and species.

The vast majority of evidence, which was reviewed recently (9), supports the following two concepts:

1. The surface of a protein is essentially a continuum of potential antigenic sites.

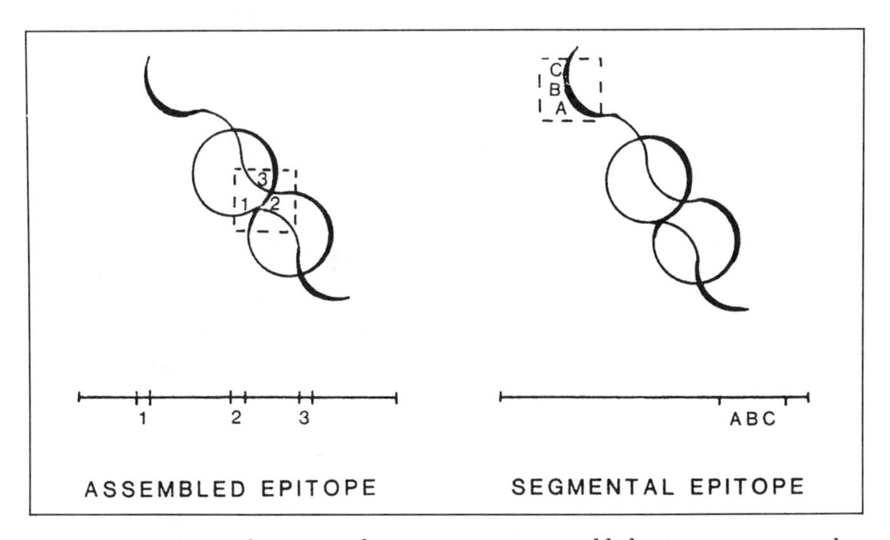

Figure 1. Types of antigenic determinants. An assembled epitope is composed of amino acid side chains (1, 2, and 3) that are well separated in the linear sequence of the protein but are brought together during folding of the protein. A segmental epitope is composed of amino acid side chains (A, B, and C) that are contiguous both in the linear sequence of the protein and on the surface of the protein in its folded form.

2. The structural differences between the antigen and the host's own version of that protein (self-protein), as well as the host's immunological regulatory mechanisms, are the important factors influencing the outcome of the immune response.

The first concept, that of a continuum of antigenic sites, is supported by studies of antibody responses to, among others, hen egg-white lysozyme (*13*) and studies on serum albumins (*14–16*). Using a panel of 60 monoclonal antibodies, Benjamin et al. (*14*) defined a minimum of 35 antigenic sites on bovine serum albumin (BSA). An extensive study of competition between pairs of monoclonal antibodies for simultaneous binding to BSA suggests complex patterns of overlap that are consistent with the entire surface of BSA being immunogenic. For example, there are numerous cases of three antibodies interacting in a pattern such that antibody A competes with B, and B competes with C, but A does not compete with C.

The second concept is based on a large number of studies on (1) the antigenicity of slowly evolving proteins such as cytochrome c (*17–19*), (2) the effect of evolutionary substitutions in the primary sequence of proteins on their immunological cross-reactivity (*20–23*), and (3) the effect of specific regulatory cells on antibody responses (*23–27*). Cytochrome c, in contrast to serum albumin, is an example of a slowly evolving protein (Table I), with few sequence differences between each cytochrome c and that of the responding individual. The rabbit antibody response to pigeon cytochrome c is directed against four sites on the pigeon cytochrome (*9*). These four sites contained all the seven sequence differences between rabbit and pigeon cytochrome c proteins.

In other studies (*20*, *21*, *28*), rabbits were immunized with a particular protein from one organism and the antisera were tested for reactivity with evolutionary variants of that protein. A high degree of correlation between immunological distance (calculated from the immunological cross-reactivity observed) and the number of amino acid differences between each pair of proteins tested was found (Table II). Such data suggest that most, if not all, of the differences in the surface amino acids of the antigen and the self-protein of the responding host can be immunologically detected.

Amino Acid Composition of Antigenic Site. Structurally, each antigenic site on the surface of a protein antigen is composed of a number of side chains of amino acids. To date, the exact boundaries of a single

Table I. Evolutionary Variance in Amino Acid Sequence

Protein	*Difference*[a]
Cytochrome c	9.6
Serum albumin	27.5

[a]Mean number of amino acid differences per 100 residues.

Table II. Correlation Between Amino Acid Sequence Identity and Immunological Cross-Reactivity

Protein	*Correlation Coefficient*
Lysozyme c	0.95
Ribonuclease	0.92
Cytochrome c	0.87

SOURCE: Adapted from Benjamin et al. (9) and based on data from reference 30.

antigenic site have not been defined. However, on the basis of what is known of the surface topography of proteins and the structure of the antibody-combining site, we estimate that the antigenic site consists of six to eight amino acids. Within this surface configuration of six to eight amino acids, a difference of a single amino acid between the antigen and a similar protein in the responding host may be sufficient to render that region immunogenic. Although there may be only one amino acid difference for a given site, X-ray crystallographic analyses of antigen–antibody complexes show that adjacent amino acids that are identical to those in the same region of the responding host's protein are involved in antibody binding to the site (29).

For sites that have been studied in detail, no single amino acid is predominant. Indeed, aliphatic, charged, neutral, and hydrophobic amino acids can all be found within one or more sites. Similarly, no particular physical feature of the surface of a protein antigen dictates antigenicity. Therefore, an antigenic site can be defined only with reference to a single antibody molecule, that is, a monoclonal antibody, which is a specific configuration of amino acid side chains on the surface of the protein antigen that possesses a configuration complimentary to the antibody-combining site.

Cross-Reactivity

Antibodies to one protein antigen will often react (cross-react) with a wide variety of similar proteins from other species. For example, 15% of the antibody in a polyclonal antiserum to bovine serum albumin will react with human serum albumin. The molecular basis for this cross-reaction is not clearly understood, but cross-reactivity is believed to be caused by the presence of a configuration of amino acid side chains on another protein molecule that has a sufficient overall similarity to the specific configuration on the original antigen that induced antibody formation.

Figure 2 shows diagrams of three protein antigens, each with a single antigenic site. The antigenic sites have similar configurations; proteins A and C share three of four amino acids, whereas protein A shares two of four amino acids with each of proteins B and C. Therefore, the efficiency of reaction of an antibody to protein B must be in the order B >C >A, whereas an antibody to protein A must react equally well with proteins B and C. However, the exact effect of any given substitution cannot be predicted.

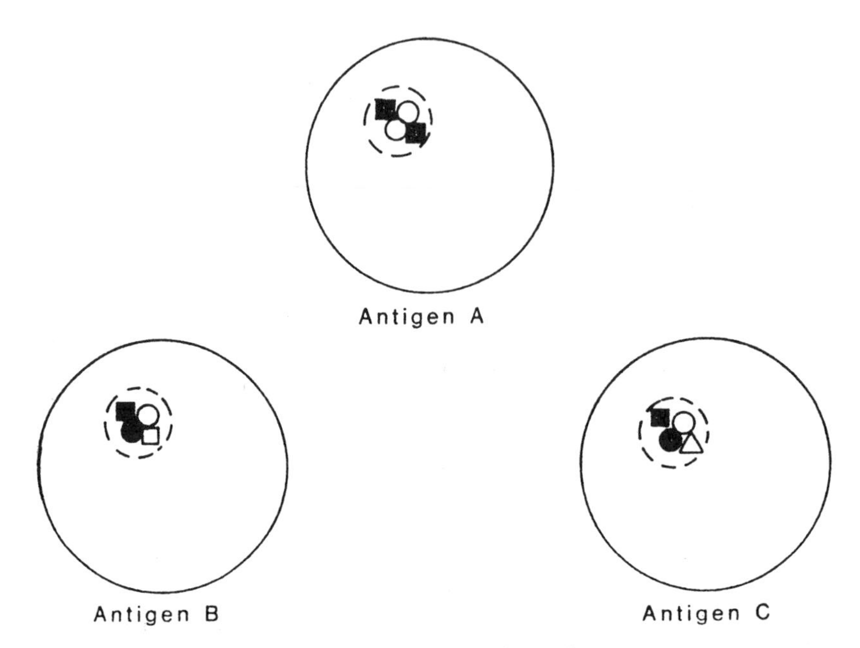

Figure 2. Diagram of three protein molecules with cross-reacting antigenic sites. Only one site is shown per protein molecule. The antigenic site on each protein is similar but not identical to the sites on the other proteins.

Berzofsky et al. (*30*) showed that one antigenic site defined by a single monoclonal antibody on the surface of sperm whale myoglobin contains the amino acids glutamic acid, alanine, and lysine at positions 83, 144, and 145, respectively. These three amino acids are brought together on the surface of the antigen by its specific folding after synthesis; that is, the antigenic site is an assembled site. Although these amino acids constitute only a part of the site, these studies provided considerable information on cross-reactivity, in addition to localizing the site on the surface of myoglobin. The data in Table III summarize the results of Berzofsky et al. (*30*) and show that a change of glutamic acid to aspartic acid at position 83 in killer whale myoglobin has a 24-fold effect on the binding of antibody, whereas a single amino acid change from alanine to glutamic acid at position 144 in bovine myoglobin essentially eliminates all reactivity. Thus, with respect to this monoclonal antibody, only killer whale myoglobin is a cross-reacting antigen.

With the divergence of two species from a common ancestor, mutations in the structural genes encoding similar proteins result in amino acid differences between the proteins produced by the now separate species. The degree of immunological similarity between any two such proteins is directly related to the evolutionary distance between the two species. Different proteins evolve at different rates. The structural basis for this difference in rate is that, for slowly evolving proteins, changes in amino acid sequence

Table III. Inhibition of Monoclonal Antibody by Myoglobin Variants

Source of Inhibitor Myoglobin	*Amino Acid at Position*			*Myoglobin Required for 50% Inhibition (nM)*
	83	*144*	*145*	
Sperm whale	Glu	Ala	Lys	10
Killer whale	Asp	Ala	Lys	240
Bovine	Glu	Glu	Lys	40,000

SOURCE: Adapted from Berzofsky et al. (*30*).

have a greater effect on function and are less well tolerated. The examples given in Table I therefore indicate that, on the average, changes in serum albumins have less effect on critical function than do changes in cytochrome c.

Some of these evolutionary changes may influence antigenicity and cross-reactivity through long-range effects. However, for most proteins, these evolutionary changes have only local effects on structure and function. Indeed, substitutions that markedly affect the binding of one monoclonal antibody have little effect on the binding of a second monoclonal antibody to an adjacent site (*30, 31*). Any hypothesis that assumes long-range effects would have predicted otherwise.

Species Differentiation by Immunological Methods

Since the turn of the century, serologists, especially in the forensic field, have used polyclonal antisera to distinguish between species (*32–34*). The reliability of the results depends on high-quality, carefully tested antisera that have been exhaustively absorbed to eliminate cross-reactivity with proteins from other species (*6*). The cross-reactivity of polyclonal antisera is the major limitation of current tests for species of origin. Thus the phylogenetic relationships between species impose intrinsic limits on the differentiation of related blood or tissue specimens. Sensabaugh (*36*) noted that forensic immunological analyses would be far more satisfactory if monospecific antisera to rapidly evolving blood marker proteins were used; this approach would take advantage of greater differences in the structures of similar proteins, even among very closely related species.

Proteins such as cytochrome c are not good target antigens for the determination of species of origin because the structures of these proteins in evolutionarily distant species are very similar. Serum albumin, on the other hand, is a rapidly changing protein (Table I) that is amenable to complete immunochemical analysis with both polyclonal and monoclonal antibodies (*14–16*). Albumins from thousands of pairs of vertebrate species have been compared immunochemically, and the results have been used to measure evolutionary distances (*22*).

A large body of literature spanning three decades has shown that albumin is a highly complex antigen with multiple antigenic sites (*9*). Studies on the antigenic structure of bovine serum albumin (*14–16*) have shown that species-specific monoclonal antibodies can be produced and that a panel of cross-reactive antibodies can be assembled for use in identifying a number of other species.

These studies suggest that monoclonal antibodies specific to human albumin may be produced; such antibodies will be of value in determining the human origin of blood, tissue, and other body fluids. We have been successful in producing a monoclonal antibody that reacts with an epitope on serum albumin that is specific to samples of human origin.

Materials and Methods

Serum Albumins. Serum albumins were either purchased from Sigma Chemical Company or prepared from serum as previously described (*37*).

Production of Hybridoma Cell Lines. Eight-week-old female BALB/c mice were injected three times intraperitoneally with 500 μg of human serum albumin (HSA) at 2-week intervals. The first injection was in complete Freund adjuvant (1:1, v/v), and the succeeding two injections were given in incomplete Freund adjuvant (1:1, v/v). The mice were allowed to rest for 30–60 days and injected intravenously with 50 μg of HSA. Four days later, spleen cells from these mice were fused with SP2/0-Ag14 myeloma cells by using polyethylene glycol, as previously described (*38*). Hybridomas of interest were cloned by limiting dilution.

Samples. Primate sera were obtained from the Yerkes Primate Center (Atlanta, GA). Human milk, saliva, and urine were obtained from donors and frozen at −70 °C within 15 min of receipt. Vaginal secretions were obtained on sterile swabs during routine pelvic examinations. Human kidney, liver, and pancreas were obtained from cadavers within 6 h of death and homogenized in carbonate–bicarbonate buffer (pH 9.2) in a Precision Scientific tissue homogenizer. The homogenates were centrifuged for 10 min at 500 × *g* to pellet particulate material, and the soluble proteins were precipitated with cold acetone at a 10:1 ratio. A pellet of the precipitated proteins was obtained after 10 min of centrifugation at 500 × *g* and was resuspended in carbonate–bicarbonate buffer.

Screening Assay. Supernatant fluids from wells that showed good hybridoma growth were screened against HSA-coated wells by enzyme-linked immunosorbent assay (ELISA) (*39*) with a goat anti-mouse immunoglobulin (anti-MIg) (heavy- and light-chain specific; Hyclone) conjugated with horseradish peroxidase (HRP–anti-MIg). Supernatant fluids from cultures of positive clones were screened extensively against a large panel of serum albumins. Hybridomas of interest were subcloned several times, and ascites fluid was produced from pristane-primed BALB/c mice.

Trapping ELISA with Two Monoclonal Antibodies. Polystyrene ELISA plates (Immulon II, Dynatech, Chantilly, VA) were coated with 10-μg/mL HSA-2 (primate-specific clone) in tris(hydroxymethyl)aminomethane (Tris) saline (pH 7.5) for 1 h at room temperature. The plates were emptied; filled with a solution containing 0.05% Tween 20 [sorbitan monododecanoate poly(oxy-1,2-ethanediyl) derivatives], 50 mM Tris, 150 mM NaCl, and 0.01% thimerosal (TTN); and incubated for 30 min at room temperature. After incubation, the plates were emptied, and samples were applied (100 μL per well) for 30 min at room temperature. The plates were rinsed with TTN, and biotinylated HSA-1 was applied (100 μL or 0.5 μg per well) for 30 min at room temperature. After three washes with TTN, 100 μL of streptavidin–HRP (0.25 μg/μL) was applied per well. After 30 min at room temperature, the plates were emptied and washed five times with TTN; 100 μL of a solution containing 1 mM ABTS, [2,2′-azinobis(3-ethyl-2,3-dihydro-6-benzothiazolesulfonic acid)], 0.03% H_2O_2, and citrate–phosphate buffer (pH 4.2) was added per well. The plates were incubated for 30 min at room temperature. Color development was measured at 410 nm with a spectrophotometer (Titertek Multiscan; Flow Laboratories, McLean, VA), after which the plates were photographed.

Ascites Production. Anti-albumin clones HSA-1 and HSA-2 were thawed, subcloned, and reassayed by ELISA on human albumin, and a rapidly growing subclone was expanded for ascites production. Forty male BALB/c mice were primed with an intraperitoneal injection of 0.5 mL pristane on day 0. On day 14, 2×10^6 cells were injected into each mouse. Ascites fluid was collected over a 1-week period and pooled. Cells and other debris were removed by centrifugation. Approximately 180 mL of ascites fluid was obtained.

Purification of Monoclonal Antibody. Immunoglobulins were precipitated from the pooled ascites by addition of an equal volume of saturated ammonium sulfate [50% $(NH_4)_2SO_4$]. The precipitate was collected by centrifugation (20 min; 10,240 × *g*), dissolved in 0.01 M sodium phosphate (pH 6.8), and reprecipitated. After the second ammonium sulfate precipitation, the pellet was dissolved in a minimum volume of 0.01 M sodium phosphate (pH 6.8) and centrifuged for 10 min at 10,600 × *g*). The resulting supernate was applied to a P6G, gel filtration polyacrylamid, column (Bio-gel; Biorad, Rockville Center, NY; 1.5 × 40 cm). Fractions containing protein were pooled and applied to a hydroxyapatite column that had been equilibrated with 0.01 M sodium phosphate (pH 6.8). Proteins were eluted with a linear gradient of 0.01 to 0.3 M sodium phosphate.

Column fractions were analyzed by sodium dodecyl sulfate–polyacrylamide gel electrophoresis (SDS–PAGE) and ELISA. Tubes containing maximal anti-albumin activity were pooled and concentrated. A portion of this concentrate was analyzed by SDS–PAGE.

Biotinylation of Monoclonal Antibody. Approximately 2 mg of monoclonal antibody HSA-1 was biotinylated with *N*-biotinyl-ω-aminocaproic acid *N*-hydroxysuccinimide ester (Enzotin; Enzobiochem, NY) at a protein-to-Enzotin ratio of 1:22 (*40*).

SDS–PAGE. Column fractions and purified immunoglobulin were analyzed on 10% or 10–20% linear gradient polyacrylamide gels in the presence of 0.1% SDS. Molecular weight standards (Sigma and LKB) were used for molecular weight estimations.

Results

Types of Monoclonal Antibodies to Human Albumin. The monoclonal antibodies to serum albumin were one of three general types (Table IV). Type A is highly specific for human albumin; type B cross-reacts extensively with primate albumins, with little or no cross-reactivity with albumins from other species; and type C shows variable reactivity with albumins from a variety of species. Examples of the reactivity of type A and type B monoclonal antibodies in an ELISA are shown in Figure 3. Serum samples from the following species were tested: human, chimpanzee, gorilla, orangutan, dog, cat, hamster, cow, sheep, pig, rat, raccoon, goat, rabbit, antelope, mule deer, elk, horse, white tail deer, guinea pig, and turkey. Only human serum reacted with monoclonal antibody HSA-1.

Two monoclonal antibodies, one highly specific for human albumin (HSA-1) and one of broader specificity for several primates (HSA-2), were selected for subsequent use. These antibodies were purified by hydroxyapatite chromatography to single heavy- and light-chain bands (Figure 4) and then biotinylated.

A double-antibody, trapping ELISA was developed that used biotinylated HSA-1 to detect the presence of albumin bound to an unsolubilized second antibody, HSA-2, which was reactive with an epitope common to

Table IV. Classes of Monoclonal Anti-Albumin Antibodies in Various Species

Antibody Type	*H*	*C*	*G*	*O*	*D*	*Ca*	*B*	*S*	*De*	*Ho*
A	+	—	—	—	—	—	—	—	—	—
B	+	+	+	+	—	—	—	—	—	—
C	+	—	+	—	—	+	+	+	—	+

NOTE: Species are abbreviated as follows: H, human; C, chimpanzee; G, gorilla; O, orangutan; D, dog; Ca, cat; B, bovine; S, sheep; De, deer; and Ho, horse.

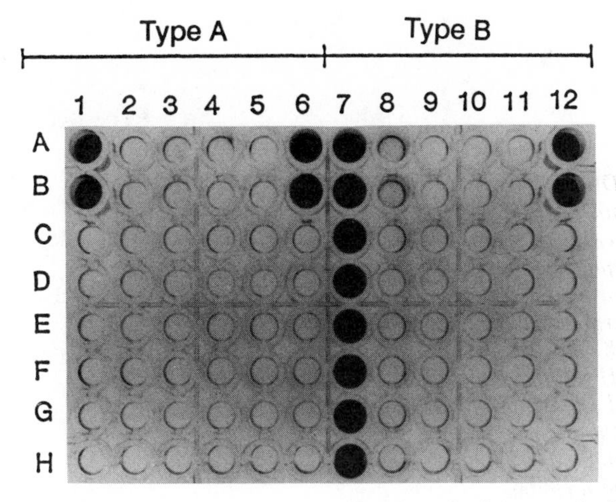

Figure 3. Results of ELISA showing monoclonal antibodies specific for human (rows 1–6; type A) and for primate (rows 7–12; type B) albumins. The plate was divided in half (rows 1–6 and 7–12 composing two groups), and purified albumins were used to coat the wells in duplicate prior to assay; the left and right sides of the plate were coated identically. Albumins were obtained from the following 20 species: human (A1, B1, A7, and B7), gorilla (C1, D1, C7, and D7), orangutan (E1, F1, E7, and F7), chimpanzee (G1, H1, G7, and H7), deer (A2, B2, A8, and B8), mule deer (C2, D2, C8, and D8), elk (E2, F2, E8, and F8), antelope (G2, H2, G8, and H8), cat (A3, B3, A9, and B9), dog (C3, D3, C9, and D9), hamster E3, F3, E9, and F9), guinea pig (G3, H3, G9, and H9), raccoon (A4, B4, A10, and B10), rabbit (C4, D4, C10, and D10), rat (E4, F4, E10, and F10), goat (G4, H4, G10, and H10), sheep (A5, B5, A11, and B11), pig (C5, D5, C11, and D11), horse (E5, F5, E11, and F11), and turkey (G5, H5, G11, and H11). Wells A6, B6, A12, and B12 were coated with a mixture of the first 10 albumins, and wells C6, D6, C12, and D12 were coated with a mixture of the second 10 albumins.

the albumins of several primate species. Solutions containing albumin were added to wells of microtiter plates that had been precoated with the cross-reactive monoclonal antibody, HSA-2, at 10 μg/mL. After extensive washing, the bound ("trapped") albumin was detected with biotinylated HSA-1 and HRP–streptavidin.

The sensitivity of the assay was determined by using various dilutions of purified serum albumin and human serum (Figure 5). This assay detected as little as 30 ng of purified albumin per mL or, expressed in another way, the albumin present in a $1:10^6$ dilution of human serum (equivalent to 1.0 nL or less of human serum) within a few minutes after the addition of substrate (Figure 5). Longer incubation times permit greater sensitivity without loss of specificity. The assay can be made more or less sensitive by varying the amount of antibody coated and the reaction times with the various reagents.

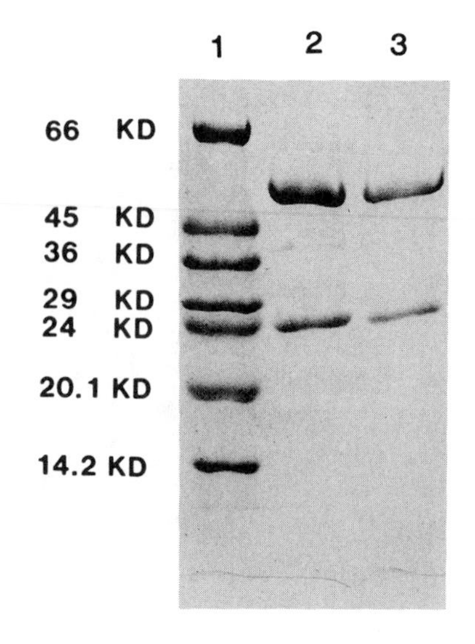

Figure 4. SDS-PAGE profile of monoclonal antibodies HSA-1 and HSA-2. After purification by hydroxyapatite chromatography, HSA-1 (lane 2) and HSA-2 (lane 3) were electrophoresed on a 10% polyacrylamide gel in the presence of 0.1% SDS. Approximately 1 μg of HSA-1 and HSA-2 was loaded in each lane. The gel shows two peptide bands corresponding to heavy and light (56- and 24-kilodalton [KD], respectively) immunoglobulin chains and no evidence of other contaminating proteins. Lane 1 was loaded with molecular weight markers.

Table V. Albumin in Tissue Extracts and Body Fluids

Fluid or Extract	*Protein at 50% Maximum OD[a] (ng)*
Serum	8
Seminal fluid[b]	630
Milk (most reactive)	1,260
Urine[b]	2,000
Kidney	3,160
Liver	10,000
Saliva[b]	30,000
Pancreas	30,000
Vaginal secretions	100,000

NOTE: Albumin was determined by the trapping assay with biotin-conjugated HSA-1.
[a]OD, optical density.
[b]Starting protein concentrations of 20, 1, and 0.5 mg/mL for seminal fluid, saliva, and urine, respectively were assumed.

Albumin in Tissues and Body Fluids. The presence of albumin in various tissues and body fluids was examined. The results are presented in Table V as the amount of protein per milliliter of fluid required to give 50% maximal reaction in the trapping assay. Albumin was easily detected in various body fluids and tissue extracts, including seminal fluid, milk, urine, saliva, vaginal secretions, and extracts of liver, kidney, and pancreas. The concentrations of albumin in these fluids, relative to serum, were such that assay conditions could be adjusted easily to readily distinguish between blood samples and other body fluids.

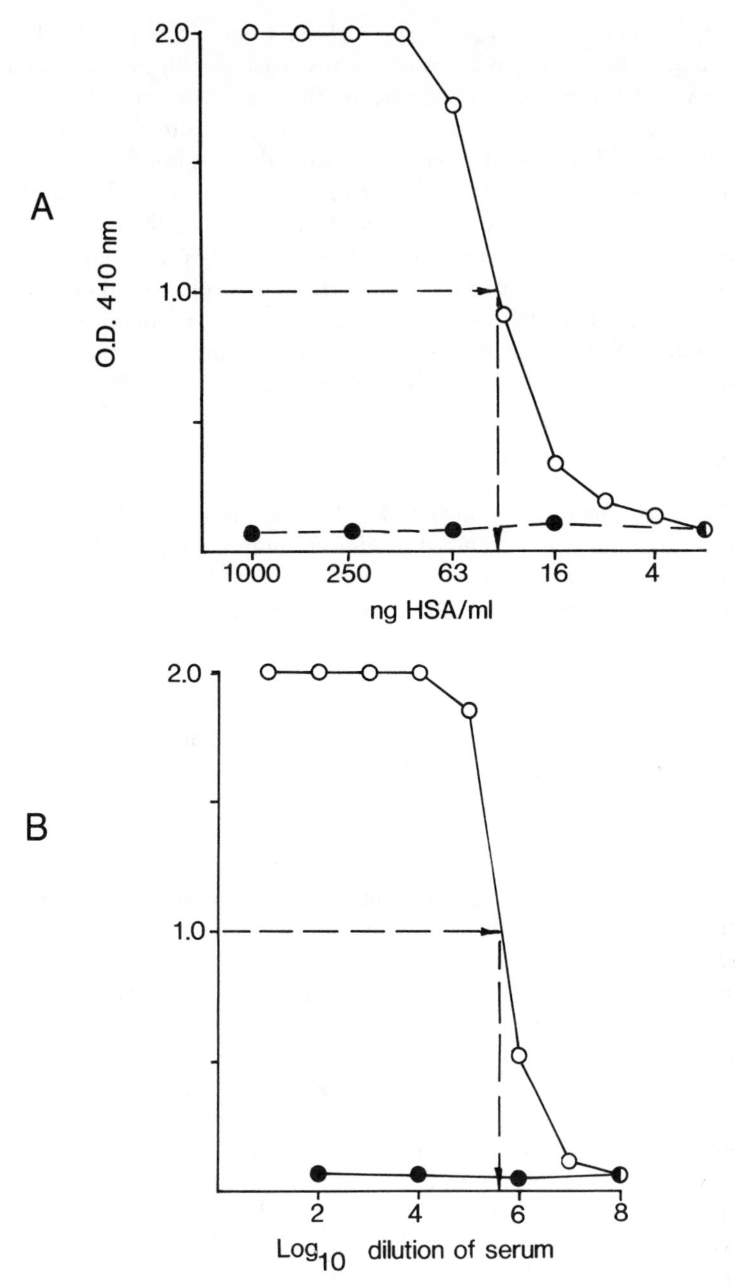

Figure 5. Sensitivity of antigen-trapping assay with monoclonal antibodies HSA-1 and HSA-2. HSA-2 (type B antibody) was coated on the plate, and then human albumin (A) or serum (B) was diluted as indicated and added to the wells. Biotin-conjugated HSA-1 served as the specific probe. The optical density at 410 nm (OD_{410}) was measured after 15 min.

Specificity of HSA-1. Every specimen from a large panel of human serum samples from various races and sexes reacted with monoclonal antibody HSA-1. No human serum was tested that did not react, a fact indicating the absence of genetic variation in the antigenic site detected by the HSA-1 antibody. The HSA-1 monoclonal antibody thus fulfills two essential criteria for an immunological probe for the identification of the species of origin: species specificity and intraspecies conservation of the antigenic site.

The unequivocal assignment of the species of origin to unknown blood stains, body fluids, or tissue fragments is critical to forensic medicine. South African forensic scientists in particular have a serious problem in differentiating human and primate bloods (*41*, *42*). Archaeologists studying artifacts with surface blood stains may face a similar challenge.

Advantages of Monoclonal Antibodies

High-titer anti-human globulin polyclonal antiserum in double-diffusion assays is the method generally used to identify human blood in forensic (*42*) and archaeological (*1*) specimens by immunological techniques. However, because of the number of antigenic sites on complex proteins, it has been difficult to produce a truly species-specific polyclonal antiserum. In addition, because of the large number of structural variations among proteins from related species, species-specific surface structures exist on proteins that may be immunologically detected with monoclonal antibodies.

Monoclonal immunoreagents offer advantages of uniformity, specificity, constant affinity, and availability in virtually unlimited supply, compared with their polyclonal counterparts. Species-specific immunoreagents, such as HSA-1, will allow the archaeological and forensic communities to standardize the identification of human blood and tissue and the direct comparison of results from different laboratories without doubt about the specificity of the antibody.

The antigen-trapping enzyme-linked immunosorbent assay that uses two monoclonal antibodies can provide quantitative determinations of unknown samples provided that the saturation level of the bound monoclonal antibody is not exceeded. Figure 6 summarizes the steps in this method. Antigen trapping assays based on this configuration should also provide the archaeological serologist with a method that is applicable over a broad range of albumin concentrations typically encountered in blood stains in samples, which may contain minute or large amounts of albumin. The trapping assay detects as little as 30 ng of purified albumin (range of concentration in serum is from 28 to 45 mg of albumin per mL). By reducing the amount of bound monoclonal antibody below the saturation of the plate-coating step of this assay or by decreasing the amount of biotinylated HSA-1 in the second step, the sensitivity of this assay may be adjusted to allow discrimination between blood samples and other body fluids.

1.

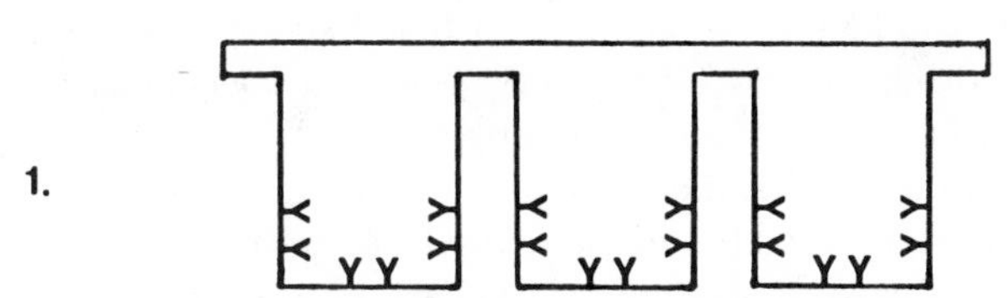

Add 0.1 ml monoclonal antibody, incubate 1 hr, 20°C.

2. Block wells with TRIS-Tween, 10 min, 20°C.

3.

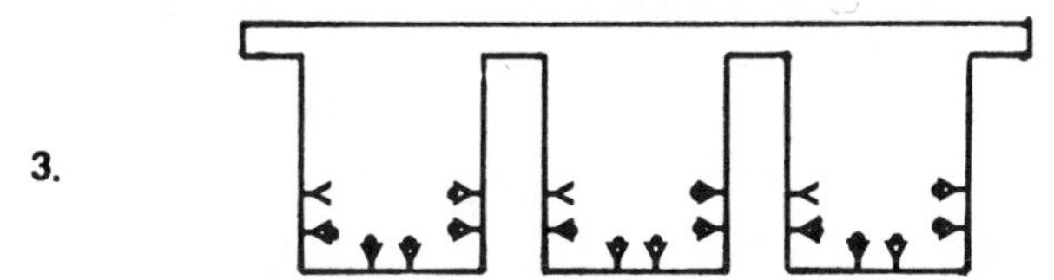

Add 0.1 ml sample of seminal fluid, incubate 1 hr, 20°C.

4. Wash 3x to remove excess sample.

5.

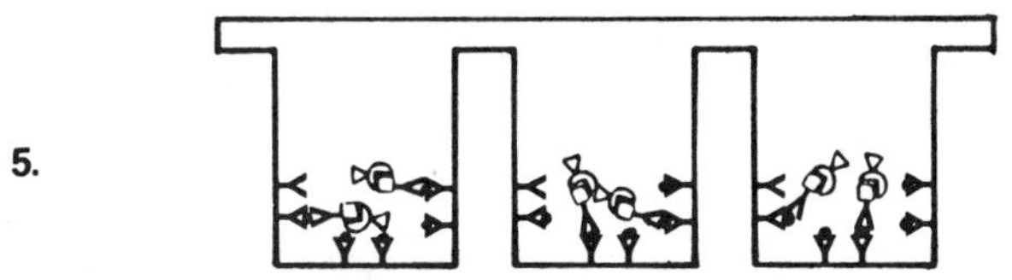

Add biotinylated monoclonal antibody-streptavidin-HRP, incubate 30 min, 20°C.

6. Wash 5x to remove excess antibody complex.

7.

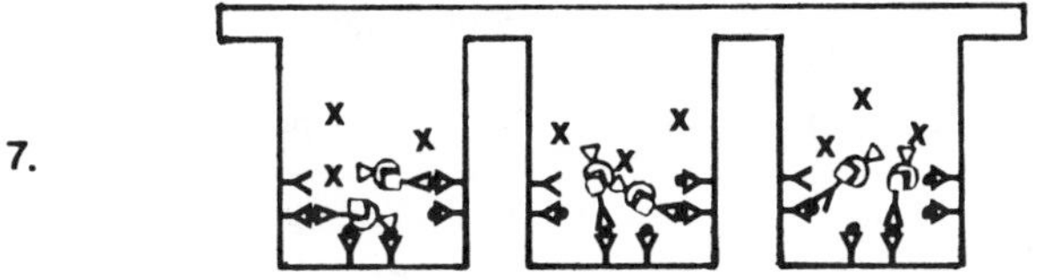

Add 0.1 ml ABTS substrate (x), incubate 30 min, 20°C.
Score visually or measure $O.D._{414}$.

Figure 6. Summary of the steps of the ELISA for detecting human albumin.

Further details on this method may be obtained from Benjamin et al. (*43*) or by obtaining product information about the commercial assay from the manufacturer (Humagen, 1500 Avon Street, Charlottesville, VA 22901). We do not know the half-life of albumin in blood stains on the surface of archaeological artifacts. Dried tissues from museum specimens or frozen samples of great age do retain albumin antigenicity (*1, 44*).

The method requires that the albumin epitopes against which the monoclonal antibodies are directed retain their native conformation. Loy (*1*) noted that blood residues on buried lithic artifacts are protected from microbiological attack and ground water removal by electrostatic interactions with soil clay particles. He states: "In some cases, up to 50% of the original blood residue may be sequestered in the first 0.1 μm of soil; thus, much of the residue may in fact be lost by extensive cleaning of artifact surfaces." This experience gives hope that albumin molecules may be maintained in their native antigenic conformation and recovered from archaeological specimens.

Conclusion

Forensic studies suggest that monoclonal antibodies specific to albumin may be used to determine the species of origin of blood stains on archaeological artifacts. A monoclonal antibody that reacts specifically with an epitope on human serum albumin has already been produced.

Acknowledgment

This study was supported by Federal Bureau of Investigation grant 115744.

References

1. Loy, T. H. *Science* **1983,** *220*, 1269.
2. Landsteiner, K. *The Specificity of Serological Reactions*; Charles C. Thomas: Springfield, IL, 1936.
3. Brown, R. K.; Delaney, R.; Levine, L.; Van Vunakis, H. *J. Biol. Chem.* **1959,** *234*, 2043.
4. Shinka, S.; Imanishi, M.; Miyagawa, N.; Amanao, R.; Inouye, M.; Tsugita, A. *Biochem. J.* **1967,** *10*, 89.
5. Gerwing, J.; Rhompson, K. *Biochemistry* **1968,** *7*, 3888.
6. Young, J. D.; Leung, C. Y. *Biochemistry* **1970,** *9*, 2755.
7. Sela, M. *Science* **1969,** *166*, 1365.
8. Sela, M.; Schechter, B.; Schechter, I.; Borek, F. *Cold Spring Harbor Symp. Quant. Biol.* **1967,** *32*, 537.
9. Benjamin, D. C.; Benzofsky, J. A.; East, I. A.; Gurd, F. R. N.; Hannum, C.; Leach, S. J.; Margoliash, E.; Michael, J. G.; Miller, A.; Prager, E. M.; Reichlin, M.; Sercarz, E. E.; Smith-Gill, S. J.; Todd, P. E.; Wilson, A. C. *Annu. Rev. Immunol.* **1984,** *2*, 67.
10. Atassi, M. Z. *Immunochemistry* **1975,** *12*, 423.

11. Twining, S. S.; David, C. S.; Atassi, M. Z. *Mol. Immunol.* **1981,** *18,* 447.
12. Atassi, M. Z. *Mol. Immunol.* **1981,** *18,* 1021.
13. Underwood, P. A. *J. Gen. Virol.* **1982,** *62,* 153.
14. Benjamin, D. C.; Daigle, L. A.; Riley, R. L. In *Protein Conformation as Immunological Signal*; Celada, F; Schumaker, V.; Sercarz, E. D., Eds.; Plenum: New York, 1983; p 261.
15. Benjamin, D. C.; DeCourcy, K.; Riley, R. L.; Wilson, L. D. *Fed. Proc. Fed. Am. Soc. Exp. Biol.* **1983,** *42,* 1237.
16. Benjamin, D. C.; Wilson, L. D.; Riley, R. L.; Holaday, B.; DeCourcy, K. In *Immune Recognition of Protein Antigens*; Laver, W. G.; Air, G. M., Eds.; Cold Spring Harbor Laboratory: Cold Spring Harbor, NY, 1985; p 98.
17. Jemmerson, R.; Margoliash, E. *J. Biol. Chem.* **1979,** *254,* 12706.
18. Urbanski, G. J.; Margoliash, E. In *Immunochemistry of Enzymes and their Antibodies*; Salton, M. R. J., Ed.; Wiley: New York, 1977; p 203.
19. Urbanski, G. J.; Margoliash, E. *J. Immunol.* **1977,** *118,* 1170.
20. Prager, E. M.; Wilson, A. C. *J. Biol. Chem.* **1971,** *246,* 5978.
21. Wilson, A. C.; Carlson, S. S.; White, T. J. *Annu. Rev. Biochem.* **1977,** *46,* 573.
22. Champion, A. B.; Soderberg, K. L.; Wilson, A. C.; Ambler, R. P. *J. Mol. Evol.* **1975,** *5,* 291.
23. Berzofsky, J. A.; Richman, L. K.; Killion, D. J. *Proc. Natl. Acad. Sci. USA* **1979,** *76,* 4046.
24. Riley, R. L.; Wilson, L. D.; Germain, R. N.; Benjamin, D. C. *J. Immunol.* **1982,** *129,* 1553.
25. Katz, M. E.; Maizels, R. M.; Wicker, L.; Miller, A.; Sercarz, E. E. *Eur. J. Immunol.* **1982,** *12,* 535.
26. Kapp, J. A.; Strayer, D. S.; Robbins, P. F.; Perlmutter, R. M. *J. Immunol.* **1979,** *123,* 109.
27. Schwartz, R. H.; Berzofsky, J. A.; Horton, C. L.; Schechter, A. N.; Sachs, D. H. *J. Immunol.* **1978,** *120,* 1741.
28. Prager, E. M.; Welling, G. W.; Wilson, A. C. *J. Mol. Evol.* **1978,** *10,* 293.
29. Amit, A. G.; Mariuzza, R. A.; Phillips, S. E. V.; Poljak, R. J. *Science* **1986,** *233,* 747.
30. Berzofsky, J. A.; Buckenmyer, G. K.; Hicks, G.; Gurd, F. R. N.; Feldmann, R. J.; Minna, J. *J. Biol. Chem.* **1982,** *257,* 3189.
31. Smith-Gill, S. J.; Wilson, A. C.; Potter, M.; Prager, E. M.; Feldmann, R. J.; Mainhart, C. R. *J. Immunol.* **1982,** *128,* 314.
32. Uhlenhuth, P. *Dtsch. Med. Wochenschr.* **1901,** *27(6),* 82.
33. Uhlenhuth, P. *Dtsch. Med. Wochenschr.* **1901,** *27(30),* 499.
34. Wasserman, A.; Schutze, A. *Berl. Klin. Wochenschr.* **1901,** *38(7),* 187.
35. Gaensslen, R. E. *Source Book in Forensic Serology, Immunology and Biochemistry*; U.S. Department of Justice. U.S. Government Printing Office: Washington, DC, 1983.
36. Sensabaugh, G. F. *Int. Microform J. Leg. Med.* **1976,** *11(2),* article 219.
37. Benjamin, D. C.; Weigle, W. O. *Immunochemistry* **1971,** *8,* 1208.
38. Herr, J. C.; Summers, T. A.; McGee, R. S.; Sutherland, W. M.; Sigman, M.; Evans, R. J. *Biol. Reprod.* **1986,** *35,* 773.
39. Volk, W. A.; Snyder, R. A.; Benjamin, D. C.; Wagner, R. R. *J. Virol.* **1982,** *42,* 220.
40. Hofmann, K.; Titus, G.; Montibeller, J. A.; Finn, F. M. *Biochemistry* **1982,** *21,* 978.
41. Taylor, J. E. D. *South Afr. Med. J.* **1952,** *26,* 81.
42. Sivrim, S.; Bhatnagar, R. K.; Bhandari, S. K.; Sijher, J. S. *Forensic Sci.* **1975,** *6,* 145.

43. Benjamin, D. C.; Herr, J. C.; Sutherland, W. M.; Woodward, M. P.; DeCourcy, K.; Condon, T. P. *Hybridoma* **1987,** 6(2), 183.
44. Prager, E. M.; Wilson, A. C.; Lowenstein, J. M.; Sarich, V. M. *Science* **1980,** *209,* 287.

RECEIVED for review September 15, 1987. ACCEPTED revised manuscript June 15, 1988.

FIBERS

23

Isotope Measurements and Provenance Studies of the Shroud of Turin

Robert H. Dinegar[1] and Larry A. Schwalbe[2]

[1]University of New Mexico—Los Alamos, Los Alamos, NM 87544
[2]Los Alamos National Laboratory, Los Alamos, NM 87545

The findings and conclusions of the 1978 investigation of the Shroud of Turin are reviewed, and a proposal for additional research that was submitted in 1984 is described. The research proposal centers primarily on two experiments: radiocarbon dating and the proposed measurements of the stable hydrogen and oxygen isotopic concentrations in the textile cellulose. The discussion of the dating experiment focuses on the history of attempts to perform the test during the past decade. That of stable isotope measurements treats the technical aspects of the method and their implications for locating the geographical origin of the Shroud.

THE SHROUD OF TURIN RESEARCH PROJECT (STURP), an American team of scientists and technicians, conducted an extensive series of nondestructive examinations of the Shroud in October 1978. About 4 years later, the data reduction and analyses were completed. Publicly released and discussed in several forms (*1*, *2*), the measurements and observations support the following conclusions:

1. The body image on the cloth is the result of an oxidation–dehydration reaction of the cellulose material of the cloth itself rather than the result of an applied pigment, stain, or dye.
2. The process for the formation of the body image remains unknown.

0065-2393/89/0220-0409$06.00/0

3. The areas traditionally identified as "bloodstains" contain blood-derived material.

In August 1984, STURP submitted a proposal to perform additional tests. The formal proposal, which consisted of 26 individual projects, addressed three broad categories of current issues: questions about how the image was formed, conservation, and provenance. Of these issues, the most complex and perplexing is how the image was produced on the surface of the cloth. Roughly half the experiments in the 1984 proposal were aimed toward solving this problem. Of the remaining experiments, nearly all are related to conservation. STURP, along with the authorities in Turin who are entrusted with the care and preservation of the relic, recognize that a comprehensive characterization of the physical and chemical state of the fabric (and its associated materials) is necessary to support valid recommendations for the safe handling and storage of the Shroud.

The final two of the 26 projects pertained directly to questions of provenance. It is perhaps ironic that so little apparent effort was given to this problem. After all, knowing the time and place of the Shroud's origin is central to what many would consider the paramount question: Was the Shroud of Turin, as pious tradition holds, the burial cloth of Jesus, or is it merely a cleverly hoaxed medieval relic? The two experiments that will provide some insight to these questions are the focus of this chapter. Both involve measurements of the isotopic composition of the fabric. The first is the familiar and widely publicized radiocarbon procedure that should give an accurate date of origin. The second is a study of the stable isotope ratios of hydrogen and oxygen that could locate the Shroud's place of origin. As we intend to show, the attention and effort given to this research has been considerable.

Determining the Age

Many accept the Shroud as the authentic burial cloth of Jesus, but those who defend this belief must explain the complete lack of records documenting the Shroud's existence throughout most of its supposed 2000-year history. For all we know today, the Shroud could have been manufactured a short time before its recorded appearance in the early 1350s. The issue of authenticity has therefore generated and sustained great interest in the radiocarbon dating technique. An objective date is possible; in fact, an attempt to date the Shroud has been completed.

Previous Proposals. Many investigators have had the idea of dating the cloth by C-14 determination ever since 1955 when Willard F. Libby showed that such analysis is both possible and meaningful. Evidently, these thoughts were not followed up, or the failure of attempts was kept secret, for there is little, if any, literature published during the 1950s–1970s on this subject. One documented attempt does exist (*3*), however, and that involved the group of investigators that later became STURP. Little came of this organization's attempt in the late 1970s. At the time, the position of the Roman Catholic Church, in whose care the cloth is entrusted, was that any type of testing should be nondestructive. In addition, the established technology (the Libby counting method) required an unacceptably large sample of cloth to be destroyed for a reliable answer. Thus, when STURP made its formal request to the Archbishop of Turin, Michelle Pellegrino, in 1977 to run tests on the Shroud, radiocarbon analysis was not mentioned.

The test program of STURP, which involved a comprehensive investigation of the cloth using many wavelengths of the electromagnetic spectrum, photography, and the removal of surface material for chemical analysis, was approved in April 1978. As STURP prepared to go to Turin in the fall, a well-known American scientist (*4*) made an unsuccessful attempt to obtain threads and swatches of cloth that had been removed in 1973 for a "quick and dirty" age determination. In addition, two other American scientists wrote a formal proposal for radiocarbon analysis of the cloth using newly developed methods involving tandem accelerators and precise miniature proportional counters. The latter request was somehow "lost" and Church authorities appeared unwilling to have the Shroud carbon dated.

Those interested in the cloth then split into three camps. One group pushed harder for immediate C-14 testing. A second said the origin of the cloth was so well-known that no part of it should be destroyed in verifying its antiquity. A third counseled making plans for an age determination in several years. STURP was the vocal exponent of the last position.

At the beginning of the last week of August 1978, the Holy Shroud of Turin was put on public display over the high altar of the Cathedral of Saint John the Baptist. The exposition marked the 400th anniversary of the Shroud's arrival in Turin from Chambéry, France. The public exposition ended the weekend of October 7–8, 1978, during which time the Second International Sindonological Conference was held. Those interested in the Shroud came from all over the world to participate. During the meeting, the newly named Archbishop of Turin, Anastasio Ballestrero, announced that when the scientific community could settle on the best way to determine the age of the cloth, he would consider asking the owner of the Shroud, the exiled King of Italy, Umberto II, to allow samples to be taken for analysis. His words were greeted with enthusiasm by those holding the third position mentioned; the other two camps redoubled their efforts to have this decision changed to their positions. A state of tension and political intrigue developed that is still very much evident.

STURP Proposal. On October 9, 1978, the Shroud was placed in the Visiting Prince's rooms of the Royal Palace of the House of Savoy, and the testing proceeded continuously for 5 days. The data and analyses of more than 100 h of cloth observation have been discussed for almost a decade. The results have been published in many journals. Most, if not all, of the articles were refereed, and, to our knowledge, not one was rejected. The general conclusions reached by STURP scientists were those outlined at the beginning of this chapter. Neither the data nor the conclusions themselves have been challenged. What has been argued, almost ad nauseam, is whether the work can be interpreted for or against a very early chronological date. No published article by any STURP scientist takes either position.

In 1979, STURP began decisive action to obtain permission to carbon date the Shroud. A C-14 committee was formed. Scientists from all over the world were contacted and asked if they were interested in joining the the effort. Some responded negatively, citing the religious nature of the project. Others failed to answer. With various levels of enthusiasm, several agreed to become involved. The Arizona, Brookhaven, and Rochester laboratories from the United States, as well as Oxford and Harwell of the United Kingdom answered affirmatively. Later, the British Museum and the Federal Institute of Technology in Zurich joined the effort. This array of scientists was deemed quite sufficient to carry out the age determination in the most objective manner. The inclusion of both methods of C-14 determination, accelerator and counter, was thought to ensure the most accurate and precise dating. The chairman of the STURP C-14 committee kept the group informed and coordinated by letter, telephone, and personal visits over the ensuing years.

In 1983, at the radiocarbon conference in Bradford, England, it was suggested that a laboratory intercomparison experiment be undertaken. The British Museum agreed to supply three samples of cloth of known dates to each laboratory. The intercomparison tests were carried out and the results reported at the radiocarbon conference held in Trondheim, Norway in June 1985. The proceedings of that meeting contain the data and the report presented by the British Museum (5). An interlaboratory agreement on the dates was clearly established. Interestingly, a serendipitous result of this test was to bring to light that the actual date of one of the samples furnished by the British Museum was not correctly known at all!

While the intercomparison experiment was being performed, a proposal for new tests on the Shroud, of which radiocarbon dating was preeminent, was drawn up by STURP. The proposal dating project was approved by the original five laboratories, and the composite proposal was sent to the Cardinal Archbishop of Turin in the fall of 1984 for endorsement and forwarding to the See of Rome, which upon the death of Umberto II had become the owner of the cloth. The Pontifical Academy of Sciences thus received the report for comment.

Within this atmosphere, STURP and the radiocarbon laboratories involved held a special meeting on the Shroud dating during the radiocarbon conference in Trondheim. It was quickly seen and made absolutely clear that one laboratory was dedicated to having the experiment come under the single and direct oversight of the Pontifical Academy of Sciences. The adamant stance and continual lobbying of this one laboratory caused a severe tension in the project that exists to this day.

Pontifical Academy of Sciences Criteria. In the last week in September 1986, at the invitation of the Cardinal Archbishop of Turin and the Pontifical Academy of Sciences, scientists from STURP, the analyzing radiocarbon laboratories (now including a Swiss team), a textile expert from Holland, and an archaeologist from Hong Kong met in Turin to discuss dating the Shroud. Also present as observers were French radiocarbon experts. The meeting was chaired by the president of the Pontifical Academy of Sciences, but under the oversight of the Archbishop of Turin and his scientific advisors. After 3 days of meetings it was decided that

1. The samples for radiocarbon dating would be taken from the relic by the Dutch cloth expert, under the direction of the scientific advisors to the Cardinal Archbishop.
2. The samples and "blinds" would first be unwoven by the British Museum and then distributed to the laboratories involved.
3. The additional tests on the cloth as outlined in the STURP proposal would then be carried out.
4. The radiocarbon data would be collated by the British Museum, the statistical department of the Polytechnic Institute in Turin, and a third group designated by the Pontifical Academy.
5. The date of the Shroud of Turin would be announced to the world by Easter 1988.

In April 1987, Turin received positive instructions from Rome on how to proceed. These directed that no more than three samples were to be taken, which required that three laboratories be chosen from the six that had offered their services. The choices were made after long deliberation and careful consultations. The criteria used for the selection were

- the specific experience in the field of archaeological radiocarbon dating
- the sample size required by the laboratory
- the international mix of the group

The criterion of sample size eliminated those laboratories that use the traditional Libby counting method. This situation was very unfortunate because one laboratory of the six best meeting the experience criterion was one of the two requiring the larger sample size. Harwell was thus eliminated, even though it had performed more radiocarbon experiments than the other five laboratories combined. Brookhaven, the other counter laboratory, was eliminated because it, too, required a large sample and was relatively inexperienced in archaeological dating. This elimination was also unfortunate, but from another standpoint: Brookhaven had done the pioneer development of the small proportional counter technique. The accelerator laboratory at Rochester required a sufficiently small sample, but they were passed over by reason of little experience in archaeological dating and their use of "non-dedicated" equipment. Arizona, Oxford, and the Federal Institute of Technology in Zurich were left after the application of all three criteria. The other scientists taking part in the overall investigation of the cloth, the British Museum for overseeing the date determination, and STURP for other experiments, remained unaffected.

The reaction of the laboratories to the proposed choices was predictable. Both Rochester and Brookhaven denounced the decision as scientifically flawed and questioned the qualifications of those who had advised the Church authorities. Harwell expressed disappointment at not being chosen and likewise disagreed with the decision to cut the number of laboratories from six to three. Arizona, Oxford, and Zurich gave the opinion that any date obtained would be more widely accepted if all the laboratories were involved instead of just three.

Current Protocol. The C-14 testing protocol at this time is essentially that submitted by STURP in 1984 with the modifications that took place in Turin in 1986. The few major changes that were made to conform to the sample-size criterion caused the loss of use of the counter method and an increase in the error associated with any date from ±100 years to about ±250 years. Although the criticisms of the plan offered by the Church authorities are valid and thought-provoking, the protocol certainly will allow a conclusion as to whether the cloth is of medieval origin or if it comes from a much earlier time.

On the night of April 21, 1988, Cardinal Ballestrero, Luigi Gonella, and Giovanni Riggi, (Vice President of STURP) with the advice of a cloth expert from Lyons, France, removed fabric samples from the Shroud and gave them to Michael Tite of the British Museum for distribution to the three laboratories. Slowly, the radiocarbon dating of the Shroud is being accomplished and the Church's decision to give free hand to the scientific community is evident.

Locating the Geographic Origin

Our discussion of the dating proposal centers more on the historical record of efforts to accomplish the work than it does on the more familiar technical aspects of the problem. In our discussion of placing, the situation is reversed. Unlike radiocarbon dating, the idea of measuring the stable isotope ratios of hydrogen and oxygen to locate the geographic origin of the Shroud is fairly recent. The method is not widely known or appreciated, and a technical discussion therefore is the most appropriate.

Previous Physical Studies. Thus far, the available physical evidence concerning the Shroud's origin is suggestive but scanty and largely inconclusive. In 1973, a fabric sample was removed from the edge of the Shroud near one end and given to a Belgian textile expert (Raes) for study. Unfortunately, despite several provocative observations, including that of the Z-twist, which is apparently unusual for linen, Raes (*6*) was unable place the origin of the material either from the twist or from the 3:1 herringbone weave type of the fabric. Still, he did report finding traces of cotton in with the flax fibers and suggested that the linen may have been spun with the same equipment used previously for cotton. The latter observation supports a Middle Eastern manufacture (because cotton was not grown in Europe at the time), but it does not prove the conjecture.

A second piece of evidence is offered by Frei (*7–9*) who studied surface debris that was removed from the Shroud with sticky tape. Among the collected materials, he observed and identified pollen from plants that are uniquely indigenous to Palestine and Anatolia (*10*). Some conclude from these observations that the Shroud originated in the Middle East and was subsequently brought to Europe via Constantinople (*4*). But we consider it just as likely that the Shroud was produced in Europe and subsequently transported to Palestine, perhaps on one of the Crusades, where it was exposed to the local pollen rain.

Stable Isotope Method. At this point, the evidence for placing the Shroud's origin is suggestive, but more work is clearly necessary if we hope to reconstruct a convincing historical scenario. So far, the evidence has been somewhat indirect; that is, it involves incidental materials associated with the fabric. The stable isotope experiment is qualitatively different from the earlier ones because, like radiocarbon, it involves direct measurements on the fabric itself.

The Shroud is a linen textile that consists of cellulose derived from flax. Each of the constituent elements of the cellulose (hydrogen, carbon, and oxygen) exists in several stable isotopic forms. The lighter isotopes ^{1}H, ^{12}C, and ^{16}O are generally more abundant, but smaller concentrations of the heavier D (^{2}H), ^{13}C, and ^{18}O are also present in measurable amounts. In any sample, the relative quantities D/H, $^{13}C/^{12}C$, and $^{18}O/^{16}O$ depend on the physical, chemical, and biological history of the material.

The isotopic compositions of oxygen and hydrogen in cellulose are particularly influenced by the climate in which the plant grew. Climate affects isotope ratios in primarily two ways. First, temperature affects cellulose composition through its effects on the isotopic concentrations of the local precipitation. In general, the magnitudes $^{18}O/^{16}O$ and D/H in rainwater decrease with cooler temperatures (*11*), and corresponding decreases in both carbon-bound D/H (*12, 13*) and $^{18}O/^{16}O$ (*14*) ratios in plant cellulose have been observed and measured. The second factor, the humidity of the growing environment, affects the isotope ratios of terrestrial plant cellulose through its influence on the transpiration process. Low humidities promote transpiration and concentrate leaf water containing the heavier isotopes, because water molecules containing ^{1}H and ^{16}O have higher vapor pressures and evaporate more readily. The effects of transpiration have also been observed experimentally (*15, 16*).

Hot, arid climates therefore cause plant cellulose to have higher relative concentrations of the heavier hydrogen and oxygen isotopes than cellulose from plants grown in cool, moist environments. The isotope ratios of cellulose have been used as reliable indicators of the climate in which the plants grew (*17*). Recently, DeNiro et al. (*18*) demonstrated how D/H and $^{18}O/^{16}O$ ratios of cellulose prepared from modern and historic linens of known provenance from Europe and the Middle East (Israel and Egypt) are related to the climates in these regions. This very study was in fact undertaken to supply a data base to which similar measurements on the Shroud could be compared to establish its geographical origin. The work's conclusion is that such a determination is indeed possible, although over a limited range, the data sets from Europe and Israel overlap slightly, thus introducing the chance of an ambiguous result.

We support STURP's proposal to measure stable isotope ratios from the Shroud fabric. The data could establish the geographical origin of the cloth in a direct and unambiguous way. Isotope ratio information is interesting in itself, but it could have an impact on other issues as well. For example, to someone familiar with the range of artistic techniques and traditions available at a specific time and location, an accurate historical placement of the Shroud could suggest how the image was formed. In conservation, the information could prove useful if the Shroud were to be uniquely classified with textiles of similar origin. The data may even be a critical supplement to the radiocarbon results if, as Meacham (*19*) notes, the presence of volcanic activity in the area where the Shroud linen was produced could affect the outcome of the ^{14}C testing.

Summary

Our understanding of the physical and chemical characteristics of the Shroud of Turin has increased enormously in the past 10 years, but the work done thus far still leaves unanswered some of the most intriguing questions asked about this most remarkable object. Foremost among these are when and where the Shroud came to be. We believe studies of the stable and unstable isotopes of the fabric will establish this information. Other questions and problems remain. In our view, the time is right to address these issues, and to see this done, we encourage the continued cooperation that we have seen between the Church and the scientific community.

References

1. Schwalbe, L. A.; Rogers, R. N. *Anal. Chim. Acta* **1982,** *135*, 3.
2. Jumper, E. J.; Adler, A. D.; Jackson, J. P.; Pellicori, S. F.; Heller, J. H.; Druzik, J. R. In *Archaeological Chemistry—III*; Lambert, J. B., Ed.; Advances in Chemistry 205; American Chemical Society: Washington, DC, 1984; Chapter 22.
3. Private communication of John P. Jackson to Willard F. Libby, December 1976.
4. Wilson, I. *The Shroud of Turin,* revised ed.; Doubleday: Garden City, NY, 1979; p 242.
5. Burleigh, R.; Leese, M.; Tite, M. *Radiocarbon* **1986,** *28(2A)*, 571.
6. Raes, G. *Report of the Turin Commission on the Holy Shroud*; translated from Italian by Screenpro Films, London, 1976. (Screepro Films, 5 Meard St., London W1V 3HQ, UK.)
7. Frei, M. *Naturwiss. Rundsch.* **1979,** *32*, 133.
8. Frei, M. *Shroud Spectrum International* **1982,** (*3*), 3. (The Indiana Center for Shroud Studies, R. 3, Box 557, Nashville, IN 47448.)
9. Wilson, I. *The Shroud of Turin,* revised ed.; Doubleday: Garden City, NY, 1979; p 293.
10. Bulst, W. *Shroud Spectrum International* **1984,** (*10*), 20.
11. Foerstel, H.; Huetzen, H. *Nature* (*London*) **1983,** *304*, 614.
12. Epstein, S.; Yapp, C. J. *Earth Planet. Sci. Lett.* **1976,** *30*, 252.
13. Yapp, C. J.; Epstein, S. *Nature* (*London*) **1982,** *297*, 636.
14. Burk, R. L.; Stuiver, M. *Science* **1981,** *211*, 1417.
15. Dongman, G.; Foerstel, H.; Wagener, K. *Nature* (*London*) *New Biol.* **1972,** *240*, 127.
16. Epstein, S.; Thompson, P.; Yapp, C. J. *Science* **1977,** *198*, 1209.
17. Brenninkmeijer, C. A. M.; van Geel, B.; Mook, W. G. *Earth Planet. Sci. Lett.* **1982,** *61*, 283.
18. DeNiro, M. J.; Sternberg, L. D.; Marino, B. D.; Druzik, J. R. *Geochim. Cosmochim. Acta*, in press.
19. Meacham, W. *Shroud Spectrum International* **1986,** (*19*), 15.

RECEIVED for review June 11, 1987. ACCEPTED revised manuscript May 9, 1988.

24

Historical Silk Flags Studied by Scanning Electron Microscopy–Energy Dispersive X-ray Spectrometry

M. Ballard[1], R. J. Koestler, C. Blair, and N. Indictor[2]

Department of Objects Conservation, Metropolitan Museum of Art, New York, NY 10028

Samples of colored silk taken from brittle flags in the National Museum of American History (Washington, DC) have been subjected to light microscopy and scanning electron microscopy–energy dispersive X-ray spectrometry (SEM–EDS). The presence of mordants, weighting materials, and colorants is discussed with reference to the embrittlement of the silk. The ash content of the samples is reported. Micrographs of individual fibers are examined. The correlation between fiber deterioration and elemental composition (as a reflection of possible manufacturing processes) is discussed.

SILK FABRICS HAVE BEEN HELD IN HIGH ESTEEM for centuries. This consideration outweighs some of the problems inherent to the silk fibers, such as susceptibility to damage from light and embrittlement with age. This esteem has encouraged museums to exhibit superb examples of the technological and aesthetic achievements associated with the silk weaving of past centuries. Even when treated poorly, some examples of silk art have survived quite well.

Among 19th-century silks, Chinese robes and silk embroideries are popular collectors' items. Far Eastern silks have often survived export, rev-

[1]Current address: Conservation Analytical Lab, Smithsonian Institution, Washington, DC 20560
[2]Current address: Chemistry Department, Brooklyn College, City University of New York, Brooklyn, NY 11210

0065–2393/89/0220–0419$06.00/0

olution, and the vagaries of use quite well. European and American silks of the same period have not. In contrast to the health of Far Eastern silks, many examples of European and American silk fabrics are shredded, split, and in unexhibitable condition. Figure 1 is an example of a deteriorated 19th-century silk.

Such degradation has been ascribed to the tin weighting (*1*, *2*), which was especially popular in the last half of the 19th century. Because the standard method of analyzing weighting processes requires gram quantities of silk, curators and conservators have been reluctant to sacrifice samples. Recently, we demonstrated the possibility of describing weighted silks by using scanning electron microscopy–energy dispersive X-ray spectrometry (SEM–EDS) on prepared modern samples (*3*, *4*). Several kinds of silk fibers, weighted in a variety of ways, were successfully analyzed in a set of blind experiments. The weighting procedures were based on recipes taken from standard processes described in the literature. Tin and iron compounds, along with other inorganic substances, were used in the weighting treatments. It was generally possible to distinguish between the weighting treatments entirely on the basis of EDS results. The methods used are similar to those used on mordanted (*5*–*8*) and metal-wrapped textiles (*9*, *10*).

Because of the success with SEM–EDS, the Division of Armed Forces History at the National Museum of American History supplied some actual, deteriorated, accessioned museum material for examination by light mi-

Figure 1. A deteriorated 19th-century silk.

croscopy and SEM–EDS. A part of each specimen was also analyzed for ash content. Some samples were tested for dye class.

The samples were taken from the vast group of military flags transferred from the Quartermaster Museum Collection in Philadelphia to the Smithsonian Institution by the War Department in 1919. This donation of military material was so large that it was shipped by railroad boxcar. All the flags sampled predate the year 1919. Unlike most silk fabrics, they can be dated quite accurately. The flags are also noteworthy for their deteriorated state. Many arrived at the Smithsonian Institution wrapped in brown paper, labeled "poor condition". Their fragmentary condition, together with their late 19th-century origin, has led the curatorial and conservation community to assume that the damage was the result of weighting during the fabric manufacture.

Table I gives catalog numbers and descriptions of the flags from which specimens were obtained at the National Museum of American History, Washington, DC.

Experimental Details

Samples (about 1–2 mm^2) were taken from the specimens and arranged on spectroscopically pure carbon stubs in a manner that would permit viewing (either with optical microscopy or scanning electron microscopy).

SEM–EDS Analyses. After observation at about 40× magnification through a light microscope (Wild M8), the samples were carbon coated and subjected to SEM–EDS analysis (AMRay 1600T equivalent with attached Kevex EDS). Generally, the operating conditions for the collection of EDS data were as follows: 15 kV, 200-s collection time, 2.56×10^5 μm^2 ($\simeq 0.25$ mm^2) excitation areas. For some samples, up to 10 replicate scans were performed at different spots. Sample preparation and data treatment have been described (*4–9, 11*). Photomicrographs and printouts of EDS scans were retained for files.

Ash Content. Samples were removed from the specimens and submitted for ash analysis (Schwartzkopf Microanalytical Laboratory, Woodside, NY). Samples were burned in oxygen at 900–1000 °C for about 0.5 h.

Color Analysis. Samples were removed from the specimens and analyzed for the colorants according to methods described in references 12 and 13.

Results and Discussion

Table II lists the qualitative EDS results of the 14 specimens. Only elements of atomic number >11 were detected. The results were obtained as normalized percentages according to the software program (ASAP) usually employed in analyses from this laboratory.

Table I. Description of Flags Sampled

Sample No.	*Catalog No.*	*Inventory No.*	*Description*	*Color*	*Date*
W1	81711W14	—	Silk repp[a], from U.S. flag, (38 stars, 13 stripes, "Corps of Cadets" on 7th stripe)	Cream	July 4, 1877– July 4, 1890
W2	81711W14	—	Silk repp, as in W1	Cream	—
W3	Unknown	—	Silk repp, Civil War	Cream	unknown
W4	81711W10	64127	Silk repp, 3rd Regiment Q.M. National Color, 36-star flag	Red	July 1863– July, 1865
W5	81711W10	64127	Silk repp, as in W4	Cream	
W6	81745W06	—	Silk repp, H.Q. of U.S.A	Blue	ca. 1884
W7	81745W07	—	Silk repp, Regimental Colors, 45-star flag, U.S. Reg. of Inf.	Red	July 4, 1896– July 3, 1908
W8	81745W07	—	Silk repp, as in W7	Cream	
W9	81759W59	—	Silk repp, U.S. flag (stars and stripes)	Cream	July 4, 1863– July 3, 1865
W10	Unknown	—	Silk repp, Flag embroid. "Hon- aux Americains," red and white stripes, Storage Unit No. 1449	Cream	WW I
W11	RWW81759W52	—	Silk repp, U.S. National Standard, post 1912	Red	July 4, 1912–1920
W12	81745W00	69143	Silk repp, painted flag	Blue	1875–1900
W13	81711W11	—	Silk repp, Sheridan, Commander of the army flag	Blue	Nov. 1, 1883– Aug. 3, 1888
W14	Unknown	—	Silk repp	Cream	unknown

[a]Repp is warp-faced plain weave fabric.

Table II. Qualitative EDS Results

Sample No.	*Na*	*Mg*	*Al*	*Si*	*P*	*S*	*Cl*	*K*	*Ca*	*Fe*	*Sn*
W1	–	–	+	+	–	++	–	(+)	+	+	–
W2	–	–	–	+	–	++	–	(+)	(+)	(+)	–
W3	–	–	–	+	–	++	–	–	+	–	–
W4	+	(+)	+	+	–	++	–	(+)	+	(+)	–
W5	+	–	+	+	–	++	(+)	+	+	–	–
W6	–	–	–	+	–	++	+	(+)	+	–	–
W7	–	–	+	+	–	++	–	–	+	–	–
W8	–	–	–	+	–	++	+	+	+	–	–
W9a[a]	+	(+)	+	+	(+)	++	(+)	+	+	(+)	–
W9b	–	–	+	+	–	++	–	–	+	(+)	–
W10	+	(+)	–	+	+	(+)	–	–	–	–	++
W11	–	(+)	(+)	+	(+)	++	–	–	+	(+)	++
W12a[b]	–	–	(+)	+	–	++	(+)	(+)	+	+	–
W12b	–	–	(+)	+	–	++	(+)	(+)	+	+	–
W13a[c]	–	–	+	+	–	++	–	(+)	+	(+)	–
W13b	–	–	+	+	–	++	–	(+)	+	(+)	–
W14	–	–	+	+	–	++	–	(+)	+	+	–

SYMBOLS: ++, present in significant amounts (>30% of elements of atomic no. >11); +, present; –, Absent (not detected); and (+), possibly present (ca. 5% of elements of atomic no. >11).

[a]Cream specimen: a is the darker (stained?) area; b is the lighter (unstained?) area.

[b]Blue specimen: a is the lighter (unpainted) side; b is the darker (painted) side.

[c]Blue specimen: a is the lighter (more faded?) side; b is the darker (less faded?) side. (Under a light microscope, at ca. 40× magnification both sides look alike.)

The wide variety and variability of elements detected for identically colored silks immediately suggested that the methods of coloring these historical flags were not at all standardized. Of the three blue specimens, two contained iron, one did not. Of the three red samples, one contained iron, one contained both iron and tin, and one contained neither iron nor tin. The cream samples showed four with iron, one with tin, and three with neither iron nor tin.

For three of the specimens EDS data were collected in two different areas:

- Specimen 9 (cream colored) showed greater discoloration in one area (9a) than in area (9b). The EDS data showed a larger variety of elements in the discolored portion, possibly because of the presence of soluble salts (Na).
- Specimen 12 (painted blue) appeared to be painted darker blue on one side (12b), and lighter blue on the other (12a). The EDS data indicated no difference in the elemental distributions on either side.

- Specimen 13 (blue) appeared similar on both sides under the light microscope at about 40× magnification. Both sides showed traces of iron, similar to the preceding result.

Table III shows raw data for specimen 12b (painted blue) from scans performed on the painted side. Variations in the numerical values illustrate the extent of nonuniformity in the sample and the general level of reproducibility achievable in performing such analyses. Two of the specimens, W10 (cream) and W11 (red), showed substantial amounts of tin. The tin in W11 is probably mordant, but in W10 it appears to be present as a weighting material, as suggested by the results presented in Table IV. Table IV shows the results of ashing the textile samples in oxygen at 900–1000 °C for about 0.5 h. This table shows that, of the 14 specimens tested, only W10 was from a weighted silk. The ash content was consistent with what might be expected from mordanting and accretions gathered during the lifetime of the silks. No attempt was made with these samples of to remove dirt or accretions that may have had a small effect on the EDS and ashing results.

A comparison of sample W10 with sample W8 using SEM (Figure 2) confirms that no identifying surface characteristics distinguish weighted from unweighted silk.

The prominence of sulfur, the major element in all scans, reflects

1. its presence in the backbone chains of the fibroin (*14*) (methionine, etc.)
2. possible residue of bleaching or other processing left from the original manufacture
3. that it was a possible component of the dying or mordanting process

Table III. EDS Data for Sample 12b

Sample No.	*Al*	*Si*	*S*	*Cl*	*K*	*Ca*	*Fe*
W12b	4.41	16.74	60.56	1.24	1.60	10.84	4.62
W12bA	5.39	16.47	51.05	1.86	3.78	10.55	10.90
W12bB	4.74	15.35	52.38	2.59	3.31	17.89	3.74
W12bC	6.21	18.15	50.87	—	1.42	12.90	10.46
W12bD	4.95	15.52	54.89	2.34	1.30	17.85	3.1
W12bE	6.71	18.13	46.99	2.86	4.08	14.01	7.23
W12bF	3.28	16.01	56.79	3.58	5.30	11.25	3.80
W12bG	4.89	17.98	52.39	3.48	3.18	14.40	3.68
W12bH	4.26	16.90	50.19	4.78	2.23	15.05	6.59

NOTE: All values are percents. All elements are of atomic number >11; data were obtained from ASAP routine, normalized; normalization factors varied from 0.86–0.91. Na, Mg, P, and Sn were not present in any sample. All samples are from the painted blue specimen, darker (painted) side.

4. that it was a possible component of the weighting process
5. that it was a possible component of soil, dirt, atmospheric pollution, etc. that accumulated at the fabric surface.

In previous reports (*4–8*), we discussed the problems of distinguishing the presence of mordant in historical textiles, the confounding effects of

Table IV. Ash Content of Flag Samples

Sample No.	*Sample Color*	*% Ash*[a]	*Ash Color*
W1	Cream	0.61	Black
W2	Cream	0.50	Rust
W3	[Cream][b]	—	—
W4	Red	2.79	Rust
W5	Cream	1.13	Rust
W6	Blue	0.42	Rust
W7	Red	1.32	White
W8	Cream	1.77	Black
W9	Cream	1.81	Rust
W10	Cream	46.49	White
W11	Red	4.32	Rust
W12	Blue	0.89	Rust
W13	Blue	0.37	Rust
W14	[Cream][b]	—	—

[a]Samples burned under oxygen at 900–1000 °C, ca. 0.5 h.
[b]No sample taken.

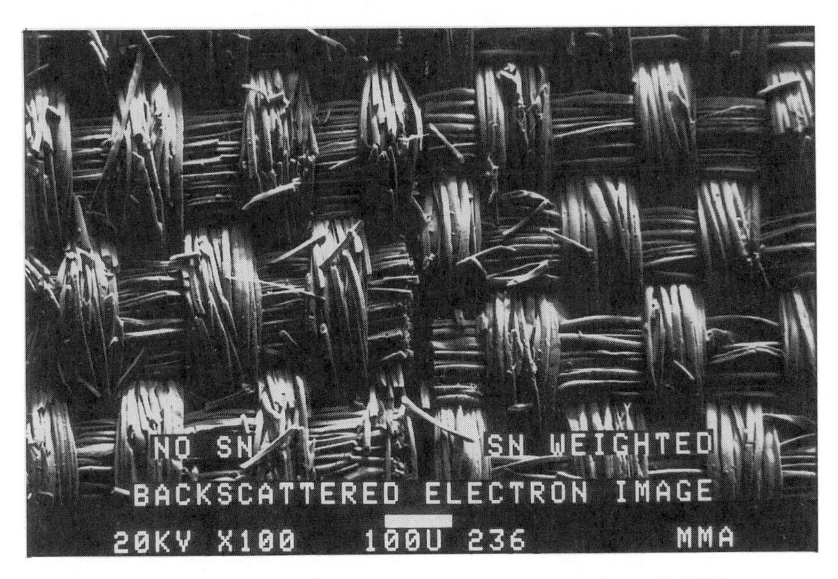

Figure 2. Scanning electron microscopic image of sample W10.

accretions, and the possible use of sulfur as an internal standard (present as part of the fiber chains in all silk and wool samples). Although the percent sulfur in pure, untreated silk is far less than in wool, it always represents a very substantial percentage of the EDS analyzable elements present. In silk samples, it is invariably observed and falls well within the range of detectability. However, because sulfur may be part of the bleaching and dyeing processes (*15–18*), and present as part of the historical accretion, the assumption of sulfur as an internal standard must be made with great caution.

In this set of EDS data, wide variations in sulfur were observed and, except for one sample (W10), sulfur had the highest percent composition of all elements of atomic number >11. In that sample (W10) in which sulfur was observed in trace amounts (about 5%), the weighting agent was so abundant that the amount of sulfur detected represented only a small amount of the total detected material. The percentage of tin observed in the weighted specimen was within the range observed among tin-weighted standards that had been previously analyzed (cf. Table V, reference 4).

Some of the colorants were analyzed with techniques developed by Schweppe (*12, 13*), and in conjunction with descriptive information delineated in the *Color Index* (*20*). The tests are essentially spot tests that are done both separately and sequentially. The spot tests indicate the presence of sulfonic acid groups in the dyestuffs. The colorants represented about 1–5% of the total weight of the sample.

Table V. Dye Analyses

Sample	*Color*	*Testing*
W4	Red	H_2O: sl. sol.; C_2H_5OH: insol.; conc. CH_3COOH: sl. sol.; conc. NH_3: soln. yellow, fiber colorless; + C_2H_5OH: pale red; conc. H_2SO_4: soln. red–violet, fiber dark red; + H_2O: soln. sl. yellow, fiber bright red. Indicative of Ponceau RR (C. I. Acid Red 26)
W7	Red	H_2O: soln. sl. yellow, fiber insol.; C_2H_5OH: soln. pink, fiber, n.r.; conc. CH_3COOH: n.r.; conc. NH_3: soln. red; conc. H_2SO_4: soln. red, fiber dark red; + H_2O: soln. yellow, fiber red. Indicative of: Ponceau RR (C. I. Acid Red 26)
W11	Red	H_2O: n.r.; C_2H_5OH: soln. pink, fiber, n.r.; conc. CH_3COOH: n.r.; conc. NH_3: sol. pink, fiber colorless; + conc. CH_3COOH: Sol. colorless, fiber colorless Indicative of Fast Red AV (C. I. Acid Red 88)
W6	Blue	H_2O: insol.; C_2H_5OH: n.r.; conc. CH_3COOH: n.r.; conc. NH_3: soln. and fiber nearly colorless; + conc. CH_3COOH: sol. blue; conc. H_2SO_4: soln. and fiber brown–red; + H_2O: sol. blue. Indicative of Water Blue IN (C. I. Acid Blue 93)

English, French, and German 19th-century dye texts (*16–19*) suggest a penchant for the use of sulfuric acid, not only in the dyeing process to maintain an acidic pH, but more importantly, as a final "sour" or clearing rinse.

Summary

1. No standard process for the use of colorants on flags was discovered:
   - Red (three examples): One had iron (traces); one had iron (traces) and tin; one had neither iron nor tin.
   - Blue (three examples): Two had iron (no tin); one had neither iron nor tin.
   - Cream (eight examples): Four had iron (trace–small amounts) and no tin; one had tin (large amount) and no iron; three had neither iron nor tin.
2. Only one silk specimen from this group (W10) appeared to be weighted. The weighting agent was tin in some form. In this sample, the most abundant element (of those elements of atomic number >11) was tin. In all other specimens, the most abundant element (of those elements of atomic number >11) was sulfur.
3. Light and SEM images of the weighted silk specimen were compared to other specimens that were not weighted. No obvious distinctions were revealed, despite the presence of about 50% ash in the case of the weighted silk and only about 1% ash in other cases.

In some cases, the deteriorated condition of historic silk flags may be the result of weighting, but other factors must be important because even these limited data show that many silk flags contain silks that are not weighted. Examples of factors other than weighting that may account for the deteriorated condition of the banners are original manufacturing processes (degumming, bleaching, etc.); storage (indoor air pollution, temperature–relative humidity cycling, etc.); and conservation practices of the past.

The widely held opinion that weighting is the cause of silk deterioration may give way, with further study, to an understanding of deterioration as a result of acid treatments or other manufacturing processes used in European and American workshops during the late 19th century.

References

1. Howitt, F. O. *Bibliography of Silk*; Hutchinson's Scientific and Technical Publications: London, 1946; Chapter 8.
2. Ross, J. E.; Johnson, R. L.; Edgar, R. *Text. Res. J.* **1936,** *6,* 207–216.
3. Ballard, M.; Koestler, R. J.; Indictor, N. *Preprints of Papers,* 13th Annual Meeting; American Institute for Conservation of Historical and Artistic Works: Washington, DC, 1985; p 155.
4. Ballard, M.; Koestler, R. J.; Indictor, N. *Scanning Electron Microsc.* **1986, 2,** 499–506.
5. Koestler, R. J.; Sheryll, R.; Indictor N. *Stud. Conserv.* **1985,** *30,* 58–62.
6. Indictor, N.; Koestler, R. J.; Sheryll, R. *J. Am. Inst. Conserv.* **1985,** *24,* 104–109.
7. Koestler, R. J.; Indictor, N.; Sheryll, R. *J. Am. Inst. Conserv.* **1985,** *24,* 110–115.
8. Williams, E. T.; Indictor, N. *Scanning Electron Microsc.* **1986, 2,** 847–850.
9. Indictor, N.; Koestler, R. J. *Scanning Electron Microsc.* **1986, 2,** 491–497.
10. Indictor, N.; Koestler, R. J.; Blair, C.; Wardwell, A. *Textile History* **1988,** *19(1),* 323
11. Goldstein, J.; Newbury, D. E.; Echlin, P.; Joy, D. C.; Fiori, C.; Lifshin, E. *Scanning Electron Microscopy and X-Ray Microanalysis*; Plenum: New York, 1984.
12. Schweppe, H. *Identification of Dyes on Textile Materials*; Course, Conservation Analytical Laboratory, Smithsonian Institution, 1985, 1986.
13. Schweppe, H. In *The Analytical Chemistry of Synthetic Dyes*; Venketararian, K., Ed.; Wiley: New York, 1977; Chapter 2.
14. Carboni, P. *Silks: Biology, Chemistry, Technology*; Walter, K., translator; Chapman and Hall: London, 1952.
15. Ganswindt, A. *Dyeing Silk, Mixed Silk Fabrics and Artificial Silks*; Salter, C., translator; Scott, Greenwood and Son: London, 1921.
16. Hummel, J. J. *Colouring Matters for Dyeing Textiles*, revised ed.; Hasluck, P. N., Ed.; Cassell: London, 1906.
17. Hummel, J. J. *Mordants, Methods, and Machinery Used in Dyeing*, revised ed.; Hasluck, P. N., Ed.; Cassell: London, 1906.
18. Hurst, G. H. *Silk Dyeing, Printing, and Finishing*; George Bell and Sons: London, 1892.
19. Knecht, E.; Rawson, C.; Loewenthal, R. *A Manual of Dyeing*, 2nd ed.; Charles Griffin: London, 1910.
20. *Color Index*, 3rd ed.; Society of Dyers and Colourists: Bradford, England, and American Association of Textile Chemists and Colorists: Research Triangle Park, NC, 1971.

RECEIVED for review June 11, 1987. ACCEPTED revised manuscript February 25, 1988.

25

Characterization of Historical and Artificially Aged Silk Fabrics

S. P. Hersh, P. A. Tucker, and M. A. Becker[1]

College of Textiles, North Carolina State University, Raleigh, NC 27695–8301

Several methods for characterizing the degradation of silk fabrics that have been subjected to natural and accelerated aging by exposure to heat and light are described. Change in color, strength degradation, and increases in amino- and ammonia-nitrogen contents were measured. Among the more notable findings were that the fabrics discolor on the order of 10 times faster when degraded thermally than when degraded by light (at equal levels of degradation as measured by strength loss) and that strength loss is proportional to the increase in ammonia and amino group contents for fabrics degraded by heat or light. The amino group and ammonia contents of 16th-, 18th-, and 19th-century silks range from values typical for artificially aged contemporary silks to 3 times greater for amino nitrogen and 4 times greater for ammonia. All of the tin-weighted historical fabrics (except three dyed with logwood) have higher amino group contents than unweighted fabric. The ammonia contents do not vary much, however.

THE BASIC KNOWLEDGE NEEDED TO SUPPORT applied research on the conservation and preservation of irreplaceable textiles is inadequate. To contribute to the knowledge needed to develop better and more acceptable

[1]**Current address: Department of Materials Science and Engineering, Johns Hopkins University, Baltimore, MD 21218**

0065–2393/89/0220–0429$06.25/0

conservation methods, we performed several studies focusing on the conservation and restoration of cotton (*1–8*). The most fragile textile, however, is generally agreed to be silk. References 9 and 10 review some of the problems associated with silk. This chapter is the continuation of an earlier study on the mechanism of silk degradation (*11*).

A major factor that contributes to the fragility of silk is the practice of weighting, which has been conducted for at least 200 years (*12, 13*). Weighting is the application of 30%–300% of inorganic salts of aluminum, iron, lead, tin, or zinc to silk fabrics to increase their body, drape, "scroop," weight per unit area, etc. By the late 1800s, weighting had become an accepted method of silk preparation. Because sensitivity to environmental stress is greatly influenced by the type and amount of weighting agents present in the silk, an analytical technique was developed to determine the presence of weighting agents. A convenient, nondestructive qualitative procedure for making such tests by X-ray fluorescence spectroscopy was described elsewhere (*14*).

Strong alkalies cause far greater damage in proteinaceous fibers than do strong acids (*15*). The damage caused by strong acids, however, is quite severe. Under milder conditions, the light stability of unweighted silk is greatest at about pH 10 but decreases rapidly as the fabric pH becomes higher or lower (*16*). Because most weighting compounds are highly acidic, weighted silks might be expected to be even less stable than unweighted silks. This sensitizing effect of weighting has indeed been shown to occur and is far more detrimental to silks exposed to light than to fabrics stored in the dark (*17, 18*). However, damage even during dark storage is severe.

Two major routes to silk degradation are oxidation and hydrolysis (*16*). Hydrolysis of an amide group splits the polymer chain to form an amino group and a carboxyl group. Determination of the increase in amino groups in the silk, therefore, should provide a measure of hydrolytic degradation. Oxidation, on the other hand, is accompanied by the formation of ammonia (*16*). Thus, it should be possible to assess the chemical breakdown by measuring the amount of amino and ammonia nitrogen present in the silk.

The ultimate objective of this study is to develop a technique for preventing, or at least retarding, the degradation of silk by applying stabilizers and consolidants to the silk. Before progress can be made, it is necessary to characterize the state of degradation of historical textiles and to establish, if possible, the mechanisms by which these textiles reached their present state. Thus, the primary purpose of this chapter is to compare the state of degradation of historical silk fabrics with those of contemporary silk fabrics that have been artificially aged by exposure to dry heat and to light from a xenon arc lamp. Properties examined for this purpose are tensile strength, color change, and the concentration of amino groups and ammonia in the water-extractable components of the fabrics.

Experimental Design

Test Materials. CONTEMPORARY SILK. As reported earlier (*11*), fabrics were chosen over yarns for treatment because of their ease of preparation and handling. Tensile properties, however, were measured on yarns extracted from the fabrics to take advantage of the ease of handling fabrics and the ease of testing yarns.

The silk fabric studied was an unweighted plain woven Chinese silk habutae (Testfabrics, Inc., Middlesex, NJ, Style No. 605) with 126 ends/in. (37.6 denier) and 117 picks/in. (32.1 denier) and weighing 1.11 oz/yd^2. The fabric had been degummed (*14*). Fabric was taken from the same bolt for all tests.

HISTORICAL FABRICS. Fabrics and garments analyzed were 11 samples from the study collection of the Division of Textiles, National Museum of American History of the Smithsonian Institution, and one sample from the Museo Poldi-Pezzoli. Many of the garments were composites consisting of several layers and decorative elements. The fabrics examined were identified as follows:

1. late 19th- or early 20th-century wedding dress, beige
2. early 20th-century green fabric with green satin stripes
3. late 19th- or early 20th-century white fabric with piping
4. 18th-century green brocade fabric background with a flower and building pattern
5. late 18th- or early 19th-century pink fabric with embroidered metallic flowers
6. late 19th-century three-layer vest, black top and lining and gold inner layer
7. ca. 1901 Edwardian collar piece, green fabric with beige lace and black velvet collar band (part of No. 8)
8. ca. 1901 Edwardian afternoon reception dress, bodice.
9. late 19th-century skirt, beige embroidered face fabric and brown lining fabric.
10. 1880–1900 dark blue and black jacquard pattern fabric.
11. late 19th-century quilted bedspread, olive with pink stripes.
12. 16th-century Flemish tapestry (being restored), deteriorated silk yarn sample. The yarn was fragmented into segments 1 to 2 mm long.

Artificial Aging by Heat. The procedure developed earlier (*11*) was followed. Fabric pieces (15 × 15 cm) were placed on racks covered with a Fiberglas screen (7- × 3-cm mesh) in a forced convection oven preheated to 150 °C. The screen was used to prevent any enhanced degradation that might be caused by direct contact with the metal rack. The control silk fabric was heated for up to 4 days in 6-h increments, and then immediately placed in a desiccator that contained silica gel to keep the silk dry while cooling.

Artificial Aging by Light. Pieces of silk fabric (20 × 7 cm) were mounted in standard specimen holders as specified in the American Association of Textile Chemists and Colorists (AATCC) Test Method 16E-1982, "Colorfastness to Light: Water-Cooled Xenon Arc Lamp, Continuous Light" (*19*). Samples of the contemporary fabric were exposed to light for 2-day intervals for up to 20 days in a water-cooled xenon arc fading apparatus Weather-ometer, model ES25 (Atlas Electric Devices, Chicago, IL). The Weather-ometer was operated at 50 °C with an arc intensity of 1500 W. The relative humidity was maintained at 30% ± 5%. The sample was exposed to 108 kJ/m^2 at 2-day intervals. Upon removal from the Weather-ometer, the samples were placed in acid-free tissue paper, then equilibrated under standard test conditions for future testing.

Parameters to Measure Degradation. BREAKING STRENGTH. Warp yarns were extracted from the fabrics, and their breaking loads were determined at a gauge length of 5.0 cm and a rate of extension of 50 mm/min on a tensile testing machine (Instron 1123) as set forth in the American Society for Testing and Materials (ASTM) Test Method D2256-80, "Breaking Load (Strength) and Elongation of Yarn by the Single-Strand Method" (*20*). Normally, 21 measurements were made from each fabric sample. The breaking strength of the yarns extracted from the control fabric was 100.0 gf (2.6 gf/denier) with a standard deviation of approximately 8 gf.

COLOR DIFFERENCE. Silk discolors when exposed to heat and light. The color change is taken as one measure of the extent of degradation. The color difference of each treated sample was evaluated on a Diano Match-Scan spectrophotometer against a standard untreated silk sample. The color differences ΔE^*_{ab} are reported in CIELAB color difference units (CDU) for Illuminant D_{65}, and were calculated by (*21*)

$$\Delta E^*_{ab} = [(\Delta L^*)^2 + (\Delta a^*)^2 + (\Delta b^*)^2]^{1/2} \quad (1)$$

where ΔL^* is the change in lightness, from lighter (+) to darker (–), Δa^* is the change in shade from red (+) to green (–), and Δb^* is the change in shade from yellow (+) to blue (–) with respect to a standard (untreated silk fabric). The variables L^*, a^*, and b^* are defined as

$$L^* = 116(Y/Y_0)^{1/3} - 16 \quad (2)$$

$$a^* = 500\,[(X/X_0)^{1/3} - (Y/Y_0)^{1/3}] \quad (3)$$

$$b^* = 200[(Y/Y_0)^{1/3} - (Z/Z_0)^{1/3}] \quad (4)$$

where X, Y, and Z are the tristimulus values for the sample, and X_0, Y_0, and Z_0 are the values for the reference white.

AMINO GROUP CONTENT. Each fabric was ground in a Wiley mill fitted with a No. 40 mesh screen. The concentration of amino groups was then determined colorimetrically by the reaction of 20 mg of the ground silk fabric with ninhydrin by using the procedure described earlier (*11*, *22*). The reaction of ninhydrin with compounds containing α-amino groups and with ammonium salts forms a compound known as Ruhmann's purple that has a maximum absorption at 570 nm. Stock solutions of *dl*-leucine and ammonium chloride were prepared for calibration. Suitably diluted aliquots of the stock solutions were then analyzed by using the ninhydrin procedure. The calibration curves obtained (shown in Figure 1) were virtually identical and fit the following linear least-squares regression equations:

$$\text{leucine } [-NH_2](\text{nmol/mL}) = 59.45\,A + 0.0585 \quad (r^2 = 0.9980) \quad (5a)$$

$$NH_4Cl\ [NH_3](\text{nmol/mL}) = 60.28\,A + 0.450 \quad (r^2 = 0.9995) \quad (5b)$$

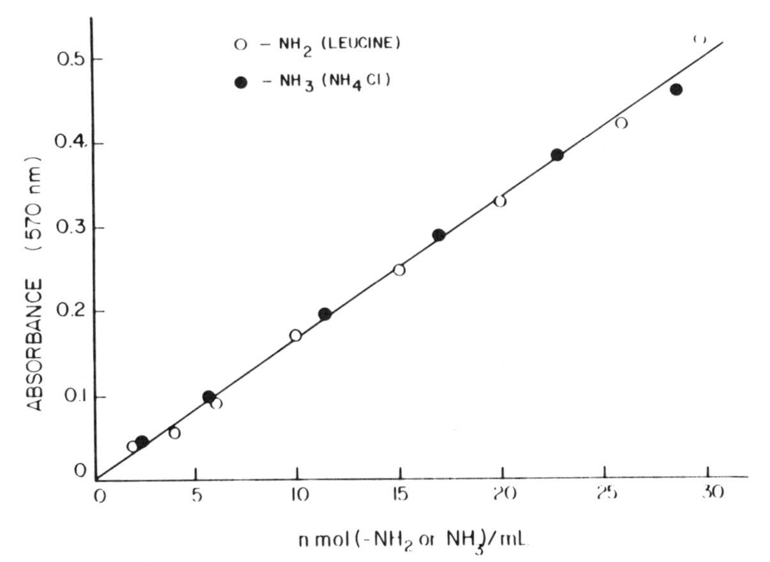

Figure 1. Calibration curve for amino group (dl-*leucine*) *and ammonium ion* (*ammonium chloride*) *analysis by ninhydrin method.*

where [–NH_2] and [NH_3] are the concentrations of amino groups and ammonia, respectively; A is the absorbance of the test solution at 570 nm; and r is the correlation coefficient. Because the analysis of the silk fabrics by the ninhydrin method gives the sum of [–NH_2] and [NH_3], the amino group concentration was calculated by subtracting the independently measured ammonia concentration.

AMMONIA CONTENT. The concentration of ammonia nitrogen in the water-extractable components of the fabric was determined colorimetrically by Nessler's reaction (23). Under alkaline conditions, Nessler's reagent (mercuric iodide–potassium iodide solution) reacts with ammonia that has been released from ammonium salts by the alkali present to produce a yellow compound by the following reaction:

$$2K_2HgI_4 + NH_3 + 3KOH \rightarrow Hg_2OINH_2 + 7KI + 2H_2O$$

A 50-mg sample of each ground fabric was introduced into a 50-mL Erlenmeyer flask, and 20–30 mL of deionized–distilled water was added. After approximately 15 min, each solution was filtered through ashless filter paper into a 100-mL volumetric flask. A 2-mL aliquot of Nessler's reagent (APHA, Fisher Scientific Co.) was added to the 100-mL solution. After at least 10 min, but not longer than 20 min, the absorbance of the solution at 425 nm was measured on a Bausch and Lomb Spectronic 20 spectrophotometer. A solution submitted to the same treatment but to which no silk was added was used as the blank. Care must be taken to prevent contamination of the water and reagents, because a number of organic materials interfere with the measurement. The colored solution formed after adding the Nessler's reagent developed some turbidity on the samples degraded 10 days or more by light. Thus far, the reason for this complication has not been established.

Two solutions were prepared for each historical silk sample. Nessler's reagent was added to one, and deionized–distilled water was added to the other, so that an independent measurement could be made of the amount of discoloration contributed to the solution by the silk and its degradation products and contaminants, if any. For these measurements, deionized–distilled water was used as the blank, and a separate absorbance measurement was made on the solution of water and Nessler's reagent.

The total absorbance A of the solution containing the ammonia–Nessler's complex is the sum of three absorbances:

$$A = A_a + A_n + A_s \qquad (6)$$

where A_a is the absorbance of the ammonia–Nessler's complex, A_n is the absorbance of the Nessler's reagent, and A_s is the absorbance of the soluble silk constituents and degradation products. A is obtained by subtracting A_n and A_s from A. The absorption of silk degradation products, A_s, is negligible for the contemporary silks, whereas a correction for A_s must be made in the analysis of the historical samples to correct for any leaching of dyes, soil, and other contaminants.

A stock solution of NH_4Cl dissolved in deionized–distilled water was prepared for calibration. Aliquots of this solution were further diluted to obtain the solutions for the calibration curve. Each solution was then subjected to the treatment just described. The calibration curve (Figure 2) was fit by the linear least-squares regression equation

$$[NH_3](\text{nmol/mL}) = 321.5\,A_a - 0.439 \quad (r^2 = 0.9979) \tag{7}$$

where $[NH_3]$ is the concentration of ammonia in nanomoles of NH_3 per milliliter.

Results and Discussion

Heat Degradation. STRENGTH. The breaking strengths of yarns extracted from the fabric after heating at 150 °C are shown in Figure 3. Considering "property kinetics" as described by Arney and Chapelaine (*24*), the data suggest a typical zero-order reaction scheme in which the strength S decreases at a constant rate with time t expressed as follows:

$$dS/dt = -k_0 \tag{8}$$

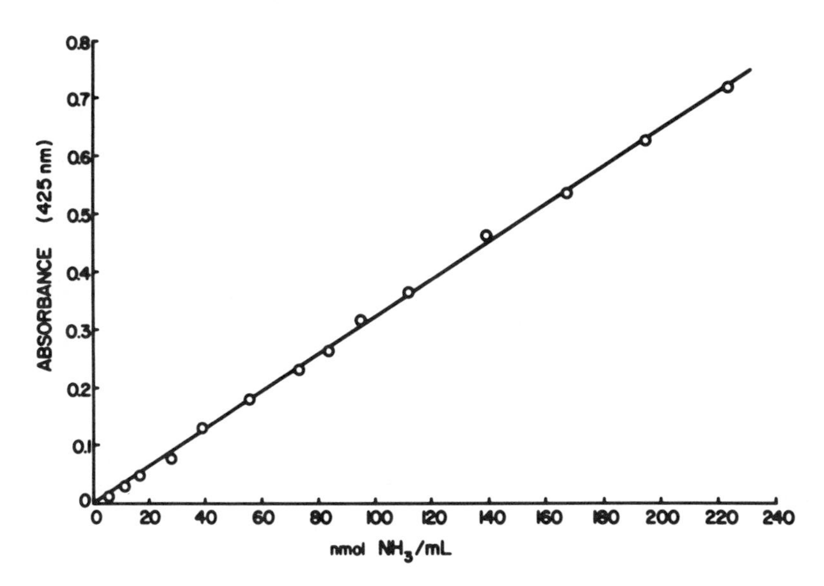

Figure 2. Calibration curve for ammonia analysis by Nessler's reaction (ammonium chloride).

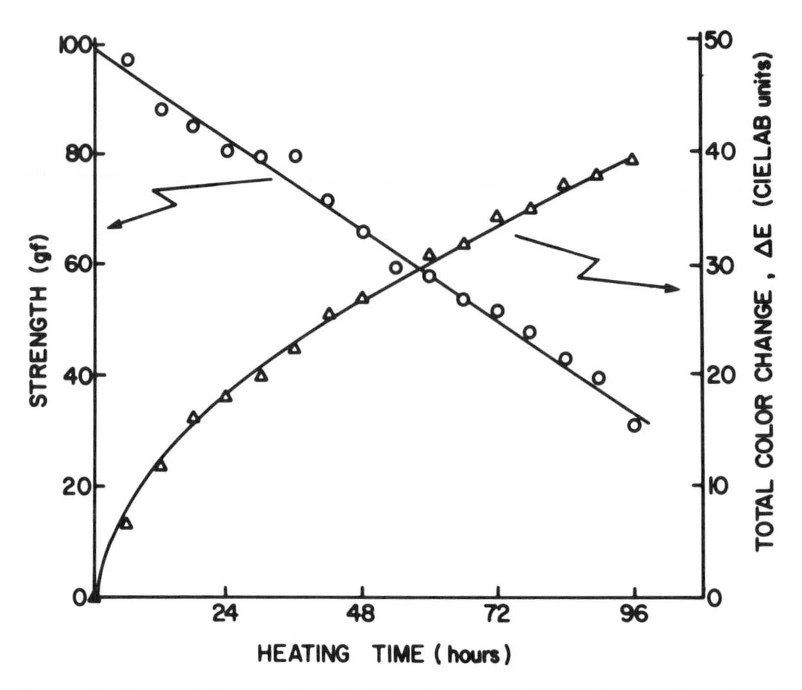

Figure 3. Strength loss and total color change of silk fabric as a function of heating time at 150 °C.

where k_0 is the zero-order rate constant. When the data are plotted as the logarithm of strength as a function of time, however, a straight line is also obtained. Such a curve indicates a first-order reaction in which the rate of strength loss (or degradation) is directly proportional to the amount of material present at any given time.

$$dS/dt = -kS \tag{9}$$

where k is the first-order rate constant. The first-order rate constant k can be estimated from the slope of the ln S vs. t curve, whose equation is given by

$$\ln S = -kt + a \tag{10}$$

where k is the slope of the line and a is the intercept on the S-axis. The slope k in equation 10 is the same as the rate constant in equation 9 as can be shown by differentiating equation 10 with respect to t:

$$(1/S)\,(dS/dt) = -k \tag{11}$$

and

$$dS/dt = -kS \tag{12}$$

A linear regression of the strengths listed in Table I and shown in Figure 3 gives the least-squares equation for the loss in strength as a function of heating time

$$S(\text{gf}) = -0.684t(\text{h}) + 99.29 \quad (r^2 = 0.9941) \tag{13}$$

where t (h) is the heating time in hours. Thus, the strength decreases at the rate of 0.684 gf/h when heated at 150 °C; 0.684 gf/h is also the zero-order rate constant. The strength of the yarn was reduced to about 30% of its original value after 4 days of heating. The regression equation that uses the natural logarithm of the strength and generates the first-order rate constant is

$$S(\text{gf}) = -0.0110 \ln t(\text{h}) + 4.67 \quad (r^2 = 0.9807) \tag{14}$$

Table I. Properties of Silk Fabric After Heating at 150 °C

Heating Time (h)	*Breaking Strength (gf)*	*Total Color Change, ΔE (CIELAB units)*	*Ammonia Concentration*[a]	*Amino group Concentration*[b]
0	100.0	0.00	11.1	53.9
6	98.3	6.71		
12	88.2	11.79	16.8	66.5
18	84.9	16.24		
24	80.4	18.20	22.1	69.3
30	79.8	20.42		
36	79.3	22.32	32.6	68.3
42	71.1	25.44		
48	65.8	26.99	44.4	72.4
54	58.9	28.78		
60	58.0	30.66	37.1	74.9
66	53.3	31.39		
72	51.4	34.15	37.8	95.5
78	48.0	34.86		
84	42.4	36.79	43.0	105.3
90	39.4	37.79		
96	30.9	39.39	55.5	107.2
Rate of Change[c]				
Zero-order	16.4	7.03	9.74	–
Range	$0 \leq t \leq 96$	$t \leq 24$	$0 \leq t \leq 96$	–
First-order	26.4%	–	–	–
Range	$0 \leq t \leq 96$	–	–	–

[a]Units are micromoles of NH_3 per gram of silk.
[b]Units are micromoles of NH_2 per gram of silk.
[c]Same units as for the corresponding column per day.

The value of k, –0.0110, indicates that the strength decreases at the rate of 1.10% per hour. Because the square of the correlation coefficients of equations 13 and 14 are virtually identical, the zero- and first-order degradation kinetics fit the data equally well. Thus, it is not possible to distinguish between the two mechanisms on the basis of the present data. If the degradation is carried further, the degradation generally follows first-order kinetics (*11*, *25*). First-order kinetics were followed in the study of light degradation described in the next section.

COLOR DIFFERENCE. As shown in Figure 3 and Table I, the color of the aged fabric changes tremendously over 4 days when heated at 150 °C. A regression analysis of the linear portion of the line from time t = 24 to 96 h (i.e., $t \geq 24$ h) gives the relationship:

$$\Delta E_{t \geq 24} = 0.293t(\mathrm{h}) + 12.362 \quad (r^2 = 0.9941) \tag{15}$$

Thus, after the first 24 h when the discoloration rate is somewhat faster, the silk fabric changes color at a rate of about 0.29 color difference units (CDU) per hour.

AMINO GROUP CONTENT. The amino group contents of heated samples are listed in Table I and are shown in Figure 4. Although the data showing

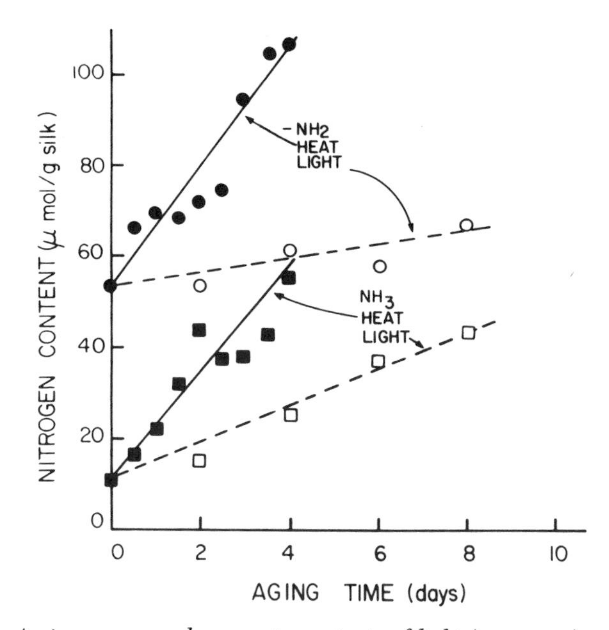

Figure 4. Amino group and ammonia contents of light (xenon arc)- and heat (150 °C)-aged silk fabrics as a function of time.

the relationship between amino concentration [$-NH_2$] and time t are scattered, evidently, amino groups are formed under these heating conditions. Because the amino groups present in silk occur only at the ends of the molecular chains and on the lysine and arginine side chains, the increase in amino groups is indicative of a decrease in chain length as a function of heating time. Such a result would also reduce the breaking strength as shown in Figure 5. The breaking strength decreases linearly with increasing amino group content and is described by the equation:

$$S(\text{gf}) = -1.148[-NH_2](\mu\text{mol/g of silk}) + 157.27 \quad (r^2 = 0.9051) \qquad (16)$$

Thus, the strength decreases at rate of about 1.15 gf per micromole of $-NH_2$ per gram of silk. Because of the complex relationship between polymer chain scission, molecular weight, and strength, and the presence of side-chain amino groups in silk, it is difficult to relate these parameters theoretically.

AMMONIA CONTENT. The concentration of ammonia in the heated samples is reported in Table I and shown in Figure 4. The formation of ammonia

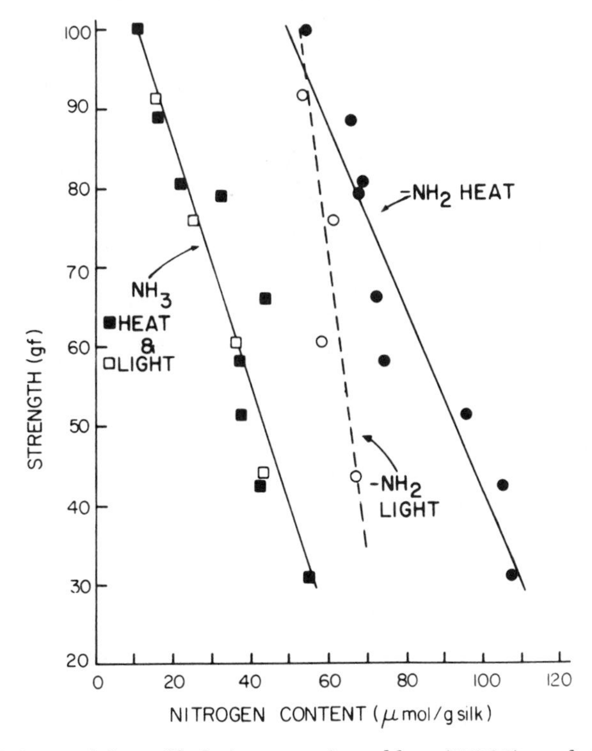

Figure 5. Strength loss of light (xenon arc)- and heat (150 °C)-aged silk fabrics as a function of nitrogen content expressed as molar concentrations of ammonia and amino groups.

in silk has been attributed to oxidation (*16*) and is also indicative of fiber degradation. A linear regression gives the relationship between ammonia concentration $[NH_3]$ and time t (in hours)

$$[NH_3](\text{mol/g of silk}) = 0.406t(\text{h}) + 13.87 \quad (r^2 = 0.9330) \qquad (17)$$

Ammonia is being formed at a rate of 0.406 μmol/g of silk/h under these heating conditions. The relationship between strength and ammonia concentration, as shown in Figure 5 and Table I, is described by the regression equation:

$$S(\text{gf}) = -1.466\,[NH_3](\mu\text{mol/g of silk}) + 115.8 \quad (r^2 = 0.9268) \quad (18)$$

Comparing equation 15 with equation 16 shows that the strength decreases more rapidly with the formation of ammonia than it does with the formation of amino groups (1.466/1.148 = 1.28; i.e., 28% faster).

Light Degradation. STRENGTH. The breaking strengths of yarns extracted from fabrics artificially aged by light are listed in Table II and are shown in Figures 6 (linear strength plot) and 7 (log strength plot). Although both plots show some curvature, the latter relationship deviates less from a

Table II. Properties of Silk Fabrics After Exposure to Xenon Arc Lamp

Irradiation Time (days)	*Breaking Strength (gf)*	*Total Color Change, ΔE (CIELAB units)*	*Ammonia Concentration*[a]	*Amino group Concentration*[b]
0	100.0	0.00	11.1	53.9
2	91.5	2.05	15.5	53.4
4	75.6	3.07	25.3	61.5
6	60.2	3.82	37.1	58.7
8	43.8	4.33	43.7	67.4
10	29.8	5.08	[c]	116.8[d]
12	21.7	5.42	[c]	116.8[d]
14	15.7	6.19	[c]	135.7[d]
16	11.3	6.48	[c]	148.3[d]
18	8.7	6.67	[c]	149.8[d]
20	6.8	6.75	[c]	163.0[d]
Rate of Change[e]				
Zero-order	–	0.221	4.35	1.48
Range	–	$t \geq 6$	$0 \leq t \leq 8$	$0 \leq t \leq 8$
First-order	14.5%	–	–	–
Range	$0 \leq t \leq 20$	–	–	–

[a]Units are micromoles of NH_3 per gram of silk.
[b]Units are micromoles of NH_2 per gram of silk.
[c]Solution was turbid.
[d]Sum of ammonia and amino group concentrations.
[e]Same units as for the corresponding column per day.

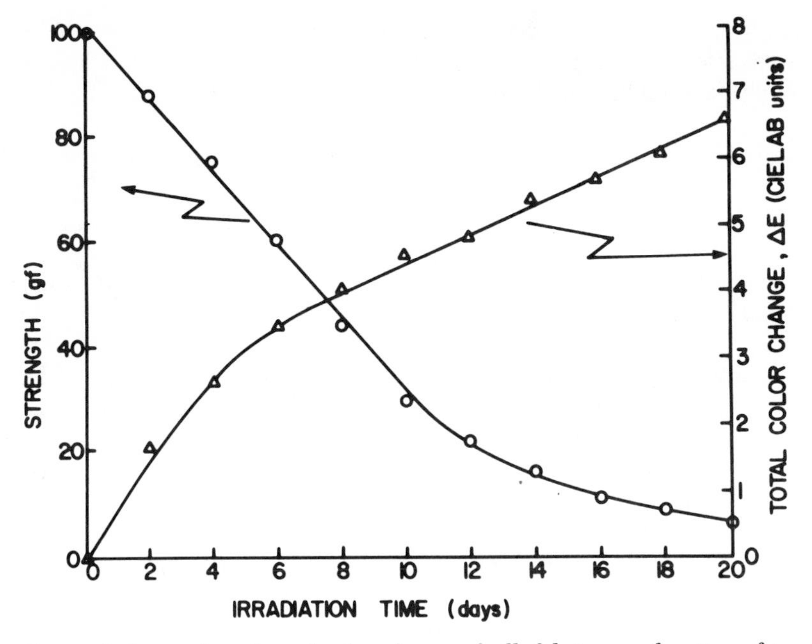

Figure 6. Strength and total color change of silk fabric as a function of irradiation time (xenon arc).

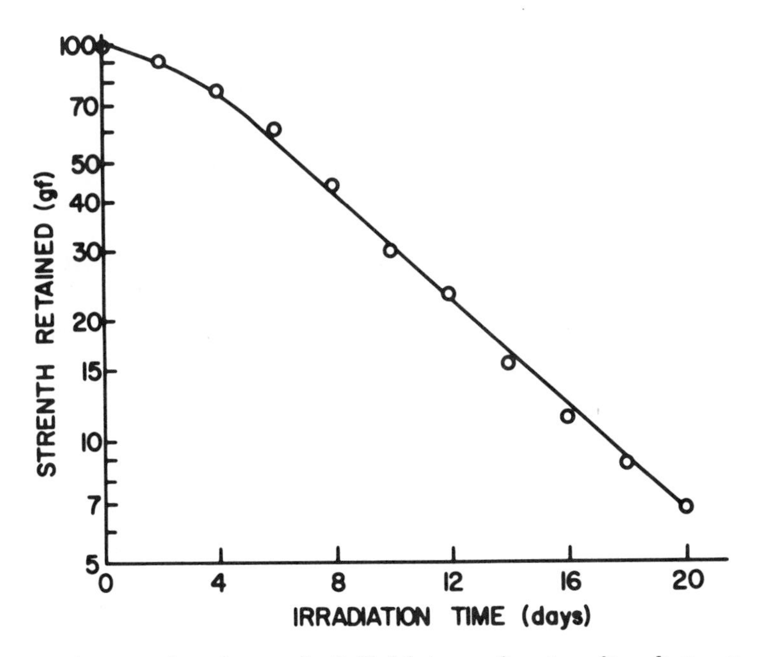

Figure 7. Logarithm of strength of silk fabric as a function of irradiation time (xenon arc).

straight line. These data suggest that more than one mechanism might be responsible for the degradation. The ln strength–time relationship, however, provides an excellent fit to first-order kinetics as follows:

$$\ln S(\text{gf}) = -0.145t(\text{days}) + 4.821 \quad (r^2 = 0.9947) \qquad (19)$$

Thus, when exposed to artificial light, the strength decreases exponentially as a function of time at a rate of 14% per day. In contrast, the first-order heat degradation rate at 150 °C (equation 14) was (0.011 × 24 h/day)/0.145) or 1.82 times faster.

COLOR DIFFERENCE. Unlike those samples aged by heat, the color of those exposed to light increased much less. The measurements are listed in Table II and are shown in Figure 6. The regression equation describing the total color difference at exposure times of 6 days or longer as a function of time is given by:

$$\Delta E_{t \geq 6} = 0.221t(\text{days}) + 2.716 \quad (r^2 = 0.9740) \qquad (20)$$

In this case, the color changed about 0.22 CDU per day compared with a rate of 7.03 CDU per day for silk heated at 150 °C. Thus, during the linear portion of the aging curve, the color changes about 32 times faster when heated than when irradiated.

AMINO GROUP CONTENT. Although the data are scattered, Figure 4 shows that the amino content increases with irradiation, but at a much lower rate than that at which amino groups are formed by heating. The heated samples (Figure 5) contain more amino groups, at constant strength, than do the light-aged samples. The relationship between strength and the amino group content of silk degraded by light is described by

$$S(\text{gf})_{t \leq 8} = -3.55[-\text{NH}_2](\mu\text{mol/g of silk}) + 283.4 \quad (r^2 = 0.8089) \qquad (21)$$

Thus, the strength decreases at a rate of 3.55 gf per mole of $-NH_2$ per gram of silk when exposed to light. This rate is about 3 times faster than that of silk exposed to heat.

AMMONIA CONTENT. The concentrations of ammonia present in the samples are listed in Table II and shown in Figure 4. The concentration appears linear up to approximately 40 μmol of NH_3/g of silk (8 days) and is described by

$$[\text{NH}_3]_{t \leq 8}(\mu\text{mol/g of silk}) = 4.345t(\text{days}) + 9.172 \quad (r^2 = 0.9906) \qquad (22)$$

As indicated in Table II, the ammonia concentrations of the samples irradiated 10 days or longer could not be measured because the solutions became turbid. The reason for the turbidity has not yet been determined. The strength vs. $[NH_3]$ relation is described by the equation

$$S(gf)_{[NH_3]<40} = -1.640[NH_3](\mu mol/g \text{ of silk}) + 117.77 \quad (r^2 = 0.9957) \tag{23}$$

Thus the strength decreases at the rate of 1.64 gf per mole of NH_3 per gram of silk with increasing ammonia content.

Additional Differences Between Heat and Light Aging

Another interesting difference in the tensile strength–color difference relationships for heat- and light-degraded samples is shown in Figure 8. Heat-aged samples discolor approximately 10 times more than light-aged samples when degraded to the same strength. Thus, different mechanisms might be responsible for the degradation caused by heat and light. Another difference

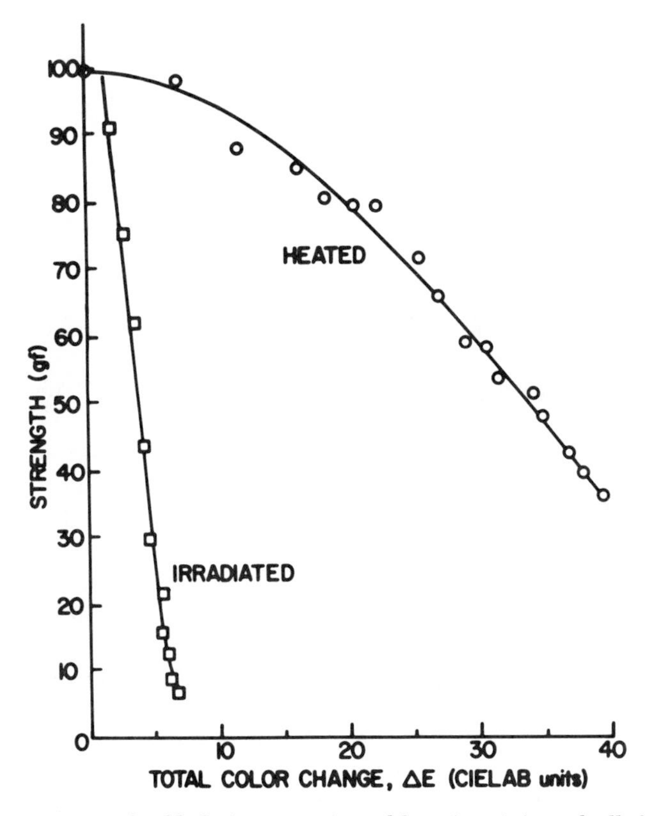

Figure 8. Strength of light (xenon arc)- and heat (150 °C)-aged silk fabric as a function of total color change.

between light- and heat-degraded samples is suggested by the plot of the relationships between ammonia content and amino-group content shown in the lower left corner of Figure 9. More detailed studies will be needed to obtain more accurate kinetic and chemical data before the nature of these hypothesized mechanisms can be established.

If the strength of silk were determined exclusively by its molecular chain length, and if the chain length (or average molecular weight) were decreased by hydrolysis with a concomitant increase in amino end groups, the strength of silk could probably be predicted by measuring the concentration of amino groups independent of the mode of degradation. The curves in Figure 5 show that strength decreases with increasing amino group content, but the relationship is different for silks degraded by light and by heat. At a given strength, heat aging generates more amino groups than does light aging (as indicated by equations 16 and 21 and Figure 5, approximately 3 times as much, much less than the 10-fold increase in color). Thus, this result is further evidence that the degradation mechanisms involved might be different for light and heat. In contrast, the relationship between strength and ammonia content as given by equations 23 and 18 and discussed earlier is almost identical for light and heat degradation (1.64 vs. 1.47 gf per micromole per gram of silk,) respectively.

Evaluation of Historical Fabric Samples

A comparison of the state of degradation of historical silk fabrics with those of artificially aged contemporary fabrics would be very useful in any effort to determine the mechanisms by which naturally aged silk is degraded. Such knowledge would also aid in the selection and evaluation of stabilizers to slow the degradation process. For these reasons, the historical fabrics were evaluated.

The change in color upon natural aging could not be determined for the historical fabrics because their original colors are unknown. The breaking strengths of yarns were extremely difficult to measure because of the types of weaves and difficulty in extracting yarns long enough to test. Figure 9 shows the relationship between the amino group and ammonia contents of the historic fabrics along with the data for the artificially aged fabrics. The presence of weighting and mordanting agents in the historical fabrics was previously determined and reported (*14*). The data plotted in Figure 9 are given in Table III.

Only three of the historical fabrics had amino group and ammonia contents within the range of the artificially aged contemporary fabrics. Most, however, were up to 3 times greater in amino group content and four times greater in ammonia content. Samples 4 and 5, both 18th-century fabrics containing no weighting agents, had relatively low concentrations of both

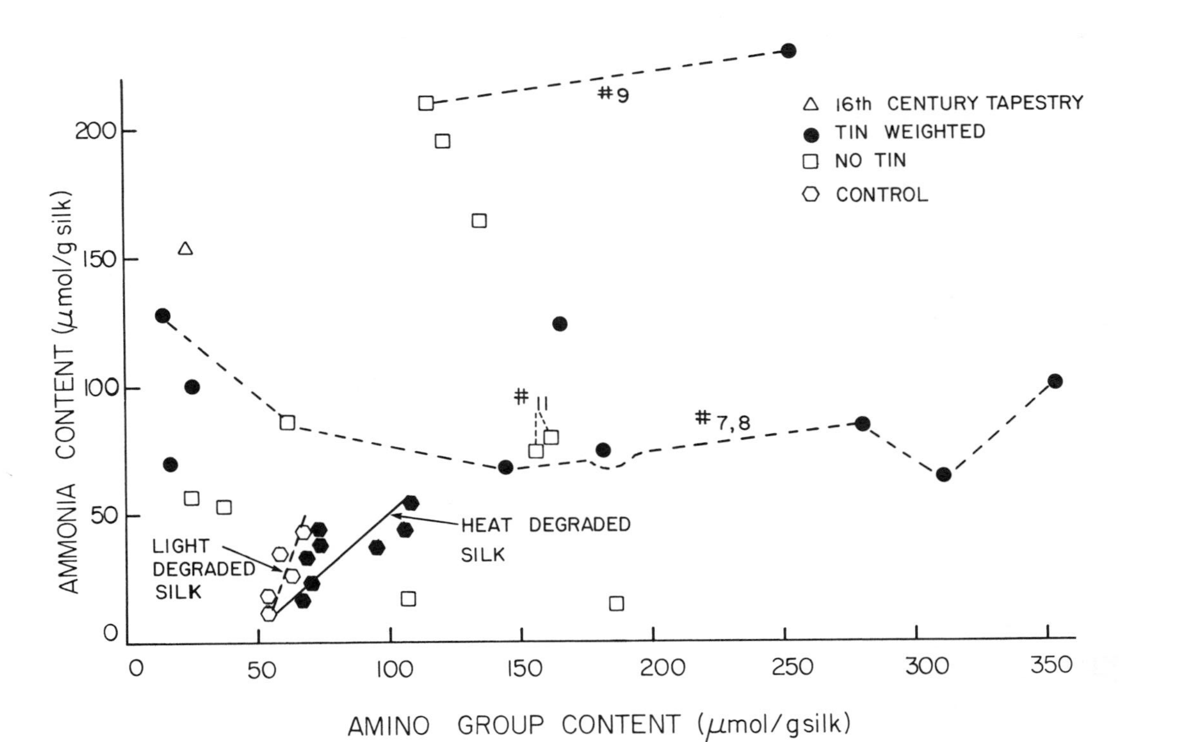

Figure 9. Relationship between ammonia and amino group contents for historical fabrics and for the artificially aged heated and irradiated silk fabrics. Points connected by dashed lines are for historical fabric samples having a common history.

Table III. Amino Group and Ammonia Contents of Historical Silk Fabrics

Time	*Sample*	*Description*	*Component*	*Metals Present*	*Ammonia Concentration*[a]	*Amino-group Concentration*[b]
1500s		tapestry		–	154.0	24.3
1700s	4	green brocade		–	56.6	25.4
	5	pink embroidered		Ag[c]	53.3	37.7
1880–	1	wedding dress		Sn	124.0	165.4
1910	2	green fabric		–	195.2	121.4
	3	white fabric		–	17.0	108.4
	6	vest	[d]black brocade	Cu, Fe, Sn	100.0	26.0
			[d]black lining	Fe, Sn	70.9	17.8
			[d]black facing	Fe[e], Sn[e]	14.4	187.6
			gold brocade	Sn	76.0	180.9
			gold lining	–	[f]–	[g]208.0
	7	collar piece	blue bow	Sn	85.1	280.9
			gold bow	Sn	69.6	143.9
			green fabric	–	87.1	61.2
			[d]black velvet	Fe, Sn	127.9	15.0
	8	Edwardian dress	face fabric	Sn	102.6	354.9
			lining	Sn	64.4	313.7
	9	brown skirt	face fabric		Sn	231.0
			lining	–	210.9	114.7
	10	Jacquard fabric		–	166.2	134.7
	11	quilted bedspread	facing		–	74.7
			backing	–	79.3	161.9
1987		habutae		–	11.1	53.9

[a]Units are micromoles of NH_3 per gram of silk.
[b]Units are micromoles of NH_2 per gram of silk.
[c]Embroidery yarn.
[d]Dyed with logwood.
[e]Trace.
[f]Turbid solution, not plotted in Figure 9.
[g]Sum of ammonia and amino group concentrations, not plotted in Figure 9.

amino groups and ammonia, approximately 30 and 55 μmol/g of silk, respectively. One particularly striking feature is that the three fabrics dyed with logwood, which contained tin and iron (and in one sample also copper), all appear to the left (low amino side) of the artificially aged property envelope. (The black facing fabric from sample 6, which was also dyed with logwood but contained only traces of iron and tin, however, had a high amino group content, 188 μmol/g of silk.)

In sample 9, a skirt in which the face fabric was weighted with tin and the lining fabric was not, the amino group concentration is 220% greater for the tin-weighted fabric than for the unweighted fabric. The ammonia concentration was about 10% greater in the weighted face fabric. (These points are the two having the highest concentrations of ammonia shown in Figure 9 and are connected with a dotted line.) The tin-weighted fabric was fractured and badly split, whereas the unweighted lining was still in one piece and relatively strong. Sample 9 was the best sample available for which a direct side-by-side comparison could be made between a weighted and a similar unweighted fabric with the same history.

Another comparison involving one nonweighted and five weighted components from samples 7 and 8 (which are part of the same costume) is also shown in Figure 9 connected by dashed lines. All of the weighted fabrics (except those dyed with logwood) have higher amino group contents than the unweighted fabric. The ammonia contents do not vary much, however. Both fabric components of sample 11, an unweighted quilt, have similar amino group and ammonia contents. Some of the measured concentrations might be artificially increased because of the presence of soil, dyes, and other contaminants.

Conclusions

The rates of silk degradation as measured by strength loss and discoloration are different for heat-aged and light-aged samples. Although the heated samples discolor much more rapidly than the irradiated samples, their accompanying strength losses are less. The heated samples discolor 10 times more than the light-aged samples at equal breaking strength. The increase in amino group and ammonia contents is an adequate indicator of degradation, although identifying the mechanism indicated by each will require further work. In order to adequately compare historical samples with contemporary samples, additional studies will be required. Additional insights might be gained by carrying out aging studies in the presence of moisture, oxygen, nitrogen, added weighting agents or combinations thereof; and by measuring additional characteristics of the fibers such as molecular weight, molecular weight distribution, and amino acid composition.

Acknowledgments

This investigation was supported in part by the National Museum Act, which is administered by the Smithsonian Institution. We are grateful for this assistance. We also express our gratitude to Katherine Dirks of the National Museum of American History for providing historical fabrics from their study collection and to Francesco Pertegato, Textile Conservator at the Museo Poldi-Pezzoli, Milan, Italy, for providing yarn samples from a 16th-century Flemish tapestry.

References

1. Berry, G. M.; Hersh, S. P.; Tucker, P. A.; Walsh, W. K. In *Preservation of Paper and Textiles of Historic and Artistic Value*; Williams, J. C., Ed.; Advances in Chemistry 164; American Chemical Society: Washington, DC, 1977; pp 228–248.
2. Berry, G. M.; Hersh, S. P.; Tucker, P. A.; Walsh, W. K. In *Preservation of Paper and Textiles of Historic and Artistic Value*; Williams, J. C., Ed.; Advances in Chemistry 164; American Chemical Society: Washington, DC, 1977; pp 249–260.
3. Tucker, P. A.; Hersh, S. P.; Berry, G. M.; Kerr, N.; McElwain, D. M. In *Physics*; Sturgess, J. M. Ed.; Microscopical Society of Canada: Toronto, 1978; Vol. 1, pp 494–495.
4. Kerr, N.; Hersh, S. P.; Tucker, P. A.; Berry, G. M. In *Durability of Macromolecular Materials*; Eby, R. K., Ed.; ACS Symposium Series 95; American Chemical Society: Washington, DC, 1979; pp 357–369.
5. Hutchins, J. K.; Hersh, S. P.; Tucker, P. A.; McElwain, D. M.; Kerr, N. In *Preservation of Paper and Textiles of Historic and Artistic Value II*; Williams, J. C., Ed.; Advances in Chemistry 193; American Chemical Society: Washington, DC, 1981; pp 315–332.
6. Hersh, S. P.; Hutchins, J. K.; Tucker, P. A.; Kerr, N. *Proceedings of the International Conference on the Conservation and Restoration of Textile, Como, Italy*; Pertegato, F., Ed.; CISST, Lombardy Section: Milan, 1982; pp 87–95.
7. Kerr, N.; Hersh, S. P.; Tucker, P. A. *Prepr. Wash. Congr. Sci. Technol. Serv. Conserv.;* Brommelle, N. S.; Thomson, G., Eds.; The International Institute for Conservation of Historic and Artistic Works: London, 1982; pp 100–103.
8. Kerr, N.; Hersh, S. P.; Tucker, P. A. *Prepr. ICOM Comm. Conserv. 7th Triennial Meet., Copenhagen, Denmark, September 1984*; de Froment, D., Ed.; International Council of Museums: Paris, 1984; pp 84.9.26–29.
9. Tucker, P.; Kerr, N.; Hersh, S. P. Presented at the 1980 Textile Preservation Symposium on Textiles and Museum Lighting of the Harpers Ferry Regional Textile Group, Washington, DC, December 1980.
10. Vigo, T. L. In *Preservation of Paper and Textiles of Historic and Artistic Value*; Williams, J. C., Ed.; Advances in Chemistry Series 164; American Chemical Society: Washington, DC, 1977; pp 189–207.
11. Kurupillai, R. V.; Hersh, S. P.; Tucker, P. A. In *Historic Textile and Paper Materials: Conservation and Characterization*; Williams, J. C., Ed.; Advances in Chemistry Series 212; American Chemical Society: Washington, DC, 1986; pp 111–127.
12. Tyler, A. G. *Text. J. Aust.* **1949,** *24,* 404-407, 464.

13. Bogle, M. M. *The Deterioration of Silks Through Artificial Weighting*; Textile Conservation Center Notes, No. 11; Merrimack Valley Textile Museum: North Andover, MA, 1979.
14. Becker, M. A.; Hersh, S. P.; Tucker, P. A.; Waltner, A. W. In *Prepr. ICOM Commit. Conserv. 8th Triennial Meet., Sydney, Australia, September 1987*, 339–344.
15. Ross, J. E.; Johnson, R. L.; Edgar, R. *Text. Res.* **1936,** *6*, 207–216.
16. Harris, M. *Am. Dyest. Rep.* **1934,** *23*, 403–404.
17. Roberts, N. M.; Mack, P. B. *Rayon Melliand Text. Mon.* **1956,** *17*, 49–53.
18. D'Olier, A. A.; Mack, P. B. *Rayon Melliand Text. Mon.* **1956,** *17*, 102.
19. *AATCC Technical Manual*; American Association of Textile Chemists and Colorists: Research Triangle Park, NC, 1987;, Vol. 52, pp 44–45.
20. *Annual Book ASTM Standards; American Society for Testing and Materials: Philadelphia, 1983; Vol. 07–01* 394–404.
21. Billmeyer, Jr., F. S.; Saltzman, M. *Principles of Color Technology*, 2nd ed.; Wiley: New York, 1981; pp 63 and 103.
22. Knott, J.; Grandmaire, M.; Thelen, J. *J. Text. Inst.* **1981,** *72*, 19–25.
23. *Standard Methods for the Examination of Water and Wastewater*, 14th ed.; American Public Health Association, Ed.; American Public Health Association: Washington, DC, 1976; pp 407–415.
24. Arney, J. S.; Chapelaine, A. H. In *Preservation of Paper and Textiles of Historic and Artistic Value*; Williams, J. C., Ed.; Advances in Chemistry 164; American Chemical Society: Washington, DC, 1977; pp 189–204.
25. Cardamone, J. M.; Brown, P. In *Historic Textile and Paper Materials*; Needles, H. L.; Zeronian, S. H., Eds.; Advances in Chemistry 212; American Chemical Society, Washington, DC, 1986; pp 41–75.

RECEIVED for review September 15, 1987. ACCEPTED revised manuscript March 18, 1988.

26

Determination of Elemental Distribution in Ancient Fibers

Kathryn A. Jakes[1] and Allen Angel[2]

[1]Department of Textiles and Clothing, Ohio State University, Columbus, OH 43210
[2]Center for Advanced Ultrastructural Research, University of Georgia, Athens, GA 30602

Archaeological textiles can provide a vast body of evidence for prehistoric environments, cultures, and technologies. They may contain clues to the processes of degradation, alteration, or mineralization that occurred after burial. Determination of elemental composition and distribution within fiber structures can contribute to the understanding of biological, systemic, and diagenetic contexts of ancient textiles. A technique suitable for brittle, fragile, and very small samples of archaeological fibers was developed, in which fibers were freeze fractured, freeze dried, mounted in carbon paste, and analyzed with a scanning electron microscope coupled to an energy-dispersive spectrometer. Elemental content and distribution in cross sections of fibers have implications for identification of fibers, description of fiber processing, determination of fabrication technology, and elucidation of the history of the fiber in long-term storage or burial.

FABRICS THAT ARE RECOVERED FROM ARCHAEOLOGICAL SITES are often so badly degraded that fiber identification on the basis of physical morphology is difficult. The changes that occur during diagenesis destroy physical and chemical evidence that is necessary to discern fiber information. On the other hand, the changes that occur during diagenesis, as well as those that occur at any stage of the fiber's lifetime, leave a record within the fiber's chemical and physical structure, such that the altered fiber reflects its history. Fibers are not inert, but interact with their surroundings in both subtle

0065-2393/89/0220-0451$06.00/0

and obvious ways. Just as a fiber exposed to an intense stress, such as immersion in a solvent, will respond to the stress by absorbing the chemical, swelling, and dissolving or decomposing, a fiber reacts in response to a less intense stress with smaller changes in chemical and physical structure. The result of the stress may not be apparent until the perturbations accumulate over a long time. Thus, a textile hanging unsupported will, over time, show evidence of molecular creep, and a fiber stored for a long time in a polluting environment may change color because of absorbed sulfur and nitrogen oxides.

Degraded fibers may be difficult to identify by the standard techniques prescribed for modern materials, but their structures offer valuable clues to the conditions of the fiber's growth (the biological context), to fiber preparation and fabric processing technology, to fabric use (the systemic or cultural context), and to the conditions of burial or long-term storage (the diagenetic context). Information must be extracted from the fiber's chemical and physical structure to interpret what the clues indicate. The goals of our work were

1. to develop a technique that used energy-dispersive analysis of X-rays that was appropriate for preparation, elemental analysis, and mapping the elemental distribution of small, fragile fiber samples, typical of those recovered from archaeological sites (or of a size small enough to be minimally destructive of artifacts in collections)
2. to determine the elemental content and distribution of a group of samples representing a broad spectrum of fiber types and histories
3. to explore the implications of the analytical results for interpretation of the biological, systemic, and diagenetic contexts of archaeological and historical materials.

Experimental Details

Samples. Fiber samples were selected from a variety of sources to represent a range of fiber types, a variety of conditions of long-term storage or burial, and a variety of ages. The samples were

1. modern wool, Testfabrics No. 522
2. modern linen, Testfabrics No. L-53
3. modern silk, Testfabrics No. 601
4. wool tunic, Coptic, ca. A.D. 200
5. hair fiber, Paracas 500 B.C.– A.D. 150, catalog No. 382-37

6. bast fibers, Etowah Mound C, Georgia, ca. A.D. 1200, catalog No. 840
7. bast fibers, Etowah Mound C, Georgia, ca. A.D. 1200, catalog No. 1145
8. feathers, Etowah Mound C, Georgia, ca. A.D. 1200, catalog No. 1145
9. metal-wrapped linen, Turkey, 15th century, catalog TSM 13.1919
10. metal-wrapped silk, Italy, 15th century, catalog MMA 46.156.134

Many of the fabric structures from which these samples were taken have been described in other publications (*1–3*). In experimenting with appropriate methods for preparation of samples for elemental analysis, fibers are of limited value because they cannot be proven to belong to the textile object in the box. For experimental purposes, these fabrics served well as representative brittle, fragile, and minute fibrous samples for preparation and analysis.

Materials. Fibers were rapidly frozen on a liquid-nitrogen-cooled block, then fractured several times with a precooled razor blade. The fibers were placed immediately in a high-vacuum evaporator (NRC) to freeze-dry overnight to sublimate any existing water and to prevent condensation over the surface of the samples after freezing. (Such condensation of water could result in elemental translocation.) Fibers were treated in this manner to produce smooth, flat cross sections, a requirement for energy-dispersive spectrometry (EDS) of X-rays and X-ray mapping. Each fractured fiber was mounted in a 1-mm hole drilled into a carbon planchet (SPI, West Chester, PA). The fibers were first affixed with pure carbon paste (SPI, West Chester, PA), then oriented in an upright position to allow at least 0.5 mm of fiber to extend out of the hole, such that the fractured plane was parallel to the surface of the planchet. The samples were then coated with approximately 40 nm of carbon by using high-vacuum evaporative techniques.

A scanning electron microscope (Philips 505) equipped with four scintillator-type backscattered electron detectors and an energy-dispersive X-ray microanalysis system (Tracor Northern 5500) were used to analyze the specimens.

Method. First, fibers were examined for morphology. If a smooth fractured plane was present and oriented to allow visualization of the entire cross section, then 50-nm spot X-ray analyses were obtained from a number of points at the periphery of the section and its center. Samples were irradiated for 50–500 s and spectra collected at 25 kV. Several fibers of each type from a given planchet were analyzed.

When spectra indicated elemental variation over the span of the cross section, or the presence of elements with atomic numbers higher than Fe, backscattered electron imaging was used to qualitatively assess the distributional variation of these elements and possible sites at which some elements had accumulated. Because the signal from backscattered electrons increased with atomic weight, accumulations of specific elements could be determined by the intensity of the signal. Backscattered electron imaging was not performed on samples that were primarily organic.

If spot analyses or backscattered imaging showed elemental variations, X-ray maps were obtained to determine the distribution of a single element in each fiber. Specific regions of spectra were defined, and maps corresponding to these energy regions were made. To ensure that data came from the sectioned fiber and not from other random locations on the planchet, maps of the whole spectrum, defined as an energy region, were obtained. X-ray maps were collected with dwell times of 0.2–0.3 s and 256- × 256-pixel resolution. A digital secondary electron or backscattered electron image usually was made for visual comparison and overlap with the elemental maps. Further details of the technique are reported in reference 4.

Results

Although freeze-fracturing, freeze-drying, and EDS analysis were generally satisfactory, some difficulties were encountered because of the nature of the fiber samples. First, not all of the samples fractured smoothly, nor did all of the fibers in one sample display smooth, flat cross sections. Because the X-ray microanalytical technique requires a flat surface geometry to obtain accurate counts, this was the surface type selected from each fiber sample examined. Although different fiber fracture tips may indicate different causes of degradation (*5*, *6*), it was assumed that any of the fibers of a yarn selected for elemental analysis were representative of all of the fibers in the yarn bundle. Because the fibers examined in each sample were sectioned from a single yarn, it was assumed that the fibers in the yarn experienced the same conditions throughout their lifetimes, except for some possible variation in growth conditions. In further work, many microsized samples could be examined by the technique described in this chapter to determine the extent of variation within a single yarn.

Additional difficulties in elemental analysis occurred with fibers that were completely or primarily organic. The elemental spectra obtained for these fibers displayed a broad band typical of hydrocarbons, and no elemental maps could be determined.

Finally, fibers with extremely fine structures, such as the feather sample, hindered elemental mapping even when they contained elements heavy enough to be counted by EDS analysis. These materials would heat up and move under the X-ray beam. Consequently, an inside spot–outside spot comparison and backscattered electron imaging had to be relied upon, rather than an elemental map, to get elemental distribution information.

Data obtained in the EDS analyses are listed in Table I. The fiber standards (primarily organic) gave broad EDS spectra. No heavy elemental composition was indicated in silk, and only a small quantity of calcium could be detected in linen. No elemental maps could be obtained for silk or linen. The EDS spectra of wool reflected its sulfur content. In some of the samples, the somewhat higher concentration of sulfur in one-half of the cross section, which is expected of this naturally bicomponent fiber, was apparent (7). In other samples, the differentiation was not clear.

Table I. Elemental Components and Their Locations in Archaeological and Historical Fiber Samples

Element	*Penetrating Surface Layer*	*Throughout Fiber*
Ca	382.37	Modern linen
		EMC 840
		EMC 1145, feather
		EMC 1145, core
		MMA 46.156.134
		TSM 13.1919
		Coptic tunic
Cu	EMC 840[a]	EMC 1145, core
	EMC 1145[a]	EMC 1145, feather
		EMC 840
Fe	EMC 840	
	EMC 1145, core	
	EMC 1145, feather	
P	EMC 1145, feather	EMC 1145, core
		EMC 840
		EMC 1145, feather[a]
K		EMC 1145, feather
		EMC 1145, core
Ag		TSM 19.19
S		Modern wool[b]
		382.37[b]
		EMC 840
		EMC 1145, feather
Si		EMC 840
		EMC 1145, feather
		EMC 1145, core

NOTES: EMC means Etowah Mound C. The samples are described under "Experimental Methods". The only elements found on the surface were Ca and Si, found only on hair fiber 382.37.

[a]Permeating throughout but somewhat larger concentration in surface layer.

[b]Somewhat heavier concentration in one-half of the cross section.

The Paracas hair fiber shown in Figure 1 fractured cleanly. In some cases, the EDS maps revealed a higher concentration of sulfur in one-half of the cross section, similar to the sulfur distribution of wool (*see* Figure 2). K, Si, and Ca were found on the fiber edges only (Figures 2 and 3). In many cases, they were clumped together on the fiber surface in a manner that indicated soil aggregation. No penetration of these elements into the fiber surface was seen. In the EDS maps of the Coptic wool fiber, the sulfur and calcium were evenly distributed throughout the fiber.

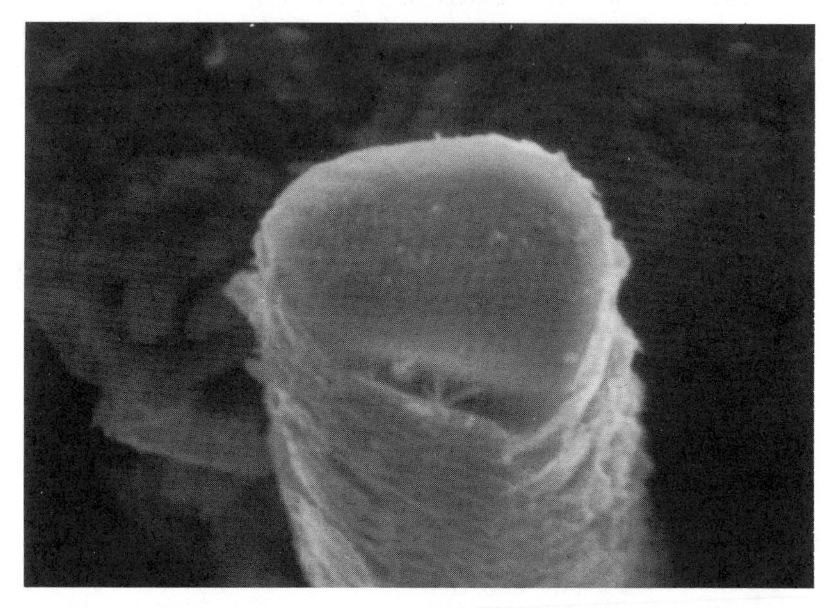

Figure 1. Hair fiber from Paracas (ca. A.D. 200); magnification: 2230×.

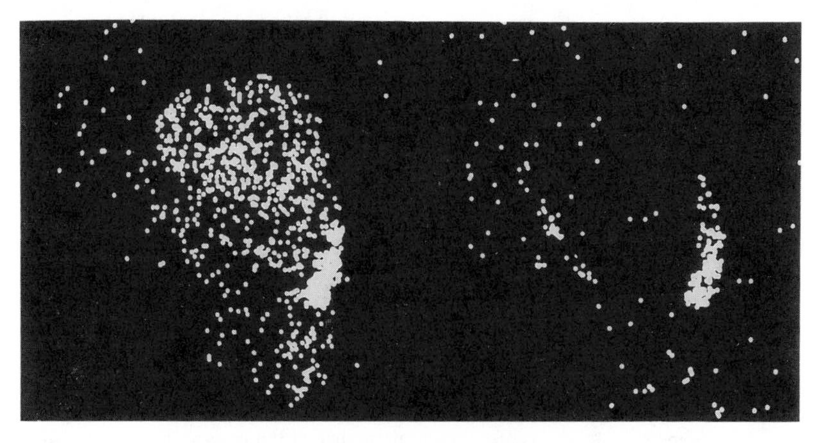

Figure 2. Elemental dot map of Paracas hair fiber indicating distribution of sulfur (left) and potassium (right).

Figure 3. Elemental dot map of Paracas hair fiber indicating distribution of silicon (left) and calcium (right).

The feather wrapping of Etowah Mound C (EMC) 1145 contained Al, Si, P, S, Cl, K, Ca, Fe, and Cu. The inside spot–outside spot elemental comparison indicated that the outer edge of each feather barbule contained more Fe and P than the inside. The backscattered electron image corroborated this point. The outer rim of the feathers appeared brighter because of the greater concentration of iron, a fact indicating that iron had penetrated only into the surface of the feathers. Al, Si, S, Cl, K, Ca, and Cu were distributed throughout the material.

A micrograph of EMC 1145 core is shown in Figure 4, the backscattered image is shown in Figure 5, and the elemental dot maps are shown in Figure 6. The bast fibers, EMC 1145 core, and EMC 840, have similar spectra, and contained Al, Si, P, S, K, Ca, Fe, and Cu. EMC 1145 also contained some Cl. The EDS maps of these materials indicate that both materials contained higher concentrations of iron in their outer surface layers and had calcium and copper distributed throughout. Phosphorus and sulfur were confined to the fiber's interior. As with the feathers, the backscattered electron image of these materials reflected the iron concentrated in the surface layers.

Elemental maps of the metal-wrapped linen (TSM 13.1919) showed that the metal wrapping was silver, but the fibers in the yarn also contained some silver and calcium. The maps of the metal-wrapped silk (MMA 46.156.134) show that the metal wrapping was gold-coated silver, and that calcium was present in the fibers but not silver. Potassium was present in both the metal wrapping and the fiber core.

The metal surface of the metal-wrapped silk showed a variation in coloration in the backscattered image, and this variation reflected the variable composition of the surface layer (Figure 7). The elemental map shows areas where silver broke through the gold coating. With no other heavy elements present, it is assumed that these layers were silver oxides.

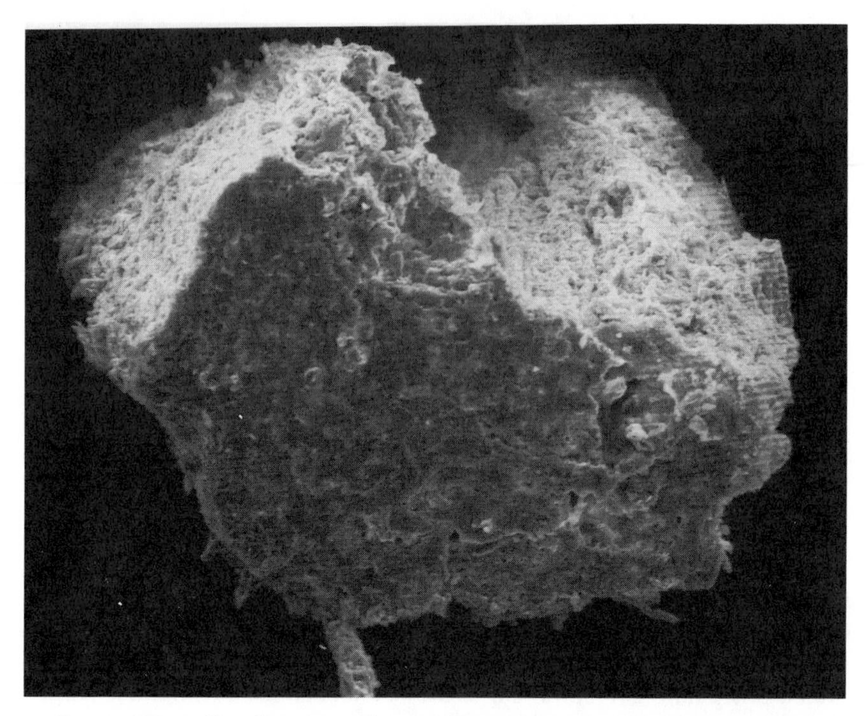

Figure 4. Bast fiber from Etowah Mound 1145 core (ca. A.D. 1200). Secondary electron image; magnification: 1200 ×.

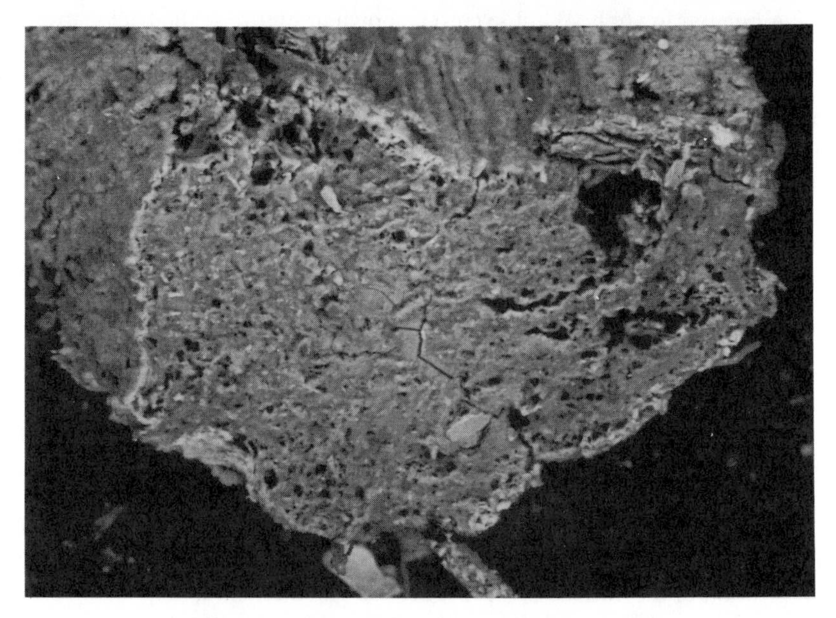

Figure 5. Bast fiber from Etowah Mound 1145 core (ca. A.D. 1200). Backscattered electron image; magnification: 345×.

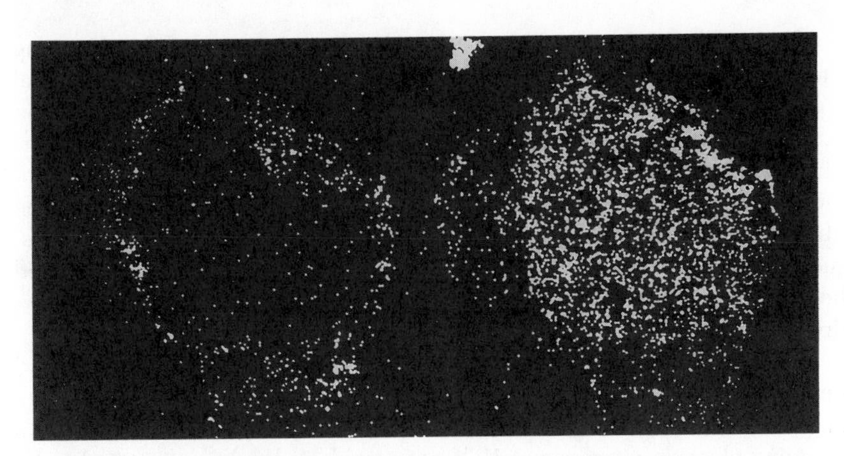

Figure 6. Elemental dot map of bast fiber, Etowah Mound C 1145 core, indicating distribution of iron (left) and copper (right).

Figure 7. Metal-wrapped silk from Italy (15th century). Backscattered electron image; magnification: 686×.

Discussion

The interaction of fibers with their environment is often studied by observing the changes in physical and chemical structure as reactions occur. Typical changes studied are increase in oxidized or unsaturated groups, decrease in degree of polymerization, decrease in tensile strength, and changes in morphology. Another way to monitor the interaction of fibers with the environment is to determine which compounds are absorbed or adsorbed into the fibers. Fibers can be altered by sorption of or reaction with hydrocarbons, such as soils; however, this work focused on compounds that contained heavier elements, because of their more obvious differences from the hydrocarbon structure of fibers.

Energy-dispersive analysis of X-rays was the chosen analytical method because of its sensitivity to elements heavier than sodium, its capability of mapping elemental distribution, and its capacity for combination with a scanning electron microscope. EDS microanalysis has been reported to be suitable for the determination of mordant treatments on historical fibers (*8–10*) and has been used to characterize metal wrappings of combination yarns (*11–13*). EDS microanalysis has also been used to determine the composition of pseudomorphs and fibers in the process of mineral replacement (*13, 14, 15*).

In this chapter, the results of past research are expanded because fiber cross sections were examined, rather than longitudinal views of fibers, and distributions of elements were obtained in addition to overall elemental spectra. Because the X-ray beam penetrates only a small distance into the surface of a sample (approximately 8–10 μm for a 25-kV excitation), examination of a longitudinally mounted fiber produces elemental spectra of surface layers only. Such spectra may not be representative of the bulk of the fiber. In addition, this work improves upon past research in that the freeze-fracturing–freeze-drying EDS technique is suited to very small, fragile fiber samples (whether single fibers or small yarn pieces), and is limited in size only in the operator's ability to see and handle the samples. By using this procedure, compression of the fiber cross section and elemental redistribution are avoided.

The ability to visualize elemental content and the distribution of those elements within a fiber's structure is potentially significant for the interpretation of the biological, systemic, and diagenetic contexts of fibrous materials. If fiber identity is questionable, EDS analysis can provide some additional information. Hair fibers contain a significant amount of sulfur, and that sulfur is somewhat more concentrated in one-half of the fibers' cross sections. This bicomponent nature sometimes is apparent in elemental maps and can be useful in identifying wool or hair. Because silica and calcium oxalate crystals form in plant materials as they grow, overall distribution of large amounts of silicon or calcium indicates bast fibers. This distribution can readily be distinguished from localized soil contamination on fiber surfaces; the larger concentration excludes the possibility that the presence of calcium is due to residual soap scum. When calcium oxalate and silica crystals are indicated by EDS, further study of these crystals and their shape may aid fiber identification. These crystals, also called phytoliths, do not degrade after burial and so provide an enduring mode of plant identification when the rest of the fiber is gone (*16–18*).

Elemental content and mapping also will contribute to the understanding of the systemic context of fibrous materials. The modern linen examined in this work contained a quantity of calcium that was too small in comparison to the organic nature of the fiber to be mapped. Possibly this calcium was the result of residual soap scum from cleaning. More significant quantities of elements, such as calcium, magnesium, and iron, within the linen or other bast fibers could indicate that the fiber was retted in hard water. Because bast fibers can act as ion-exchange columns and attract minerals, and because these ions could be more concentrated in the protoplasm lining the lumen, elemental mapping could provide more evidence concerning bast fiber processing. Other aspects of fiber–yarn–fabric treatment that can be indicated by EDS analysis are dyeing with mordants, whitening and delustering with agents such as titanium dioxide, and weighting of silk. The elemental maps will reveal depth of penetration of these treatments.

Examination of the metal-wrapped yarns shows that EDS analysis and mapping can be used to infer yarn structure (systemic contextual information) and long-term storage conditions (diagenetic contextual information). Fibers surrounded with silver metal showed absorbed silver ions; fibers surrounded with gold-coated silver did not. Although some silver oxides were present on the surface of the metal wrapping, the inert gold coating or other environmental conditions hindered silver ion migration into this yarn. The fibers, thus, reflect their long-term storage conditions. These fabrics have not been buried, and have not been subjected to long-term immersion in water. Rather, conditions of normal humidity and temperature have been sufficient for elemental transfer to the inner fibers and for corrosion layers to form on the metal wrappings. Hardin and Duffield (*11*) found sulfur-containing corrosion layers on the silver wrapping of 16th-century Turkish combination yarns and attributed the sulfur source to burning fossil fuels.

EDS analysis of yarn cross sections could provide additional information, and document, for example, whether sulfur from the air permeates the fibrous core as well. Such is the case in the "fume fading" of textiles enclosed in environments with sulfur and nitrogen oxides. Elemental mapping of the fiber cross sections would show the depth of penetration of the absorbed or adsorbed species, and perhaps indicate the length of exposure, as well as of exposure to gaseous rather than aqueous environments.

The results of the analysis of materials examined in this work particularly lend themselves to inferences of diagenetic context. Just as the metal-wrapped yarns showed that metal ions can transfer to fibers in contact with silver, fibers buried in association with copper reflect that association by their large copper content. At present, it is believed that the copper is bound to fiber cellulose or protein polymers. Under appropriate conditions, however, the copper localized within the fiber structure will oxidize to form copper minerals such as malachite and cuprite. The resulting fiber pseudomorph maintains the physical shape of the fiber but no longer has its organic composition (*19*, *20*).

The presence of Si, Al, P, Ca, and K throughout the fibers from Etowah Mound is evidence of the interaction between the fiber and soil in a wet environment. These soil elements, as well as copper ion from the corroding copper metal, readily dissolve in an oxidizing aqueous environment and migrate with water percolating through the ground. By absorbing this ground water, buried fibers also absorb the dissolved elements.

Iron ions migrate only under reducing conditions, and therefore do not behave in a manner similar to the other soil elements. Rather, microparticulate iron is carried by the waters percolating through the site and deposits in the fiber surface in a manner comparable to rust staining of modern textiles. Whether the iron particles are mechanically entrapped or are complexed in some manner with the fiber's organic structure has not been determined, but the backscattered electron images of the fibers from Etowah Mound clearly show this thin layer of iron penetration (Figure 5).

In contrast to the wet, oxidizing conditions of the Etowah Mound, with its iron-containing claylike soil, the dry conditions and limestone-type soil of the Paracas peninsula have resulted in fibers with soil encrustations of silicon and calcium, but with no absorbed soil elements.

An unresolved question remains concerning the presence of significant quantities of sulfur in the materials from Etowah Mound. Not only do the fibers contain sulfur, but copper sulfate corrosion layers are also present on the associated copper metal from these burials (*21*). Because sulfur is present in the soil only in trace quantities, some other sulfur source must be found. The source of the sulfur may be a treatment of the textile or may be a product of the decomposing body nearby. Further examination other textiles from these burials should provide answers to the questions of the source of sulfur.

Conclusions

The freeze-fracturing, freeze-drying, EDS analysis, and mapping procedure was tailored for the examination of brittle, fragile, and small fiber samples. Evaluation of a group of fibers with this technique produced results that suggest a great potential for producing information that may contribute to the understanding of the biological, systemic, and diagenetic contexts of archaeological and historical textiles. In each of the cases described, many implications for contextual information could be inferred from the data, but no conclusions could be drawn for any of the materials examined because many more samples from well-documented sites need to be studied.

Elemental content and distribution may be useful in determining biological context (discussed here only in terms of fiber identification), but further work may suggest growth conditions of the fibers as well. Inferences concerning systemic contexts were discussed, including fiber processing, dyeing, and finishing. Further work, particularly examining treated model fibers, will confirm the speculations made here. The definition of the diagenetic contexts of archaeological materials made here can be verified and expanded with the analysis of samples from fully documented sites. Examination of soil, metal, wood, or any other associated material from the sites must be included as well as evaluation of moisture content, acidity, and oxidation potential. After study of fully documented materials, patterns can be deduced, including definition of those factors that persist through diagenesis and those that change in predictable manner. From these patterns, it will be possible to examine fibers of unknown origin and to infer some information concerning their biological, systemic, and diagenetic contexts. Determination of elemental content and mapping of elemental distribution can make an important contribution to the understanding of archaeological and historical materials.

Acknowledgments

We are indebted to L. R. Sibley of the Department of Textiles and Clothing, Ohio State University; L. Larson, State Archaeologist of Georgia; A. Paul of the University of Texas, Arlington; and D. Mott of the Fine Arts Department of the University of Georgia for the donation of fiber samples. We gratefully acknowledge the financial support of the Center for Advanced Ultrastructural Research and the Center for Archaeological Sciences of the University of Georgia. K. Jakes also thanks J. D. Taylor and R. L. Hauss for their encouragement during the progress of this work.

References

1. Paul, A.; Niles, S. A. *Text. Mus. J.* **1985,** *23,* 5–15.
2. Sibley, L. R. *Muse* **1983,** *17,* 81–95.
3. Sibley, L. R.; Jakes, K. A. In *Historic Textile and Paper Materials, Conservation and Characterization*; Needles, H. L.; Zeronian, S. H., Eds.; Advances in Chemistry 212; American Chemical Society: Washington, DC, 1986; pp 253–275.
4. Angel, A.; Jakes,, K. A. In *Proceedings of the 45th Annual Meeting of the Electron Microscopy Society of America*; Bailey, G. W., Ed.; San Francisco Press: San Francisco, 1987; pp 410–411.
5. Zeronian, S. H.; Alger, K. W.; Ellison, M. S.; Al-Khayaff, S. M. In *Historic Textile and Paper Materials, Conservation and Characterization*; Needles, H. L.; Zeronian, S. H., Eds.; Advances in Chemistry 212; American Chemical Society: Washington, DC, 1986; pp 79–94.
6. Bresee, R. R.; Goodyear G. E. In *Historic Textile and Paper Materials, Conservation and Characterization*; Needles, H. L., Zeronian, S. H., Eds.; Ad-

vances in Chemistry 212; American Chemical Society: Washington, DC, 1986; pp 95–110.
7. Morton, W. E.; Hearle, J. W. S. *Physical Properties of Textile Fibres*; The Textile Institute: London, 1975.
8. Indictor, N.; Koestler, R. J.; Sheryll, R. *J. Am. Inst. Conserv.* **1985,** *24,* 104–109.
9. Koestler, R. J.; Indictor, N.; Sheryll, R. *J. Am. Inst. Conserv.* **1985,** *24,* 110–115.
10. Koestler, R. J.; Sheryll, R.; Indictor, N. *Stud. Conserv.* **1985,** *30,* 58–62.
11. Hardin, I. R.; Duffield, F. J. In *Historic Textile and Paper Materials, Conservation and Characterization*; Needles, H. L.; Zeronian, S. H., Eds.; Advances in Chemistry 212; American Chemical Society: Washington, DC, 1986; pp 231–252.
12. Stodulski, L. P.; Mail H. F.; Nauman, A; Kennedy, M. In *Application of Science in Examination of Works of Art*; England, P. A.; Van Zelst, L., Eds.; Museum of Fine Arts; Boston, MA, 1985; pp 76–91.
13. Hoke, E.; Petrascheck-Heim, I. *Stud. Conserv.* **1977,** *22,* 49–62.
14. Jakes, K. A.; Sibley, L. R. In *Archaeological Chemistry—III*; Lambert, J. B., Ed.; Advances in Chemistry 205; American Chemical Society: Washington, DC, 1984; pp 403–24.
15. Jakes, K. A.; Sibley, L. R. In *Application of Science in Examination of Works of Art*; England, P. A.; Van Zelst, L., Eds.; Museum of Fine Arts: Boston, MA 1985; pp 221–224.
16. Catling, D.; Grayson, J. *Identification of Vegetable Fibers*; Chapman and Hall: London, 1982.
17. Rovner, I. *Quat. Res.* **1971, 1,** 345–359.
18. Jakes, K. A. In *Eighth National Textile Preservation Symposium; Harpers Ferry Regional Textile Group: Washington, DC, 1986; pp 23–25.*
19. Jakes, K. A.; Howard, J. H. In *Historic Textile and Paper Materials, Conservation and Characterization*; Needles, H. L., Zeronian, S. H., Eds.; Advances in Chemistry 212; American Chemical Society: Washington, DC, 1986; pp 277–290.
20. Jakes, K. A.; Howard, J. H. In *Proceedings of the 24th International Archaeometry Symposium*; Olin, J. S.; Blackman, M. J., Eds., Smithsonian Institution: Washington, DC, 1986; pp 165–178.
21. Jakes, K. A.; Sibley, L. R. In *Proceedings of the 25th International Archaeometry Symposium*; Smithsonian Institution: Washington, DC; in press.

RECEIVED for review June 11, 1987. ACCEPTED revised manuscript April 18, 1988.

27

Photomicrography and Statistical Sampling of Pseudomorphs after Textiles

L. R. Sibley[1], Kathryn A. Jakes[1], J. T. Kuttruff[1], V. S. Wimberley[2], D. Malec[3], and A. Bajamonde[4]

[1]Department of Textiles and Clothing, Ohio State University, Columbus, OH 43210
[2]Department of Home Economics, University of Texas, Austin, TX 78712
[3]National Center for Health Statistics, Hyattsville, MD 20782
[4]Department of Statistics, University of California—Berkeley, Berkeley, CA 94720

The nature and extent of pseudomorphs after textiles on a Shang bronze spearpoint are explored by using mapping of the evidence by photomicrography and simple random sampling. The two techniques are designed to increase the rigor of the methodology used to study certain forms of the archaeological textile record that require extensive analysis. Complete linkage cluster analysis and classification and regression tree (CART) of technical fabrication attribute data reveal the presence of unbalanced plain weave silk fabric with areas of float.

THE RECONSTRUCTION OF PAST HUMAN BEHAVIOR from early textile remains continues to challenge the analyst. Despite recent strides made in understanding the basic chemistry and degradation of textile fibers in the diagenetic context (*1*, *2*), there is not yet a rigorous methodology capable of yielding valid and reliable data with which to infer human behavior. Because textiles were used widely in nearly all facets of the cultural system, they provide important evidence of past decision-making. Yet their composition has resulted in differing degrees of survival, in forms often difficult to analyze.

0065–2393/89/0220–0465$06.00/0

Without better research designs, textile analysts run the risk of having potentially powerful data being misapplied in the development of inferences about the past use of textiles. The purpose of the research described in this chapter was to apply two techniques of analysis to the investigation of pseudomorphs after textiles, one form of the archaeological textile record. Specifically these techniques are mapping textile evidence by photomicrography and statistical sampling of that evidence.

Archaeological textiles occasionally occur in forms that mislead the analyst. The mineralized fabric evidence known as pseudomorphs after fabric is one such form. Fabric remains may be recovered in pieces crumpled together and may exhibit pronounced structural variation. In such instances, one cannot rely on a series of tests from one area of the fabric to identify characteristics of the entire fabric.

The fragmented and fragile condition of archaeological fabrics has led to bias because scientists have evaluated parts of fabrics that were accessible and of interest. Without sampling designed to avoid bias, inferences relating to the structure of textile remains or the human behavior associated with their production and use may be in question. To infer the whole from the part with any degree of rigor requires an assumption that the part (a single observation) represents the whole (textile remains). That assumption should not be made unless data gleaned from analysis of the fragment are obtained through probabilistic sampling procedures, because the alternative of using a large nonprobabilistic sample is difficult to accomplish (*3*). The potential error in haphazard sampling could negate the results of very precise instrumental and technical fabrication analyses.

Mapping the textile evidence by photomicrography is another technique used to study the archaeological textile record. Mapping is the preparation of an enlarged visual record of adjacent areas of fabric from photomicrographs, and it allows the investigator to plot the movement and interaction of textile components and to study the fabric in detail. Mapping also provides the means to trace the spatial relationships of a fragment's fiber–yarn–fabric evidence with respect to adjacent fragments or to a metal host. The techniques of mapping by photomicrography and statistical sampling improve the reliability of the data obtained. Used together, they strengthen the research design developed to evaluate archaeological textile remains.

Examination of archaeological textile evidence includes analysis of two classes of evidence to yield what is referred to as *attribute data*. These data are derived from (1) technical fabrication examinations of the fiber–yarn–fabric evidence and (2) the physical and chemical analyses of the fiber. Both classes of information are necessary to characterize fully the fabric evidence, and both require photomicrography as an initial step in the analytical procedure. The nature of the information sought leads to different avenues of testing. Because both are essential to the complete understanding of the textile, the sets of results obtained in the testing complement each other.

Two problems should be considered when conducting an investigation that incorporates mapping and sampling techniques. One problem is that the irreplaceable and fragile nature of the textile evidence demands non-destructive or minimally destructive procedures. Removal of enough microsized samples to provide a statistical sample would destroy much of the textile. Another consideration is that the cost of techniques such as as X-ray diffraction (XRD) or Fourier transform infrared spectroscopy (FTIR) would be prohibitive when applied to a large number of samples. Because of these constraints, chemical and physical data should be obtained from a limited number of locations. These areas can best be identified once a survey of the evidence has been conducted and technical fabrication analyses of evidence obtained from a random sample of locations have been performed.

We used mapping by photomicrography and statistical sampling to evaluate pseudomorphs after fabric from one Shang bronze weapon, a spearpoint (Sp1), dated about 1300 B.C. (Figure 1). The spearpoint has been the focus of earlier investigations (*1, 2, 4, 5*), which constitute a survey of the evidence. This investigation emphasizes technical fabrication data through sampling and mapping, and the implications for future chemical and physical analyses.

Methodology

The objective of the study is to determine the nature and extent of pseudomorphs on Sp1 by mapping, with photomicrography and statistical sampling techniques. The "population" to be studied consisted of all pseudomorphs after fabric located on the spearpoint. A survey previously conducted with microscopy identified the general location of the mineralized fabrics. A research hypothesis, derived from the survey of evidence, governed the study and is as follows: Pseudomorphs after fabric located on Sp1 are fragments of an unbalanced plain-weave type of fabric.

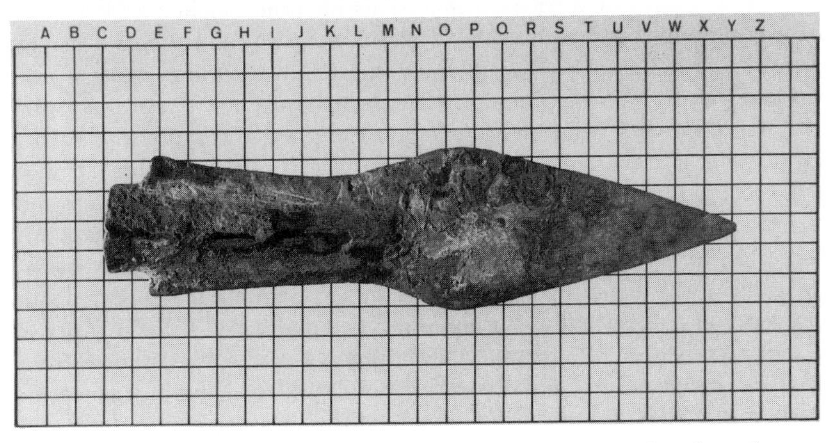

Figure 1. Shang bronze spearpoint (Sp1) from the Museum of Anthropology, University of Missouri— Columbia. Reduction: 45×.

Data Collection

Mapping. No attempt was made to map the entire area of pseudomorphs after fabric; rather, photomicrographs of adjacent location points were reviewed, and two groups were selected for further study. Criteria for inclusion in the map were the proximity to other location points and 10.2× magnification for the photomicrograph. Slides were developed as 5- × 7-in. photographs. The photographs were placed upon a neutral-colored mat board and fitted together so that yarns could be tracked across several location points.

Sampling. A sample was drawn at random from a list of randomly combined x and y coordinates that indicated location points for pseudomorphic evidence. The *location point* is defined as the field of vision at 10× magnification that allowed attribute data occurring in the field to be observed. The spearpoint was placed upon a special microscope stage that had a centimeter grid system. The x axis was identified by letters A–Y, and the y axis was identified by numbers 1–10. These combinations of coordinates led to the spatial location points or subunits of the site. Photomicrography was used to record the mineralized fabric attributes at 101 location points.

Before sampling, a list of attribute dimensions based upon available technical fabrication information was compiled. The dimension of an attribute differs from the attribute and is defined as a trait or characteristic capable of subdivision into numerous subcategories (attributes). For example, color is a dimension of an attribute; red is an attribute. For the evidence being studied, fiber entity or the number of fiber entities occurring together naturally is an attribute dimension. Attributes of the dimension "fiber entity" include singles, pairs, and bundles. The entities form the basis for converting fibers into yarn. The attribute dimensions identified included fiber entity, fiber crimp, fiber surface, yarn type, number of yarn components, yarn twist direction, number of yarn systems, yarn system direction, type of interworking, layers identified, and fabric deformation.

Attribute data were identified from the photomicrographs of each location point in the sample, and each potential categorical variable (attribute) was recorded as present or absent. Because it was desirable to determine whether the evidence at the location points was related, the data were subjected to hierarchical clustering. The measure of dissimilarity used in the project was the number of matches among attribute measurements that two location points shared. For example, two points had a dissimilarity of 0 if they matched on all attribute measurements, and at the other extreme, the two location points had a dissimilarity of 13 (the total number of measured attributes) if they did not match on any of the measurements. Each match was weighed as equally important. In addition to this intuitive measure of dissimilarity, we used complete linkage (also called nearest neighbor), a procedure that tends to form clusters in which all points within a cluster are relatively homogeneous with regard to their attributes.

Once clusters were determined, the next step was to identify which measures were important in defining the clusters. A classification procedure similar to discriminant analysis was used to determine which attributes actually placed a point in a particular cluster. Because all measurements are categorical (presence or absence), a nonparametric procedure called classification and regression tree (CART) was used.

Of the 13 available attributes, three have direct relevance to the research hypothesis concerning incidence of unbalanced plain-weave fabric. These attributes are the presence of float in one set of yarns of a fabric structure, the presence of a plain weave (defined as a repeated simple alternating) interworking pattern (1/1), and the presence of layers of fabric. If all three are present, there are at least three possible interpretations:

1. Two layers of a single patterned fabric are present.
2. Layers of different fabrics are present.
3. Layers of plain fabric and patterned fabric are superimposed on one another.

A structural patterned fabric is suggested by the presence of float and plain weave together without any layers. If layers of fabric are observed in conjunction with float but without 1/1 interworking, then two layers of float fabric are indicated. The presence of float by itself would suggest a single layer of float fabric, and the presence of 1/1 by itself would support a plain weave. Layers of fabric would only occur where some interworking of fabric evidence is present either, in the form of repeated simple alternating systems or the repeated float progression of alternation. These combinations are summarized in Table I. The remaining 11 attributes would define fiber and yarn characteristics, but taken by themselves, do not indicate fabric structure.

Table I. The Six Attribute Combinations for Pseudomorphs After Fabric on Shang Spearpoint

Combination	*Float*	*1/1*	*Layers*
1	Present	Present	Present
2	Present	Present	Absent
3	Present	Absent	Present
4	Absent	Present	Present
5	Present	Absent	Absent
6	Absent	Present	Absent

Results

Mapping of the two areas chosen for magnification shows the interplay of two different systems of yarns in a regular plain-weave interlacing pattern. One system consistently has yarns that are 1.5 times larger than those of the other system. The difference in the yarn systems confirms the unbalanced plain-weave structure, which is observable without magnification as ribbed fabric (Figure 2). One of the areas shows the presence of layers of mineralized remains on the blade. A section of red, identified as cuprite in the plain-weave formation, rests on top of a green or malachite section (*6*). Both formations share the same type of structure, and the two layers do not have the same orientation with respect to the spearpoint, despite the similarity in weave.

The pseudomorphs after fabric were located in an area roughly between H and U on the x axis, and between 3.6 and 9.0 on the y axis. This area corresponds to approximately 52% of the spearpoint's length and 81% of its width. Examination of the spearpoint during the survey phase had already indicated shifts in the geometry of the fabric with respect to the weapon's form (*6*).

Of 101 sampling units drawn, 11 revealed no pseudomorphs after fabrics present, and two points had pseudomorphic evidence that could not be identified because of blurring of the photomicrograph. Only the remaining 88 location points yielded textile information.

Figure 2. Mapping by photomicrography of pseudomorphs after fabric on Shang spearpoint (Sp1). Magnification: 51×.

Inspection of the fabric attribute dimensions for all location points revealed that certain attributes were either absent or unrecognizable. These attributes are fiber bundle, fiber single, fiber crimp, fiber surface, fiber pattern, yarn type, and yarn twist direction. Therefore, these attributes were deleted from the statistical analysis. The remaining attributes were either present at all location points or exhibited variation. These attributes included the three (layers, 1/1, and float) whose interactive effects had direct relevance to the research hypothesis, and the following fiber and yarn attributes: paired fibers, 0 twist, combined yarns, yarn system A, yarn systems A + B, yarn systems A + B + C + D, fabric distortion, red, green, and black.

Statistical Analysis

In the first complete linkage analysis, the interaction term (layers, 1/1, float) was assigned the same weight as any of the other relevant attributes. The algorithm yielded three clusters (Figure 3) when there was only one match (or equivalently, when the maximum distance between clusters was 12 "no matches"). Two of these clusters accounted for only 13 of the 101 location points. Interestingly, 11 of these 13 points were precisely those and only those points that revealed no pseudomorphs. The other two points were those for which pseudomorphic evidence could not be identified. As such, these two small clusters were taken together to form the cluster of "indeterminates". When matches were increased to three, the big cluster of 88 location points was subdivided into three smaller clusters. A cluster could represent either a fabric type or a fabric pattern.

When there are only two clusters, a CART demonstrates that yarn system direction A is the most important classification attribute. This information was not surprising because yarn system A indicated the presence of some kind of pseudomorph after fabric evidence. When the one big cluster is subdivided into three smaller ones, attributes important in forming the clusters are as follows: The presence of layers, whether in combination with 1/1 interworking, with float interworking, or with both, completely determines one cluster. Another cluster is dominated by no-layer locations having at least float present. The third attribute comprises locations where only 1/1 interworking is present among the three interacting attributes (Table II).

The two latter clusters are not completely distinct from each other, and could reasonably be taken together. The nondetection of layers and the presence of at least one layer of 1/1 interworking and float would characterize this cluster. The classification tree, when the number of clusters is four, shows that the more important classification attributes, in decreasing order of significance, are yarn system direction A, float, red yarn, layers, and fabric distortion. Float is the only attribute that is found in a single cluster. The remaining attributes were dispersed among the three clusters.

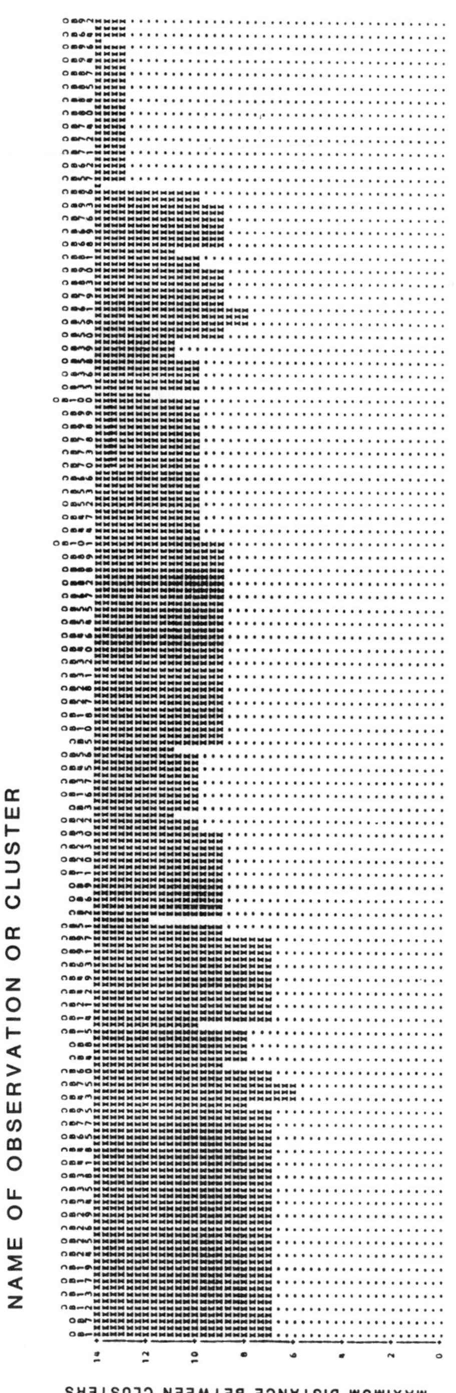

Figure 3. Complete linkage cluster analysis: unweighted interaction.

Table II. Unweighted Attribute Interaction for Four Clusters

Attributes	*Cluster 1*	*Cluster 2*	*Cluster 3*	*Cluster 4*
Float, 1/1, layers	2.00	–	–	–
Float, 1/1	–	7.00	–	–
Float, layers	2.00	–	–	–
Float	–	8.00	1.00	–
1/1, layers	28.00	–	–	–
1/1	–	1.00	39.00	–
All absent	–	–	–	13.00

The second cluster analysis involved attaching twice as much importance to the interaction term than to any of the other 12 attributes. We obtained basically the same cluster tree as with the first method, with only a few modifications: some cluster breakups were more severe than before, indicating more cluster-to-cluster distinctiveness, and some were less severe. Inspection of the tree indicated that there were either two clusters, in which case one cluster was approximately twice as big as the other cluster, or there were three clusters, with the larger cluster subdivided into two clusters. (*See* Figure 4.) Further examination of spatial plots revealed no clear separation of cluster, whether the number of clusters was designated two or three (Figures 5a and b).

When the number of clusters was two, the classification tree showed that the interaction term perfectly determined to which cluster a location point should belong. When layers were detected and either 1/1 interworking or float is present, the location point goes to Cluster 2. When layers is not present, the location goes to Cluster 1. With our data set, the cases for which both 1/1 interworking and float were absent were also those cases for which layers were either absent or could not be identified. Statistically, the interaction term and the variable corresponding to the detection of layers measure essentially the same effects. Table III shows how layers completely determine cluster membership.

When the larger cluster was broken into two (i.e., a total of three clusters), one subcluster corresponded to the indeterminates, and the other had no layers with either 1/1 interworking or float present (*see* Table IV). Moreover, the corresponding classification listed the interaction term, or equivalently the layer variable, and yarn system direction A as the more important classification variables.

When we considered four clusters under the first setup, it was suggested that two of these four clusters could reasonably be seen to constitute only one cluster. Indeed, the analysis verified this suggestion. Exhibiting either 1/1 or float as the type of interworking in only one layer is a strong enough measure of homogeneity to weld the corresponding location points.

In either setup, there is adequate evidence that grouping the location points into three clusters is a useful summary of the data. Two of these clusters can be determined by the presence or absence of layers. The third

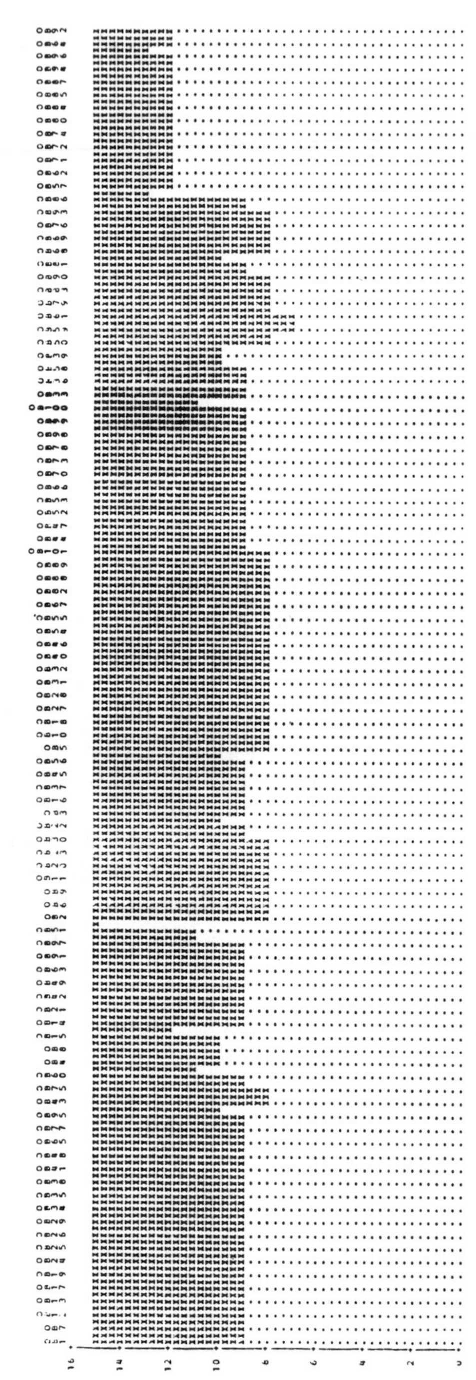

Figure 4. Complete linkage cluster analysis: weighted interaction.

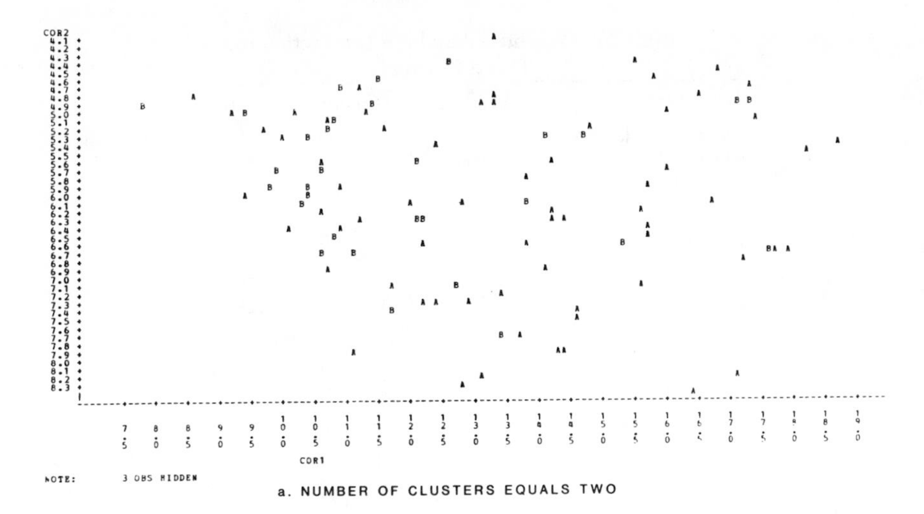

a. NUMBER OF CLUSTERS EQUALS TWO

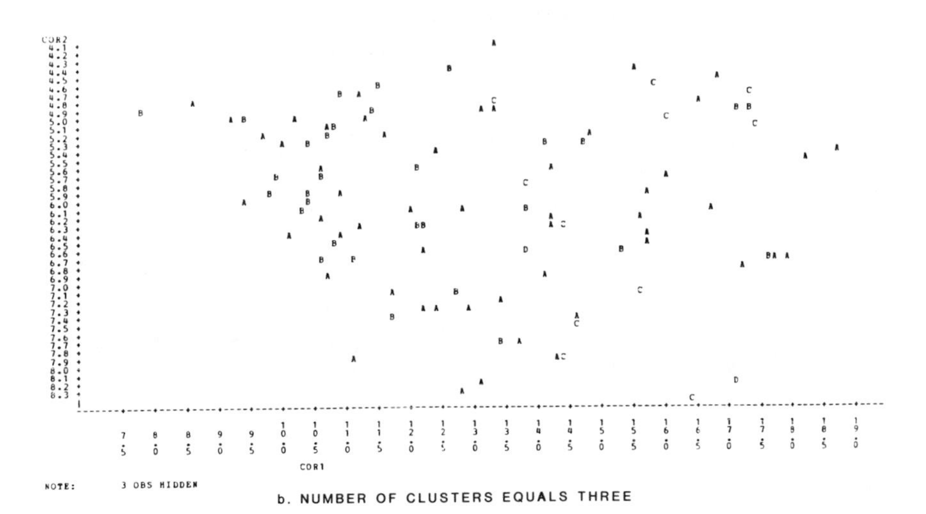

b. NUMBER OF CLUSTERS EQUALS THREE

Figure 5. Spatial plots of weighted interaction.

Table III. Weighted Attribute Interaction for Two Clusters

Attributes	*Cluster 1*	*Cluster 2*
Float, 1/1, layers	–	2.00
Float, 1/1	7.00	–
Float, layers	–	2.00
Float	9.00	–
1/1, layers	–	28.00
1/1	40.00	–
All absent	13.00	–

Table IV. Weighted Attribute Interaction for Three Clusters

Attributes	*Cluster 1*	*Cluster 2*	*Cluster 3*
Float, 1/1, layers	–	2.00	–
Float, 1/1	7.00	–	–
Float, layers	–	2.00	–
Float	9.00	–	–
1/1, layers	–	28.00	–
1/1	40.00	–	–
All absent	–	–	13.00

layer consists of the indeterminates. Because the indeterminates are of no practical importance, there are essentially only two clusters.

Of the 88 sampling units with pseudomorphs studied, 77 (88%) contained evidence of 1/1 interworking of the two yarn systems. Float interworking without any other type of interworking was observed at nine locations (10%), and seven locations (8%) displayed evidence of 1/1 and float together. Two locations contained layers in conjunction with float interworking, and two locations contained layers, float, and 1/1 interworking. The attribute "layers of fabric" was found at 32 location points (36%).

The incidence of float and float with 1/1 interworking occurred in the L–Q area of the x axis and between 4.9 and 7.7 on the y axis. Two adjacent points [(N.2, 6.2) and (N.2, 6.3)] differ in the "interworking" attribute dimension. The point (N.2, 6.2) is classified as float only and (N.2, 6.3) is classified as float plus 1/1 interworking. All of the micrographs displaying float were in the area where at least the main layer had a similar orientation across the blade. This area also had raised pseudomorphs after fabric that suggested a crumpled fabric.

Interpretation

Of the 32 location points where layers were identified, only two points also had both 1/1 interworking and float; however, this finding did not necessarily indicate that one layer contained 1/1 only and the other layer contained float only. Thus, for only two points within this cluster (Table IV, cluster 2) can we possibly suspect that there were two different types of fabrics. For the other points in this same cluster, we concluded that 1/1 was superimposed on 1/1, or float was superimposed on float, which is strongly indicative of the presence of only one fabric. For the other cluster for which no layers were identified, there was only one type of fabric with some structural variation.

A cluster is strong evidence that there is only one fabric type. One question remains: Are there two types of fabric present, or do the differences between clusters only signify a certain type of patterning within a single fabric? The latter hypothesis will be supported if the two clusters tend not to be separated spatially. Examination of the spatial plot suggests there was no spatial separation (Figure 5 a, b).

Strong evidence suggests that only one type of fabric was present; it was an unbalanced plain weave. The evidence also indicates there were areas of float interworking that would support the possibility of structural patterning. The mineralized fabric identified as an unbalanced plain weave located on the spearpoint should be considered a variant with areas of float.

Discussion

The sampling technique applied in the research project unexpectedly exposed areas of float interworking. The spinning of more than one yarn in a consistent manner in conjunction with 1/1 interlacing is indicative of a structurally patterned fabric. Such patterning is achieved by changing the system of interlacing at certain points to provide visual variation. However, consistency and pattern in float interworking became keys to the identification of a fabric as structurally patterned, because flaws or discrepancies in the interlacing order can lead to float. In the case of the pseudomorphic evidence, there is not enough consistency to confirm a structurally patterned fabric. The evidence collected supports the possibility of a structurally patterned interpretation, however. Because the pseudomorphs after fabric are difficult to evaluate, caution should be used in identification of fabric structure.

Statistical sampling combined with cluster analysis is a useful investigative tool because it revealed an important characteristic, that is, the presence of float interworking. Not only did the technique reduce the possibility of bias in selecting points to measure, but it also allowed identification of a structural feature hitherto unobserved.

Despite the differential survival of material culture in the archaeological record, the record is more extensive than previously realized, albeit incomplete. The archaeologist faced with constraints of time and cost must choose what to study. Sampling based upon probability is a tool that is used increasingly to reconstruct the past. This tool may be applied to a number of situations or populations. The population to be sampled may be a geographic region or a culture area (*7, 8*), it may consist of subunits of a specific site (*9–11*), it may comprise attribute traits or variables of artifacts (*12, 13*), or it may be composed of a specific class of artifacts from a given site (*11, 14*). The classes of data recoverable from the archaeological record (populations) seem infinite but tend to group themselves into cultural, environmental, and spatial types of information (*15*).

If the attribute analysis had revealed that the pseudomorphs after fabric located upon the spearpoint indicated only an unbalanced plain weave, then the probability of the presence of different types of fabric would have been low. If, however, enough variation in fabric attributes had existed, then the possibility of multiple fabric types would have been strengthened. Variation, of course, may have been the result of other possibilities such as structural patterning in one fabric, layers of fabric, or even deformation of fabric structure. Given extensive and pronounced variation in fabric attributes and breaks in the regularity of pseudomorph patterning, the presence of more than one type of fabric would have to be considered. The presence of multiples of one type of fabric with regularity in attributes cannot be discounted, particularly if there are breaks or discontinuities in the pseudomorphic evidence. Careful analysis of the evidence is essential.

The research hypothesis is partially supported because the float occurred in one fabric type as demonstrated by cluster analysis. The new definition of the fabric expands our knowledge of Shang fabric production substantially. The unbalanced plain weave structure produced a ribbed effect on the surface of the fabric, with the size of the rib related to the size and number of the yarns (*16*). If the number of yarns in one direction is great enough to cover the yarns from the other system, then the fabric is called either warp-faced or weft-faced. Without a selvage, we cannot identify which of these two possibilities was present on the spearpoint. However, one widely used structure found in the Han Dynasty fabrics (206 B.C.–A.D. 220) is the warp-faced compound silk fabric, which achieved its structural pattern by float (*17*). If the Shang pseudomorphs after fabric located upon the spearpoint are part of a warp-faced, unbalanced plain-weave fabric with patterning, then the Shang people apparently used the same weaving techniques as the Han, but some 1000 years earlier. On the other hand, if the mineralized fabric was part of an unbalanced plain-weave fabric that contained weaving flaws, then the earlier Shang were developing the weaving tradition that later evolved into the Han warp-faced silk identified by Burnham (*17*).

The presence of layers of mineralized fabric suggests that the silk fabric may have been wrapped around the blade, but the fabric was altered in the diagenetic context. All evidence of fabric was destroyed on the reverse side of the spearpoint. Furthermore, the shift in the geometric relation of the layers with the same attributes, as indicated in the mapped area, tends to support the wrapping of one fabric rather than the possibility of double cloth or some other compound fabric variant. The Han warp-faced compound tabby (unbalanced plain weave) has two sets of warp that are integrated into one layer and may be considered a compound fabric with complementary warp (*16*, *17*). The reason for the placement of fabric and weapon in the burial context remains a crucial question for reconstruction of past decision making.

Mapping by photomicrography and sampling have assisted in the technical fabrication analysis of pseudomorphs after fabric from the Shang spearpoint. The two techniques have strengthened the investigative efforts and should be useful in analyzing other misleading forms of archaeological textile evidence. Certain recommendations for the application of mapping should be considered.

1. A vertical dimension must be considered at each location point. The investigator should record and evaluate evidence at all levels that are visible, rather than observing only the upper level of evidence in the field of vision.
2. Attribute dimensions need to be more clearly defined and linked more closely to the initial survey in order to focus upon variation.
3. Selection of sites for the mapping by photomicrography should occur after the sampling and cluster analysis are accomplished. Results of sampling could suggest certain areas to be studied in greater detail.
4. The use of a stratified random sampling technique would be useful.

The next stage in the research project will focus upon those location points where significant variation exists, or where the data collected are puzzling. Selection of a limited number of points for the physical and chemical testing would preserve at least part of the evidence and assist in the search for information about the character of the fiber. An array of instrumental analyses would be designed to yield essential data about the textile evidence and its mineralization processes.

Summary

The objective of this study was the determination of the nature and extent of pseudomorphs after fabric located on one Shang Dynasty bronze spear-point. Reid, Schiffer, and Neff (9) noted that the very nature of archaeological research involves sampling, but the real concern is with "how to secure a sample that best provides data to answer questions about past behavioral systems." If the ultimate goal of studies of pseudomorphs after textiles is to infer the use of textiles in a cultural system where other forms of textile evidence do not survive, then the investigator must determine first the nature and extent of the mineralized evidence on a metal host/object. The techniques used in this chapter give us the opportunity to expand our knowledge of one form of the archaeological textile record. The presence of layers of an unbalanced plain-weave silk fabric with areas of float was confirmed. A profile of the fabric–metal object relation is a beginning, but many more objects and pseudomorphs would need to be examined before generalizations can be made about cultural context. A careful and systematic study of one object and its pseudomorphs will provide the basis for more extensive work in the future.

Acknowledgments

The work was conducted at the Ohio State University. Initial support for the project was received from Wenner Gren Foundation for Anthropological Research. We express our gratitude to the Museum of Anthropology of University of Missouri for loan of the Shang spearpoint. Additional support of project was received from The Statistical Consulting Laboratory and the Department of Textiles and Clothing of the Ohio State University, and the Center for Archaeological Sciences of the University of Georgia.

References

1. Jakes, K. A.; Sibley, L. R. In *Archaeological Chemistry—III*; Lambert, J. B., Ed.; Advances in Chemistry 205; American Chemical Society: Washington, DC, 1984; pp 403–424.
2. Jakes, K. A.; Howard, J. H. In *Historic Textile and Paper Materials, Conservation and Characterization*; Needles, H. L.; Zeronian, S. H., Eds.; Advances in Chemistry 212; American Chemical Society: Washington, DC, 1986; pp 277–287.
3. Watson, P. J.; LeBlanc, S. A.; Redman, C. L. *Archaeological Explanation*; Columbia University: New York, 1984; p 179.
4. Sibley, L. R.; Jakes, K. A. *Clothing Text. Res. J.* **1982,** *1,* 24–30.
5. Sibley, L. R. In *Proceedings of the Twenty-Fourth International Archaeometry Symposium*; Olin, J.; Blackman, J., Eds.; Smithsonian Institution: Washington, DC, 1986; pp 153–163.
6. Jakes, K. A.; Sibley, L. R. In *Proceedings of the Application of Science in Examination of Works of Arts*; England, P.; Van Zelst, L., Eds.; Museum of Fine Arts: Boston, MA, 1983; pp 221–224.
7. Berry, B. *Flood Plain Data: Sampling, Coding, and Storing*; Agriculture Handbook 237; U.S. Department of Agriculture, Farm Economics Division and Economic Research Service: Washington, DC, 1962.
8. Redman, C. L. *Archaeological Sampling Strategies*; Addison-Wesley: Phillipines, 1974; 134.
9. Reid, J. J.; Schiffer, M. B.; Neff, J. M. In *Sampling in Archaeology*; Mueller, J. W., Ed.; University of Arizona: Tucson, 1975; pp 209–229.
10. Rootenberg, S. *Am. Antiq.* **1964,** *30,* 181–188.
11. Cowgill, G. L. *Am. Antiq.* **1964,** *29,* 467–473.
12. Pollnac, R. B.; Rowlett, R. M. In *Experimental Archaeology*; Ingersoll, D.; Yellen, J. E.; Macdonald, W., Eds.%lumbia University: New York, 1977; pp 167–190.
13. Christenson, A. L.; Read, D. W. *Am. Antiq.* **1977,** *42,* 163-179.
14. Benfer, R. A. In *Sampling in Archaeology*; Mueller, J. W., Ed.; University of Arizona: Tucson, 1975; pp 170–191.
15. Mueller, J. W. In *Sampling in Archaeology*; Mueller, J. W., Ed.; University of Arizona: Tucson, 1975; pp 33–43.
16. Emery, I. *The Primary Structures of Fabrics: An Illustrated Classification*; The Textile Museum: Washington, DC, 1966.
17. Burnham, H. B. Centre International d'Etudes des Textiles Anciens, Lyons. *Bull. Liaison* **1965,** *22,* 25–45.

RECEIVED for review June 11, 1987. ACCEPTED revised manuscript May 6, 1988.

INDEXES

Author Index

Affiliation Index

Subject Index

C

D

F

G

H

M

N

Q

U

V

W

X

Y

Z

Copy editing: Linda Romaine Ross and A. Maureen R. Rouhi
Indexing: Linda Romaine Ross
Production: Barbara J. Libengood
Managing Editor: Janet S. Dodd

Typesetting of text by Techna Type, Inc., York, PA
Typesetting of front matter and index elements
by Hot Type Ltd., Washington, DC
Printing and binding by Maple Press Company, York, PA

Titles of Related Interest

Archaeological Chemistry
Edited by Curt W. Beck
Advances in Chemistry Series 138
ISBN 0-8412-0211-7

Archaeological Chemistry II
Edited by Giles F. Carter
Advances in Chemistry Series 171
ISBN 0-8412-0397-0

Archaeological Chemistry—III
Edited by Joseph B. Lambert
Advances in Chemistry Series 205
ISBN 0-8412-0767-4

Historic Textile and Paper Materials: Conservation and Characterization
Edited by Howard L. Needles and S. Haig Zeronian
Advances in Chemistry Series 212
ISBN 0-8412-0900-6

Preservation of Paper and Textiles of Historic and Artistic Value
Edited by John C. Williams
Advances in Chemistry Series 164
ISBN 0-8412-0360-1

Preservation of Paper and Textiles of Historic and Artistic Value II
Edited by John C. Williams
Advances in Chemistry Series 193
ISBN 0-8412-0553-1